AF801470

Challenges and Solutions Against Visceral Leishmaniasis

Angamuthu Selvapandiyan • Ruchi Singh •
Niti Puri • Nirmal K. Ganguly
Editors

Challenges and Solutions Against Visceral Leishmaniasis

Springer

Editors
Angamuthu Selvapandiyan
Department of Molecular Medicine
Jamia Hamdard
New Delhi, India

Ruchi Singh
ICMR-National Institute of Pathology
New Delhi, India

Niti Puri
School of Life Science
Jawaharlal Nehru University
New Delhi, India

Nirmal K. Ganguly
Institute of Liver and Biliary Sciences
New Delhi, India

ISBN 978-981-99-6998-2 ISBN 978-981-99-6999-9 (eBook)
https://doi.org/10.1007/978-981-99-6999-9

© The Editor(s) (if applicable) and The Author(s), under exclusive license to Springer Nature Singapore Pte Ltd. 2023
This work is subject to copyright. All rights are solely and exclusively licensed by the Publisher, whether the whole or part of the material is concerned, specifically the rights of translation, reprinting, reuse of illustrations, recitation, broadcasting, reproduction on microfilms or in any other physical way, and transmission or information storage and retrieval, electronic adaptation, computer software, or by similar or dissimilar methodology now known or hereafter developed.
The use of general descriptive names, registered names, trademarks, service marks, etc. in this publication does not imply, even in the absence of a specific statement, that such names are exempt from the relevant protective laws and regulations and therefore free for general use.
The publisher, the authors, and the editors are safe to assume that the advice and information in this book are believed to be true and accurate at the date of publication. Neither the publisher nor the authors or the editors give a warranty, expressed or implied, with respect to the material contained herein or for any errors or omissions that may have been made. The publisher remains neutral with regard to jurisdictional claims in published maps and institutional affiliations.

This Springer imprint is published by the registered company Springer Nature Singapore Pte Ltd.
The registered company address is: 152 Beach Road, #21-01/04 Gateway East, Singapore 189721, Singapore

Paper in this product is recyclable.

Foreword

It is my great pleasure to write the foreword for the book, *Challenges and Solutions Against Visceral Leishmaniasis,* edited by Prof. Nirmal K. Ganguly (Global Health Strategies, New Delhi) and colleagues, including Dr. A. Selvapandiyan (Jamia Hamdard, New Delhi), Dr. Ruchi Singh (ICMR-National Institute of Pathology, New Delhi), and Dr. Niti Puri (Jawaharlal Nehru University, New Delhi). Having experience in organizing an international conference on "Innovations for the Elimination and Control of Visceral Leishmaniasis (IECVL)" in 2018, sponsored by various national and international funding organizations, including the WHO, this elite group of editors brought together an impressive set of experts and authors to contribute to this pivotal work.

Microbes are perhaps the most intelligent living form, which rapidly adapts to environmental cues by virtue of their evolutionary dynamics. Tropical countries are subjected to infection by viral, bacterial, fungal, and parasitic organisms. When it comes to eliminating infectious diseases, smallpox eradication is a valuable lesson that must be strictly followed, since the pathogen considered eliminated may re-emerge and re-establish infection. The re-emergence of schistosomiasis in China and poliomyelitis are examples before us. Constant surveillance and continued interventions are of paramount importance in disease eradication strategies. Fighting against infectious diseases has constantly become an uphill task. Although several countries claim the total eradication of some of the infectious diseases in their territories, more often, those diseases shift from higher to a stealthy lower incident level. Such a dynamic shift to a lower level poses greater challenges.

Leishmaniases range from the mild self-healing cutaneous form to the severe fatal form of visceral leishmaniasis (VL, black fever or the well-known kala-azar) and warrant greater efforts to eliminate the disease on a global scale. There are around 20 known species and variants of *Leishmania* that cause a spectrum of different forms of diseases, with different pathology, epidemiology, and vector specificities. In addition, the challenges include loss of immunity, buildup of susceptible regions, emergence of parasite resistant to treatments, population heterogeneities, noncompliance to control measures, and so on.

From the time that Prof. C.P. Thakur was the Minister of Health of India, attention was given to the epidemics of kala-azar in India, particularly in Bihar and West Bengal. Professor V. Ramalingaswami as Director of ICMR supported control efforts of kala-azar, but there were no efficacious tools to combat the epidemics, which took the lives of tens of thousands of reported cases, but most likely with many more unreported cases of death caused by VL. During the period of Prof. N.K. Ganguly and through his vision and support of open international and national collaborations, new treatment modalities and diagnostic tests were developed. By implementing the new tools, the incidence of VL has fallen to a level that in comparison to the old data can be considered that kala-azar has been almost controlled as a major public health problem and cause of death in these endemic regions. The century-old treatment of daily injections of toxic antimonial drugs for a month is history and no longer in use. It was the pioneering work on AmBisome by Dr. S. Sundar that opened the road to new treatment. However, the cost of this drug was prohibitive. The WHO negotiated the price with the company for India, and DNDi conducted low-dose AmBisome in combination with other drugs available that provided a powerful tool for the control of VL in the Indian subcontinent. There is now the challenge of maintaining the implementation of control, when the incidence is low and will fall under the radar of public health priorities. The cyclic nature of the epidemiology of kala-azar has been documented repeatedly. One tool for the ultimate elimination of all forms of leishmaniasis is having a safe, efficacious, and affordable prophylactic universal vaccine. The new approaches for developing vaccines were discussed by prominent Indian and international scientists and reported in this book. The recurrence of leishmaniasis in an individual, who has been cured following immunosuppression (HIV infection or treatment by immunosuppressive drugs), attests to the presence of live *Leishmania*, which is controlled by the effective immune response of the host. Indeed, even over 65 years after the cure from CL, in my own case, Prof. Sima Rafati using RT-PCR showed the presence of live *Leishmania* (unpublished). The lifelong immunity after cure from leishmaniasis may be due to the presence of a live parasite kept in control by the effective immune response of the host. Hence, the importance of the use of a therapeutic vaccine and new adjuvants as an adjunct to drug treatment has been recognized and is presented in this valuable book.

The chapters in this book are from eminent personalities working tirelessly toward the goal in various fields, including those involved in epidemiology, clinical studies, entomology, diagnosis, development of drugs and vaccines, treatment, etc. The chapters address the current state of leishmaniasis infections worldwide, with a focus on the Indian subcontinent. The book also highlights the challenges of identifying asymptomatic infections and susceptible regions and improving disease control efforts. Additionally, it discusses the economic impact of disease control and the importance of sustainability.

This book has a total of 17 chapters covering all the relevant areas in the current scenario, including leishmaniasis and challenges, omics and advancement in diagnosis, genetics, new drug discovery opportunity, immunity, and vaccine development, written by eminent scientific groups nationally and internationally, including from the Food and Drug Administration, USA; London School of Hygiene & Tropical Medicine, London, UK; Ohio State University, Ohio, USA; University of Colombo, Colombo, Sri Lanka; Jamia Hamdard, New Delhi; Jawaharlal Nehru

University, New Delhi; ICMR-National Institute of Pathology, New Delhi; Indraprastha Apollo Hospitals, New Delhi; Sharda University, Noida, UP, India; Central University of Punjab, Bathinda, Punjab, India; Central Drug and Research Institute, Lucknow, UP, India; CSIR-Indian Institute of Chemical Biology, Kolkata, WB, India; and Banaras Hindu University, Varanasi, UP, India. As the last-mile walk information from the authors, I am sure that this volume will serve as an informative resource for practitioners, students, and researchers alike.

Drugs for Neglected Diseases Initiatives (DNDi) Farrokh Modabber
Geneva, Switzerland

Preface

The fatal disease, visceral leishmaniasis (VL, kala-azar), and its sequel infection as post-kala-azar dermal leishmaniasis, due to the single-celled parasite *Leishmania donovani*, are prevalent in the tropical world, mainly in the Indian subcontinent and Africa. Drug resistance due to available medications and nonavailability of vaccines for the disease add to the misery. Due to the tireless efforts of the Governments of India, Nepal, Bangladesh, Bhutan, and Thailand in association with the World Health Organization, since 2005, there has been a considerable reduction in the kala-azar cases in the Indian subcontinent. However, there is still a long way to go for its total elimination as per the renewed pledge by the WHO to eradicate it by 2025 or even earlier as per the National Health Policy (India). The major objective of this edition, *Challenges and Solutions Against Visceral Leishmaniasis,* is to bring together leading scientists from academia and industry, and nongovernment organizations, as important patrons around the world to write upon new tools, new approaches, and routes to implementation of innovations for elimination. The book seriously discusses the status of kala-azar, its epidemiology, transmission role of innate immunity, diagnostics, and treatment options, i.e., the development of drugs and vaccines and last but not least the models for partnerships, programs, and integration to combat the disease. Discussions on the complexities and challenges of disease eradication, including sustainable elimination, would yield for the development of important target-oriented innovations and control measures. We appreciate the entire team of authors of the chapters that include persons from our institute and other national and international bodies who have teamed up to write for this unique and important endeavor.

There have been significant advances in the discovery and development of new drugs, diagnostics, vaccines, and approaches to vector control for VL—new tools that can be used for more rapid and effective treatment of patients as well as those that are deployable in control and elimination programs. The challenge will be deployment: How and where will these new tools best be used? How do we work with national governments, health systems, and other stakeholders to prepare them for these changes so that they are introduced in a timely and accessible manner to the greatest effect? The recently observed inverse trend of declining VL incidents and increase of PKDL cases in the Indian subcontinent needs systematic deliberation. The last few years have seen an increasing number of cases coming from previously non-endemic areas in the hilly regions. This poses a serious challenge to sustain the successes, which have been achieved so far. The other challenges include the occurrence of outbreaks, which could increase the incidence to above the elimination target, especially in districts with low populations. With the epidemic continuing in neighboring Bihar, lack of effective cross-border coordination is another important area of concern. Perhaps, the key aspect is to recognize that both surveillance and control of VL rely on prompt identification of cases and that the target should reflect the continued effort that will be required to find cases, even as they become rarer. Efforts are continuing to understand the epidemiology, visualizing the potential and genetic variation of *L. donovani* in Sri Lanka, which causes cutaneous leishmaniasis instead of VL in Sri Lanka, Himachal Pradesh, India, etc. Additionally, no high-quality evidence-based treatment regimens exist for HIV-VL coinfection in the Indian subcontinent. Apart from this, a timely diagnosis and treatment of post-kala-azar dermal leishmaniasis cases assume paramount importance in the context of achieving kala-azar elimination in a sustainable manner. The development of new treatment options with improved profiles is thus a priority for VL. Efficient programs to deliver safe, highly efficacious oral therapy that can be deployed at grassroots levels in the health system should be a priority. Since there is no vaccine yet commercially available for VL, several gene-deleted intracellular stage-attenuated whole *Leishmania* parasite vaccines are under development. Such second-generation leishmanization—live vaccine with a more acceptable and safer non-lesion-producing attenuated *L. major* strain—is therefore a rational vaccine strategy and is being demonstrated in detail by the USFDA. Importantly, this book brings on the various models getting developed to address these challenges.

The book comes with 17 chapters distributed in 5 major categories, viz. ***A. Leishmaniasis and Challenges, B. Omics and Advancement in Diagnosis, C. Genetics, D. New Drug Discovery Opportunity,*** and ***E. Immunity and Vaccine Development.*** Chapter 1, being introductory, describes the challenges, innovations, and solutions to VL. Chapter 2 brings out the activities of a UK-based program KalaCORE—to tackle VL in South Asia and East Africa, initiated in 2014 and lasted till recently. This program served as a model to function toward the elimination of any kind of disease. Chapter 3 deals with the essentials needed for the elimination of VL including diagnostics, drugs, and vaccines. Chapter 4 mainly brings out the factors of the sand fly, the vector of leishmaniasis that plays an important role in the life of the parasite in its gut and salivary portion. Chapter 5 lists the proteome part of the parasite that can help in the parasite's diagnosis and the development of therapeutics. Chapter 6 describes the diagnosis and treatment aspects of post-kala-azar dermal leishmaniasis. Chapter 7 is all about the emergence of novel *Leishmania* genetic variants, as a new challenge to focus on the ongoing leishmaniasis elimination program in the Indian subcontinent. Chapter 8 focuses on the atypical leishmaniasis in Sri Lanka as another challenge to the disease elimination program. Chapter 9 narrates the critical roles of microRNAs in the pathogenesis and immunoregulation of the parasite infection. Chapter 10 is dedicated to the heat-shock proteins of *Leishmania* as emerging therapeutic and vaccine targets. The advancements in antileishmanial chemotherapy have been discussed in detail in

Chaps. 11 and 12 that follows on the drug miltefosine's unresponsiveness in the VL treatment, a major challenge to VL treatment. Chapter 13 introduces the development and use of the live attenuated whole-parasite vaccine as a safe and efficacious pan-*Leishmania* vaccine. Chapter 14 is about the needed understanding of the heterogeneity in mast cell role in the host defense during the disease. The next chapter (15) is on the feasibility of therapeutic vaccines for the management and control of VL. Chapter 16 tells the worldwide efforts for the prevention of VL using vaccinations, including the commercially available vaccines to canine populations as model therapeutics. And the last Chap. 17 brings out the emerging concepts in *Leishmania* vaccine adjuvants and its use.

This book features the collective wisdom of authors/resource persons as important stakeholders from various corners globally, who look at the challenges and advise us on the key issues to be addressed in the VL disease elimination strategies.

New Delhi, India Angamuthu Selvapandiyan
New Delhi, India Ruchi Singh
New Delhi, India Niti Puri
New Delhi, India Nirmal K. Ganguly

Contents

Editors and Contributors

About the Editors

Angamuthu Selvapandiyan Ph.D. is a senior investigator in the Department of Molecular Medicine, Jamia Hamdard, New Delhi, India. He has been involved in teaching and research for the past 33 years. He has been teaching molecular biology and immunology courses to the visiting participants from the member countries at the International Centre for Genetic Engineering and Biotechnology (a UNIDO organization), New Delhi, for 10 years and to the visiting trainees at the USFDA for 10 years. He has worked in the molecular and biochemical aspects in several fungi, bacteria, and parasites and cloned and functionally characterized several of their genes. His current major thrust areas are in the diagnostics of microbial pathogens in the clinical isolates and development of drug/vaccine candidates against visceral leishmaniasis. He was instrumental in the development of one live attenuated leishmania vaccine candidate against the fatal "visceral leishmaniasis" disease, which is currently under plan for a clinical trial after its successful test as vaccine in the experimental animals. At USFDA, as a regulatory officer, he was also involved in the reviewing of certain Biological License Applications on diagnostic devices submitted by the industries, through review committees. He has published over 70 research articles and organized several national and international conferences. He is also a recipient of 2002 NIH-FARE award; 2007 "Group Recognition Award" to Chagas Disease Donor Screening Biological License Application Review Group by CBER/USFDA; and 2009 USFDA Center Director's Scientific Achievement Award.

Ruchi Singh Ph.D. is a Scientist F at ICMR-National Institute of Pathology, New Delhi, India. Her research is focused on tropical diseases, leishmaniasis, and malaria. Dr. Ruchi has investigated basic and clinical aspects of leishmaniasis, making outstanding contributions in genomics, diagnostics, and mechanism of drug resistance in kala-azar and post-kala-azar dermal leishmaniasis. Her work on drug resistance (antimony, miltefosine, paromomycin, and artemisinin) in leishmaniasis yielded new insights into mechanisms operative in clinical isolates. She has also authored more than 60 research articles and several book chapters. She has been at the forefront of clinical research in visceral leishmaniasis. She has received several prestigious national and international awards, including ICMR Awards: Shakuntala Amir Chand (Best Published Work, 2006) and Maj Gen Sahib Singh Sokhey (Communicable Diseases, 2015) Awards and UNESCO L'Oréal FWIS fellowship for Asia Pacific, 2006. She is Associate Editor at Frontiers in Epidemiology.

Niti Puri Ph.D. joined School of Life Sciences, Jawaharlal Nehru University (JNU), New Delhi, as Assistant Professor in Immunology in May 2008, after postdoctoral experience in India at JNU and National Institute of Immunology, Delhi, and as a Fogarty Fellow and Visiting Fellow Employee at the National Cancer Institute, National Institutes of Health (NIH), USA. Her extensive research in cellular and molecular immunology is focused on studying the molecular mechanisms of innate immune cell effector mechanisms during infectious and noninfectious diseases with the intent to find novel therapeutics. Her research has revealed the molecular mechanisms and novel effector responses involving mast cells during infectious diseases like leishmaniasis, tuberculosis, and candidiasis and noninfectious inflammatory diseases like an allergic response, anemia, and cancer. Her studies of mast cell interactions with intracellular pathogens like leishmania have provided evidence for a very important direct role for mast cells in clearance of these pathogens by both intracellular and extracellular mechanisms. These studies highlight the importance of mast cells in these infectious disease outcomes and therapeutic strategy development. She has carried out several highly competitive externally funded research projects and also set up productive collaborations with other researchers nationally and internationally. She has also published more than 50 research articles in high-impact peer-reviewed international journals and several book chapters and presented her work in national and international conferences. She has been a reviewer for various national and international journals and extramural projects. She has also designed several research protocols for studying various aspects of mammalian immune system and host-pathogen interactions in vitro and in vivo. She also teaches basic and advanced immunology to postgraduate and higher level students at JNU and serves as guest lecturer at South Campus, DU, and also earlier at NIH, USA. She has been a member of the IBSC, Animal House Advisory Committee, and the Academic Council of JNU and the core faculty-in-charge for the flow cytometry facility and M.Sc. laboratory at SLS, JNU. She is also currently part of the Institutional Animal Ethics Committee at South Asian University and member of Research Advisory Committee for Ph.D. students at All India Institute of Medical Sciences, National Institute of Immunology, and Delhi Technological University, New Delhi. She has supervised so far 14 Ph.D. theses.

Nirmal Kumar Ganguly M.D., Ph.D. is a Former Director of the Indian Council of Medical Research, the National Academy of Medical Sciences, and the National Institute of Biologicals. He is a microbiologist and Chairman of the Advisory Committee to the Minister of Health on COVID-19. He is also the Chairman of the Research Council of the Institute of Advanced Virology, National Academy of Medical Sciences; Chairman, Indian Pharmacopoeia Commission, Expert

Committee of the National Institute of Biologicals guiding R&D activities; and President, Immunology Foundation of India. Prof. Ganguly is at present the Senior Scientific Advisor, Global Health Strategies, New Delhi, and President, Apollo Hospitals Educational and Research Foundation. He is Emeritus Professor and Director, PGIMER, Chandigarh. He is on the Advisory Board of NIH Fogarty International Center, the Health Vaccine Center, the U.S. Centers for Disease Control (CDC) Advisory Board for Global Health, and the Public Health Advisory Board of the University of California (UCLA). He is a Senior Scientific Advisor to the Bill and Melinda Gates Foundation (BMGF) and a Fellow of Imperial College and the Royal College of Pathologists and the Tropical School, London, and he is on the Board of the Public Health Innovative Fund, the Canada Innovative Fund, and Grand Challenges, Canada. Prof. Ganguly is Chairman of the Advisory Committee for Health Research of the World Health Organization-SEARO, the International Vaccine Institute Cholera Board (CHOVI), and the United Nations Children's Fund SAG TDR. He has published more than 775 research papers and supervised or co-supervised 130 Ph.D. candidate dissertations and more than 20 book chapters. Prof. Ganguly has been honored with 7 international and 113 national awards along with the prestigious "Padma Bhushan" Award in the field of "Medicine" for the year 2008.

Contributors

Nahid Ali Infectious Diseases and Immunology Division, CSIR-Indian Institute of Chemical Biology, Kolkata, West Bengal, India

Apeksha Anand Division of Biochemistry and Structural Biology, CSIR-Central Drug Research Institute, Lucknow, India
Academy of Scientific and Innovative Research (AcSIR), Gaziabaad, India

Mudsser Azam ICMR-National Institute of Pathology, Safdarjung Hospital Campus, New Delhi, India

Surbhi Badhwar Cellular and Molecular Immunology Laboratory, School of Life Sciences, Jawaharlal Nehru University, New Delhi, India

Mirza A. Beg Department of Molecular Medicine, Jamia Hamdard, New Delhi, India

Parna Bhattacharya Division of Emerging and Transfusion Transmitted Diseases, Office of Blood Research and Review, Center for Biologics Evaluation and Research, Food and Drug Administration, Silver Spring, MD, USA

Sakib Burza London School of Hygiene and Tropical Medicine, London, UK

Yogesh Chauhan Department of Biochemistry, Central University of Punjab, Bathinda, Punjab, India

Garima Chouhan Department of Biotechnology, School of Engineering and Technology, Greater Noida, Uttar Pradesh, India

Simon Croft London School of Hygiene and Tropical Medicine, London, UK

Amrita Das Infectious Diseases and Immunology Division, CSIR-Indian Institute of Chemical Biology, Kolkata, West Bengal, India

Saswata Das Department of Biotechnology, School of Engineering and Technology, Greater Noida, Uttar Pradesh, India

Tirtharaj Datta Department of Biotechnology, School of Engineering and Technology, Greater Noida, Uttar Pradesh, India

Margriet den Boer Medicines sans Frontieres, Amsterdam, Netherlands

Rajika Dewasurendra Department of Parasitology, Faculty of Medicine, University of Colombo, Colombo, Sri Lanka

Ranadhir Dey Division of Emerging and Transfusion Transmitted Diseases, Office of Blood Research and Review, Center for Biologics Evaluation and Research, Food and Drug Administration, Silver Spring, MD, USA

Anuradha Dube Division of Molecular Microbiology and Immunology, CSIR-Central Drug Research Institute (CSIR-CDRI), Lucknow, India

Itibaw Farooq Department of Molecular Medicine, Jamia Hamdard, New Delhi, India

Rebecca Fultz Department of Pathology, The Ohio State University Medical Center, Columbus, OH, USA

Nirmal K. Ganguly Institute of Liver and Biliary Sciences, New Delhi, India

Sreenivas Gannavaram Division of Emerging and Transfusion Transmitted Diseases, Office of Blood Research and Review, Center for Biologics Evaluation and Research, Food and Drug Administration, Silver Spring, MD, USA

Sushmita Ghosh ICMR-National Institute of Pathology, Safdarjung Hospital Campus, New Delhi, India

Neena Goyal Division of Biochemistry and Structural Biology, CSIR-Central Drug Research Institute, Lucknow, India
Academy of Scientific and Innovative Research (AcSIR), Gaziabaad, India

Charu Gupta Department of Molecular Medicine, Jamia Hamdard, New Delhi, India
Clinical Research Division, Galgotias University, Greater Noida, Uttar Pradesh, India

Niharika Gupta Biochemistry and Structural Biology, Division, CSIR-Central Drug Research Institute (CSIR-CDRI), Lucknow, India

Ryan H. Huston Department of Pathology, The Ohio State University Medical Center, Columbus, OH, USA

Manju Jain Department of Biochemistry, Central University of Punjab, Bathinda, Punjab, India

Nadira D. Karunaweera Department of Parasitology, Faculty of Medicine, University of Colombo, Colombo, Sri Lanka

Anuja Krishnan Department of Molecular Medicine, Jamia Hamdard, New Delhi, India
Clinical Research Division, Galgotias University, Greater Noida, Uttar Pradesh, India

Vinay Kumar ICMR-National Institute of Pathology, Safdarjung Hospital Campus, New Delhi, India

Priyanka Madaan Department of Biochemistry, Central University of Punjab, Bathinda, Punjab, India

Greg Matlashewski Department of Microbiology and Immunology, McGill University, Montreal, QC, Canada

Dharmendra Kumar Maurya Department of Biochemistry, Institute of Science, Banaras Hindu University, Varanasi, Uttar Pradesh, India

Stefanie Meredith Mott MacDonald, London, UK
Global Health Consulting, Divonne, France

Shyam Lal Mudavat Infectious Disease Biology Laboratory, Chemical Biology Unit, Institute of Nano Science & Technology, Mohali, Punjab, India

Hira L. Nakhasi Division of Emerging and Transfusion Transmitted Diseases, Office of Blood Research and Review, Center for Biologics Evaluation and Research, Food and Drug Administration, Silver Spring, MD, USA

Parma Nand School of Engineering and Technology, Greater Noida, Uttar Pradesh, India

Nilofer Naqvi Cellular and Molecular Immunology Laboratory, School of Life Sciences, Jawaharlal Nehru University, New Delhi, India
Department of Microbiology, University of Chicago, Chicago, Illinois, IL, USA

Thalia Pacheco-Fernandez Departments of Pathology and Microbiology, Wexner Medical Center, The Ohio State University, Columbus, OH, USA

Niti Puri Cellular and Molecular Immunology Laboratory, School of Life Sciences, Jawaharlal Nehru University, New Delhi, India
School of Life Sciences, Jawaharlal Nehru University, New Delhi, India

V. Ramesh Department of Dermatology, ESIC Hospital, Faridabad, India

Enam Reyaz Department of Molecular Medicine, Jamia Hamdard, New Delhi, India

Amogh A. Sahasrabuddhe Biochemistry and Structural Biology, Division, CSIR-Central Drug Research Institute (CSIR-CDRI), Lucknow, India

Poonam Salotra ICMR-National Institute of Pathology, Safdarjung Hospital Campus, New Delhi, India

Abhay R. Satoskar Department of Pathology, The Ohio State University Medical Center, Columbus, OH, USA
Department of Microbiology, The Ohio State University Medical Center, Columbus, OH, USA

Angamuthu Selvapandiyan Department of Molecular Medicine, Jamia Hamdard, New Delhi, India

Shital Clinical Research Division, Galgotias University, Greater Noida, Uttar Pradesh, India

Shaikh Shuhail Department of Biotechnology, School of Engineering and Technology, Greater Noida, Uttar Pradesh, India

Shyamali Department of Biochemistry, Institute of Science, Banaras Hindu University, Varanasi, Uttar Pradesh, India

Om Prakash Singh Department of Biochemistry, Institute of Science, Banaras Hindu University, Varanasi, Uttar Pradesh, India

Riya Singh Amity Institute of Biotechnology, Amity University, Noida, Uttar Pradesh, India

Ruchi Singh ICMR-National Institute of Pathology, Safdarjung Hospital Campus, New Delhi, India

Sanjay Singh Gennova Biopharmaceuticals, Pune, India

Rahul Srivastava Cellular and Molecular Immunology Laboratory, School of Life Sciences, Jawaharlal Nehru University, New Delhi, India

Shyam Sundar Department of Medicine, Institute of Medical Sciences, Banaras Hindu University, Varanasi, Uttar Pradesh, India

Mohd Tasleem ICAR-NIPB, Pusa Campus, New Delhi, India

Priya Tyagi Department of Biotechnology, School of Engineering and Technology, Greater Noida, Uttar Pradesh, India

Aditya Verma ICMR-National Institute of Pathology, Safdarjung Hospital Campus, New Delhi, India

Chaitenya Verma Department of Pathology, The Ohio State University Medical Center, Columbus, OH, USA

Greta Volpedo Department of Pathology, The Ohio State University Medical Center, Columbus, OH, USA
Department of Microbiology, The Ohio State University Medical Center, Columbus, OH, USA

Abigail R. Wharton Department of Pathology, The Ohio State University Medical Center, Columbus, OH, USA

Alok K. Yadav Biochemistry and Structural Biology, Division, CSIR-Central Drug Research Institute (CSIR-CDRI), Lucknow, India

Shailendra Yadav Division of Biochemistry and Structural Biology, CSIR-Central Drug Research Institute, Lucknow, India
Academy of Scientific and Innovative Research (AcSIR), Gaziabaad, India

Samer Zidan Department of Pathology, The Ohio State University Medical Center, Columbus, OH, USA

The Burden of Visceral Leishmaniasis: Need of Review, Innovations, and Solutions

1

Itibaw Farooq, Ruchi Singh, Angamuthu Selvapandiyan, and Nirmal K. Ganguly

Abstract

Leishmaniasis is an ongoing public health crisis and asks for unabated addressal by all stakeholders. Efforts to eradicate the disease have been underway for over a decade. While the number of new cases each year has gone down globally, the goal of total disease elimination has roughly been achieved. The parameters (poverty, malnutrition, poor sanitation and health care, climate change, globalization, etc.) that drive the endemics of the disease have proven critical in slowing down the elimination process. Effective approaches to tackling this neglected tropical disease lie in revamping the epidemiological system so as to eliminate the discrepancy in the reported versus the actual number of cases, improving the diagnostic techniques for timely intervention among the patients and go beyond to uplifting the socioeconomic conditions of the affected populations, keeping a check on the existing and emerging disease reservoirs including but not limited to PKDL patients along with sand fly vector control in the endemic regions. The development of modernized drug therapy for treatment, which is the core facet of disease management, has also been in the pipeline. However, the challenge of emerging and increasing drug resistance in the parasite in countries like India has necessitated the generation of a successful vaccine at the earliest. Several investigations are underway worldwide, taking benefits of both conventional

I. Farooq · A. Selvapandiyan
Department of Molecular Medicine, Jamia Hamdard, New Delhi, India
e-mail: selvapandiyan@jamiahamdard.ac.in

R. Singh
ICMR-National Institute of Pathology, Safdarjung Hospital Campus, New Delhi, India
e-mail: ruchisingh.nip@gov.in

N. K. Ganguly (✉)
Institute of Liver and Biliary Sciences, New Delhi, India

© The Author(s), under exclusive license to Springer Nature Singapore Pte Ltd. 2023

A. Selvapandiyan et al. (eds.), *Challenges and Solutions Against Visceral Leishmaniasis*, https://doi.org/10.1007/978-981-99-6999-9_1

and new-age tools like CRISPR-Cas 9–based gene editing, to develop a safe and efficacious vaccine against leishmaniasis. Combining the advancements in science and technology with the efforts of the researchers and educators at work and backed by monetary support from the funding organizations, the target of WHO to achieve the elimination of leishmaniasis by the year 2030 can be foreseeable.

Keywords

Visceral leishmaniasis · Innovations · Diagnostics · Drug development · Vaccine development · Epidemiology · Kala azar

1.1 Leishmaniasis: Problems Worldwide

The protozoan parasite *Leishmania* is the causative agent of the broad-spectrum neglected tropical disease (NTD) called leishmaniasis. Their discovery as human pathogens is attributed to William Boog Leishman and Charles Donovan, who, in the year 1901, independently visualized these cells under the microscope, living inside the infected human tissues. The parasite belongs to the Kinetoplastida order of Trypanosomes and follows a digenetic life cycle, shuttling between an insect vector and a mammalian host. The parasite resides extracellularly in the phlebotomine sand fly gut, where it presents itself in a flagellated form known as promastigote. Vector-to-host transmission occurs when an infected female sand fly bites a healthy human, allowing the parasite to invade its stamping ground—the cellular macrophages, and establish infection, leading to disease manifestation in the human host. The free-living promastigote form within the poikilothermic vector shows adaptive changes, losing its flagella to become an amastigote—the obligate intracellular form in the homoeothermic mammalian host. However, the basic cellular architecture is conserved between the two *Leishmania* cell shapes (Sunter and Gull 2017). The change from promastigote to amastigote helps avoid the host's immune response (Elmahallawy et al. 2021).

Leishmaniasis is often related to poverty, illiteracy, weak immune system, malnutrition, gender inequality, lack of resources, and poor housing. Generally, high morbidity and low mortality of infectious diseases are well-recognized determinants of poverty (Okwor and Uzonna 2016). As such, while the disease doesn't contribute significantly to global deaths, it continues to pose threat to the livelihood of the poorer populations, preventing them from escaping poverty by impacting their agriculture and livestock and affecting cognitive, developmental, and educational outcomes. Leishmaniasis is a debilitating healthcare problem spread across 96 countries on all continents except Oceania. Current estimates show an annual global incidence of 0.2–0.4 million cases of visceral leishmaniasis (VL) and 0.7—1.2 million cases of cutaneous leishmaniasis (CL). The number of new cases may vary or change over time and are difficult to estimate (Mann et al. 2021). Limited and poor reporting of new cases in the affected areas hints that the actual disease burden often goes undetermined (Okwor and Uzonna 2016). Over 90% of new VL cases occur in the poor rural and suburban areas of Bangladesh, Ethiopia, India, Nepal, South Sudan, Sudan, and Brazil. Cutaneous leishmaniasis is much more widespread, particularly in South and Central America, the Middle East, and Central Asia (Panahi et al. 2021; Shita et al. 2022).

Approximately 20 species of the parasite are known to cause leishmaniasis in humans. Different species, together with host factors, determine the diverse clinical manifestations and sequelae (Mann et al. 2021). *L. tropica*, *L. major*, *L. aethiopica*, *L. mexicana,* and *L. braziliensis* are associated with the most commonly occurring type, i.e., cutaneous leishmaniasis (CL), which involves skin lesions, mainly ulcers, at the bite site. Mucocutaneous leishmaniasis (MCL) is more serious, causing destructive lesions in the mouth and nose. *L. donovani* and *L. infantum* are responsible for causing visceral leishmaniasis (VL), a more progressive form of the disease and is fatal if left untreated (Mann et al. 2021).

New epidemics keep occurring in the endemic areas due to the lack of effective therapeutics. Factors such as tourism, migration, and military activities contribute to the spread of the disease to nonendemic regions. In addition, the international traffic of blood products has resulted in *Leishmania* infections of patients, who never traveled to leishmaniasis endemic regions. There is also evidence that global warming will lead to an extension of the distribution of sand flies more northward, which could result in the transmission of leishmaniasis in hitherto nonendemic regions in the future (Steverding 2017).

1.2 Visceral Leishmaniasis

Visceral leishmaniasis, also known as Black Fever (or Kala Azar in the endemic regions), is considered the most severe form of the disease. In some patients who recover from VL, the causative parasite presents itself in the fourth form of the disease, which is called Post kala-azar dermal leishmaniasis (PKDL).

The visceral disease spreads over a large part of the South and East Asia (mainly in India and China), a large part of Africa, the Mediterranean (affecting children and adults), and South America (where children are affected). It is caused by *L. donovani* (India and Eastern Africa), *L. infantum* (Mediterranean area), (South America) (Mann et al. 2021; Torres-Guerrero et al. 2017).

The transmission characteristics of VL are zoonotic in areas of *L. infantum*-caused infection. Here, the parasite alternates between a reservoir animal (mainly dogs), the sand fly vector and human hosts. The other type of VL is anthroponotic, transmitted from human to vector to human and found in areas of *L. donovani* transmission (Khosravi et al. 2017). Nevertheless, due to the variety of the *Phlebotomus* species, leishmanial species, and foci specificity of these, identifying the type of vector that transmits the parasite species and their role in transmission to animal reservoirs and man is often difficult. Congenital and parenteral transmissions (through needle sharing among drug addicts) have also been reported (Cruz et al. 2002). There is, however, no direct transmission from person to person observed (Singh et al. 2020, 2021).

It is widely accredited that the frequency and transmission dynamics of VL are closely interrelated to socioeconomic, climatic, and environmental factors, including land use/land cover, topography, rainfall, temperature, and vegetation coverage (Bhunia et al. 2013). Sand flies breed in high relative humidity, warm temperatures, high subsoil water, and an abundance of vegetation. VL incidence is seen to have seasonal peaks in the spring that are likely due to temperature, humidity, and vector habits (de Souza Fernandes et al. 2022; Deb et al. 2018). The transmission risk is also highest from dusk to dawn because this is when sand flies are generally the most active (Pareyn et al. 2020). The at-risk population in the endemic regions mainly includes preschool children and immunocompromised and undernourished individuals (Torres-Guerrero et al. 2017). VL contributes significantly to household economic loss, as shown by individual studies across countries. The socioeconomic bias of the disease prevalence largely in the poorer sections of society is indeed why leishmaniasis is called as a "poor man's disease."

An estimated 50,000–90,000 new cases of VL occur worldwide annually, with only 25–45% reported to World Health Organization (WHO). In 2020, more than 90% of new cases reported to WHO occurred in 10 countries: Brazil, China, Ethiopia, Eritrea, India, Kenya, Somalia, South Sudan, Sudan, and Yemen (Yimer et al. 2022). The infection is characterized by a broad clinical spectrum, ranging from mild (oligo symptomatic) to moderate and severe clinical manifestations. The incubation period can range from 10 days to 34 months, with an average of 3-–8 months. If symptoms develop and the full-blown disease is left untreated, Visceral Leishmaniasis is usually fatal.

The classic manifestations of VL include fever, splenomegaly (enlargement of the spleen, manifested in the great majority of patients), hepatomegaly (enlargement of the liver), pallor (caused by severe anemia), leukopenia (low white blood cell count), and weight loss or even severe wasting of the body (cachexia). Other signs and symptoms include respiratory problems or gastrointestinal disturbances such as vomiting and diarrhea; in severe cases, malnutrition and lower limb edema may progress to anasarca (extreme generalized edema). Other important signs are bleeding from the nose or mouth, jaundice, and fluid build-up in the abdomen (Varma and Naseem 2010). VL symptoms often persist for several weeks to months before patients seek medical care (Chappuis et al. 2007). The fatality of untreated cases, usually within 2 years, results from organ failure, anemia, or secondary infections.

The major symptoms caused by *L. donovani* and *L. infantum* are not generally distinguishable (Ready 2014). However, the clinical presentation of VL sometimes differs in the endemic areas. For example, hyperpigmentation, which probably led to the name kala azar, has only been described in VL patients from the Indian subcontinent, not commonly observed nowadays (Chappuis et al. 2007). Similarly, enlarged lymph nodes are rarely found in Indian VL patients but are frequent in Sudanese VL patients (Siddig et al. 1990).

VL is an opportunistic infection associated with HIV infection and is emerging as a major threat, having been reported in 45 countries as of 2021. Most of the patients were initially reported from Southwestern Europe, but the number of coinfected patients is increasing, especially in Ethiopia, Brazil, and Bihar in India (Sundar 2015).

PKDL is a postrecovery sequela of VL. It is characterized by indurated nodules or depigmented macules, which start from the mouth and then spread to other body parts depending upon the severity of the infection. It is largely restricted to areas where *L. donovani* is the causative parasite; cases are mainly reported from Sudan and India. PKDL serves as a reservoir of infection, and its effective treatment is essential for VL elimination (Zijlstra et al. 2003).

VL is a systemic illness. Once the *Leishmania* promastigotes are deposited into the skin of the human host by the sand fly vector, they are taken up by the dendritic cells and macrophages of the dermis, where they differentiate into the infectious amastigotes and proliferate extensively by binary fission (Chappuis et al. 2007). Entry into the macrophages is mediated primarily through immunoglobulin-mediated phagocytosis (Torres-Guerrero et al. 2017). Evasion of the host immunity and persistence in the host are achieved through a combination of strategies, including neutralization of complement components, preventing the release of macrophage superoxide and nitric oxide, and suppressing induction of antigen-specific CD4+ T helper lymphocytes (Bogdan 2008; Selvapandiyan et al. 2022). After successfully evading the defensive attack of the host immune system, *Leishmania* amastigotes sneak into the vascular and lymphatic systems to attack other cells in the reticuloendothelial system, thereby infecting the liver, spleen, bone marrow, and other viscera.

The immunological basis of the disease involves an imbalance between TH1 and TH2 responses, and people with primary TH1 responses are shown to have excellent parasite control. TH2 responders are more likely to develop disseminated disease, which leads to visceral disease (Selvapandiyan et al. 2009; Volpedo et al. 2021). A TH2 response leads to increased parasite load as antibody neutralization is ineffective against the intracellular parasite (Mann et al. 2021).

The review here identifies the various reasons that could contribute to establishing VL (Fig. 1.1). It also identifies the concomitant challenges and the solutions.

1.3 Innovations in Epidemiology

VL is the second largest parasitic killer in the world (after malaria), responsible for an estimated 0.7–1 million new cases and 26,000–65,000 deaths each year globally (DebRoy et al. 2017). However, its complex epidemiology makes the assessment of the disease burden challenging (Bern et al. 2008). It has been reported that the total number of estimated cases could be 2–2.5 times higher than the actual incidence and maybe even five times higher than the officially reported figures. In the absence of accurate statistics, it is difficult for health planners and policymakers to evolve a suitable control strategy for elimination of VL (Singh et al. 2010). Therefore, scrutiny of this disease remains necessary for its control, eradication, and prevention.

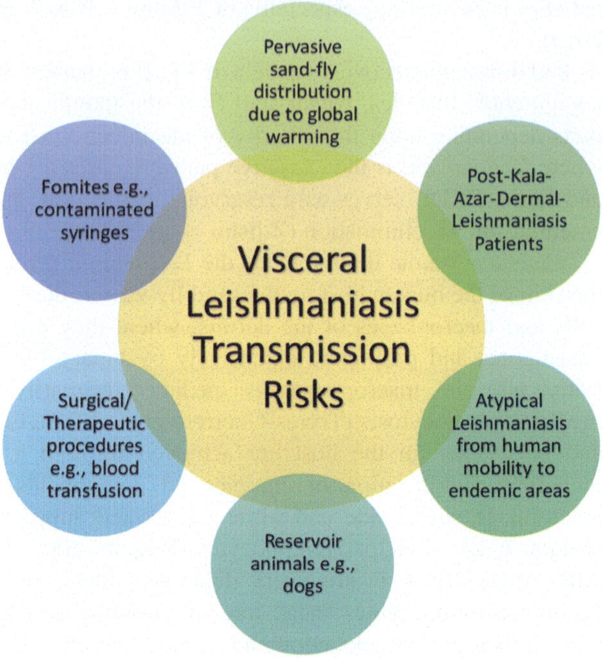

Fig. 1.1 Visceral leishmaniasis transmission risks that describe the areas where a major focus is needed to identify the challenges and the subsequent solutions

Following the 60th World Health Assembly in the year 2007, the WHO gathered support from all the member states and laid out a minimum set of indicators to monitor the global burden of leishmaniasis. The organization's Global Health Repository (GHO) lists six of these indicators including (1) status of endemicity of VL; (2) number of cases of VL reported; (3) number of imported cases of VL reported. The rest three indicators similarly describe the CL cases globally (Ruiz-Postigo et al. 2020).

Strategies for examining the worldwide distribution of disease cases have evolved through the years. The first comprehensive overview of leishmaniasis incidence published by the WHO's *Leishmania* control program was based on an empirical study (Alvar et al. 2012). The surveillance method included data collection of local VL cases from 98 countries over a period of 5 years, assisted with a literature review. Mapping technology (GIS) used the compiled epidemiological data to develop final incidence estimates (Bhunia et al. 2013). The WHO World Epidemiological Record (WER) provides data of the disease burden with time trends from 1998 to 2016. In a more recent report from 2017 to 2018, the WHO mentions PKDL burden and VL case fatality rate (from all countries) as well as gender and age distribution and HIV coinfection rate for VL (from high-burden countries) as additional indicators of disease surveillance (Ruiz-Postigo et al. 2020).

These reported data, however, came with limitations, mainly due to underreporting and surveillance gaps. Since VL occurs mainly in rural and remote populations, most disease-related deaths occur outside medical facilities. Therefore, uncertainty in the mortality range poses yet another limitation (Kruk et al. 2018; Wamai et al. 2020).

Recent innovations in disease epidemiology also involve the use of mathematical models for predicting disease transmission and dynamics. Such models combined with computer-aided simulations help researchers directly predict future VL prevalence and subsequently identify disease control strategies to fight a VL pandemic (Pigott et al. 2014). The researchers used a boosted regression tree modeling framework to create a comprehensive database of both CL and VL occurrences worldwide. Their study highlights areas and countries requiring more leishmaniasis treatment and prevention methods and identifies the focality of the distribution of cases (Wamai et al. 2020). These reports collectively indicate three global endemic foci, or hotspots, for VL: in East Africa (Ethiopia, Kenya, Somalia, South Sudan, Sudan, and Uganda), in the Indian subcontinent (Bangladesh, India, and Nepal), and in Brazil. The WER, which is the most recent of the three studies, indicates a shift in the global distribution of VL. For example, the proportion of global VL burden from East Africa rose from 40% to 50% between 2015 and 2016. On the other hand, the proportion of global burden from the Indian subcontinent decreased from 39% to 30% within the same time frame. Brazil, however, did not change and consistently represented 14% of the global VL burden. Between 2017 and 2018, East Africa and India weighed 45% and 28%, respectively, whereas Brazil showed a leap to 20%.

Since WHO designated VL as an NTD in 2015, an abundance of research studies have focused on developing environmental mathematical models of VL. Various researchers have made incremental contributions. For example, Subramanian et al. proposed a compartment-based mathematical model of VL transmission to explain disease transmissions in symptomatic VL, asymptomatic VL, and PKDL-infection classes (Subramanian et al. 2015). Sensitivity analysis of model parameters found that the biting and birthing rates of sand flies and the recovery rate of symptomatic humans are dominating factors for VL epidemic control (Bi et al. 2018). Thompson et al. studied relationships between climate and VL epidemics by establishing a statistical regression model. Their research found that rainfall is the most significant parameter statistically correlated to VL incidences, and the foothill populations are at higher risk of disease infections (Thompson et al. 2002). Karagiannis-Voules et al. built another model that predicted that regions with humid climates and dense vegetation distributions are more vulnerable to VL than other regions (Karagiannis-Voules et al. 2013). In making the modeling process increasingly dynamic, statistical and machine learning models can utilize real-world data to ensure accuracy while predicting possible trends of an ongoing epidemic (Bi et al. 2018). Montenegro or Leishmanin test is an intradermal skin test (developed on the lines similar to the Mantoux tuberculin skin test for TB), which is also being explored for use in epidemiological surveys to identify high-risk populations (Mann et al. 2021).

By and large, the prevalence of VL has been proven to be influenced by environmental, meteorological, and socioeconomic factors. Changes in these factors are likely to lead to fluctuation in the prevalence and distribution of the disease by affecting the habitat of vectors and animal hosts, as well as the interaction of humans and vectors or animal hosts (Zhao et al. 2021).

1.4 Innovations in Diagnosis

The clinical presentation of VL lacks specificity, and its symptoms resemble common infections like typhoid fever, tuberculosis, brucellosis, and malaria (Mann et al. 2021), as well as some hematologic malignancies (Safavi et al. 2021). As such, the diagnosis of VL is tricky. In light of the toxicity of the available drugs and the degree of fatality of VL, the diagnostic tests should be highly specific and sensitive and should ideally be able to distinguish between an acute disease and an asymptomatic infection. Moreover, such tests should be simple and affordable.

The history, epidemiology, symptoms, and signs on physical examination should alert the clinician of the possible diagnosis of leishmaniasis (Mann et al. 2021). However, species identification of *Leishmania* and confirmatory tests to identify which patients need to be treated are important to prevent misdiagnosis (Chappuis et al. 2007; Torres-Guerrero et al. 2017).

Traditionally, the gold-standard VL diagnosis is the direct visualization of the parasite via microscopy or culture on invasive samples (spleen, bone marrow, or lymph node aspirates or liver biopsy). The diagnostic yield is highest (93–99%) for spleen aspirates, which is routinely done only in Eastern Africa. (van Griensven and Diro 2019).

However, spleen aspiration can be complicated by life-threatening hemorrhages in ~0.1% of individuals and therefore requires considerable technical expertise, as well as facilities for nursing surveillance, blood transfusion, and surgery (Chappuis et al. 2007). Therefore, bone marrow is the preferred first source despite lower sensitivity (50–80%) (Mann et al. 2021) and is more commonly done in Europe, Brazil, and the United States (van Griensven and Diro 2019). Blood specimens are useful in diagnosing HIV-VL coinfections and immunocompromised patients with higher levels of parasitemia (Mann et al. 2021).

The detection of parasite DNA by PCR in blood or bone marrow aspirates is substantially more sensitive than microscopic examination and has found use in Europe and North America. Owing to high sensitivity, PCR detects more asymptomatic infections (Scarpini et al. 2022). However, the cost of testing limits its suitability in resource-restricted countries. Moreover, the lack of standardized primers and protocols used in laboratories makes the diagnostic accuracy less reliable (van Griensven and Diro 2019).

Serological tests, which include ELISA, IHA, IFAT, and Immunoblotting, also provide fair sensitivity and specificity. With the advent of new tests, some of these are rarely used in routine practice (van Griensven and Diro 2019). Considerations that strongly limit the accuracy of serological methods are: first, relapses cannot be reliably diagnosed because previous antibodies remain detectable for several years after cure; and secondly, a significant proportion of healthy people living in endemic areas with no history of VL are positive for antileishmanial antibodies owing to asymptomatic infections.

Two serological tests developed and sufficiently validated for field use are the direct agglutination test (DAT) and the rK39 dipstick test. The rK-39 antibody test involves the detection of antibodies against a 39-amino acid repeat that is part of a kinesin-related protein in *Leishmania chagasi* and which is conserved within the *L. donovani* complex (Canavate et al. 2011). These tests have over 90% sensitivity and specificity for diagnosing VL in regions like India and Nepal. They, however, show variable sensitivity and limited use for PKDL and VL-HIV cases (Selvapandiyan et al. 2019).

Approaches to develop more reliable, cost-effective, and noninvasive diagnostic methods have been in the pipeline. Urine antigen detection by ELISA, real-time PCR, and loop-mediated isothermal amplification (LAMP) assays have been shown to improve diagnosis and measure treatment outcomes in VL, PKDL, and relapsed VL cases that are more difficult to test (Selvapandiyan et al. 2019). Dendrogram-based kinetoplast DNA sequencings that differentiate the important various species of *Leishmania* seen globally were developed and validated using clinical samples from India and US soldiers (Selvapandiyan et al. 2008). In addition, a novel multiplex fluorescence-PCR-based detection of certain blood-borne pathogens (bacterial and parasitic) simultaneously in blood including for *Leishmania* was demonstrated (Selvapandiyan et al. 2005). As accuracy looks comparable to conventional methods, molecular tests are likely to be increasingly used in resource-constrained endemic areas. Noninvasive methods of cure like quantitative PCR and antigen tests may also prove beneficial with respect to immunocompromised patients on whom the treatment fails to work (van Griensven and Diro 2019). Irrespective of the infallibility of the old and new techniques, the test specificity primarily depends on the level of the health system at which it is administered.

1.5 Innovations in Therapeutics

Visceral leishmaniasis manifests as a systemic infection in affected individuals; hence, timely intervention is pivotal to the survival of the individual. A panel of the Infectious Diseases Society of America (IDSA) and the American Society of Tropical Medicine and Hygiene (ASTMH) recommends that persons with clinical abnormalities compatible with VL and laboratory evidence of VL should be given treatment. Their guidelines suggest close monitoring of asymptomatic individuals with the initiation of treatment only if symptoms develop (Ready 2014). Treatment strategies differ among patients depending on disease presentations, patient's age and status of immunocompetency, geographic region, and systematic therapy side effects (Safavi et al. 2021). Effective treatment and disease cure require an immunocompetent system because medicines will not get rid of the parasite from the body, thus the risk of relapse if immunosuppression occurs (Safavi et al. 2021).

Currently, the treatment of VL relies largely upon antileishmanial drugs. The intravenous pentavalent antimonial monotherapy being cheap, effective, and well tolerated has been considered the standard of care for VL in many areas for more than seven decades (Chappuis et al. 2007; Sundar and Agarwal 2018). In the Indian subcontinent, parasite susceptibility to the antimonials has decreased substantially. Consequently, sodium stibogluconate is no longer recommended in India (Burza et al. 2018). Amphotericin B deoxycholate and pentamidine have been used as second-line medicines. However, these have many adverse effects, and the drug is expensive, too. Therefore, this drug does not offer any obvious advantages for areas outside India over pentavalent antimonial agents (Sundar and Agarwal 2018). Recent clinical research has focused on shorter regimens and avoiding resistance (Burza et al. 2018). Lipid formulation of amphotericin, commercially sold as AmBisome, is considered by many experts as the best existing drug against VL and is used as first-line treatment in Europe and the United States (Chappuis et al. 2007). The drug also results in a good response in immunocompromised patients, but the percentage of recurrences in these patients is high (Torres-Guerrero et al. 2017). Other alternative antileishmanial agents are miltefosine, the only oral antileishmanial drug used to date, and paromomycin (van Griensven and Diro 2019). However, a comprehensive evaluation of the role of these drugs in reducing disease incidence is yet to be determined. A rising limitation of miltefosine use in the Asian Elimination Program is evidence of reduced effectiveness in both visceral leishmaniasis and PKDL. There is also a need for follow-up studies to show whether the single-course AmBisome regime is associated with higher PKDL incidence (Burza et al. 2018; Selvapandiyan et al. 2019).

Compounds with synergistic or additive activity acting at different sites shorten the duration of therapy and decrease dose requirement, thereby reducing the chances of toxic side effects and cost and preventing the emergence of drug resistance (Sundar and Agarwal 2018). This rationale has paved the way for using these drugs in combination regimens and has proven remarkably effective. The AmBisome and Paromomycin/Miltefosine combination therapy is now included in WHO recommendations for the Indian subcontinent.

HIV-coinfected patients present unique problems. While AmB is the research-confirmed drug for such patients, their treatment response is still poor. Moreover, relapse rates are about 60% in the first year, regardless of the medication employed (Lindoso et al. 2018). Antiretroviral therapy and secondary prophylaxis targeting the specific parasite may improve the response to antileishmanial treatment, avoid or delay relapses of leishmaniasis, and improve overall survival (Aderie et al. 2017).

Different modifications and innovations in conventional therapeutics can still not ameliorate these drugs' adverse effects. It implies that there is a need to investigate new targets for the chemotherapy of visceral leishmaniasis. Various biomolecules and pathways unique to the parasite and important for its survival and virulence are being studied as druggable targets for the design of novel antileishmanial drug molecules or potential antileishmanial vaccines. These include, for example, proteins and enzymes regulating DNA replication and cell cycle, thiol metabolism, kinases, and synthetases. Several nanotechnology-based formulations for drug delivery and optimization are also in different phases of clinical trials (Jain and Jain 2018).

However, the limitations of drug therapy go beyond efficacy, drug resistance, and drug toxicity. Injectable administration is another problem of treatments since it requires patients to go to health centers, hospitalization, and professional administration, which are conditions that are not adapted to the reality of the poverty conditions of patients with the disease (de Souza et al. 2020). Thus, overall, the available treatment options for VL are limited and not up to satisfactory standards; hence, vaccination seems as the best hope for controlling leishmaniasis (Mann et al. 2021; Singh et al. 2016).

The development of a vaccine will benefit an estimated 350 million people, who are at risk of developing the disease worldwide (Okwor and Uzonna 2016). The quest for developing a successful vaccine candidate is still underway. Nevertheless, the strong immunity conferred to leishmaniasis patients after primary infection supports vaccination as an effective prevention measure against the disease. Protection from subsequent infections has also been observed through leishmanization, which is the voluntary inoculation of live, virulent *L. major* promastigotes into healthy humans. While the practice is not encouraged under normal conditions, leishmanization has been carried out in Uzbekistan, Iran, and Israel (Chappuis et al. 2007).

Various investigations and trials have employed different strategies in vaccine development. While results from whole-killed parasite vaccines have been inconclusive, second-generation vaccines consisting of recombinant *Leishmania* proteins and genetic vaccines have proven effective in dogs (one of the primary parasite reservoirs) and are being reviewed. LEISH-F3 + GLA SE is one such promising candidate being developed by Infectious Disease Research Institute (IDRI, Washington, Seattle using recombinant proteins from *L. donovani* and *L. infantum* (Mann et al. 2021). Other groups, including the Sabin Vaccine Institute in collaboration with the National Institutes of Health, are exploring prototype combination vaccines comprising recombinant *L. donovani* antigens together with a recombinant protein encoding one of several different sand fly salivary gland antigens (Gillespie et al. 2016).

In addition, a gene (a growth regulating "centrin" gene) deletion-based development of parasite live-attenuated vaccine as a concept in *L. donovani* was undertaken against VL and tested extensively including safety, efficacy, and toxicity analysis in the animals and in human cells using funds from Indo-US vaccine development, Department of Biotechnology, Government of India by USFDA, National Institute of Pathology, New Delhi, India, and Department of Molecular Medicine, Jamia Hamdard, New Delhi (Avishek et al. 2016; Bhattacharya et al. 2016; Dey et al. 2014; Fiuza et al. 2015; Fiuza et al. 2013; Selvapandiyan et al. 2004, 2009, 2014). In the same line of thought, another more promising centrin-deleted *L. major* and *L. mexicana* live-attenuated vaccine candidates have been developed, by USFDA, Ohio State University, Ohio, USA, and McGill University, Montreal, QC, Canada, OH, using marker-free CRISPR Cas9 technology with funds from Global Health Innovative Technology Fund and CBER Intramural Research Program, FDA (Volpedo et al. 2021, 2022a, 2022b; Zhang et al. 2020). The *L. major* vaccine candidate is currently in clinical testing against leishmaniasis. More recently, a third-generation vaccine candidate, ChAd63-KH, has completed a first-in-human clinical trial for VL/PKDL with beneficial therapeutic outcomes (Younis et al. 2021). The study demonstrated the safety of this adenoviral vaccine in healthy volunteers and induced potent innate and cell-mediated immune responses against the two *Leishmania* antigens—KMP-11 and HASPB.

Critical to advancing *Leishmania* vaccine research and development is adequate funding and support from international agencies. Major public sector and philanthropic funders like the Bill & Melinda Gates Foundation, the Carlos Slim Foundation, the European Commission, the German Federal Ministry of Education and Research (BMBF), the Indian Council of Medical Research (ICMR), Institut Pasteur, and the U.S. NIH, and Wellcome Trust have been directing their funding toward prophylactic as well as therapeutic *Leishmania* vaccine development (Gillespie et al. 2016).

1.6 Visceral Leishmaniasis: Elimination Strategies

Disease elimination and eradication are the ultimate goals of public health, evolving naturally from disease control, and require the confluence of biological, political, and socioeconomic factors (Dowdle 1998). The WHO launched the VL elimination campaign in Southeast Asia in 2005. The elimination target was defined as the reduction to less than one case for 10,000 inhabitants in these districts by 2015. However, the set goal was extended to 2017 and then to 2020 (Wamai et al. 2020). The VL elimination program is currently underway, and the roadmap has set 2030 as the target year. The global estimate from WHO shows that the total cases were under 13,000 in 2020 compared to nearly 60,000 in 2011, with a 75% decrease in *Leishmania* cases from 2015 to 2020 in India alone. Although the progress toward disease elimination appears to be growing, the success of the program still remains dubious in the light of already existing and newly erupting challenges like sporadic outbreaks, asymptomatic infections, and changing foci.

Nonetheless, the campaign has proven fruitful in many regions. The declining trend in the number of VL cases from 2011 till now can be largely attributed to the VL elimination program in Southeast Asia region. China is a notable example where the total cases had fallen from 530,000 in 1951 to only 378 cases in 2012. Even more plausible were Nepal and Bangladesh achieving the elimination threshold in 2013 and 2016, respectively (Rijal et al. 2019).

To boot, the Seventh World Congress on Leishmaniasis—WorldLeish7 held in Colombia in 2022 has brought positive updates from the SEAR, where the VL reported cases in three endemic regions have dropped from 2335 cases in 2020 to 1577 in 2021. Bangladesh in particular has remained successful in sustaining the elimination threshold. In view of its success, the WHO SEARO aims at launching a strategic plan of integrating the regional initiatives across South East Asia to scale up the elimination of VL in the near future (Balboni 2022).

In the Americas, a 2017 epidemiological analysis reported by a regional information system—SisLeish, has shown an increased incidence in Brazil of 26.4% from

2016. Unlike Southeast Asia, East Africa does not have a regional approach to achieving the elimination of visceral leishmaniasis as a public health problem. The goal of the WHO 2021–30 NTDs roadmap is to reduce mortality caused by the disease to less than 1% (Alvar et al. 2021). Emphasis should be laid on strategic planning and execution of measures to help contain the disease transmission through reservoir and vector control, along with improved surveillance and case detection to achieve the goal. The following bullet points bring out such specific strategies.

- Measures to control vectors have primarily been indoor residual spraying of insecticides using insecticide-treated nets, environmental management, and personal protection. Newer tools being evaluated include wall paint containing three insecticides, including a larvicide, and an insecticide-repellent combination for canine leishmaniasis (Rijal et al. 2019).
- Effective disease surveillance will entail vigilant and sustainable disease monitoring and a strengthened healthcare system. Since passive case detection has previously led to underreporting cases in Bihar state of India, the surveillance strategy has recently shifted to the active case detection method (ACD). ACD not only reduces the time to diagnose and thus the risk of transmission but also ensures a double check on the proportion of cases getting captured (Dubey et al. 2021). The WHO-driven web-based surveillance of the disease burden has also improved the quality and quantity of the reported data in the recent years.
- Control of animal reservoir hosts, primarily dogs, is a practiced strategy, though debatable. Vaccination of dogs would nevertheless be the best strategy if an efficacious vaccine can be developed (Dubey et al. 2021).
- Together with complete case management, improved diagnostics and therapeutics will help reduce the disease prevalence, control the cyclical transmission patterns, and prevent disabilities or deaths caused by leishmaniasis (Rijal et al. 2019).
- In the case of anthroponotic leishmaniasis, PKDL patients may continue to have symptoms for years without systemic illness and treatment, so they probably constitute the main interepidemic reservoir. The measures mentioned above, accompanied with steps for preventing reinfection, should reduce or eliminate the parasite load and reduce transmission (Zijlstra et al. 2020).
- Resource mobilization, public–private partnerships, and community mobilization are equally important and must be prioritized (Narain et al. 2010).
- Studies looking at the national and global socioeconomic impact of leishmaniasis also need to be enhanced as they will help in redesigning the existing mitigation strategies or formulating new ones (Okwor and Uzonna 2016).

In reaching the disease eradication target, it is necessary for the already existing elimination threshold achieved in South-East Asian countries to sustain. An unflagging commitment from all implementing and funding partners also remains a priority. The fear of losing political and donor interests without attaining a lasting outcome is challenging (Burza et al. 2018). Be that as it may, it is worth mentioning that the world has marked the first NTDs Day ever on January 30, 2020, which promises hope and good things to come for VL and these other diseases to 1 day remove the "neglected" from the term (Wamai et al. 2020).

Acknowledgments AS is supported by research grant from Indian Council of Medical Research (ICMR) (No. GIA/2/VBD/2021/ECD-II) and Biotechnology Industry Research Assistance Council (BIRAC) (No. BT/CRS0214/CRS-10/16).

References

Aderie EM, Diro E, Zachariah R et al (2017) Does timing of antiretroviral treatment influence treatment outcomes of visceral leishmaniasis in Northwest Ethiopia? Trans R Soc Trop Med Hyg 111:107–116

Alvar J, Velez ID, Bern C et al (2012) Leishmaniasis worldwide and global estimates of its incidence. PLoS One 7:e35671

Alvar J, den Boer M, Dagne DA (2021) Towards the elimination of visceral leishmaniasis as a public health problem in East Africa: reflections on an enhanced control strategy and a call for action. Lancet Glob Health 9:e1763–e1e69

Avishek K, Kaushal H, Gannavaram S et al (2016) Gene deleted live attenuated leishmania vaccine candidates against visceral leishmaniasis elicit pro-inflammatory cytokines response in human PBMCs. Sci Rep 6:33059

Balboni A (2022) Worldleish7. Lancet Microbe 3:e734

Bern C, Maguire JH, Alvar J (2008) Complexities of assessing the disease burden attributable to leishmaniasis. PLoS Negl Trop Dis 2:e313

Bhattacharya P, Dey R, Dagur PK et al (2016) Live attenuated leishmania donovani centrin knock out parasites generate non-inferior protective immune response in aged mice against visceral leishmaniasis. PLoS Negl Trop Dis 10:e0004963

Bhunia GS, Kesari S, Chatterjee N et al (2013) Spatial and temporal variation and hotspot detection of kala-azar disease in Vaishali district (Bihar), India. BMC Infect Dis 13:64

Bi K, Chen Y, Zhao S et al (2018) Current visceral leishmaniasis research: a research review to inspire future study. Biomed Res Int 2018:9872095

Bogdan C (2008) Mechanisms and consequences of persistence of intracellular pathogens: leishmaniasis as an example. Cell Microbiol 10:1221–1234

Burza S, Croft SL, Boelaert M (2018) Leishmaniasis. Lancet 392:951–970

Canavate C, Herrero M, Nieto J et al (2011) Evaluation of two rk39 dipstick tests, direct agglutination test, and indirect fluorescent antibody test for diagnosis of visceral leishmaniasis in a new epidemic site in highland Ethiopia. Am J Trop Med Hyg 84:102–106

Chappuis F, Sundar S, Hailu A et al (2007) Visceral leishmaniasis: what are the needs for diagnosis, treatment and control? Nat Rev Microbiol 5:873–882

Cruz I, Morales MA, Noguer I et al (2002) Leishmania in discarded syringes from intravenous drug users. Lancet 359:1124–1125

de Souza Fernandes W, de Oliveira Moura Infran J, Falcao de Oliveira E et al (2022) Phlebotomine sandfly (diptera: Psychodidae) fauna and the association between climatic variables and the abundance of lutzomyia longipalpis sensu lato in an intense transmission area for visceral leishmaniasis in Central Western Brazil. J Med Entomol 59:997–1007

de Souza ML, Gonzaga da Costa LA, Silva EO et al (2020) Recent strategies for the development of oral medicines for the treatment of visceral leishmaniasis. Drug Dev Res 81:803–814

Deb RM, Stanton MC, Foster GM et al (2018) Visceral leishmaniasis cyclical trends in Bihar, India - implications for the elimination programme. Gates Open Res 2:10

DebRoy S, Prosper O, Mishoe A et al (2017) Challenges in modeling complexity of neglected tropical diseases: a review of dynamics of visceral leishmaniasis in resource limited settings. Emerg Themes Epidemiol 14:10

Dey R, Natarajan G, Bhattacharya P et al (2014) Characterization of cross-protection by genetically modified live-attenuated leishmania donovani parasites against leishmania mexicana. J Immunol 193:3513–3527

Dowdle WR (1998) The principles of disease elimination and eradication. Bull World Health Organ 76(Suppl 2):22–25

Dubey P, Das A, Priyamvada K et al (2021) Development and evaluation of active case detection methods to support visceral leishmaniasis elimination in India. Front Cell Infect Microbiol 11: 648903

Elmahallawy EK, Alkhaldi AAM, Saleh AA (2021) Host immune response against leishmaniasis and parasite persistence strategies: a review and assessment of recent research. Biomed Pharmacother 139:111671

Fiuza JA, Santiago Hda C, Selvapandiyan A et al (2013) Induction of immunogenicity by live attenuated leishmania donovani centrin deleted parasites in dogs. Vaccine 31:1785–1792

Fiuza JA, Gannavaram S, Santiago Hda C et al (2015) Vaccination using live attenuated leishmania donovani centrin deleted parasites induces protection in dogs against leishmania infantum. Vaccine 33:280–288

Gillespie PM, Beaumier CM, Strych U et al (2016) Status of vaccine research and development of vaccines for leishmaniasis. Vaccine 34:2992–2995

Jain V, Jain K (2018) Molecular targets and pathways for the treatment of visceral leishmaniasis. Drug Discov Today 23:161–170

Karagiannis-Voules DA, Scholte RG, Guimaraes LH et al (2013) Bayesian geostatistical modeling of leishmaniasis incidence in Brazil. PLoS Negl Trop Dis 7:e2213

Khosravi A, Sharifi I, Fekri A et al (2017) Clinical features of anthroponotic cutaneous leishmaniasis in a major focus, Southeastern Iran, 1994-2014. Iran J Parasitol 12:544–553

Kruk ME, Gage AD, Joseph NT et al (2018) Mortality due to low-quality health systems in the universal health coverage era: a systematic analysis of amenable deaths in 137 countries. Lancet 392:2203–2212

Lindoso JAL, Moreira CHV, Cunha MA et al (2018) Visceral leishmaniasis and hiv coinfection: current perspectives. HIV AIDS (Auckl) 10:193–201

Mann S, Frasca K, Scherrer S et al (2021) A review of leishmaniasis: current knowledge and future directions. Curr Trop Med Rep 8:121–132

Narain JP, Dash AP, Parnell B et al (2010) Elimination of neglected tropical diseases in the South-East Asia region of the world health organization. Bull World Health Organ 88:206–210

Okwor I, Uzonna J (2016) Social and economic burden of human leishmaniasis. Am J Trop Med Hyg 94:489–493

Panahi S, Abbasi M, Sayehmiri K et al (2021) Prevalence of cutaneous leishmaniasis in Iran (2000-2016): a systematic review and meta-analysis study. Infect Disord Drug Targets 21:173–179

Pareyn M, Kochora A, Van Rooy L et al (2020) Feeding behavior and activity of phlebotomus pedifer and potential reservoir hosts of leishmania aethiopica in Southwestern Ethiopia. PLoS Negl Trop Dis 14:e0007947

Pigott DM, Bhatt S, Golding N et al (2014) Global distribution maps of the leishmaniases. Elife 3: e02851

Ready PD (2014) Epidemiology of visceral leishmaniasis. Clin Epidemiol 6:147–154

Rijal S, Sundar S, Mondal D et al (2019) Eliminating visceral leishmaniasis in South Asia: the road ahead. BMJ 364:k5224

Ruiz-Postigo JA, Grouta L, Jaina S (2020) Global leishmaniasis surveillance, 2017–2018, and first report on 5 additional indicators josé antonio ruiz-p. WHO: Weekly Epidemiological Record 95: 265–280

Safavi M, Eshaghi H, Hajihassani Z (2021) Visceral leishmaniasis: kala-azar. Diagn Cytopathol 49: 446–448

Scarpini S, Dondi A, Totaro C et al (2022) Visceral leishmaniasis: epidemiology, diagnosis, and treatment regimens in different geographical areas with a focus on pediatrics. Microorganisms 10:1887

Selvapandiyan A, Debrabant A, Duncan R et al (2004) Centrin gene disruption impairs stage-specific basal body duplication and cell cycle progression in leishmania. J Biol Chem 279: 25703–25710
Selvapandiyan A, Stabler K, Ansari NA et al (2005) A novel semiquantitative fluorescence-based multiplex polymerase chain reaction assay for rapid simultaneous detection of bacterial and parasitic pathogens from blood. J Mol Diagn 7:268–275
Selvapandiyan A, Duncan R, Mendez J et al (2008) A leishmania minicircle DNA footprint assay for sensitive detection and rapid speciation of clinical isolates. Transfusion 48:1787–1798
Selvapandiyan A, Dey R, Nylen S et al (2009) Intracellular replication-deficient leishmania donovani induces long lasting protective immunity against visceral leishmaniasis. J Immunol 183:1813–1820
Selvapandiyan A, Dey R, Gannavaram S et al (2014) Generation of growth arrested leishmania amastigotes: a tool to develop live attenuated vaccine candidates against visceral leishmaniasis. Vaccine 32:3895–3901
Selvapandiyan A, Croft SL, Rijal S et al (2019) Innovations for the elimination and control of visceral leishmaniasis. PLoS Negl Trop Dis 13:e0007616
Selvapandiyan A, Puri N, Kumar P et al (2022) Zooming in on common immune evasion mechanisms of pathogens in phagolysosomes: potential broad-spectrum therapeutic targets against infectious diseases. FEMS Microbiol Rev 47:fuac041
Shita EY, Nibret E, Munshea A et al (2022) Burden and risk factors of cutaneous leishmaniasis in Ethiopia: a systematic review and meta-analysis. Int J Dermatol 61:1336–1345
Siddig M, Ghalib H, Shillington DC et al (1990) Visceral leishmaniasis in Sudan. Clinical features. Trop Geogr Med 42:107–112
Singh VP, Ranjan A, Topno RK et al (2010) Estimation of under-reporting of visceral leishmaniasis cases in Bihar, India. Am J Trop Med Hyg 82:9–11
Singh OP, Singh B, Chakravarty J et al (2016) Current challenges in treatment options for visceral leishmaniasis in India: a public health perspective. Infect Dis Poverty 5:19
Singh OP, Hasker E, Boelaert M et al (2020) Xenodiagnosis to address key questions in visceral leishmaniasis control and elimination. PLoS Negl Trop Dis 14:e0008363
Singh OP, Tiwary P, Kushwaha AK et al (2021) Xenodiagnosis to evaluate the infectiousness of humans to sandflies in an area endemic for visceral leishmaniasis in Bihar, India: a transmission-dynamics study. Lancet Microbe 2:e23–e31
Steverding D (2017) The history of leishmaniasis. Parasit Vectors 10:82
Subramanian A, Singh V, Sarkar RR (2015) Understanding visceral leishmaniasis disease transmission and its control—a study based on mathematical modeling. Mathematics 3:913–944
Sundar S (2015) Visceral leishmaniasis. Tropenmed Parasitol 5:83–85
Sundar S, Agarwal D (2018) Visceral leishmaniasis-optimum treatment options in children. Pediatr Infect Dis J 37:492–494
Sunter J, Gull K (2017) Shape, form, function and leishmania pathogenicity: from textbook descriptions to biological understanding. Open Biol 7:170165
Thompson RA, de Oliveira W, Lima J, Maguire JH et al (2002) Climatic and demographic determinants of american visceral leishmaniasis in Northeastern Brazil using remote sensing technology for environmental categorization of rain and region influences on leishmaniasis. Am J Trop Med Hyg 67:648–655
Torres-Guerrero E, Quintanilla-Cedillo MR, Ruiz-Esmenjaud J et al (2017) Leishmaniasis: a review. F1000Res 6:750
van Griensven J, Diro E (2019) Visceral leishmaniasis: recent advances in diagnostics and treatment regimens. Infect Dis Clin N Am 33:79–99
Varma N, Naseem S (2010) Hematologic changes in visceral leishmaniasis/kala azar. Indian J Hematol Blood Transfus 26:78–82
Volpedo G, Pacheco-Fernandez T, Bhattacharya P et al (2021) Determinants of innate immunity in visceral leishmaniasis and their implication in vaccine development. Front Immunol 12:748325

Volpedo G, Bhattacharya P, Gannavaram S et al (2022a) The history of live attenuated centrin gene-deleted leishmania vaccine candidates. Pathogens 11:431

Volpedo G, Pacheco-Fernandez T, Holcomb EA et al (2022b) Centrin-deficient leishmania mexicana confers protection against new world cutaneous leishmaniasis. NPJ Vaccines 7:32

Wamai RG, Kahn J, McGloin J et al (2020) Visceral leishmaniasis: a global overview. J Glob Health Sci 2:e3

Yimer M, Nibret E, Yismaw G (2022) Updates on prevalence and trend status of visceral leishmaniasis at two health facilities in Amhara regional state, Northwest Ethiopia: a retrospective study. Biochem Res Int 2022:3603892

Younis BM, Osman M, Khalil EAG et al (2021) Safety and immunogenicity of chad63-kh vaccine in post-kala-azar dermal leishmaniasis patients in Sudan. Mol Ther 29:2366–2377

Zhang WW, Karmakar S, Gannavaram S et al (2020) A second generation leishmanization vaccine with a markerless attenuated leishmania major strain using crispr gene editing. Nat Commun 11:3461

Zhao Y, Jiang D, Ding F et al (2021) Recurrence and driving factors of visceral leishmaniasis in Central China. Int J Environ Res Public Health 18:9535

Zijlstra EE, Musa AM, Khalil EA et al (2003) Post-kala-azar dermal leishmaniasis. Lancet Infect Dis 3:87–98

Zijlstra EE, Kumar A, Sharma A et al (2020) Report of the fifth post-kala-azar dermal leishmaniasis consortium meeting, Colombo, Sri Lanka, 14–16 may 2018. Parasit Vectors 13:159

KalaCORE: A Programme to Tackle Visceral Leishmaniasis in South Asia and East Africa (2014–2019)

2

Stefanie Meredith, Margriet den Boer, Sakib Burza, and Simon L. Croft

Abstract

"Tackling Visceral Leishmaniasis in South Asia and East Africa" (also known as KalaCORE), a UK AID funded programme, worked in two regions with the greatest burden of visceral leishmaniasis (VL)—East Africa (South Sudan, Sudan and Ethiopia) and South Asia (India, Bangladesh and Nepal) between April 2014 and June 2019. The overall aim of the KalaCORE programme was to sustainably reduce the economic and health impact of VL in the two regions. KalaCORE's key strategies were to: (a) increase access to diagnosis, treatment and prevention, (b) improve surveillance and control of outbreaks of VL, (c) improve data on VL mortality and morbidity, (d) increase quality of diagnosis and treatment and (e) engage with communities impacted by VL. In this chapter, we present the outcomes of the programme, the challenges faced in achieving these objectives, their implementation and sustainability, as well as lessons learned.

Keywords

Visceral leishmaniasis · Kala-azar · Control · Elimination · East Africa · South Asia

S. Meredith
Mott MacDonald, London, UK

Global Health Consulting, Divonne, France

M. den Boer
Médecins Sans Frontières, Amsterdam, Netherlands

S. Burza · S. L. Croft (✉)
London School of Hygiene and Tropical Medicine, London, UK
e-mail: simon.croft@lshtm.ac.uk

© The Author(s), under exclusive license to Springer Nature Singapore Pte Ltd. 2023

A. Selvapandiyan et al. (eds.), *Challenges and Solutions Against Visceral Leishmaniasis*, https://doi.org/10.1007/978-981-99-6999-9_2

2.1 Executive Summary

"Tackling Visceral Leishmaniasis in South Asia and East Africa" (also known as KalaCORE), a UK AID funded programme, worked in two regions with the greatest burden of VL—East Africa (South Sudan, Sudan and Ethiopia) and South Asia (India, Bangladesh and Nepal) between April 2014 and June 2019. The overall aim of the KalaCORE programme was to sustainably reduce the economic and health impact of visceral leishmaniasis (VL) in the two regions. KalaCORE's key strategies were to: (a) increase access to diagnosis, treatment and prevention, (b) improve surveillance and control of outbreaks of VL, (c) improve data on VL mortality and morbidity, (d) increase quality of diagnosis and treatment and (e) engage with communities impacted by VL.

2.1.1 South Asia

At the time the programme started, the introduction of AmBisome® single-dose treatment, although adopted by the national programmes throughout the region, was not yet implemented, and KalaCORE was critical to establishing it as first-line therapy within the target countries. This involved facility assessments and improvements, cold chain strengthening, as well as clinical training for doctors and healthcare staff. Ninety-eight percent coverage of the target population with AmBisome® treatment was achieved. AmBiCal Assist, a simple user-friendly mobile application for both android and IOs, was developed to assist healthcare providers in charge of preparation and administration of single-dose AmBisome for administration to VL and HIV-VL coinfected patients. Crucial to this programme was the sustained use of evidence about implementation realities and demonstrating to governments how problems could be tackled effectively to produce positive results. This large-scale example of delivering a treatment to hard-to-reach populations for a long-neglected disease offers important lessons for equity and Universal Health Coverage.

India has observed a steep decline in disease burden since the VL elimination initiative was launched in the year 2005. In 2019, only 3128 cases were reported, which was all-time low. This represented a 90% decline in disease incidence since 2005 (32,803 reported cases) (World Health Organization 2020).

Nepal also reported a significant decrease in VL burden from the 2005 level of 1463, to 185 reported cases in 2019, the decrease being particularly significant in the regions close to the border with India (Bihar state). However, most of these 2019 cases were in the new foci in rural higher-altitude regions in western Nepal; this presents new challenges.

Bangladesh represents the greatest success in the region with a reduction in VL burden from a reported 6892 in 2005, 544 at the start of the KalaCORE programme in 2015 to 97 at the end of the programme in 2019. The elimination target was achieved although it remains to be validated by WHO. Political commitment and coordination from experts in government and research organisations were important for this success.

2.1.2 East Africa

In Africa, VL puts high demands on the health system due to the complex diagnostic tree and long treatment courses requiring hospitalisation. VL typically occurs in remote populations in areas with poor health systems and where access to care is also poor because of the lack of infrastructure and conflict. KalaCORE supported the national programmes in Ethiopia, Sudan and South Sudan in capacity building aimed at the development of a pool of trained health workers to deliver quality diagnosis, treatment and care of VL patients.

A context-specific strategy based on building laboratory capacity, improved diagnosis, treatment, case management, surveillance and outbreak investigation and management were developed for Ethiopia, Sudan and South Sudan. Treatment centres were assessed and upgraded where needed. Clinical mentoring teams consisting of a clinician, one or two nurse(s) and a laboratory technician were trained and deployed in South Sudan, Sudan and Ethiopia to visit VL treatment facilities. Due to the security issues and ethnic sensitivities in South Sudan, two teams of different ethnicities responded to outbreak rumours and provided VL care, training to healthcare providers and health education to the populations in remote and often highly insecure settings where in some cases only the most basic health services were available.

2.2 Introduction

Over the past two decades, there have been substantive changes in our approach to the control and elimination of human visceral leishmaniasis (VL), a neglected tropical disease caused by two species of the protozoan parasite (*Leishmania donovani*, *Leishmania infantum*) (Burza et al. 2018). These changes result from: (1) a clearer picture of the numbers and distribution (in Asia, Africa, South & Central America, Europe) of VL (Alvar et al. 2012), (2) the development of resistance to the long used pentavalent antimonial drugs in the Indian subcontinent (ISC) (Croft et al. 2006), (3) the introduction of the first oral drug for VL (miltefosine) and the single course infusion of liposomal amphotericin B (AmBisome) in the ISC (Alves et al. 2018), (4) the implementation of novel diagnostics, including the RDT rk39, for patient detection and epidemiology, (5) a full evaluation of the impact of HIV/VL co-infections (Alvar et al. 2008; Burza et al. 2014) and (6) in 2005, the establishment of a regional alliance to eliminate VL in India, Nepal and Bangladesh supported by the WHO (WHO 2005). The target for VL elimination for this region was originally 2015, but as the elimination target was not reached by that date, this was extended initially to 2020. With the establishment of the WHO NTD Road Map (2021–2030), the goals and measurements for VL control and elimination have been further modified to include both S Asia and East Africa and extended to 2030[1] (WHO 2020a).

Despite the recent developments in VL diagnosis and treatment (Alves et al. 2018; Burza et al. 2018), VL remains one of the most neglected of the WHO-defined neglected tropical diseases (WHO 2020b). VL patients are often among the poorest of the poor and frequently suffer from malnutrition and concomitant diseases. In addition, VL places high demands on health systems due to the complex diagnostic tree, long treatment courses requiring hospitalisation and sometimes nutritional supplements (except in ISC, where diagnosis can be done with a simple rapid test and the current first-line treatment is a single infusion of AmBisome) and frequent clinical complications due to the high prevalence of co-morbidities: malnutrition, anaemia, diarrhoea, pneumonia, malaria, HIV and tuberculosis (Burza et al. 2018). VL typically occurs in low-resource settings where a lack of well-trained personnel and a high staff turnover are common.

Despite these known challenges, there had been limited donor support for VL control programmes, and unlike many of the other NTDs, there was no specific partnership to address this. The London Declaration on NTDs in January 2012 marked the launch of an expanded and coordinated international effort against NTDs by the end of the decade (Uniting to Combat NTDs 2012). It was within this context that the UK government Department for International development (DFID) committed to large-scale funding for VL through the "Tackling Visceral Leishmaniasis in South Asia and East Africa" Programme. This programme, known as the KalaCORE Programme, was established in April 2014 and ran until June 2019 (FCDO 2020).

Here we describe the KalaCORE programme its inception, challenges, achievements and outcomes in today's wider context and look at what this may mean in terms of political commitment for sustainability and health system strengthening for disease control and elimination programmes.

2.3 KalaCORE Programmes in the South Asia and East Africa Regions: Two Approaches Taken

The KalaCORE programme worked in two regions with the greatest burden of VL—East Africa (South Sudan, Sudan and Ethiopia) and South Asia (India, Bangladesh and Nepal) with the overall aim of sustainably reducing the economic and health impact of VL. The overall programme strategies to achieve this aim were to: (a) increase access to diagnosis, treatment and prevention, (b) to improve

[1] The new WHO NTD roadmap has proposed a new target for elimination of KA (VL) as a public health problem. It is defined as achieving less than 1% case fatality rate due to primary KA. It has two more sub-targets for South East-Asia region (SEAR), namely: Number of countries in SEAR validated for elimination <1 case (new and relapses) per 10,000 population at block level in India at 100% by 2030; and PKDL cases detected (VL post-treatment follow-up for 3 years) and treated 100% by 2030.

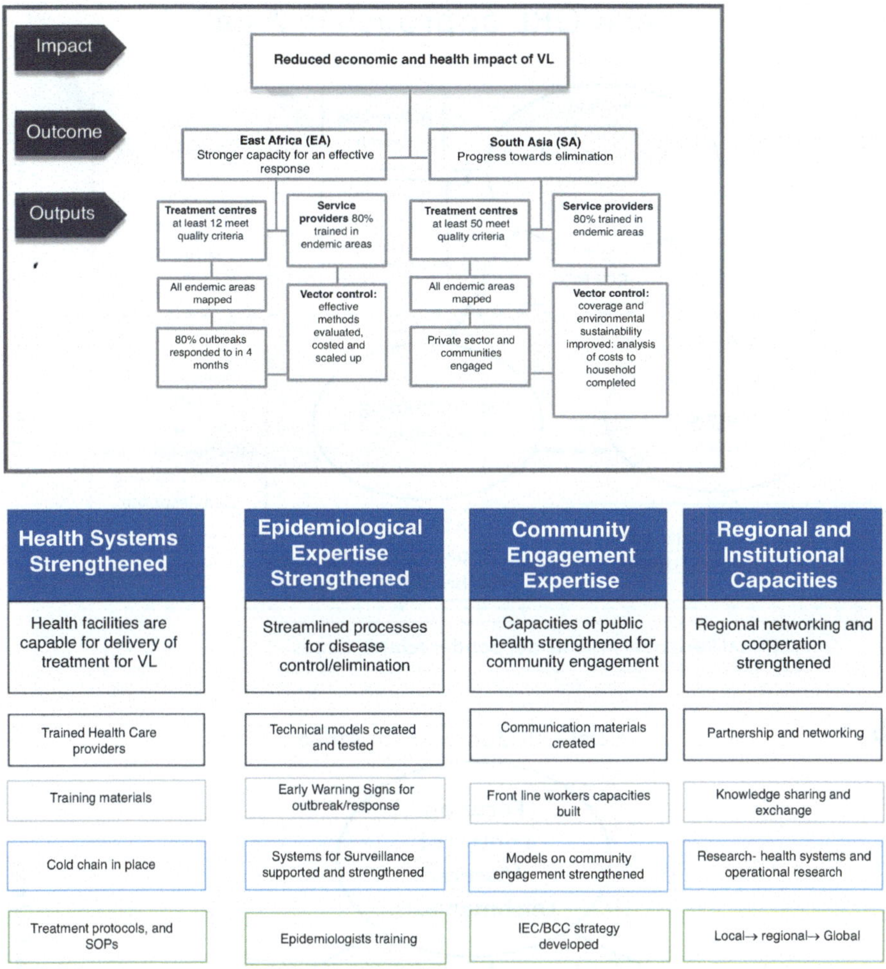

Fig. 2.1 Summary of the overall approach of the KalaCoRE programme

surveillance and control of outbreaks of VL, (c) improve data on VL mortality and morbidity, (d) increase quality of diagnosis and treatment and (e) engage with communities impacted by VL as summarised in Fig. 2.1.

However, the VL situation was different in the two regions and required different approaches as shown in Boxes 2.1 and 2.2.

In South Asia, KalaCORE aimed to support governments in their goal to eliminate VL as a public health problem (defined by WHO at less than 1 case in 10,000 population each year in each endemic sub-district), by providing technical assistance, strengthening surveillance and expanding access to diagnosis and treatment. Essentially, the approach focused on a three pronged approach as illustrated in Fig. 2.2a.

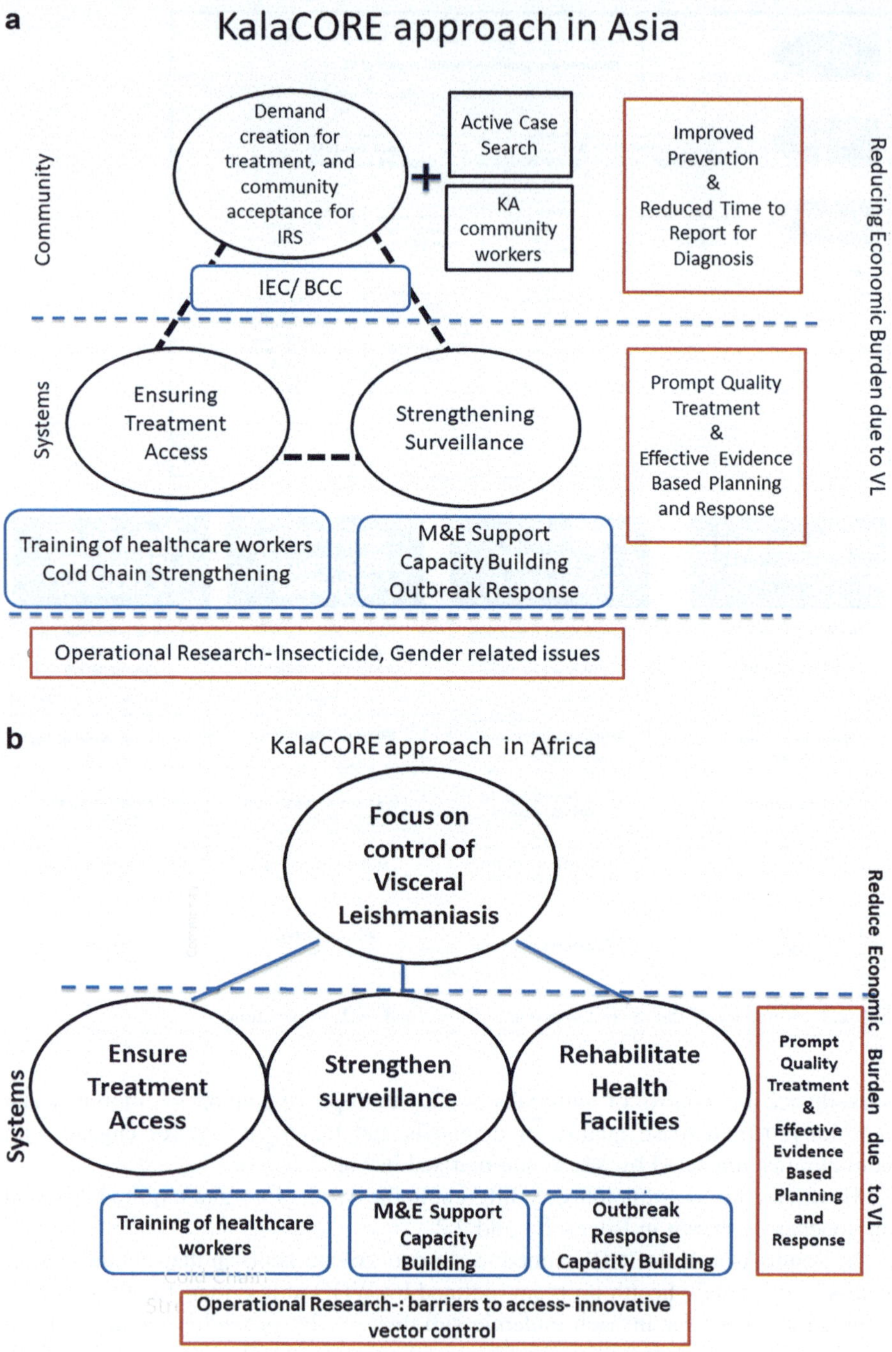

Fig. 2.2 The Kalacore approaches in Asia and East Africa

In East Africa, where less is known about how to prevent transmission of VL, the programme had a focus on VL control through support for: (a) enhanced disease surveillance, (b) capacity building for response to outbreaks, in order to ensure early detection of disease outbreaks and to minimise their spread and impact, (c) improved procurement of drugs and diagnostics, (d) operational research to identify more effective prevention methods (e) improvement of quality of care by training and clinical mentoring and (f) the strengthening and expansion of hospital capacity within endemic areas (Fig. 2.2b).

2.4 Overall Achievements

2.4.1 Towards the Elimination of VL in South Asia

2.4.1.1 Elimination Targets

During the period of the programme, Bangladesh and Nepal reported significant progress to achieve their elimination targets as measured by the proportion of sub-districts and districts, respectively, reaching elimination as a public health problem as shown in Fig. 2.3a, b. The exception to this was India, where the States of Bihar and Jharkhand remained a challenge. In Jharkhand by 2019, only 47% of the blocks achieved the elimination threshold. In Jharkhand, the populations at risk for VL are mainly from the Scheduled Castes and Scheduled Tribes[2] who were not only in hard-to-reach areas but were also suspicious of government officials including health workers and outside organisations like NGOs. Thus, successful strategies for reaching these communities needed to be based on working with local, trusted groups.

Another important criterion for success, the duration of illness, time between onset of symptoms and receiving treatment, was reduced in all countries in Asia and across genders. However, the results from the KalaCORE Monitoring & Evaluation (M&E) and theory of change studies showed that in all countries, patients still visited a number of traditional/private practitioners before they finally received correct diagnosis and treatment at government facilities. Although KalaCORE made a very significant investment in a Behaviour Change Communication (BCC) strategy which focused on treatment-seeking behaviour, it did not fully effect the required change. The involvement of the private and informal health sector in future strategies will be paramount.

A major component in progress towards elimination was the use of single infusion treatment with AmBisome, for 98% of primary VL cases, shortening the diagnosis and treatment time and reducing hospitalisation time and attendant issues. The use of AmBisome, however, created specific demands to the health system which KalaCORE addressed, including (1) the roll-out of a cold chain by Ice Line

[2]The Scheduled Castes and Scheduled Tribes (STs) are officially designated groups of people and among the most disadvantaged socio-economic groups in India.

a

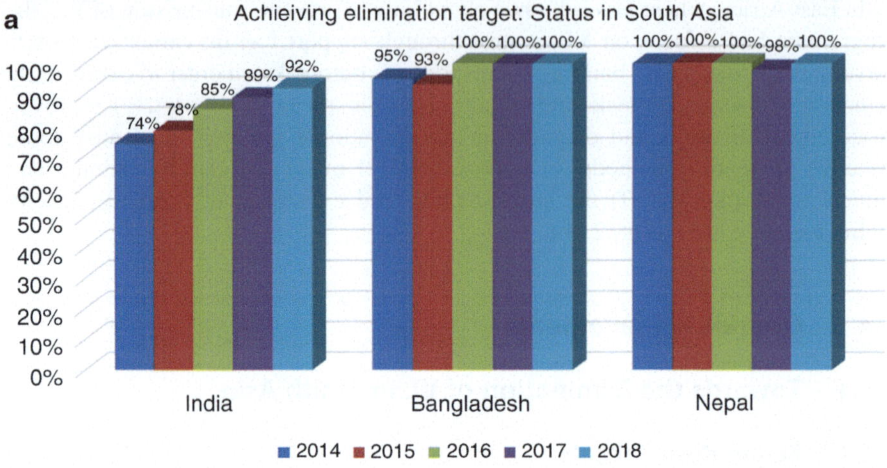

b

BANGLADESH	Year	Number of sub-districts (upazillas) with ≥1 case / 10,000 population	Number of sub-districts (upazillas) with <1 case / 10,000 population	Number of endemic upazillas	% of sub-districts (upazillas) above the elimination threshold
	2014	8	92	100	8%
	2015	4	96	100	4%
	2016	0	100	100	0%
	2017	0	100	100	0%
	2018	0	100	100	0%

INDIA	Year	Number of sub-districts (blocks) with ≥1 case / 10,000 population	Number of sub-districts (blocks) with <1 case / 10,000 population	Number of endemic blocks	% of sub-districts (blocks) above the elimination threshold
	2014	157	439	596	26%
	2015	136	488	624	22%
	2016	94	548	642	15%
	2017	72	561	633	11%
	2018	53	580	633	8.30%

NEPAL	Year	Number of endemic districts with ≥1 case / 10,000 population	Number of endemic districts with <1 case / 10,000 population	Number of endemic districts	% of endemic districts above the elimination threshold
	2014	0	12	**12**	0%
	2015	0	12	**12**	0%
	2016	0	18	**18**	0%
	2017	0	18	**18**	0%
	2018	0	18	**18**	0%

Fig. 2.3 (**a**) Progress towards reaching and sustaining elimination targets in S. Asia. Note: Nepal 2017 98% is because of a non-endemic district having reported above the threshold (**b**) Progress on VL Elimination targets for S. Asia, 2014–2018 (source: WHO, March 2019)

Refrigerators at select health centres, with trained technical monitors in place and (2) the Ambical assist App to facilitate dosing, important as AmBisome preparation and administration involves several critical calculations. Although healthcare providers referred to dosage charts or technical materials provided during the KalaCORE training, the development of AmBiCal Assist, a simple user-friendly mobile application addressed these treatment challenges and assisted healthcare providers in charge of preparation and administration of single dose AmBisome as well as preparation of AmBisome for administration to VL-HIV coinfected patients. The main features of the application include: information on how to test a patient for VL and VL-HIV; the suitable procedure based on whether the patient has VL or VL-HIV; calculation of the dosage; and the materials needed for administration. AmBICal Assist is compatible with android and iOs, it was made freely available and is login-based.

It is noteworthy that in 2017, one of the non-endemic districts in Nepal reported VL cases above the elimination threshold and in 2018, 32 non-endemic districts reported VL cases; in 2019, as the KalaCORE programme had ended, assessment as to whether they were actually endemic was still pending. In Bangladesh, a similar situation was observed where, despite reports showing less than 1 case/10,000 population in all (100) endemic sub-districts in 2018, there were cases in an additional 27 sub-districts where further clarification of endemicity was needed.

During the course of the KalaCORE programme, Bangladesh and Nepal made substantial progress towards reaching the elimination target, a target that has to be maintained for 3 consecutive years for validation. Progress towards achievement and validation of the elimination target is assessed through a formal process involving preparation and submission of a dossier to WHO (2016).

2.4.1.2 Contribution of Active Case Finding

The KalaCORE programme provided strategic technical assistance for active case finding using digital mapping and spatial risk factor analysis targeting village-level populations at risk with the aim of increasing patient access to high-quality care through early detection and treatment of Post-Kala-Azar-Dermal-Leishmaniasis (PKDL) and VL, with a focus on addressing gender-based barriers, to reduce transmission and help achieve the elimination target.

This support for active case finding, improved disease surveillance and reporting contributed to a steady decline (Fig. 2.4) in the number of VL cases reported over the life of the programme in all endemic states in India and across all countries as shown in Figs. 2.5–2.7. However, challenges in reducing VL in Bihar and Jharkhand remained due to access to schedule casts, remoteness and distrust of the government as mentioned in 3.1.1. In contrast, the number of reported PKDL cases increased, representing the effectiveness of community-based detection of a previously hugely neglected manifestation of leishmaniasis.

Over the programme period, cost per positive VL diagnosis decreased by 47.8%. Although the costs remained high, ACD had several unintended positive outcomes, for example health services reaching more marginalised members of the population and helping to ensure no one was left behind. In addition, the three phases of ACD continued to identify PKDL cases which had added both to the overall cost-effectiveness and potential reduction in parasite transmission.

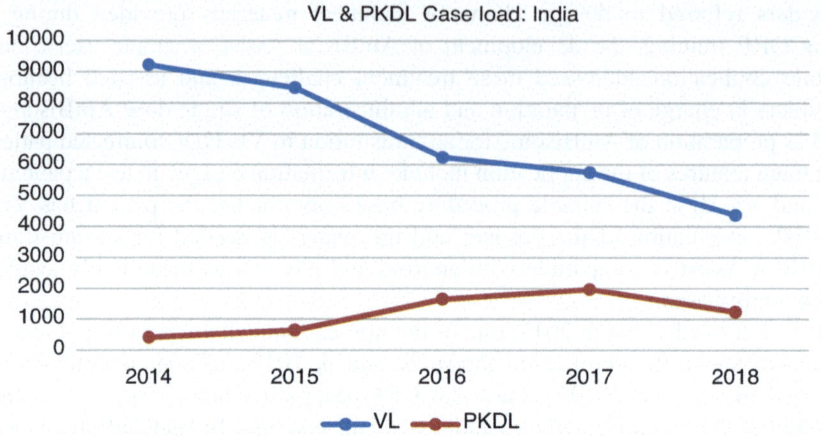

Fig. 2.4 VL and PKDL case load in India 2014–2018

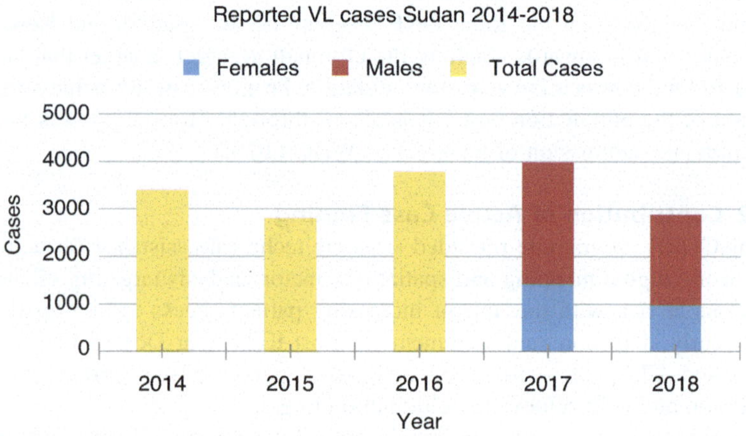

Fig. 2.5 Reported VL cases In Sudan 2014–2018

Comprehensive outbreak response strategies were developed in Bangladesh and Nepal under KalaCORE funding, where the elimination targets have been reached,[3] while KalaCORE provided technical support in the development of these guidelines in India.

[3]Nepal was the first country in the South-East Asia Region to reach the elimination threshold at its implementation level in 2013, However, other cases were reported from "non-endemic" districts during the KalaCORE programme.

2.4.1.3 Case Fatality Rates

Overall, the case fatality rate regional targets for the programme countries were achieved. However, it is important to note that these reported mortality rates and use of CFR as an indicator must be interpreted with caution as: (1) the quality of reporting was not optimal, (2) there was underreporting as some patients never reached a VL treatment centre and (3) the definition of VL mortality was not then standardised. Importantly, in India, the National Centre for Vector Borne Disease Control Programme (NVBDCP) adapted the definition of death due to VL in 2014, and despite ongoing discussions, there remained a lack of agreement regarding the definition of death due to VL; this resulted in zero deaths being officially recorded between 2018. However, the current national database suggests 3% case fatality rate recorded (Postigo, WHO, personal communication), consistent with previous data analysis for a data set in 2012–2013.

2.4.1.4 Economic Impacts

The programme successfully contributed to the reduction of both economic and health impacts of VL in South Asia and East Africa and overall met the target of a 12% reduction in economic costs to households where the VL patient was a wage earner in the three countries where this was measured. The target of 12% reduction in economic costs to households was met in all three countries where it was possible to conduct both a baseline and endline survey (India, Ethiopia and Sudan) for households where the VL patient was a wage earner and was 9% in Ethiopia (Paintain et al. 2019).

2.4.2 Strengthening Capacity for Control of VL in East Africa

2.4.2.1 Treatment

The first important achievement towards stronger capacity for effective response in East Africa was the introduction and implementation of effective, safe treatment and control measures. This was implemented through the country-wide introduction of the SSG plus paromomycin combination therapy and the availability of rapid diagnostic tests, resulting in a major impact on control of VL in the three East African countries. KalaCORE's support greatly improved access to treatment, and death due to VL decreased. The KalaCORE programme rolled out country-wide training on the use of diagnostics and treatments in all three countries, supported quality procurement and increased access to safe and effective drugs and diagnostics. In all three countries, the numbers of VL diagnosis and treatment facilities were extended, refurbished and supplied with drug, diagnostics and therapeutic food. KalaCORE defined readiness criteria for these facilities for provision of first- and second-line treatment. At the end of the programme in 2019, a total of 50 VL treatment centres were in place in Ethiopia, 12 of which met the readiness criteria and 38 hospitals and health centres provided diagnostic services. A total of 44 VL treatment centres were in place in South Sudan, 38 of which met basic readiness criteria defined as at least one person trained in the last year and diagnostic services and equipment available with no stock ruptures in the last 3 months and a basic laboratory. A total of 44 VL treatment centres were in place and met the readiness criteria in 12 endemic states in Sudan.

2.4.2.2 Training and Mentoring

Treatment of VL patients requires complex clinical management. In addition to central training of medical staff, KalaCORE implemented a clinical mentoring strategy involving mobile teams in Ethiopia and South Sudan to provide on-site training, bedside mentoring and problem-solving. These teams followed up suspected outbreaks and new endemic areas. This model was also adopted by the WHO-managed programme in Sudan. The clinical mentoring model was important for training the medical and laboratory staff in the new treatment regimens and diagnostics and helped to address the high staff turnover in treatment facilities. However, as this proved to be an expensive and labour-intensive activity to manage for long-term and sustainable national programmes, mentoring needs to be adapted to a more cost-efficient and sustainable model, for example through local mentees in universities and teaching hospitals that provide VL treatment.

Through its implementing partners, the KalaCORE programme carried out health education, awareness raising and behaviour change activities in target communities. In addition, the programme had a major role in supporting development of outbreak response surveillance systems and provided direct support to national programmes for monitoring and evaluation and technical assistance to national programmes to sustain VL control.

2.4.2.3 VL Cases

The caseload in East Africa countries varied over the programme and remained unpredictable as shown in Figs. 2.5, 2.6 and 2.7). It has been speculated that this could be related to environmental variables such as rainfall and temperatures, but no reliable prediction models exist. Population movements play an important role, as is evidenced by the outbreak in 2014 in South Sudan which coincided with the civil war (Al-Salem et al. 2016). In 2018, the number of cases was relatively low in both Sudan and South Sudan compared to previous years, while in Ethiopia cases rose slightly after 2016, for reasons unknown. The mortality rates varied as well over the years in all three countries, but with incomplete reporting during the programme period, this variation remained difficult to interpret.

With an absence of evidence-based prevention methods for East Africa, the control of VL relied on early detection and treatment. Over the course of KalaCORE, better reporting, increased drug and test availability and an increased number of VL treatment facilities with better access to treatment as a result will have positively influenced the ratio of numbers of cases reported and treated versus those remaining unseen. However, KalaCORE's studies in Ethiopia and Sudan described in Sect. 3.2.4, and situational reports from South Sudan, demonstrated that access to treatment was still problematic. It cannot be assumed that all VL cases in these countries were detected and treated, nor that the KalaCORE Programme's interventions had an impact on the number of new VL infections.

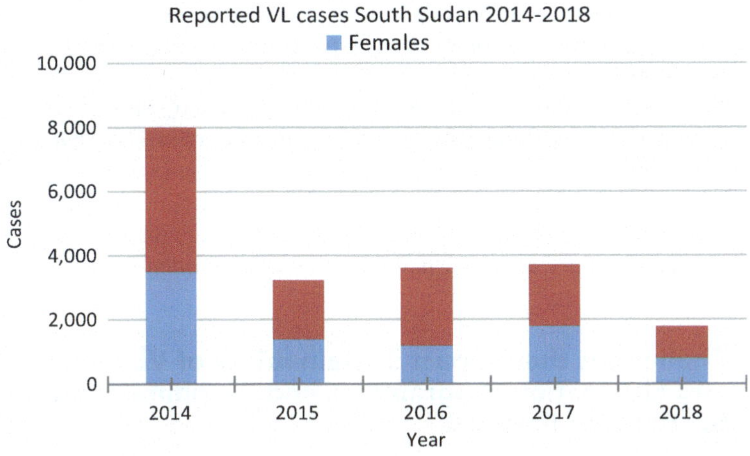

Fig. 2.6 Reported VL cases in South Sudan 2014–2018

Fig. 2.7 Reported VL cases in Ethiopia 2014–2018

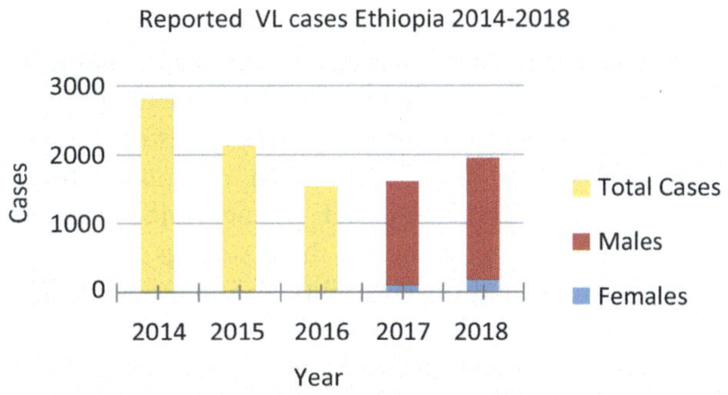

2.4.2.4 Operational Research on Effective Prevention Methods

Most of the planned operational research (OR) was delayed due to in part to contracting issues and the time taken to obtain national ethical approvals. Hence, the results of the studies were only available towards the end of the programme.

With hindsight, the programme was overly optimistic regarding those OR projects that aimed to identify effective vector control methods for VL in the target countries and scaling up a pilot that would yield results and inform country disease control policies within the life of the programme. The proof-of-concept study that showed real promise in Sudan involving spraying fences around household compounds to reduce sand fly populations inside compounds (Elnaiem et al. 2020), and hopefully transmission of VL, could not be scaled up to a community trial within the programme due to the changes in the protocol requested by the WHO experts, these would have required extending the study beyond the life of the programme funding. It was clear from the other OR studies that the only real option was to encourage and facilitate the use of personal protection measures through use of insect repellents which were not available in the endemic areas and to advocate for integrated vector control with other disease programmes (Alemayehu et al. 2020; Berhe et al. 2022).

The OR study by Coulborn et al. (2018) in Ethiopia and by Sunyoto et al. (2018) in Sudan which assessed barriers to access to care observed that the study participants reported very poor access to diagnosis and, consequently, significantly delayed access to treatment. Interventions tailored to the socio-economic and health needs of the migrant workers in Ethiopia and remote communities in Sudan are urgently needed to reduce health disparities and the VL burden.

2.5 Challenges that Impact Sustainability of VL Control and Elimination Programmes and Recommendations for the Way Forward

To date, the major focus of donor support for NTDs is preventive chemotherapy through mass drug administration of diseases that blind, maim, disfigure and debilitate (WHO 2019). But unlike many of the preventive chemotherapy (PCD) NTDs, VL is fatal within a relatively short time of onset of disease if not diagnosed and treated promptly. Where access to treatment is poor, the death toll can be high.

In East Africa, VL diagnosis and treatment are highly complex, and in addition, medical doctors and nurses do generally not receive practical training in VL during their studies, and there were often no trained health personnel present in the treatment sites. The KalaCORE partners MSF and DNDi had direct experience in managing VL programmes in these settings, and there was an acute awareness of these problems and responsibility and urgency to implement measures that would address them immediately and thus 'save lives'. Interventions such as mobile teams that visited treatment sites and train and mentor health workers on-site, treat patients, collect data, address drug shortages and follow up on indications of outbreaks, were highly effective but very costly and not sustainable in the long run without dedicated funding. Systemic health system-related issues such as poor supply chains and a high turn-over of human resources in remote hospitals could not be systematically resolved during the time span of the programme. Long-term health system strengthening therefore took place mostly in the form of leaving behind hospital equipment and structures and training and health education materials. Several of the challenges encountered in the programme to implement these aims are described below.

2.5.1 Integration

Although integration of cross-cutting activities with other NTD and disease programmes was seen as essential by DFID,[4] WHO and implementing partners, this was a challenge operationally. There was no common understanding of what level of integration there should be and if it was actually achievable. In the discussions on integration, and as exit strategies were being developed, this led to tensions in the VL programmes and with the other NTD programmes of what could really be integrated and what that would mean for the disease elimination targets. For example, Behaviour Change Communication (BCC) is an important component of disease control programmes and an integrated BCC strategy for the four PCT diseases plus VL could mean savings on training, communications costs, travel, transport, etc. However, the timing of the BCC for health workers and the community is critical so both impact and effectiveness may be lost if it is not implemented at the right time with respect to seasonality, supply chain management, etc.

So, the question remained, how does VL fit into a possible integrated NTD programme approach?

Integrated vector control could involve integration of LF and VL with the malaria programmes as bednets, IRS and some of the novel vector control methods have an impact on LF vectors and some impact on VL vectors. An integrated strategy for advocacy and market creation for personal protection measures (e.g. insect repellents and even some innovative repellent clothing) could target several diseases with value for money for the health systems.

VL certainly remains one of the most neglected diseases; however, lumping it in with the PCT NTDs does not help countries in East Africa in particular, with 'mainstream' training and case management. VL patients require diagnosis, in-patient treatment with monitoring as the drugs are toxic, severely ill patients require supplemental feeding as well as some patients have VL/HIV co-infections. Thus, VL has more in common with HIV and TB in terms of hospital facilities, case management, etc. KalaCORE attempted, but did not manage, to initiate an effective collaboration with malaria, HIV and TB programmes at country level.

Advocacy and bringing the various donors together to look at how to ensure VfM across the different programmes and how to support establishment of effective and functioning health systems are needed.

2.5.2 Procurement and the Supply Chain

The supply chain for both diagnostics and drugs remains an issue for VL in most countries. There were country-specific issues, for example, the procurement and bidding process in India resulted in stock-outs of RDTs and miltefosine; Ethiopia

[4] In 2020, the UK Government merged DFID and FCO to become FCDO and dispersed much of the £15.2 bn budget to other government departments.

suffered stock-outs at treatment facilities due to forecasting and distribution processes, and supply chain management in South Sudan has major systemic issues.

However, it is important to understand that supply chain systems in countries are a sensitive issue. In the course of the KalaCORE programme, situations were encountered that even when countries acknowledged that there were problems, often due to systemic problems in the health systems, countries were not willing to have an external, disease-specific, donor-driven and short-term programme address them.

Donors need to be realistic about what can be achieved when they want programmes to support supply chain improvements. Where donors have country staff, it is important that they provide support at country level to all their programmes whether they are managed from the country office or from headquarters.

2.5.3 Training

The high turn-over and attrition of trained health workers were an ongoing challenge in all countries, necessitating continual training programmes. For example, regular trainings were carried out at State level in India on diagnosis and treatment and on surveillance through the M&E component of KalaCORE. In Ethiopia, Sudan and South Sudan, clinical mentoring teams supported by KalaCORE carried out on site trainings whilst WHO/KalaCORE and FMOH carried regular trainings for all levels at central and regional locations in Ethiopia. To sustain a fully effective programme for VL control training remains essential and should be institutionalised in some way (e.g. through medical/nursing schools/national Public health institutes) with cooperation between ministries of Health, Education and planning. To support this training effort, an MOOC (massive online open access course) on VL was developed for healthcare workers by one of the KalaCORE partners, made available by 2018 and completed by over 500 participants by the end of the programme. It was subsequently taken over by WHO and accessible via https://openwho.org/courses/NTDs-visceral-leishmaniasis-east-africa.

2.5.4 Case Fatality Rate

CFR as an impact indicator for disease programmes should be used and interpreted with caution, as discussed in Sect. 3.1.3. Programme Log frames often call for such targets without taking the context into account. As per today, death due to VL is still not adequately defined and poorly reported in most countries.

2.6 Overall Summary of Challenges and Remaining Issues at the Closure of the KalaCORE Programme in April 2019

2.6.1 Asia

A major challenge for elimination was, and remains, not only achieving but subsequently the validation and thereafter, sustaining the targeted elimination threshold in all Asian countries at district and sub-district levels (Fitzpatrick et al. 2017). It is essential that national programmes do not become complacent once the threshold has been reached. Sustained advocacy from key stakeholders will continue to be critical to build and sustain country ownership.

The major burden of kala-azar in India at the end of the programme (and to date) is in Bihar and Jharkhand states. Jharkhand remains a major challenge in relation to the VL elimination agenda. Some 75% of the endemic blocks reported more than 1/10,000 cases over the last 5 years of the KalaCORE programme. Reasons for this included difficult topography, a tribal population living in difficult to reach areas and lack of accessibility of the health system to marginalised groups. Advocacy by key stakeholders including KalaCORE, CARE, BMGF, PATH and DNDi resulted in a special focus by the national programme and all stakeholders on Jharkhand, and active case detection was initiated on a regular basis with sensitisation of the health system. This resulted in a 50% reduction in the number of blocks reporting more than 1 case per 10,000 population in Jharkhand to 8 in 2018 in comparison to 16 in 2017.

The changing pattern in VL case distribution in all three countries and the geographical expansion of cases in "non-programme" (synonym of non-endemic) districts of Nepal and Bangladesh highlight the need for continued monitoring by the national programmes and key stakeholders. Different strategies may need to be developed, for example for the mountainous, hard-to-reach areas in western Nepal.

PKDL cases in people previously treated for VL is also a key challenge in both Asia and Africa. As mentioned above, there is strong evidence that PKDL cases are reservoirs for VL transmission and so pose a real threat to elimination if a better strategy for the identification, diagnosis and treatment is not developed and implemented (Mondal et al. 2019; Le Rutte et al. 2019).

In Asia, the KalaCORE programme focused on the roll-out of quality first-line treatment with AmBisome and consequently, the entire Kala-azar Elimination Programme relied upon one single drug, AmBisome, and one single producer, Gilead. Given the WHO recommendations for use of combination therapies (to slow down the eventual development of drug resistance), countries need to be encouraged to make available the miltefosine-paromomycin combination treatment for primary VL in the field as a treatment option to complement AmBisome. Miltefosine plus AmBisome is also available for HIV/VL infections.

Going forward, continued capacity building to assess and respond to emerging outbreaks is important and continued capacity development of health workers on case finding and management as case numbers drop, diagnostic and treatment skills are lost. In Asia in particular, private providers and informal sector providers are the first port of call in health seeking. The KalaCORE programme carried out awareness and training on VL for private practitioners, but for sustainable impact and elimination of VL, this needs to be expanded and continued in order to improve and shorten the treatment pathway for VL patients. As already mentioned, the key challenge of staff turn-over necessitates regular and continuous plans for training.

2.6.2 Africa

In all three target countries in East Africa, there remained a lack of evidence-based and feasible prevention measures, so that at the end of the KalaCORE programme, the only way to control VL was still through early detection and treatment and awareness raising.

One of the issues VL treatment centres in Africa face, particularly in Ethiopia, Sudan and South Sudan, is that many patients present when they were already severely ill and required therapeutic food. In Ethiopia, the country policy at the time was that children under 5 were eligible for therapeutic food, but provision for older children and adults was at the doctor's discretion. In South Sudan, access to therapeutic food provided by UNICEF was highly challenging, and most VL patients were severely malnourished, and their treatment outcome likely to be affected. Similarly, in Sudan, the issue of the supply of therapeutic food was a problem. The KalaCORE programme procured and supplied the plumpy nut therapeutic food for severely ill adults to facilities in Ethiopia and South Sudan. However, this was not a sustainable solution for the provision of therapeutic food to VL patients, and advocacy for a policy that addresses this issue at global, regional and national levels is much needed.

BCC strategies and other intervention strategies should be highly specific to the affected groups. For example, the migrant seasonal workers in Ethiopia, who are the high risk group for VL, but also at high risk for malaria, LF, cholera, snake bite, TB, HIV and malnutrition, a multisectoral approach is called for. Migrant workers need access to bednets, food, shelter, water and healthcare at farms and compensation for their sick leave (Coulborn et al. 2018). This multisectoral approach would require high-level advocacy, political commitment and collaboration (with the health, education, agricultural and planning sectors.)

The absence of effective treatment available for HIV/VL coinfection was a major challenge. These patients relapse multiple times which poses a challenge in clinical management and hospital resources. Although the AmBisome/miltefosine combination therapy is effective (Burza et al. 2022; Diro et al. 2019), access to miltefosine remains highly problematic. A collaborative approach with WHO, the national authorities, MSF, IDA Foundation, DNDI and future programmes will need to address the issues regarding miltefosine availability and implementation of the 2022 WHO guidelines for treatment for HIV/VL using this combination (WHO 2022b).

In Sudan, several challenges and issues were encountered as a result of the international sanctions that caused difficulties for DFID to provide funding to the

National programme. DFID funding for the VL programme was channelled through the WHO office in Khartoum. However, there were many bureaucratic and political hurdles encountered that resulted in delayed implementation for all the elements of the programme. Once implementation started, interventions could be quickly rolled out, to a large extent due to a close collaboration with the Ministry of Health.

In South Sudan, without NGO support and International funding, the Ministry of Health had limited or no capacity to continue any of the interventions that had been provided by the KalaCORE partner, IMA. There remained problems in transport of drugs and tests to health facilities, remuneration for Health care workers, difficulties in obtaining reliable VL reports because of the lack of communication in remote sites and access to treatment centres from insecure and remote areas.

2.7 Conclusions

The KalaCORE programme faced two distinct challenges: (1) to improve treatment access for VL patients and reduce disease related mortality and (2) to work with ministries and health systems which varied from region to region, country to country and in India from state to state, to support the implementation of VL control and elimination programmes. KalaCORE found, as have other programmes, 'You can't have one without the other'. Implementation of improved approaches to treatment and reduction of mortality through use of single infusion AmBisome and RDTs was effective in South Asia, and KalaCORE ensured a cold chain for the drug and contributed to the reduction in time between illness, diagnosis and treatment. But this approach was not possible in East Africa where efficacy of treatment and sensitivity of diagnostics were different. Even in South Asia, embedding sustainable improved treatment involved the commitment of the health systems and education/ communication to ensure awareness of VL in the endemic populations; this commitment varied between countries. In East Africa, the interaction with NTD staff in ministries was essential to ensure procurement, delivery, access and training of key staff; but very different approaches to this were necessary for Ethiopia, Sudan and South Sudan. The overall investment of £27 M contributed in both regions to improved treatment and reduction in mortality, to better trained staff and to necessary interactions between clinical/field staff and the health systems and increased visibility of VL within and outside the target countries. The largest challenge was and remains the need for a sustained effort to eliminate VL in both regions and to ensure that monitoring, surveillance, staff trained in outbreak response and the tools (drugs, diagnostics, data banks) are available.

Since the end of KalaCORE in 2019, there have been a series of developments which are having a major impact on targets for the future control and elimination of VL in both South Asia and East Africa (Dahl et al. 2021). These include the recent elevation of control to elimination status in East Africa (Makau-Barasa et al. 2022; Alvar 2021; WHO 2023) and the move from elimination as a public health problem to zero transmission in Bangladesh (WHO 2020–2030). This is all re-enforced by the WHO commitment to NTDs, as seen the WHO NTD (2021–2030) Road Map (WHO

2020a), which contains new targets for VL elimination, and in South Asia by the WHO task force report on VL status in India (WHO 2020) and a detailed WHO SEARO framework for sustaining elimination of VL in South east Asia, 2022–2026 (WHO 2022). The impact of COVID-19 on these ambitious targets is still being assessed (Uwishema et al. 2022; WHO 2021); at least there is now a general awareness of how pandemics can impact NTD programmes and precautionary measures can be planned.

Acknowledgements We thank members of the KalaCORE Expert Steering Group, Koert Ritmeijer (MSF), Jorge Alvar (DNDi 2022), the Mott MacDonald managerial and administrative teams, in particular Lucy Palmer, Sinead Rowan, Sara Hanka, Duncan Tossel, Aska Wisniewska-Pickering and the late Bob Grose; Daniel Dagne and Jose Ruiz Postigo at WHO, key staff in countries, Vikas Aggarwal, Kingsuk Misra (India), Atia Atia, Moussab Elhag (Sudan), Cherinet Adera, Paco Moreno, Merce Herrero (Ethiopia) Duncan Ochol and Mounir Lado, IMA,(South Sudan) for their considerable efforts, expertise and knowledge to ensure the success of the KalaCORE programme. The LSHTM team Jayne Webster and Lucy Pantain and Vanessa Yardley for the M& E and technical oversight.

The KalaCORE was a UK Aid programme and partnership funded by the Department for International Development, UK government.

Box 2.1: VL Endemic Areas in South Asia

- Sand fly vector considered to be an indoor biter, prevention strategies therefore possible with indoor insecticide spraying.
- Comparatively easier access to populations than East Africa.
- Densely populated endemic areas.
- Relatively good infrastructure.
- Easy diagnosis with well performing rapid tests, specifically rK39 RDT.
- Effective safe treatment with AmBisome, single intravenous infusion.
- No or limited hospitalisation.
- Limited testing of VL/HIV; both diseases found in VL endemic areas.
- Large investments and many partners in context of elimination program.

Box 2.2: VL Endemic Areas in East Africa

- Vector bites outdoors.
- Unproven animal reservoirs.
- Difficult access to populations, remote and isolated.
- Sparsely populated.
- Poor health infrastructure.
- Insecurity.
- Regular population movements from and into endemic areas, further spreading the disease (economic migrants, refugees, nomadic populations).
- Complex diagnostic tree as rK39 RDT not highly sensitive.
- Effective but toxic and painful first-line treatment with 17-day injections with SSG + paromomycin.
- High cost of long obligate hospitalisation.

(continued)

Box 2.2 (continued)

- High dose of AmBisome for vulnerable and HIV co-infected populations.
- World's highest VL/HIV coinfection rate in Ethiopia.
- VL tends to occur in outbreaks.
- Supply chain issues.
- Elimination of VL not considered possible at the time (N.B: In 2023, an East African elimination programme has launched).

References

Alemayehu M, Paintain L, Adera C, Berhe R, Gebeyehu A, Gizaw Z, Herrero M, Mulugeta Beshah A, Den Boer M, Nigusie A, Alem M, Yechale E, Elnaiem DA (2020) Impact of education on knowledge and practice of Kala Azar preventive measures among seasonal and migrant agricultural Workers in Northwest Ethiopia. Am J Trop Med Hyg 102(4):758–767. https://doi.org/10.4269/ajtmh.19-

Al-Salem W, Herricks JR, Hotez PJ (2016) A review of visceral leishmaniasis during the conflict in South Sudan and the consequences for east African countries. Parasit Vectors 9:460. https://doi.org/10.1186/s13071-016-1743-7

Alvar J, Aparicio P, Aseffa A, Den Boer M, Cañavate C, Dedet JP (2008) The relationship between leishmaniasis and AIDS: the second 10 years. Clin Microbiol Rev 21(2):334–359

Alvar J, Vélez ID, Bern C, Herrero M, Desjeux P, Cano J et al (2012) Leishmaniasis worldwide and global estimates of its incidence. PLoS One 7(5):e35671. https://doi.org/10.1371/journal.pone.0035671

Alves F, Bilbe G, Blesson S, Goyal V, Monnerat S, Mowbray C, Muthoni Ouattara G, Pécoul B, Rijal S, Rode J, Solomos A, Strub-Wourgaft N, Wasunna M, Wells S, Zijlstra EE, Arana B, Alvar J. Recent Development of Visceral Leishmaniasis Treatments: Successes, Pitfalls, and Perspectives. Clin Microbiol Rev. 2018 Aug 29;31(4):e00048–18. https://doi.org/10.1128/CMR.00048-18. PMID: 30158301; PMCID: PMC6148188

Berhe R, Spigt M, Schneider F, Paintain L, Adera C, Nigusie A et al (2022) Understanding the risk perception of visceral leishmaniasis exposure and the acceptability of sandfly protection measures among migrant workers in the lowlands of Northwest Ethiopia: a health belief model perspective. BMC Public Health 22:989. https://doi.org/10.1186/s12889-022-13406-3

Burza S, Sinha PK, Mahajan R, Lima MA, Mitra G, Verma N, Balasegaram M, Das P (2014) Risk factors for visceral leishmaniasis relapse in immunocompetent patients following treatment with 20 mg/kg liposomal amphotericin B (Ambisome) in Bihar, India. PLoS Negl Trop Dis 8(1): e2536. https://doi.org/10.1371/journal.pntd.0002536

Burza S, Croft SL, Boelaert M (2018) Leishmaniasis. Lancet 392(10151):951–970. https://doi.org/10.1016/S0140-6736(18)31204-2. Epub 2018 Aug 17

Burza S, Mahajan R, Kazmi S et al (2022) AmBisome Monotherapy and Combination AmBisome-Miltefosine Therapy for the Treatment of Visceral Leishmaniasis in Patients Coinfected With Human Immunodeficiency Virus in India: A Randomized Open-Label, Parallel-Arm, Phase 3 Trial. Clin Infect Dis 75(8):1423–1432. https://doi.org/10.1093/cid/ciac127

Coulborn RM, Gebrehiwot TG, Schneider M, Gerstl S, Adera C, Herrero M, Porten K, den Boer M, Ritmeijer K, Alvar J, Hassen A, Mulugeta A (2018) Barriers to access to visceral leishmaniasis diagnosis and care among seasonal mobile workers in Western Tigray, northern Ethiopia: a qualitative study. PLoS Negl Trop Dis 12(11):e0006778. https://doi.org/10.1371/journal.pntd.0006778. PMID: 30408042; PMCID: PMC6224040

Croft SL, Sundar S, Fairlamb AH (2006) Drug resistance in leishmaniasis. Clin Microbiol Rev 19(1):111–126. https://doi.org/10.1128/CMR.19.1.111-126.2006. PMID: 16418526; PMCID: PMC136027

Dahl EH, Hamdan HM, Mabrouk L, Matendechero SH, Mengistie TB, Elhag MS, Lado M, Adera C, Atia AA, Potet J, den Boer M, Ritmeijer K, Alvar J, Alves F, Alcoba G (2021) Control of visceral leishmaniasis in East Africa: fragile progress, new threats. BMJ Glob Health 6(8):e006835. https://doi.org/10.1136/bmjgh-2021-006835. PMID: 34413075; PMCID: PMC8365799

Diro E, Blesson S, Edwards T, Ritmeijer K, Fikre H, Admassu H, Kibret A, Ellis SJ, Bardonneau C, Zijlstra EE, Soipei P, Mutinda B, Omollo R, Kimutai R, Omwalo G, Wasunna M, Tadesse F, Alves F, Strub-Wourgaft N, Hailu A et al (2019) A randomized trial of AmBisome monotherapy and AmBisome and miltefosine combination to treat visceral leishmaniasis in HIV co-infected patients in Ethiopia. PLoS Negl Trop Dis 13(1):e0006988. https://doi.org/10.1371/journal.pntd.0006988

DNDi 2022.: https://dndi.org/newsletters/n18/3_1.php

Elnaiem D-EA, Dakein O, Alawad AM-A, Alsharif B, Khogali A, Jibreel T et al (2020) Outdoor residual insecticide spraying (ODRS), a new approach for the control of the Exophilic vectors of human visceral Leishmaniasis: Phlebotomus orientalis in East Africa. PLoS Negl Trop Dis 14(10):e0008774. https://doi.org/10.1371/journal.pntd.0008774

FCDO 2020.: https://devtracker.fcdo.gov.uk/projects/GB-1-203148/documents Accessed Jan 2023

Fitzpatrick A, Al-Kobaisi NSMS, Beitman Maya J, Ren Chung Y, Duhan S, Elbegdorj E, Jain S, Kuhn E, Nastase A, Ahmed BN, Olliaro P (2017) Sustaining visceral leishmaniasis elimination in Bangladesh-could a policy brief help? PLoS Negl Trop Dis 11(12):e0006081. https://doi.org/10.1371/journal.pntd.0006081

Le Rutte EA, Zijlstra EE, de Vlas SJ (2019) Post-Kala-Azar dermal Leishmaniasis as a reservoir for visceral leishmaniasis transmission. Trends Parasitol 35(8):590–592. https://doi.org/10.1016/j.pt.2019.06.007. Epub 2019 Jun 29. PMID: 31266711; PMCID: PMC6667731

Makau-Barasa LK, Ochol D, Yotebieng KA, Adera CB, de Souza DK (2022) Moving from control to elimination of visceral Leishmaniasis in East Africa. Front Trop Dis 3:965609. https://doi.org/10.3389/fitd.2022.965609. Front. Trop. Dis, 19 Aug 2022. Sec. Neglected Tropical Diseases, Volume 3 - 2022 | doi:10.3389/fitd.2022.965609

Mondal D, Bern C, Ghosh D, Rashid M, Molina R, Chowdhury R, Nath R, Ghosh P, Chapman LAC, Alim A, Bilbe G, Alvar J (2019) Quantifying the infectiousness of post-Kala-Azar dermal Leishmaniasis toward sand flies. Clin Infect Dis 69(2):251–258. https://doi.org/10.1093/cid/ciy891. PMID: 30357373; PMCID: PMC6603265

Paintain L, Boudarene L, Pearson M, Edwards T, Alatiaby AA, Mondal D, Adera C, Prakash M, Ruiz Postigo J, Argaw Dagne D, den Boer M, Burza S, Webster J (2019) The catastrophic economic burden of visceral leishmaniasis in Bangladesh, India, Ethiopia and Sudan: financial risk protection is not enough. Trans R Soc Trop Med Hyg 113(Supplement_1):S1–S98. https://doi.org/10.1093/trstmh/trz094

Sunyoto T, Adam GK, Atia AM, Hamid Y, Babiker RA, Abdelrahman N, Vander Kelen C, Ritmeijer K, Alcoba G, den Boer M, Picado A, Boelaert M (2018) "Kala-Azar is a dishonest disease": community perspectives on access barriers to visceral Leishmaniasis (Kala-Azar) diagnosis and care in Southern Gadarif, Sudan. Am J Trop Med Hyg 98(4):1091–1101. https://doi.org/10.4269/ajtmh.17-0872

Uniting to Combat NTDs 2012.: https://unitingtocombatntds.org/resource-hub/who-resources/london-declaration-neglected-tropical-diseases. Accessed Jan 2021

Uwishema O, Sapkota S, Wellington J, Onyeaka CVP, Onyeaka H (2022) Leishmaniasis control in the light of the COVID-19 pandemic in Africa. Ann Med Surg (Lond) 80:104263. https://doi.org/10.1016/j.amsu.2022.104263. Epub 2022 Jul 31. PMID: 35936565; PMCID: PMC9339101

WHO 2005.: https://www.who.int/neglected_diseases/diseases/en/ Accessed Sept 2022

WHO 2016.: https://www.who.int/news/item/11-11-2016-visceral-leishmaniasis-who-publishes-validation-document-as-countries-approach-elimination

WHO 2019: Evaluation of the WHO neglected tropical diseases programme volume 1: report. https://www.who.int › docs › evaluation-ntd-report
WHO National Strategic Plan for Kala-azar elimination in Bangladesh: 2020–2030
WHO 2020a: Ending the neglect to attain the sustainable development goals: a road map for neglected tropical diseases 2021–2030. https://www.who.int/publications/i/item/9789240010352 Accessed Nov 2022
WHO 2020b. https://www.who.int/neglected_diseases/diseases/en/ Accessed Sept 2022
WHO 2021 Global leishmaniasis surveillance: assessing the impact of the COVID-19 pandemic weekly epidemiological record. 11 Nov 2022
WHO 2022 Regional strategic framework for accelerating and sustaining elimination of kala-azar in the South-East Asia region: 2022–2026. ISBN: 978-92-9020-981-2
WHO 2022b. https://www.who.int/publications/i/item/9789240048294
WHO 2023. https://www.who.int/news-room/events/detail/2023/01/24/default-calendar/stakeholders-meeting-for-the-development-of-a-strategic-plan-for-the-elimination-of-visceral-leishmaniasis-in-east-africa
World Health Organization 2020.Independent assessment of kala-azar elimination programme India. ISBN: 978-92-9022-796-0

Essentials in Leishmaniasis Elimination 3

Surbhi Badhwar, Angamuthu Selvapandiyan, and Niti Puri

Abstract

Leishmaniasis is a neglected tropical infectious disease caused by multiple species of an obligate intracellular protozoan pathogen of the genus *Leishmania*. The disease, with a worldwide prevalence of 12 million cases, is endemic in tropical and subtropical countries and affects mainly the poor in remote areas. Being a vector-borne disease, the parasite is transmitted through the bite of *a Phlebotomine* female sand fly resulting in a spectrum of diseases with symptoms ranging from self-healing ulcers to fatal infections affecting the liver and spleen depending upon the species infected. The current treatment approach for leishmaniasis includes antiparasitic drugs, combinational drugs, and immunotherapy. However, these therapeutics are not successful in eradicating Leishmaniasis due to adverse side effects, resistance development against drugs, high cost of production, etc. Despite being an infectious disease, there is no vaccine approved currently against any type of Leishmaniasis for use in humans. Recent studies have reported novel approaches to vaccine development and the important role of the innate immune system and its interaction with *Leishmania* for the development of an effective vaccine against leishmaniasis.

Keywords

Leishmania · Vaccine · Innate immune cells · Leishmaniasis · Therapy · Drug

S. Badhwar · N. Puri (✉)
Cellular and Molecular Immunology Laboratory, School of Life Sciences, Jawaharlal Nehru University, New Delhi, India
e-mail: purin@mail.jnu.ac.in

A. Selvapandiyan
Department of Molecular Medicine, Jamia Hamdard, New Delhi, India
e-mail: selvapandiyan@jamiahamdard.ac.in

© The Author(s), under exclusive license to Springer Nature Singapore Pte Ltd. 2023
A. Selvapandiyan et al. (eds.), *Challenges and Solutions Against Visceral Leishmaniasis*, https://doi.org/10.1007/978-981-99-6999-9_3

3.1 Introduction

Leishmania is an obligate intracellular pathogen that belongs to the Trypanosomatidae family and leads to the development of leishmaniasis in vertebrates (Mann et al. 2021). *Leishmania*, being endemic in tropical and subtropical regions, can lead to a spectrum of diseases varying from self-healing ulcers to disseminated and often fatal infections, based on the species infecting and the host immune response to it (Wheeler et al. 2011; Naqvi et al. 2020). Humans and 69 other animal species are known to act as natural reservoir hosts of *Leishmania* parasites (WHO 2023a). Nearly 30 different species of Phlebotomine sand flies are known to transmit this parasite (Ghorbani and Farhoudi 2018), and approximately, 21 *Leishmania* species are known to infect humans (Kobets et al. 2012). Leishmaniasis, being a vector-borne disease, is transmitted to the host cells by the bite of the *Phlebotomus* genus (Old World *Leishmania*) or *Lutzomyia* (Arenas et al. 2017) genus (New World *Leishmania*) female sand flies and invades the host phagocytes. The sand fly can transmit the parasite either anthropologically or zoonotically (Ghorbani and Farhoudi 2018). It is considered one of the top ten neglected tropical diseases (Makoni 2021).

3.2 Life Cycle of *Leishmania*

Leishmania possesses a digenetic life cycle, being capable of colonizing an invertebrate (sand fly) and vertebrate host (human, dog, etc.). In both the hosts, it exhibits a distinct morphology with "promastigote" form in sand fly and "amastigote" form in vertebrates, respectively. The sand fly injects infective flagellated metacyclic promastigotes into a susceptible vertebrate host during blood feeding (CDC 2023). The promastigotes along with the sand fly saliva lead to the recruitment of macrophages, neutrophils, and mononuclear cells at the site of the insect bite (Fig. 3.1). The peptides present in the saliva of sand fly promote the initiation of infection in the macrophages located in the dermis by causing erythema and vasodilation (Kaye et al. 2020). The neutrophils are able to either phagocytose the parasite or form neutrophil extracellular traps (NETs) for the parasite clearance (Bhattacharya et al. 2021). However, if the promastigotes survive by evading immune surveillance, they indirectly or directly find their way to reach the macrophages (diagnostic stage) within which they transform into nonmotile amastigotes and divides by binary fission in a parasitophorous vacuole (CDC 2023). These amastigotes further infect macrophages in the skin or secondary lymphoid organs depending upon the parasitic and host factors, thus, causing a symptomatic or asymptomatic form of the disease. During these interactions, the parasites also interact with the extracellular matrix (ECM) and the proteins associated with it on the cell surface. The presence of glycosaminoglycans is one of the factors found responsible for the modulation of immune response of the cell to the parasite. Glycosaminoglycan, for example, hyaluronic acid leads to the reduction

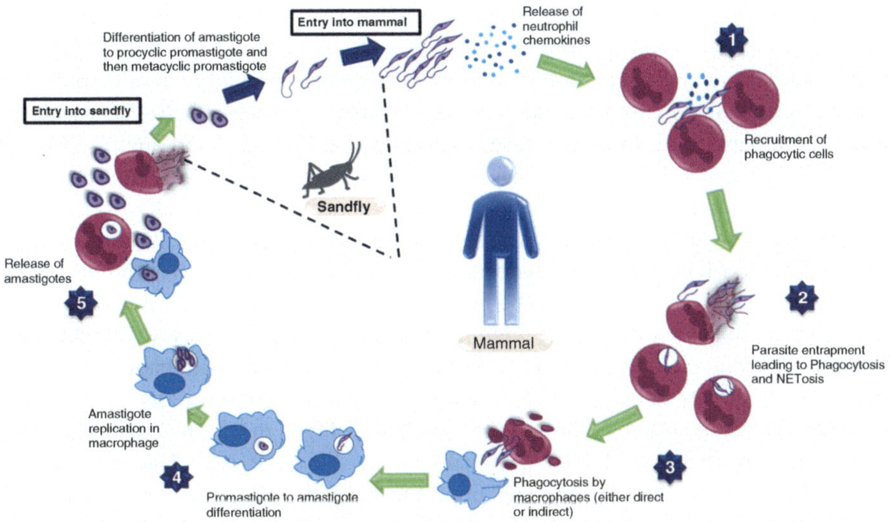

Fig. 3.1 Life cycle of *Leishmania*. The life cycle of *Leishmania* involves two hosts. Flagellated promastigote stage is present in the sand fly (vector), which injects the parasite by biting the vertebrate host and causes the recruitment of phagocytic cells and release of chemotactic factors. The parasite is trapped inside neutrophils either leading to phagocytosis or NETosis. Neutrophils recruit macrophages where promastigotes convert into nonmotile amastigotes and divide. Amastigotes are released from the phagocytic cells and enter the sand fly after another bite. Inside sand fly midgut, the amastigotes again transform into promastigotes. [Blue arrows indicate stages in sand fly, and green arrows denote stages in mammals]

in the growth and motility of the promastigotes of *Leishmania*, thus promoting its uptake by the host (Merida-De-Barros et al. 2018).

When the female sand fly bites the infected vertebrate (Fig. 3.1), it also takes up the amastigotes along with the blood. Once inside the sand flies, the amastigotes again transform into promastigotes within the midgut in low pH (Endrew Souto Maior Torres Bonfim et al. 2022) conditions and reproduce followed by migration to the foregut and transformation into infectious metacyclic promastigotes. These infectious promastigotes reach the proboscis of the sand fly and reach the healthy vertebrate host once the sand fly takes the next blood meal resulting in the continuation of the infectious cycle.

3.3 Clinical Manifestations

The clinical manifestations in the infection caused by *Leishmania* vary due to the interactions between the type of virulence characteristic present in the infecting species and the type of immune responses generated by the host toward it (Srivastava et al. 2016). The three main clinical manifestations of leishmaniasis as caused by different species include Mucocutaneous Leishmaniasis (MCL), Cutaneous

Leishmaniasis (CL), and Visceral Leishmaniasis (VL). VL, also known as kala-azar, is caused by *L. donovani, L. chagasi* (New World), and *L. infantum* (Old World). It is the most severe and fatal among the three forms with high outbreak potential. The symptoms associated with it include irregular bouts of fever (WHO 2023a), swelling in the spleen and liver, hypergammaglobulinemia, weight loss, and anemia (Ready 2014). The patient with HIV infection is also prone to opportunistic infection by VL, which shows a high mortality rate and relapse as per the 2021 WHO data (WHO 2023a). Another rare form of leishmaniasis observed includes Post kala-azar Dermal Leishmaniasis (PKDL), which can occur 6 months to 1 year after VL is cured with a papular rash on different areas of the body (Zijlstra 2019). CL is the most common form caused by *L. mexicana, L. viannia and L. amazonensis* (New World) and *L. major, L. tropica, L. ethiopia* (Old World) and is associated with papule development, which further develops into nodules (lesions) in exposed parts of the body resulting in atrophic scars. Although the lesions developed in CL are sometimes self-healing (as in Old World CL) (Volpedo et al. 2022), it may lead to lifelong disfiguration and social stigma. The third form, MCL affects the oral–nasal and the pharyngeal cavity leading to the partial or total destruction of these regions (WHO 2023a) due to the infection by *L. braziliensis, L. panamensis, and L. guyanensis (*Abadías-Granado et al. 2021*)*. Diffuse cutaneous Leishmaniasis (DCL), an anergic form of CL, occurs due to the absence of a cell-mediated immune response to the parasite. It is caused by *L. amazonensis, L. aethiopica, L. mexicana*, etc., and mainly occurs in patients with HIV infection. The symptoms include the development of skin lesions mimicking the sores of lepromatous leprosy along with nonulcerating nodules (Sinha et al. 2008) with a high parasitic load that may spread all over the body. Atypical leishmaniasis is a form occurring due to heterogeneity in parasite genes. This evolution led to the species causing a form of the disease with different phenotypes (Thakur et al. 2018). For instance, *L. tropica*, which is normally linked to CL, causes mild VL and DCL in atypical form in Iran while *L. donovani*, which usually causes VL, leads to the sporadic CL phenotype in the Mediterranean basin.

3.4 Epidemiology

The characterization of *Leishmania* and particular sand fly species, ecological characteristics at transmission areas, past and current exposure of the human population to *Leishmania*, and human behavior define the epidemiology of leishmaniasis (WHO 2023a). Though *Leishmania* was initially discovered in India, it was also present in other countries. Leishmaniasis has a worldwide prevalence of 12 million cases spread in approximately 98 countries (Akhoundi et al. 2016). Being one of the neglected diseases, it is estimated that around 1 billion people in tropical and subtropical countries are susceptible to leishmaniasis (Zhang et al. 2020) and 70,000–1,000,000 cases arise annually. Out of 88 endemic countries, 81% of the cases arise in developing countries. Leishmaniasis is sensitive to climate changes, which also affects its epidemiology as changes in temperature affect the distribution of vector and host and their survival. This also affects the life cycle of *Leishmania*, thus, allowing it to survive and transmit infection in nonendemic areas. Conditions such as drought and flood lead to the displacement of people in areas that might transmit *Leishmania* and cause infection due to poor immune response. According to the 2020 data published by WHO, VL has a worldwide prevalence of 50,000–90,000 cases annually with 90% of the new cases of VL observed in Brazil, India, Yemen, Somalia, Ethiopia, China, Kenya, Sudan, Eritrea, and South Sudan (WHO 2021). While in CL, 95% of the cases arise in America, Central Asia, the Middle East, and the Mediterranean basin with around a 1 million annual rise in cases. More than 90% of MCL cases are endemic to Bolivia, Peru, Brazil, and Ethiopia. There's a decline in the number of annual cases of VL globally in 2021 (1275 cases) as compared to 1977 (77,102 cases) due to a 98% decrease in cases in India owing to better sanitation, proper housing facilities, etc., but is still a cause of concern in poorer countries (WHO 2023a; Selvapandiyan et al. 2019). However, the cases of CL are increasing each year due to increased global warming, climate change, and globalization (Sasidharan and Saudagar 2021).

India accounts for 18% of the VL cases globally in 2020 with West Bengal, Jharkhand, Bihar, and Uttar Pradesh being the main endemic areas. The new add-ons to these states include Gujarat, Tamil Nadu, and Assam along with sporadic cases occurring in Madhya Pradesh and Kerala (WHO 2023b). According to recent reports, the VL parasite has also spread to Himachal Pradesh, Uttarakhand, and Jammu-Kashmir, the high-altitude regions. For. CL, Bikaner and Thar desert were known to be endemic regions earlier, but the parasite for CL has also been reported to develop sporadic cases in Delhi, Punjab, Haryana, Assam, and Varanasi (WHO 2023b). People suffering from VL are also reported to be coinfected with HIV in 5–7% cases, thus increasing the chances of relapse and mortality (WHO 2022). VL and PKDL are known to be caused by *L. donovani*; however, there are several reports that confirm the coexistence of another parasite *Leptomonas seymouri* that resembles *L. donovani*, which could affect the vaccine development and increase the mortality (Ahuja et al. 2015, 2020; Selvapandiyan et al. 2015).

3.5 Therapeutics Available for the Treatment of Leishmaniasis

The current treatment approach for leishmaniasis includes chemotherapy involving antimonials for the treatment of CL and MCL and Amphotericin B for the treatment of VL. The therapeutics available for the treatment of Leishmaniasis are depicted in Fig. 3.2. Some of the drugs in practice for the treatment of leishmaniasis are:

Antimonials: These are the drugs used all over the world to produce an antileishmanial effect. The drug once consumed is reduced from pentavalent to trivalent active form in macrophages, thus making the amastigotes sensitive to it. Sodium stibogluconate was used as a drug for the treatment of VL for around six decades since 1945 (Hendrickx et al. 2019). However, over the last few decades, *Leishmania* began to show resistance to the drug by a decrease in its uptake by the parasite, increased efflux, etc. (Taslimi et al. 2018), leading to a ban on its use in

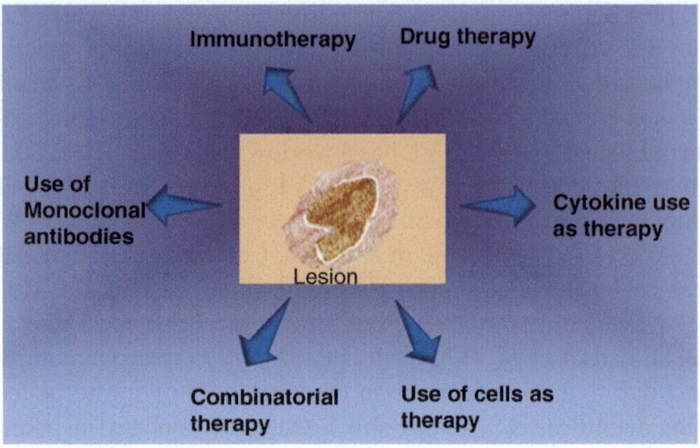

Fig. 3.2 Therapeutics in practice for the treatment of leishmaniasis. The different therapeutic approaches for the treatment of leishmaniasis depend on its different forms. Some of the treatment methods include drug therapy, combinatorial therapy for VL, immunotherapy by using biological molecules, use of cytokines like IL-12 for a targeted therapy, use of monoclonal antibodies to treat specific form of leishmaniasis, etc.

India. The drug also posed renal toxicity, pancreatitis, gastro-intestinal (GI) tract problems along with pain at the site of injection (Sasidharan and Saudagar 2021). Other limitations of this drug include its mode of administration. The drug is not active orally and needs to be injected either intravenously or intramuscularly to cure the disease (Vennerstrom 1997).

Amphotericin B: Amphotericin B is an antifungal drug obtained from *Streptomyces nodusus (*Sasidharan and Saudagar 2021*)* and is majorly used for the treatment of VL by controlling leishmaniasis *via* two distinct mechanisms. This includes either the production of free radicals by Amphotericin B auto-oxidation (Taslimi et al. 2018) or the interaction of Amphotericin B with ergosterol in parasites to create aqueous pores in the membrane leading to ion imbalance and parasitic death. Its side effects like muscle pain, nausea, headache, and vomiting lead to the development of its modified form with a liposomal formulation called Ambisome (Sasidharan and Saudagar 2021), which although expensive has an increased bioavailability and fewer side effects on other tissues. In patients with VL, paromomycin sulfate (Humantin), approved for use, also showed cure rates and side effects similar to Amphotericin B (Shukla et al. 2010).

Miltefosine: Miltefosine is used against cancer and leishmaniasis and is the only known oral drug for treating VL and CL with a cure rate of more than 90% in both. It works by three mechanisms, namely inhibition of protein kinase leading to apoptosis; changing the plasma membrane structure of *Leishmania* like its GPI synthesis; and causing an immunomodulatory effect in macrophages. However, it is associated with GI tract disorder, teratogenicity, and the development of parasitic resistance in some species.

Paromomycin: Paromomycin is a broad-spectrum aminoglycoside approved for the treatment of VL and mostly for CL. It can be injected intramuscularly or can be used as an ointment for skin infections. The drug inhibits ribosomal function and thus, affects the protein synthesis in parasite and mitochondrial membrane potential, which affects ATP production (Taslimi et al. 2018). Some of the side effects associated with the drug include high Alanine Transaminase (ALT), Aspartate Transaminase (AST) levels in the liver, nausea, diarrhea, pain, and resistance (Sasidharan and Saudagar 2021).

Combinatorial therapies: Combinatorial approach to disease treatment is adopted to reduce the chances of drug resistance, decrease the treatment time and toxicity, and increase the patient response to the drug. VL patients treated with miltefosine and amphotericin B showed a higher cure rate when resistant to antimonials. Similarly, allopurinol and antimonial when given together showed a high cure rate with fewer relapses. However, combination therapy did not prove to be a 100% effective treatment as in some cases it could not decrease the *Leishmania* parasite burden (Sasidharan and Saudagar 2021).

Immunotherapy: Immunotherapy involves the use of biological molecules to stimulate the immune system by using them individually or in combination with drugs. It has lesser side effects and reduces drug resistance. Berberine chloride is found to be effective in enhancing the apoptosis of *Leishmania*-infected neutrophils by altering the mitogen-associated protein (MAP) kinase pathway (Sasidharan and Saudagar 2021). In nonhealing infections of *Leishmania*, immunotherapy works as a better approach for recovery.

Other treatment approaches in practice against Leishmaniasis include the use of cytokines like interferon-gamma (IFN-γ), interleukin-12 (IL-12), and granulocyte-macrophage colony-stimulating factor (GM-CSF) and blocking IL-10 receptor to inhibit the type 2 T helper (Th2) response and reduction in parasitic burden. Monoclonal antibodies (mAb) are also used similarly to enhance a particular response and block the other. Another approach used for treating leishmaniasis is the use of cells, mostly dendritic cells, as therapeutics. Dendritic cells (DCs) prevent infection by suppressing IL-10 release. So, the use of bone marrow–derived dendritic cells (BMDDCs), when pulsed with *L. donovani* along with treatment of antimonial, showed parasitic clearance from the liver and was an effective treatment strategy. Also, BMDDCs when pulsed with *L. amazonensis* antigen and injected showed an increase in IL-12 and suppression of infection (Taslimi et al. 2018). Sereno et al. observed the use of nicotinamide (NAM) as a potential drug for the treatment of leishmaniasis as it specifically affects the amastigote proliferation in infected macrophages without any toxicity to the host (Santos et al. 2008). The use of genomics, transcriptomics, proteomics, etc., is also being employed to identify small molecules as drugs against different *Leishmania* species by predicting protein structures through Molecular Dynamics (MD) simulations (Ochoa et al. 2016).

3.6 Limitation of Available Therapeutics

There are many drugs and therapies available by now to cure leishmaniasis that are advantageous in the sense that they do not depend on parasite heterogeneity and can be easily produced in new formulations (Handman 2001), but the use of these is limited due to high toxicity, less efficacy, side effects, high cost of production, difficult administration, evolution of resistant strains leading to relapse of leishmaniasis infection (Sasidharan and Saudagar 2021). In underdeveloped countries too, there's increased resistance to these drugs due to long and tedious treatment duration and availability (Ghorbani and Farhoudi 2018). Thus, these therapies including drug therapy do not seem to be an effective approach to eradicating leishmaniasis. To reduce transmission and a long-term response against Leishmaniasis, vaccination would be a better approach. In view of this, there are some vaccines produced and used to treat and prevent leishmaniasis in dogs, and some are under trials to be used for humans. Vaccination programs could reduce the disease incidence and would prove to be a better cure.

3.7 Vaccine Against Leishmaniasis

Vaccination against leishmaniasis is an economically feasible method for disease prevention and long-term immunity. Patients recovering from leishmaniasis either naturally or following drug therapy usually develop life-long immunity against reinfection suggesting the feasibility of vaccine development (Zhang et al. 2020). The knowledge gained from leishmanization in the early century, which involves the development of immunity against reinfection by initial infection using a low dose of *L. major*, provides evidence that vaccine development against leishmaniasis would be fruitful (Cecílio et al. 2018). The different forms of the vaccine developed for treating leishmaniasis include: first to third-generation vaccines.

First-generation vaccines: These include whole-killed parasites in promastigote form. The use of killed promastigotes along with adjuvants like *Mycobacterium bovis* BCG or Alum provides a therapeutic value and low cost. Two first-generation vaccines developed in the 20th century in Brazil include Leishvacin (a pentavalent preparation made without adjuvant in which the parasite was inactivated by merthiolate) and other similar autoclaved vaccines (Ghorbani and Farhoudi 2018). Both provided similar immunogenicity. *L. mexicana* adjuvanted with BCG was also developed in Venezuela as immunotherapy to cure CL. However, these vaccines had low efficacy and didn't provide long-term immunity. Also, trials in humans with three forms of a vaccine comprising *L. major, L, mexicana*, and *L. amazonensis* have not been much effective as it showed immunogenicity but not protection (Ghorbani and Farhoudi 2018).

Second-generation vaccine: Second-generation vaccine includes the use of live attenuated vaccines, recombinant antigen vaccines, purified *Leishmania* protein vaccines, and vaccines using recombinant bacteria or virus as a delivery agent. Generation of a live vaccine involves the knock-out of important genes like dihydrofolate reductase (DHFR), thymidylate synthase, centrin (Ghorbani and Farhoudi 2018) that results in an inactivated infection causing pathogen, thus generating an adaptive immune response. Similar to knock-out-based attenuation, the addition of drug-sensitive cassettes like the addition of cytidine deaminase gene of *Saccharomyces cerevisiae*, which being sensitive to 5-fluorocytosine results in the death of the parasite post inoculation (Duthie and Reed 2014), can also be a vaccine approach. Recently, bacterial and viral vaccines were also developed by cloning the *Leishmania* antigen in either of them and using it as a delivery vehicle to initiate an immune response. For example, cloning of Gp63 surface protease of *L. major* in *Salmonella typhimurium* mutant bacteria or expressing *L. infantum Leishmania*-receptor for activated kinase (LACK) antigen in virus showed protection against *L. major* and generated $CD4^+$ and $CD8^+$ T cell response with increased IL-2 and IFN-γ in vaccinated mice. Thiol-specific antioxidant, *L. major* stress–inducible protein 1 is also being formulated into vaccines with adjuvants to generate recombinant vaccines and provide an immune response against the disease (Ghorbani and Farhoudi 2018). Lipophosphoglycan (LPG) conjugated with polyelectrolytes like Polyacrylic acid (PAA) has also been characterized and tested for its toxicity for its use as a potential vaccine candidate against leishmaniasis in future (Topuzogullari et al. 2013).

Third-generation vaccine: Third-generation vaccines include DNA and mRNA vaccines. DNA vaccines are known to be more stable than second-generation vaccines along with their low cost of production and provide flexibility in using a combination of different genes. These vaccines generate a response through activation of the innate immune system, which is engaged by sequences of nonmethylated CpG of bacteria, thus allowing recombinant protein expression for a longer duration (Ghorbani and Farhoudi 2018). DNA vaccines have been found to be more effective in rodent and simian models (Kaye et al. 2021). Thiol-specific antioxidant (TSA) gene in the DNA vaccine was able to show maximum protection against leishmaniasis. However, these results have been reported in Phase I and Phase II clinical trials only. The first human trial of a third-generation vaccine to protect against VL and PKDL had two *Leishmania* proteins (kinetoplastid membrane protein-11 (KMP-11) and hydrophilic acylated surface protein B (HASPB) encoded in Simian adenovirus (ChAd63) conferring protection and generating potent immune response (currently in Phase II clinical trials) (Coutinho De Oliveira et al. 2020). Administration of DNA vaccine along with mTOR inhibitors like sirolimus (SIR), also called as rapamycin, has been reported to reduce the parasitic load of *L. major* in the draining lymph nodes in mouse model (Martínez-Flórez et al. 2020). This coadministration has also been reported to enhance the differentiation of $CD8^+$ T cells and the release of IFN-γ (Martínez-Flórez et al. 2020).

3.8 Vaccination in Canine

Canine leishmaniasis (CanL) is caused by *L. infantum* with dogs being the most important reservoir responsible for maintaining the parasite in the endemic region and thus, is one of the important targets for controlling the disease. Also, leishmaniasis in canines can cause a chronic disease ultimately leading to death. The symptoms of CanL include weight loss, lymph node enlargement, and dermatological disorders (Almeida et al. 2021). CanL prevention is achieved by immunotherapy and vaccination. Four vaccines approved for the treatment of CanL include Leishmune® and Leish-Tec® in Brazil in 2004 and CaniLeish® and LetiFend® in Europe (Velez and Gállego 2020), respectively.

Leishmune®: It was the first vaccine licensed for use against CanL in Brazil in 2004. Being a second-generation vaccine, it comprises saponin adjuvant along with fucose mannose ligand (FML) from *L. donovani* and is administered subcutaneously three times with a gap of 21 days and annual boosters thereafter. It was first tested on murine models that induced seroconversion of IgG2, showed delayed-type hypersensitivity response to promastigote, and was successful in protecting against *L. donovani* and cross-infection by *L. chagasi*. Further studies demonstrated that Leishmune® upregulated $CD8^+$ T cells and immunization increased the nitric oxide (NO), IFN-γ, and anti-*L. chagasi* IgG2 levels after one and six months of vaccination and returned to normal levels after 12 months. Low IL-10 and IL-4 and high IL-17 were observed in the first 6 months. Leishmune® also provided a protective immune response by stimulating neutrophils and monocytes (Velez and Gállego 2020). Vaccination with Leishmune® reduced canine and human leishmaniasis cases in Brazil along with no detection of clinical signs of infection and *Leishmania* DNA, which led to the assumption and consideration of this vaccine as a transmission-blocking vaccine. However, the use of Leishmune® was withdrawn by Brazil in 2014 due to insufficient effectiveness evidence during phase III trials (de Oliveira Aguiar-Soares et al. 2020).

Leish-Tec: Leish-Tec®, licensed in 2007 in Brazil, is the only authorized vaccine at present for CanL in Brazil. It comprises recombinant A2 protein of *L. donovani* amastigotes with saponin adjuvant and is given to dogs aged 4 months or older with three doses at an interval of 21 days and annual boosters thereafter. The A2 protein is an amastigote-specific virulence protein essential for the survival of *Leishmania* in the host. The immunization in murine models showed low parasitic burden in the liver and a specific humoral and cell-mediated response with high levels of IFN-γ in the vaccinated as compared to unvaccinated. Immunization in dogs showed a protective response against a high dose of *L. chagasi* and high anti-A2 IgG2 levels after vaccination. High IFN-γ levels were detected in vaccinated dogs on stimulation with A2 antigen or total protein extract of *L. chagasi*. However, it showed adverse reactions compared to Leishmune®, and efficacy evidence could not be obtained (Velez and Gállego 2020).

CaniLeish: CaniLeish® and LetiFend® are the two CanL vaccines currently approved for use in Europe (Lima et al. 2019). CaniLeish®, licensed in Europe in 2011, is a combination of excreted–secreted proteins of *L. infantum* (LiESP) and

Quilaja saponaria saponin (QA-21) as the adjuvant. The initial vaccination dose was given to measure the humoral and cellular marker levels in the immunity of dogs. The results showed a high IgG2 response and mainly type 1 T helper (Th1) response to LiESAp with a 99% protection in field study and hence was found to be safe to use with only mild side effects. A high IFN-γ was also observed in vaccinated dogs as compared to the control dogs along with increased NO_2 production and parasite-killing capacity on exposure of macrophage of vaccinated dog with *L. infantum*. IFN-γ ELISpot and macrophage leishmanicidal assay also showed a high level of IgG2 in the case of vaccinated dogs. The vaccine's side effects included local edema and crusting. Also, the vaccination prevented the migration of *Leishmania* to deep tissues and led to the reversion of some of the vaccinated dogs to the *Leishmania*-free population (Velez and Gállego 2020).

LetiFend®: It is a second-generation recombinant vaccine licensed in Europe in 2016, comprising protein Q derived from four *L. infantum* antigenic proteins (LiP0, LiP2a, LiP2b, and H2A) without adjuvants. It is administered to dogs aged 6 months or older as a single dose with annual boosters. Initial experiments on mice and dogs by vaccinating protein Q with BCG as an adjuvant to check the immune potential followed by *L. infantum* infection showed protection and prevention of parasite establishment. However, treatment with other adjuvants led to the development of infection. Vaccination with either BCG or protein Q alone elicited a protective response. LetiFend®, consisting of only protein Q, showed vaccine efficacy of 72% in phase III clinical trials and prevented signs and symptoms of CanL in vaccinated dogs along with no side effects or false-positive results of vaccination (Velez and Gállego 2020).

3.9 Novel Approaches in Vaccine Development Against Leishmaniasis

Among the vaccines under clinical trials, only two of the vaccines designed are currently approved for human administration trial, with one based on killed *L. amazonensis* for CL immunotherapy while the second one consisting of live *L. major* (Coutinho De Oliveira et al. 2020). However, none of the vaccines designed or under trial have been licensed for human use owing to their inconsistency, ineffective results, and recurrence of disease in some cases. The roadblocks to effective vaccine development include target antigen selection, lack of knowledge about unknown *Leishmania* proteins, lack of funding for clinical trials, vaccine demand, affordability, and lack of epidemiological modeling for a better impact of vaccine (Kaye et al. 2021). Hence, there's a need for novel approaches to vaccine design that would be effective and provide long-lived immunity against leishmaniasis (Fig. 3.3). Some of these novel approaches include:

New-generation leishmaniasis live-attenuated vaccines consisting of either centrin gene deleted *L. donovani* parasites *via* double homologous recombination (Selvapandiyan et al. 2004, 2009, 2014, 2019; Avishek et al. 2016) or the same homologous gene deleted *L. major*/*L. mexicana* parasites *via* CRISPR Cas9 have

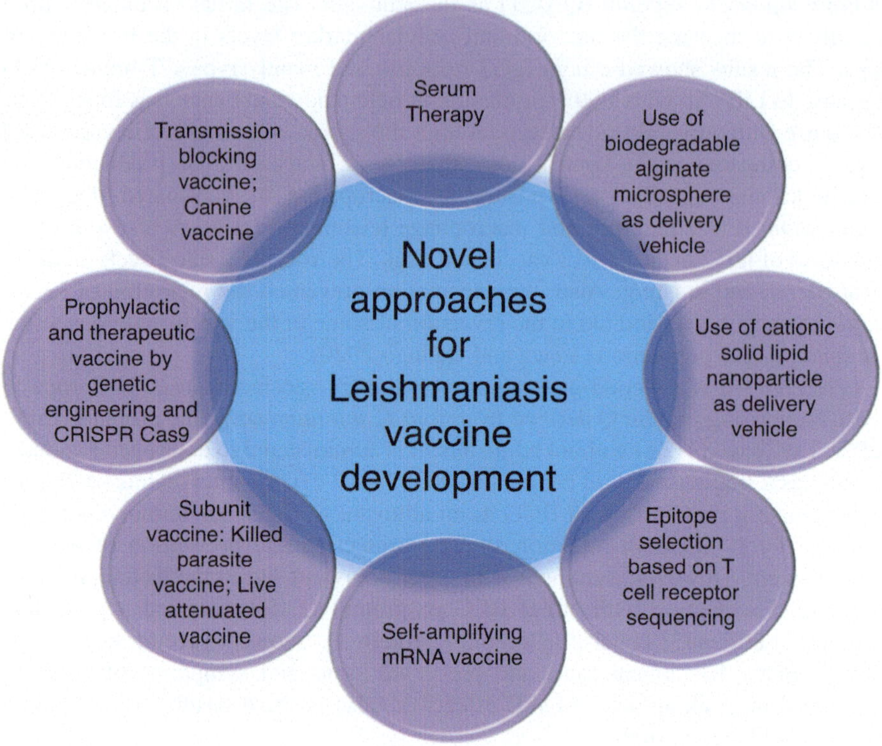

Fig. 3.3 Novel approaches to vaccine development against leishmaniasis. To provide a safe, effective, and long-lasting immune response, there's a need to use novel approaches in vaccine design against leishmaniasis. Some of the novel approaches that can be used to develop leishmaniasis vaccine include use of CRISPR Cas9 to develop a live-attenuated vaccine, use of biodegradable molecules as delivery vehicle, use of lipid nanoparticle as delivery vehicles, use of mRNA for vaccine design, T-cell receptor sequencing for the selection of specific epitope, etc.

been developed (Volpedo et al. 2022; Zhang et al. 2020) and tested for the treatment against homologous and heterologous strains of *Leishmania* sp. in animals. Such mutant *Leishmania* strains were tested for their safety and efficacy in several animal models including mice (including aged mice), hamsters, and dogs and in human cells (Volpedo et al. 2022; Zhang et al. 2020; Avishek et al. 2016; Fiuza et al. 2013, 2015; Selvapandiyan et al. 2009) . The use of CRISPR Cas9 generates mutants without any antibiotic resistance gene marker activity and thus, does not pose any hindrance during clinical trials. Recombinant antigens from other species like using nonpathogenic *L. tarentolae* as a vehicle to deliver the A2 gene of *L. donovani* in mice showed a high IFN-γ generating immune response on vaccination (Taslimi et al. 2018).

Selection of epitope on the basis of T-cell receptor sequencing could provide more insights into human antigen recognition. Another approach involves the use of reverse genetics through transcriptomic studies to identify novel targets that are correlated to the infective stage of the parasite life cycle. mRNA vaccine, similar to that used in COVID-19, against leishmaniasis is one of the approaches that need to be explored in in vivo studies as only in vitro studies have been performed to date (Kaye et al. 2021). Self-amplifying mRNA vaccine that is being explored for the malaria vaccine can also be studied for development against leishmaniasis. This approach requires less amount of antigen but provides an enhanced immune response, thereby reducing the requirement for a booster dose. It induces differentiation of both $CD4^+$ and $CD8^+$ T cells and has high suitability due to the non-requirement of freezer storage (Laurens 2021). The use of alginate microspheres as a delivery vehicle for *Leishmania* antigen to act as an immunoadjuvant and induce a humoral and cell-mediated immune response could also be one of the approaches. It is also biodegradable, low-cost, and safe to use. Another approach is to use cationic solid lipid nanoparticles (cSLNs) as a delivery system owing to its better stability than liposomes. The antigen is first absorbed on the nanoparticle surface and then trapped inside. Studies using lipid nanoparticles to deliver parasite antigens in mice showed protection against *L. infantum* by generating a Th1 immune response and high IFN-γ levels (Ghorbani and Farhoudi 2018). To target leishmaniasis, in silico designing and packaging of self-amplifying mRNA derived from an alphavirus in the form of viral replicon particles are also being studied (Savar et al. 2022). Transmission-blocking vaccines involving the use of mAb against epitopes expressed in malaria parasites are known to block transmission to humans. Similar strategies can be applied for leishmaniasis as well using *Leishmania* antigens or epitopes. Self-assembling protein nanoparticles, known to induce cellular and humoral immunity in malaria vaccines (Laurens 2021), can be tested for leishmaniasis as well.

The reverse vaccinology approach is also used widely to identify novel vaccine antigens. The proteome of parasite species is studied to identify proteins whose transmembrane helices and subcellular localization is predicted. Selected proteins are then analyzed for their binding efficiency to MHC-I and II alleles (John et al. 2012). Therefore, it can be used to discover potential vaccine antigens through in silico analysis of the pathogen's genome or proteome instead of culturing the whole organism (John et al. 2012).

3.10 Role of Different Innate Immune Cells in Vaccine Development

The infection with *Leishmania* parasites leads to the activation of a spectrum of host immune responses involving both innate and adaptive immune responses (Selvapandiyan et al. 2023). Once the parasite infects, the innate immune cells tend to create an environment for the host, which could be either permissive or hostile depending on the effector functions and ultimately modulates the adaptive system for the susceptibility or resistance of the host (Volpedo et al. 2021). *Leishmania*, through evolution, has devised strategies to survive phagocytosis or complement activation by the innate immune cells. Innate immune cells that react to *Leishmania* infection include Macrophages, Natural killer (NK) cells, Mast

Cells (MCs), Neutrophils, Dendritic cells (DCs), and Monocytes along with T cells, NO, and cytokines (IL-12, IFN-γ) to elicit a host response (Elmahallawy et al. 2021). Studies have provided evidence for the role of innate immune cells in interacting with the parasite to determine the outcome and could play an important role in the development of a potential vaccine against leishmaniasis.

The innate immune cells and their role in vaccine development are as follows:

Neutrophils: Upon parasite inoculation, the sand fly, parasite, or the host factors lead to the initial recruitment of neutrophils at the dermal region of infection, which promotes phagocytosis or extracellular trap formation along with the release of reactive oxygen species (ROS) and cytokines like IL-12, IFN-γ, and α tumor necrosis factor-α (TNF-α) to induce T-cell activation or transforming growth factor-β (TGF-β) and IL-10 to inhibit T-cell response at the site of infection. Besides interaction with the parasite, neutrophils recruit macrophages, NK cells, DCs, and adaptive cells like B cells and T cells (Volpedo et al. 2021). Neutrophil interaction with other immune cells also affects the infection caused by *Leishmania.*

Neutrophils also act as "Trojan horses" by carrying *Leishmania* in apoptotic neutrophils and transferring them to the macrophages (Ger van Zandbergen et al. 2004), which act as the principal host of *Leishmania.* The role of neutrophils in vaccine immunity is relevant as they can present the antigen directly to the naïve T cells and activate them. Studies have also shown an impairment in protection against *L. donovani* infection in the absence of neutrophils in centrin-deleted *L. donovani* (LdCen$^{-/-}$) immunized mice. Recent studies provided evidence that due to epigenetic and metabolic changes, the neutrophils may be able to acquire a nonspecific memory against the parasite called "trained immunity." Therefore, the importance of neutrophils in the anti-*Leishmania* vaccine may be attributable to their antigen presentation ability and the development of trained immunity for which more research needs to be done (Volpedo et al. 2021).

Monocytes: Leishmania infection causes the release of proinflammatory cytokines that leads to the recruitment of inflammatory monocytes (iMO) mediated by C-C motif chemokine receptor 2 (CCR2). Monocytes, through toll-like receptors (TLRs), secrete anti- and proinflammatory cytokines that may either prevent ROS production and antigen presentation by activating IL-10 in case of *L. donovani* infection or it may increase TLR9 and TLR2/4 expression with increased TNF-α in the case of *L. tropica*, leading to decrease in inducible nitric oxide synthase (iNOS). iNOS is capable of reducing the parasitic burden during secondary infections by producing ROS and NO, thus clearing the parasite and initiating a Tissue-resident CD4^{+} memory (TRM) T cells response, which could make iMOs crucial in improving the efficacy of the anti-*Leishmania* vaccine. Leishmune® studies have also shown an increase in phagocytic activity, TLR expression, IFN-γ levels, and NO by monocytes and neutrophils in vaccinated dogs. The ability of monocytes to activate the adaptive immune response is also a factor in their role in vaccine design (Volpedo et al. 2021). BCG and *Candida albicans* are known to induce trained immunity in monocytes and macrophages by epigenetic reprogramming. A similar mechanism can be looked upon during *Leishmania* infection to develop a pan-*Leishmania* vaccine that could be targeted to induce robust trained immunity along with adaptive immune response (Volpedo et al. 2021).

Macrophages: Macrophage, known to be the ultimate host for *Leishmania*, responds to the infection depending upon the molecules present on the surface of *Leishmania* like GP63, GPI, LPG and their recognition by the TLRs, mannose receptors, and Fc receptors. Macrophages serve to act as effector cells against *Leishmania* as well as a permissive host for *Leishmania* depending upon the microenvironment of the host tissue in response to the parasite. Proinflammatory macrophages (M1) promote parasite clearance by increasing IL-12, TNF-α, and ROS levels. Anti-inflammatory macrophages (M2) allow amastigote growth by increasing IL-10 and IL-4 levels and decreasing NO levels. The role of macrophages as antigen presenting cells (APCs) to induce an antigen-specific T-cell response is crucial for its role in vaccine development and has been known in the preparation of protein-based and DNA-based vaccines. There's a possibility that the vaccine against leishmaniasis targeting macrophages could complement the T-cell response and allow the macrophages to enhance the internalized parasite killing. Macrophages are known to enhance the Th1 immune response in vaccine immunity against *L. donovani*. The role of macrophages in producing TRM cells could provide some of their effects in the development of the *Leishmania* vaccine or determination of efficacy (Volpedo et al. 2021).

Dendritic Cells: DCs recognize *Leishmania* through TLRs and are activated to produce IL-12 and chemoattractants in *L. infantum* and IL-12, IL-23, IL-27 in *L. donovani* infections, respectively. DCs migrate to secondary lymphoid organs upon activation by parasites and induce the activation of T cells for a cellular response. Some of the parasitic proteins like excreted or secreted proteins decrease the IL-12p7 production by DCs. DCs are being used as potential vaccine candidates due to their property of acting as an APC on which DC-derived exosomes or other antigens can be loaded. Along with this, another vaccine strategy using DC is pulsing of DCs ex vivo with N terminal region of *Leishmania* elongation factor 2 (EF-2) and then using them to vaccinate mice along with CpG oligodeoxynucleotides. This generated a potent T-cell response with IL-2 and TNF-α being produced by $CD4^+$ T cells and IFN-γ produced by $CD8^+$ and $CD4^+$ T cells. Thus, DCs play a pivotal role in a potential vaccine response due to the activation of T-cell response and production of IL-12 whose expression reduces the *Leishmania* burden in the host (Volpedo et al. 2021).

Mast Cells: MCs are the sentinel immune cells packed with secretory granules that include mediators like histamine, TNF-α, serotonin, and some enzymes like β-hexosaminidase and tryptase that are released at the site of infection (Varma and Puri 2019). MCs are predominantly present in the skin within the superficial dermis and recruit other innate immune cells and activate the adaptive immune response. MCs are also involved in the immune response against leishmaniasis as the parasite also deposits in the dermis post the bite of an infected sand fly. MCs are known to control *L. major* infection by recruiting DCs to activate the adaptive immune response (Volpedo et al. 2021). Thus, MCs chaperone the immune response against *L. major* by recruiting neutrophils, macrophages, and DCs sequentially (Naqvi et al.

2020). MCs show heterogeneity in the uptake and killing of *Leishmania* depending on the type of species (Naqvi et al. 2020). Interaction of MCs with *L. tropica* leads to higher phagocytosis and killing as compared to *L. donovani*. However, both these species cause the release of MC extracellular traps (MCETs). There's a hypothesis that MCs, on encountering *Leishmania*, not only induce local inflammation but also recruit other immune cells to the site of infection ensuring a protective and long-lasting response along with the development of memory against the infection, suggesting its importance in the development of an efficacious vaccine. Centrin deleted *L. donovani* ($LdCen^{-/-}$) parasite infection followed by challenge with *L. mexicana* led to a decrease in parasitic burden, less infiltration, and no degranulated MCs, hence could protect against *Leishmania* infection from different species and serve as a prophylactic vaccine (Dey et al. 2014). Molecules secreted by MCs are known to show adjuvant-like activity in many vaccines, and their administration may result in Th1 polarization and release of Th2 cytokines resulting in a heightened immune response. Further research is required into the function of MC secretory molecules as immunomodulators and their potential as a *Leishmania* vaccine (Volpedo et al. 2021).

Natural Killer and NKT cells: NK Cells are known to induce an immune response against *Leishmania*. NK cells are activated by the binding of LPG of *Leishmania* to the TLR-2 and induce the secretion of IFN-γ and TNF-α along with the translocation of NF-κB transcription factor to the nucleus. In spite of their role in immune response, they are not able to generate an efficient Th1 immune response against *L. tropica* or *L. major* amastigote but are crucial for *L. donovani* clearance. The role of NK cells in vaccine development has been identified in recent studies where scientists have observed an increase in the production of IFN-γ by the NK cells when living *L. major* is used as a vaccine along with the CpG DNA (Volpedo et al. 2021).

$CD1d^{+}$ NK-T cells are known to show some protection against leishmaniasis during the early stages by detecting the glycoconjugates on the parasite surface (Volpedo et al. 2021). $CD8^{+}$ NKT cells also confer protection against *L. donovani* by enhancing IFN-γ and killer cell immunoglobulin–like receptors (KIRs) expression.

Hence, the role of NK and NKT cells in the anti-*Leishmania* vaccine along with the trained immunity needs to be further investigated for their use in vaccine success (Volpedo et al. 2021).

3.11 Population Relation to Vaccine

In the case of vaccine development for commercial use, many factors like size of the target population, economy, regulatory feasibility, and development feasibility are crucial especially for Neglected Tropical Diseases (NTDs) as it mainly affects lower- and low-middle-income countries. The size of the target population and the demand for the vaccine are also looked upon by policymakers and people in the regional health system to determine the health impact of the implementation of the vaccine program with respect to financial resources required, mortality, and morbidity rates. The exact estimate of the burden of leishmaniasis is difficult due to the less knowledge on the geographical distribution and epidemic nature of this disease (Malvolti et al. 2021).

By the time of vaccine introduction, leishmaniasis vaccines will be able to target a total population of approx. 1.1 billion in 2030. Based on 2018 data, the overall target for vaccine would be nearly equal to the total population at risk for VL (approx. 404 million) and CL (approx. 765 million), while the total population of people at risk for VL and PKDL is much less. Therefore, the use of the vaccine in a huge population could provide significant protection and benefit. To implement the immunization program against VL and CL over 10 years, the vaccine required would be around 1.1 billion doses in the case of CL and 0.5 billion doses in the case of VL. This large number of doses also attracts the manufacturers of the vaccine to finance the immunization program along with the clarity on the recommendation made by world health organization (WHO) for the use of the vaccine. Hence, there's a growing demand for the need of a vaccine against leishmaniasis to reduce the parasitic burden. Based on the prioritization made by countries, the vaccine against CL and VL could reduce the disease burden of NTD and brings a commercial prospect for the developers of the vaccine (Malvolti et al. 2021).

3.12 Discussion

Vaccines are the most effective way of controlling disease and should be of global public health priority for the diseases for which it is not available. However, before the development, there are many challenges like the research transition from laboratory to the field, differences in animal models and humans, lack of sufficient knowledge of parasitic pathogenesis and the immune response generated along with the variation in old-world and new-world *Leishmania* that needs to be addressed.

The use of combination drug therapy has been known for the treatment of leishmaniasis; however, due to the associated side effects and high production cost, there's a need for the development of an effective vaccine that would be more affordable and able to protect billions of affected people annually. The use of the vaccine for the treatment of leishmaniasis has been approved only in canines currently. The vaccine used would also provide long-term immunity against infection and help in reducing transmission.

The innate immune cells play an important role in shaping the adaptive response, which is crucial for protective immunity against infection. The host response against *Leishmania* is initially initiated by the innate immune cells that also activate the adaptive immune system for the clearance of the parasite. However, *Leishmania* has evolved to develop strategies for the evasion of the innate immune response. The *Leishmania* infection also induces a change in the metabolome of the host's innate immune system that could provide biomarkers of disease progression and can be looked upon for the development of a vaccine. The innate immune cells are also known to acquire trained immunity due to metabolic and epigenetic changes that could also be studied in the development of an effective vaccine. The vaccine candidates currently under trial are insufficient to provide long-term immunity against the parasite. Hence, there's a need for novel approaches to the development of an effective therapeutic vaccine.

Acknowledgments NP is supported by research grant from Department of Science and Technology-Science and Engineering Board (DST-SERB), Govt of India—Project (No. CRG/2019/003651). AS is supported by grants (No. 6/9-7(241)/2020-ECD-II; No. GIA/2/VBD/2021/ECD-II) from Indian Council of Medical Research. SB is supported by junior research fellowship from Department of Biotechnology, India.

References

Abadías-Granado I, Diago A, Cerro PA, Palma-Ruiz AM, Gilaberte Y (2021) Cutaneous and mucocutaneous leishmaniasis. Actas Dermosifiliogr 112:601–618

Ahuja K, Arora G, Khare P, Selvapandiyan A (2015) Selective elimination of Leptomonas from the in vitro co-culture with Leishmania. Parasitol Int 64:1–5

Ahuja K et al (2020) High resolution melting based method for rapid discriminatory diagnosis of co-infecting Leptomonas seymouri in Leishmania donovani-induced leishmaniasis. Parasitol Int 75:102047

Akhoundi M et al (2016) A historical overview of the classification, evolution, and dispersion of Leishmania parasites and sandflies. PLoS Negl Trop Dis 10:e0004349. https://doi.org/10.1371/journal.pntd.0004349

Almeida GG et al (2021) FML/QuilA-vaccinated dogs naturally infected with Leishmania infantum: serum cytokines, clinicopathological profile, and parasitological parameters. Biomed Res Int 2021:3192960

Arenas R, Torres-Guerrero E, Quintanilla-Cedillo MR, Ruiz-Esmenjaud J (2017) Leishmaniasis: a review. F1000Res 6:750. https://doi.org/10.12688/f1000research.11120.1

Avishek K et al (2016) Gene deleted live attenuated Leishmania vaccine candidates against visceral leishmaniasis elicit pro-inflammatory cytokines response in human PBMCs. Sci Rep 6:33059

Bhattacharya P et al (2021) HHS public access, vol 205, pp 3333–3347

CDC (2023) The life cycle of Leishmania species. In: Sandflies inject infective. Download scientific diagram https://www.researchgate.net/figure/The-life-cycle-of-Leishmania-species-Sandflies-inject-infective-promastigotes-into-a_fig 10_234087025

Cecílio P, Oliveira F, da Silva AC (2018) Vaccines for human leishmaniasis: where do we stand and what is still missing? In: Leishmaniases as re-emerging diseases. https://doi.org/10.5772/intechopen.75000

Coutinho De Oliveira B, Duthie MS, Alves Pereira VR (2020) Vaccines for leishmaniasis and the implications of their development for American tegumentary leishmaniasis. Hum Vaccin Immunother 16:919–930

de Oliveira Aguiar-Soares RD et al (2020) Phase I and II clinical trial comparing the LBSap, Leishmune®, and Leish-Tec® vaccines against canine visceral leishmaniasis. Vaccine 8:1–17

Dey R et al (2014) Characterization of cross-protection by genetically modified live-attenuated Leishmania donovani parasites against Leishmania mexicana. J Immunol 193:3513–3527

Duthie MS, Reed SG (2014) The emergence of defined subunit vaccines for the prevention of leishmaniasis. Curr Trop Med Reports 1:154–162

Elmahallawy EK, Alkhaldi AAM, Saleh AA (2021) Host immune response against leishmaniasis and parasite persistence strategies: a review and assessment of recent research. Biomed Pharmacother 139:111671. https://doi.org/10.1016/j.biopha.2021.111671

Endrew Souto Maior Torres Bonfim N, Lígia Barbour Scott A, de Azevedo Calderon L (2022) Leishmaniasis: molecular aspects of parasite dimorphic forms life cycle. In: Leishmaniasis-

general aspects of a stigmatized disease. IntechOpen. https://doi.org/10.5772/intechopen.102370
Fiuza JA et al (2013) Induction of immunogenicity by live attenuated Leishmania donovani centrin deleted parasites in dogs. Vaccine 31:1785–1792
Fiuza JA et al (2015) Vaccination using live attenuated Leishmania donovani centrin deleted parasites induces protection in dogs against Leishmania infantum. Vaccine 33:280–288
Ghorbani M, Farhoudi R (2018) Leishmaniasis in humans: drug or vaccine therapy? Drug Des Devel Ther 12:25–40
Handman E (2001) Leishmaniasis: current status of vaccine development. Clin Microbiol Rev 14: 229–243
Hendrickx S, Caljon G, Maes L (2019) Need for sustainable approaches in antileishmanial drug discovery. Parasitol Res 118(10):2743–2752
John L, John GJ, Kholia T (2012) A reverse vaccinology approach for the identification of potential vaccine candidates from leishmania spp. Appl Biochem Biotechnol 167:1340–1350. Springer
Kaye PM, Cruz I, Picado A, Van Bocxlaer K, Croft SL (2020) Leishmaniasis immunopathology—impact on design and use of vaccines, diagnostics and drugs. Semin Immunopathol 423(42): 247–264
Kaye PM et al (2021) Overcoming roadblocks in the development of vaccines for leishmaniasis. Expert Rev Vaccines 20:1419–1430
Kobets T, Grekov I, Lipoldova M (2012) Leishmaniasis: prevention, parasite detection and treatment. Curr Med Chem 19:1443–1474. https://doi.org/10.2174/092986712799828300
Laurens MB (2021) Novel malaria vaccines. Hum Vaccin Immunother 17:4549–4552
Lima C et al (2019) The use of specific serological biomarkers to detect CaniLeish vaccination in dogs. Front Vet Sci 6:373
Makoni M (2021) New threats to visceral leishmaniasis control. The Lancet Microbe 2:e574
Malvolti S, Malhame M, Mantel CF, Le Rutte EA, Kaye PM (2021) Human leishmaniasis vaccines: use cases, target population and potential global demand. PLoS Negl Trop Dis 15:e0009742
Mann S et al (2021) A review of leishmaniasis: current knowledge and future directions. Curr Trop Med Rep 8:121–132. https://doi.org/10.1007/s40475-021-00232-7
Martínez-Flórez A et al (2020) Sirolimus enhances the protection achieved by a DNA vaccine against Leishmania infantum. Parasit Vectors 13:294
Merida-De-Barros DA, Chaves SP, Belmiro CLR, Wanderley JLM (2018) Leishmaniasis and glycosaminoglycans: a future therapeutic strategy? Parasit Vectors 111(11):1–12
Naqvi N et al (2017) Role of mast cells in clearance of Leishmania through extracellular trap formation. Sci Rep 7:1–16
Naqvi N, Srivastava R, Selvapandiyan A, Puri N (2020) Host mast cells in leishmaniasis: friend or foe? Trends Parasitol 36:952–956
Ochoa R et al (2016) Drug search for leishmaniasis: a virtual screening approach by grid computing. J Comput Aided Mol Des 30:541–552
Ready PD (2014) Epidemiology of visceral leishmaniasis. Clin Epidemiol 6:147–154. https://doi.org/10.2147/CLEP.S44267
Santos DO et al (2008) Leishmaniasis treatment - a challenge that remains: a review. Parasitol Res 103:1–10
Sasidharan S, Saudagar P (2021) Leishmaniasis- where are we and where are we heading. Parasitol Res 120:1541–1554
Savar NS, Shengjuler D, Doroudian F, Vallet T, Kain AM, Arashkia A, Khamesipour A, Lundstrom K, Vignuzzi M, Niknam HM (2022) An alphavirus-derived self-amplifying mRNA encoding PpSP15-LmSTI1 fusion protein for the design of a vaccine against leishmaniasis. Parasitol Int 89:102577. https://www.researchgate.net/publication/359330144_An_alphavirus-derived_self-amplifying_mRNA_encoding_PpSP15-LmSTI1_fusion_protein_for_the_design_of_a_vaccine_against_leishmaniasis
Selvapandiyan A et al (2004) Centrin gene disruption impairs stage-specific basal body duplication and cell cycle progression in Leishmania. J Biol Chem 279:25703–25710

Selvapandiyan A et al (2009) Intracellular replication-deficient Leishmania donovani induces long lasting protective immunity against visceral leishmaniasis. J Immunol 183:1813–1820
Selvapandiyan A et al (2014) Generation of growth arrested Leishmania amastigotes: a tool to develop live attenuated vaccine candidates against visceral leishmaniasis. Vaccine 32:3895–3901
Selvapandiyan A, Ahuja K, Puri N, Krishnan A (2015) Implications of co-infection of Leptomonas in visceral leishmaniasis in India. Parasitology 142:1657–1662
Selvapandiyan A, Croft SL, Rijal S, Nakhasi HL, Ganguly NK (2019) Innovations for the elimination and control of visceral leishmaniasis. PLoS Negl Trop Dis 13:e0007616
Selvapandiyan A et al (2023) Zooming in on common immune evasion mechanisms of pathogens in phagolysosomes: potential broad-spectrum therapeutic targets against infectious diseases. FEMS Microbiol Rev 47:fuac041
Shukla AK, Singh BK, Patra S, Dubey VK (2010) Rational approaches for drug designing against leishmaniasis. Appl Biochem Biotechnol 160:2208–2218
Sinha S, Fernández G, Kapila R, Lambert WC, Schwartz RA (2008) Diffuse cutaneous leishmaniasis associated with the immune reconstitution inflammatory syndrome. Int J Dermatol 47: 1263–1270
Srivastava S, Shankar P, Mishra J, Singh S (2016) Possibilities and challenges for developing a successful vaccine for leishmaniasis. Parasit Vectors 9:277
Taslimi Y, Zahedifard F, Rafati S (2018) Leishmaniasis and various immunotherapeutic approaches. Parasitology 145:497–507
Thakur L et al (2018) Atypical leishmaniasis: a global perspective with emphasis on the Indian subcontinent. PLoS Negl Trop Dis 12:e0006659
Topuzogullari M et al (2013) Conjugation, characterization and toxicity of lipophosphoglycan-polyacrylic acid conjugate for vaccination against leishmaniasis. J Biomed Sci 20:35
van Zandbergen G et al (2004) Cutting Edge: Neutrophil Granulocyte serves as a Vector for *Leishmania* Entry into Macrophages. *J Immunol.* 2004 Dec 1; 173(11):6521–6525
Varma R, Puri N (2019) Dampening of mast cell secondary responses to allergen involves specific signalling and epigenetic changes. Cell Immunol 344:103944
Velez R, Gállego M (2020) Commercially approved vaccines for canine leishmaniosis: a review of available data on their safety and efficacy. Tropical Med Int Health 25:540–557
Vennerstrom JL (1997) In: Anand N (ed) Approaches to design and synthesis of antiparasitic drugs by Satyavan Sharma. Elsevier Science B.V, Amsterdam. xi + 511 pp. 17 × 25 cm. ISBN 0-444-89476-4. $265.75. J. Med. Chem. 41, 3100–3101 (1998)
Volpedo G et al (2021) Determinants of innate immunity in visceral leishmaniasis and their implication in vaccine development. Front Immunol 12:1–18
Volpedo G et al (2022) Centrin-deficient Leishmania mexicana confers protection against new world cutaneous leishmaniasis. NPJ Vaccines 71(7):1–14
Wheeler RJ, Gluenz E, Gull K (2011) The cell cycle of Leishmania: morphogenetic events and their implications for parasite biology. Mol Microbiol 79:647–662
WHO. (2021). Visceral leishmaniasis elimination: India gears-up to overcome last-mile challenges. https://www.who.int/news/item/29-07-2021-visceral-leishmaniasis-elimination-india-gears-up-to-overcome-last-mile-challenges
WHO. (2022). Visceral leishmaniasis and HIV coinfection: WHO publishes new guideline with region-specific treatment recommendations. https://www.who.int/news/item/08-06-2022-visceral-leishmaniasis-and-HIV-coinfection-WHO-publishes-new-guideline-with-region-specific-treatment-recommendations
WHO. (2023a). Leishmaniasis. https://www.who.int/news-room/fact-sheets/detail/leishmaniasis
WHO. (2023b). Leishmaniasis - India. https://www.who.int/india/health-topics/leishmaniasis
Zhang WW et al (2020) A second generation leishmanization vaccine with a markerless attenuated Leishmania major strain using CRISPR gene editing. Nat Commun 11:1–14
Zijlstra EE (2019) Biomarkers in Post-kala-azar Dermal leishmaniasis. Front Cell Infect Microbiol 9:228

Factors Affecting *Leishmania* Infection in Sand Fly

4

Shital, Charu Gupta, Anuja Krishnan, and Angamuthu Selvapandiyan

Abstract

Leishmania is a causative agent of leishmaniasis infection. It requires two hosts to complete its life cycle, mammals and sand flies. The parasite has two forms: the amastigotes that reside in the phagocytic macrophages and the promastigotes that reside in the midgut of the vector, sand flies. To achieve a successful infection in sand flies, the parasite has to encounter unfavorable circumstances, including digestive enzymes, detachment from the midgut, and the other immune responses shown by the vector, which are detrimental or lethal to the parasite. In contrast, certain factors are produced by the vector fly, which parasites later use for survival. These factors and components are deciding factors of infection rate, parasitic load, transmission and survival in another host. Lots of studies have been done on various species of sand flies to investigate such factors in their exact role in infection and their own survival. This review briefly describes such factors affecting parasitic infection and survival.

Keywords

Promastigote · Amastigote · Phlebotomus · Procyclic · Metacyclic · Midgut

Shital · A. Krishnan · A. Selvapandiyan (✉)
Department of Molecular Medicine, Jamia Hamdard, New Delhi, India
e-mail: selvapandiyan@jamiahamdard.ac.in

C. Gupta
Clinical Research Division, Galgotias University, Greater Noida, Uttar Pradesh, India

© The Author(s), under exclusive license to Springer Nature Singapore Pte Ltd. 2023

A. Selvapandiyan et al. (eds.), *Challenges and Solutions Against Visceral Leishmaniasis*, https://doi.org/10.1007/978-981-99-6999-9_4

4.1 Introduction

Leishmania, a protozoan parasite, is a pathogenic agent of leishmaniasis. It is a digenetic parasite, with one form, flagellated promastigotes multiplying in the vector sand fly, and the other round and rudimentary-flagellated amastigotes replicating in the mammalian macrophages. The vector mostly is Phlebotomine sand flies of the genera either *Phlebotomus* or *Lutzomyia* (Coutinho-Abreu et al. 2020).

On ingesting infected blood meal, sand fly also consumes the amastigotes (present within macrophages). Twenty-four hours post infected blood meal (PIBM), the parasites released from the phagocytes transformed into promastigotes (Pimenta et al. 1997). Recent studies show that a few intermediate stages accompany alternation between these two forms. Immediately after the release from the macrophage, the amastigotes transform into procyclic promastigotes within the peritrophic matrix. During this transformation, they undergo many morphological changes, including lengthening cell bodies and nucleating their flagella from one of the two basal bodies. Two to four days PIBM, these procyclic promastigotes transformed into Nectomonads, which are more of elongated banana shape (Githure et al. 1990; Walters 1993; Walters et al. 1989). Eventually, Nectomonads migrate from ectoperitrophic space and colonize the midgut microvilli. These parasites further differentiate into the third form known as Leptomonads. They have longer flagellum and are smaller in body size than Nectomonads. Six days PIBM, Leptomonads undergo a differentiation process known as Metacyclogenesis. During this transformation, the parasites undergo many morphological changes, which will give rise to the infective form of the parasite, Metacyclopromastigotes, which exhibit the factors specific for infection. As the parasites reach the metacyclic stage, they detach themselves from the midgut epithelia and rush toward the foregut and the cardia. There these parasites regulate the chitinase expression, which causes stomodeal valve damage. This ultimately causes regurgitation of gut contents whenever a vector fly attempts to blood feed (Rogers et al. 2008; Schlein et al. 1992). The summarized form of the *Leishmania* life cycle is shown in Fig. 4.1. The transition between different stages is crucial for parasitic development, including persevering the digestive enzyme action, escaping the peritrophic membrane, attachment to the epithelial membrane of gut microvilli, etc. (Coutinho-Abreu et al. 2020). These parasites are exposed to many harmful enzymes and substances released by a vector to counteract the infection and survival of parasites within the gut. This review briefly describes such factors (Table 4.1), which are turned around by the parasite for successful infection and survival within the sand fly.

4.2 Digestive Enzymes

To achieve a successful infection in the sand fly, *Leishmania* parasites must overcome a series of barriers offered by the host system. One among them is exposure to digestive enzymes (mainly trypsin-like enzymes), which have proven to be detrimental to the parasites. The parasite infection suppresses the secretion of trypsin in

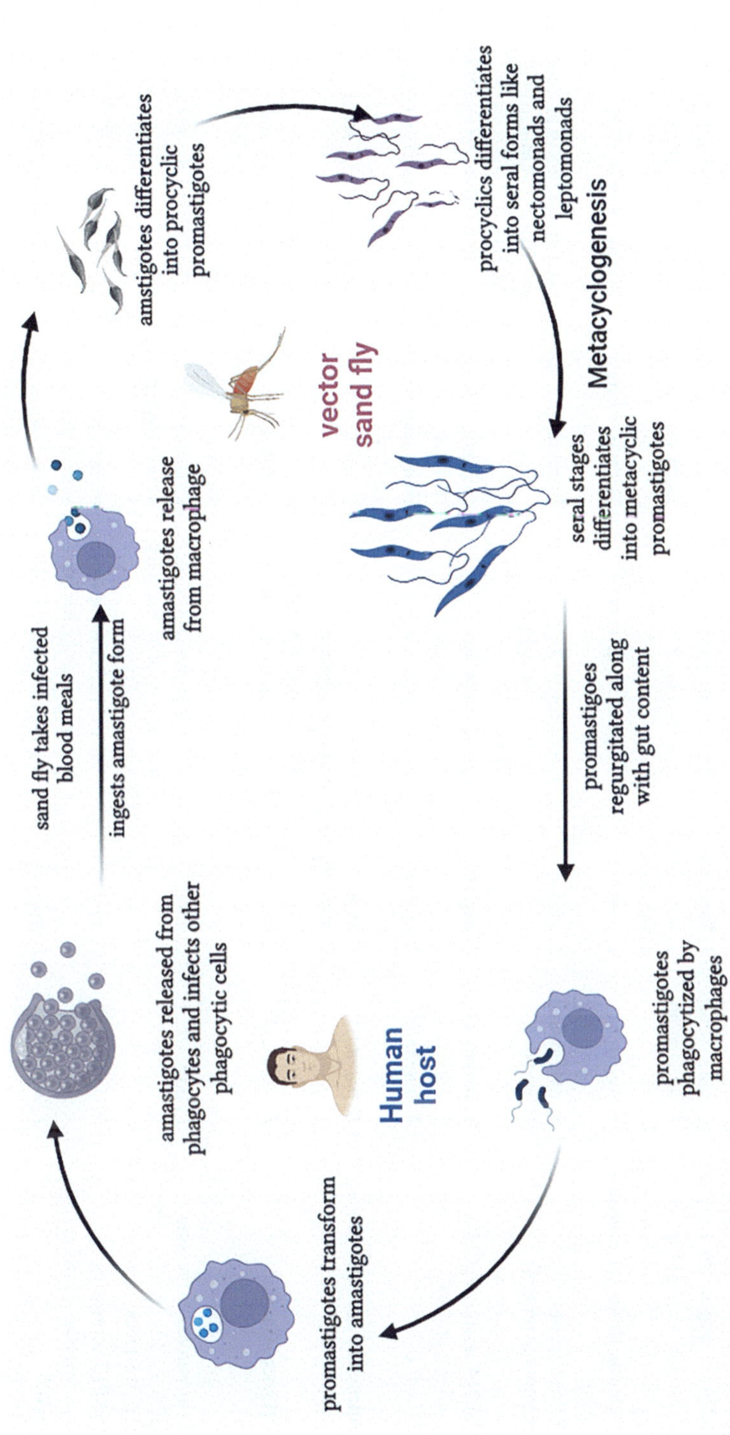

Fig. 4.1 *The life cycle of leishmania with specific forms in sand fly:* Vector sand fly takes the amastigote infected blood meals, and these amastigotes differentiate into promastigotes. These metacyclic infective forms along with gut content regurgitate into human blood where they are phagocytized by the macrophage. These promastigotes transform into amastigotes and replicate. These newly released amastigotes infect other macrophages. Created in BioRender.com

Table 4.1 Factors of sand fly regulating the parasite infection

Sand fly factor (antigen/ therapeutic)	Reported sand fly species	Reported parasite species	Effects on host due to antigen	Parasitic load	References
RNAi of trypsin like enzymes	*L. longipalpis*	*L. mexicana*	Reduces trypsins	Increases parasites	Sant'Anna et al. (2009)
RNA inhibition of Caspar	*L. longipalpis*	*L. mexicana*, *L. infantum*	Suppression of Caspar	Decreases parasites	Telleria et al. (2012)
Defensin (from gut epithelial cells)	*P. duboscqi*	*L. major*	Antiparasitic activity	Decreases parasites	Boulanger et al. (2004)
PpPer1 (peritrophic-membrane protein)	*L. longipalpis*, *P. papatasi*	*L. major*	Knockdown of PpPer1 in the fly reduces PpPer1 in the gut	Increases parasites	Coutinho-Abreu et al. (2013) and Ramalho-Ortigão et al. (2007)
dsRNA against PpChit1 (mid-gut protein)	*L. longipalpis*, *P. papatasi*	*L. major*	Reduces PpChit1 expression in the gut	Decreases parasite survival	Coutinho-Abreu et al. (2010)
LlTGF-β	*L. longipalpis*	*L. chagasi*	Signal transduction in the host	Helps parasite survival	Di-Blasi et al. (2019)
Adenosine or AMP (saliva compounds)	*P. papatasi*	*L. major*, *L. amazonensis*	Suppress metacyclogenesis, inflammatory response	Decrease	Carregaro et al. (2015)
PpGalec (Galectin)	*P. papatasi*	*L. major*	Parasite survival in the gut of vector	Anti-PpGalec ab reduced parasite survival	Kamhawi (2006) and Kamhawi et al. (2004)
L. tarentolae expressing CPA + CPB proteins with PpSP15 DNA	*P. papatasi*	*L. major*	Cell and humoral-mediated immunity in BALB/c mice	Protects mice against parasite	Zahedifard et al. (2014)
PpSP15 expressing *L. tarentolae + CpG*	*P. papatasi*	*L. major*	Cell and humoral-mediated immunity in BALB/c mice	Protects mice against parasite	Katebi et al. (2015)
PpSP9	*P. sergenti*	*L. tropica*	Th1 immunity in BALB/c mice	Protects mice against parasite	Gholami et al. (2019)

the sand fly midgut. When trypsin1 expression was silenced by RNAi, there was an increase in *L. mexicana* parasitic burden (Sant'Anna et al. 2009). The expression of trypsin1 in *L. longipalpis* infected with *L. mexicana* was downregulated in infected flies, and such modulation by the parasites brings the successful infection (Sant'Anna et al. 2009).

4.3 Role of Caspar and Defensin of Sand Fly in Parasite Infection

Caspar is a negative regulator of the immune deficiency (Imd) pathway associated with the TNF/NF-kB pathway. Activation of the sand fly immune system was initiated by the downregulation of caspar expression in females between 3 days and 6 days after blood feed containing amastigotes of *L. mexicana*. The caspar expression when depleted by the RNAi method greatly reduced the prevalence of *L. infantum* and *L. mexicana*. Depletion of this gene disrupts the transmission of the parasite by vector sand fly (Telleria et al. 2012). This study demonstrated that silencing a single caspar gene could disrupt the development of the parasite and its transmission. Poor details are available regarding the immune response of the sand fly toward *Leishmania* infection. One such study showed 'defensin' production after the parasite and bacterial challenge. This antiparasitic activity is against the promastigote form of *L. major* (Boulanger et al. 2004).

4.4 Role of Peritrophins and Chitinase of Sand Fly in Parasite Infection

There are three major peritrophins PpPer1, PpPer2, and PpPer3 that display 4, 1, and 3 chitin-binding domains (CBDs), respectively (Ramalho-Ortigão et al. 2007). The temporal and spatial analysis shows the restricted expression of PpPer1 in the midgut after the blood feeding, whereas others express consecutively in the midgut and the hindgut. The expression of PpPer3 was also observed in the Malpighian tubules. It is observed that knocking down of PpPer1 increases the parasitic load up to 39% at 48 h and 22% at 96 h (Coutinho-Abreu et al. 2013). Transcriptome analysis revealed the upregulation of PpPer1 mRNA levels at 24 h postinfection (PI) of *L. major* in *L. longipalpis* and *P. papatasi*. However, it is still debatable whether the upregulation of PpPer1 by *L. major* is to protect itself or to sand fly to reduce the permeability of plasma membrane (PM) to prevent parasitic load. On the other hand, a similar set of experiments were performed with other peritrophins. However, no significant changes were observed in PpPer2, whereas downregulation of PpPer3 mRNA was observed at 24 h and 48 h PI (Coutinho-Abreu et al. 2013). These results indicate the importance of PpPer1 in the *Leishmania* pathogenesis.

PpChit1 is a midgut-specific chitinase, which mainly plays a role in PM maturation and degradation in *P. papatasi*. Studies revealed that the sand fly's chitinase might be responsible for the parasite's escape. Coutinho-Abreu et al. 2010 prepared the knockdown of PpChit1 by RNAi procedure and observed its effect on the parasitic load at different time intervals. They injected dsRNA targeting the PpChit1 gene, which reduced the gene's mRNA levels to 72% at 96 h, and then analyzed the parasitaemia at different time intervals. The parasitic load within the sand fly is observed to be reduced to 46% at 48 h PIBM and 63% at 120 h PIBM. The downregulation of PpChit1 opens two explanations (Coutinho-Abreu et al. 2010). First, the impermeability of PM causes more prolonged exposure of parasites to the digestive enzymes, and the second is the inaccessibility of the nutrients in the endoperitrophic space. Interestingly, the expression profile of PpChit1 resembles PpPer1 expression profile.

4.5 Role of TGF-β in Suppression of the Immune Response

Transforming growth factor-β (TGF-β) is a conserved cytokine involved in various signal transduction pathways of embryonic development, immune response, and stress tolerance in an organism. These cytokines consist of 40 members, which are based on their functional and structural similarity and classified into various subgroups (Di-Blasi et al. 2019). To investigate the role of TGF-β in the parasite survival, the homolog of TGF-β was identified in *L. longipalpis*, LlTGF-β and checked for the role it played in the immune response. Western-blot analysis results revealed that the expression of LlTGF-β is maintained throughout different stages of life but 3 days after feeding. The abrogation of the signaling pathway by RNAi silencing and TGFβ inhibitor *SB431542* decreases the parasitic load within *L. longipalpis* significantly at 72 h and 144 h after infection. The effect of NO on the TGF-β signal transduction was assessed by measuring iNOS expression. Results revealed a significant increase in NO, while signaling is abrogated when TGF-β is silenced by RNAi method; however, negligible change is observed when the chemical TGF-β inhibitor *SB431542* was used (Di-Blasi et al. 2019).

4.6 Suppression of Metacyclogenesis by Purines

Metacyclogenesis is the process of conversion of procyclic promastigotes into infective metacyclic promastigotes. Lesser details are available regarding the in vivo mechanism (Kamhawi 2006) and factors affecting this process. A study reported that stress condition promotes metacyclic stage formation (Serafim et al. 2012). The presence and absence of adenine control the metacyclic stage. The induction of metacyclogenesis occurs strongly when CGS, the adenosine receptor's antagonist added to the culture system. This antagonist not only increases the metacyclic growth by sixfold but also the rate of infection in macrophages than the control. Reports revealed that CGS-treated cultures are more resistant to complement-mediated lysis than control promastigotes. This suppression in the presence of purine might be explained by the trigger known as the "purine sensor." Deficiency in purine is suggested to trigger the metacyclogenesis process. The metacyclogenesis in *Leishmania* is also accompanied by a few changes, including an increase in major surface proteinase (gp63) and modification in lipophosphoglycan (LPG), which ultimately leads to resistance to complement-mediated lysis.

4.7 Role of *Leishmania* Lipophosphoglycan (LPG) and Its Receptor Galactin in Infection

Lipophosphoglycan (LPG) is a tripartite molecule, a major glycoconjugate on the surface of *the Leishmania* promastigote. LPG mediates the parasite's attachment with the epithelia of the sand fly midgut, making it one of the crucial factors in parasite infection (Mahoney et al. 1999). Any structural changes in the LPG molecules result in the loss of parasites through excretion, reducing the parasitic load or infection rate. Studies of phosphoglycan (PG) of the LPG show the disaccharide repeats of PO4-6Gal (β1–4) Manα1 (Kamhawi et al. 2004). In the case of *L. major*, LPG consists of beta 1,3-glucose residues that branch off the disaccharide phosphate repeat units in the procyclic stage. However, this glucose substitution is downregulated in metacyclic parasites, thus making it unable to bind to the insect midgut. Similar results were also reported in *L. donovani* (Mahoney et al. 1999).

To discover the changes that lead to the infective form of the parasites, the parasite populations isolated from both the log and the stationary phases were tested, and it was found that the ratio of Gal (β 1,4) Man repeats with no side chains was increased in metacyclics, which results in 2.5 times more attachment of procyclics than metacyclics. In another set of experiments, PGs (isolated from LPG of the above metacyclics and procyclics) were incubated with the gut, followed by incubation with culture-derived procyclics, and it was found that the PGs strongly inhibited the binding of procyclics to the gut. The above study revealed the role played by LPG in the attachment of the parasite to the gut epithelia and apprised the importance of its sugar side chains (Soares et al. 2002).

Galectins are soluble proteins that bind specifically to β-galactoside with high affinity. It plays a vital role in various processes, including embryonic development, immune response, and homeostasis (Colin Hughes 2001; Pace and Baum 2002; Rabinovich et al. 2002; Yang and Liu 2003; Young and Meeusen 2002). Recent studies revealed that it has a role in parasite survival. It acts as a receptor for the parasite's LPGs and hence plays an equally crucial role in the attachment of the parasite to the midgut. Distribution of the tandem repeats of galectin (PpGal) was restricted to the species belonging to the *Phlebotomus* subgenus, so responsible for species-specific vector competence. Anti-PpGal antibodies, when preincubated with midgut followed by incubation with procyclics of *L. major*, completely inhibited LPG and promastigote binding, indicating that parasites use PpGal as the receptor to bind with LPG.

4.8 Nucleoside as a Promoter of Infection and Its Role in Immunosuppression

Sand fly saliva consists of many components with pharmacologic properties, making it a crucial factor in parasite infection and its survival. Two such components, adenosine and adenine monophosphate (AMP) of saliva, are known to suppress the immune response against the parasite (Carregaro et al. 2015). Adenosine downregulates the production of proinflammatory cytokines and costimulatory markers (Novitskiy et al. 2008). Adenosine specifically targets dendritic cells—to block their antigen presentation activity, interrupting the Th17 activation and ultimately inhibiting the inflammatory response's activation (Carregaro et al. 2011). When pair of these components, along with the parasites (*L. major* and *L. amazonensis*), were injected into the *L. papatasi*, it showed the immunosuppressive response and the enhancement of IL-10 expression.

Furthermore, similar experiments with mice treated with infected salivary gland extract led to an increase in parasitic load and produced larger-sized lesions. The immunosuppressive effects might be induced by an adenosine $A2_A$ receptor ($A2_AR$)- dependent mechanism. It was found that $A2_AR^{-/-}$ mice resulted in lesser parasitic load and decreased iTreg (induced regulatory T cell) production. The immunosuppressive effects seemed to be reversed when adenosine deaminase (ADA) was added. Immunosuppressive factors were still upregulated, but no enhancement in lesion size and the parasitic load was observed. The $CD4^+$ $CD25^-$ phenotypic cells express Treg-specific markers suggesting the T-effector cell induction accompanied by the high-level expression of COX2 and IL-10 by the nucleoside-induced tolerogenic dendritic cells (Carregaro et al. 2015). It is concluded from the above information that the presence of nucleoside in the saliva suppresses the immune response, whereas the addition of ADA enzyme can do the reversal.

4.9 Contribution of Blood and Its Derivatives in the Parasitic Infection in Sand Fly

Other than the above factors, parasitic infection is also affected by the presence or absence of different blood components. In a series of experiments, sand flies inoculated with the different blood derivatives (including defibrinated blood, heparinized blood, and blood with trypsin inhibitor) spiked with *Leishmania* promastigotes reveal reduced parasite infection. Whereas erythrocytes in saline with either promastigotes or amastigote inoculation increases infection in the sand fly (Schlein and Jacobson 1998). Hence, it is concluded that the favorable nature of saline maintains the viability of the parasites, but whereas unfavorable condition like heparin and fibrin removal results in the lethality of parasites.

4.10 Effects of Second Blood Meals on Infected Flies

Many studies revealed the effect of a second blood meal on the infected flies (Elnaiem et al. 1994; Stamper et al. 2011; Vivenes et al. 2001). Results reveal that the second feed not only enhances the rate of infection but also affects the oviposition in vector flies. It has been shown that flies that laid eggs prior to the second feed were more competent, indicating that this second feed also enhances the parasite's competence in vector flies. To unmask the effects of second feed in parasite infection, the evaluation of the parasite infection in *L. longipalpis* has been done. Results clearly show that there is no significant change in average parasitic load in sand flies, but the proportion of heavily infected flies was increased. Reports also revealed that the migration of the parasites after the second feed is enhanced (Moraes et al. 2018), thus suggesting the role of multiple feeds on the development of *Leishmania* in the sand fly.

4.11 Role of Sand Fly's "Relish" Protein in Imd Pathway

The Imd pathway is well-studied immune response pathway in insects against bacterial and fungal challenges. The pathogen-associated molecular pattern (PAMP) of the bacteria or fungus was recognized as an antigen leading to the production of antimicrobial peptides by the activation of "relish," the only transcription factor of the Imd pathway (Louradour et al. 2019). To uncover the whole contribution of the relish (*rel*), three lines of knockout mutants were generated by CRISPR/CAS9 system and marked as rel^{B}, rel^{E}, and rel^{L}. To check whether the *rel* protein affects the gut microbes, Louradour et al. 2019 compared the gut microbiota of the wild type and mutant rel^{B} in the flies infected with *L. major* by qPCR of 16S rRNA gene. They found that the infected rel^{B} mutant flies have 5.45 times higher amount of amplicon than the wild type flies (Louradour et al. 2019). Furthermore, to test the effect of rel on vectoral competence, they carried out another set of experiments in which *P. papatasi* was challenged with *L. major* parasite. The study showed a significant increase in the promastigotes population in the midgut of infected rel^{B} mutant files than the wild type files on day 9 after infected blood feeding. From the above study, it can be easily concluded that *rel* mutant sandflies were more prone to parasitic infection and harbored more microbes in their guts.

4.12 Impact of Glucose Metabolism on Parasites

The positive correlation between glucose metabolism and the other *Leishmania*-infected model organism is well known from prior studies (Anstead et al. 2001; Crampton and Luckhart 2001; Ibrahim et al. 2013; Losada-Barragán et al. 2019; Nweze et al. 2020; Serafim et al. 2010). The decrease in glucose level decreases the immune response of the host, hence increasing the host's susceptibility to the infection. On comparing the glucose level with the respective parasite burden, a moderate correlation between the liver and spleen is found, whereas no such correlation is found in the bone marrow. Further, the report reveals that promastigotes remain unaffected by blood glucose and insulin growth factor (IGF) levels (Hendrickx and Caljon 2022). The study showed that the optimum growth of parasites was only seen at the normoglycemic levels when compared with hypo- and hyperglycemic vector flies. Despite that, glucose concentration seems to affect the metacyclogenesis process. The metacyclic stage developed earlier than the control in the presence of a high glucose concentration (Hendrickx and Caljon 2022).

Furthermore, the enhanced survival rate of parasites in the sand flies was reported in the hyperglycemic infected flies, whereas the lowest was in the hypoglycemic infected flies. No impact of IGF1 was reported in the development of *Leishmania*, but prominent positive effects of the higher IGF1 or insulin concentration were reported in the survival of the flies. These studies clearly presented the effect of host blood chemistry impacting the survival of the parasites in the sand fly.

4.13 Modulation of Midgut Genes by *Leishmania* Parasites

For survival, parasites also alter the midgut expression pattern. Coutinho-Abreu et al. 2020 investigated the differential expression pattern of these transcripts (Coutinho-Abreu et al. 2020). They divided transcripts into differential functional groups such as immune-related genes, digestive enzymes, PM modification and degradation genes, metabolism genes, and detoxification-associated genes. They found that the modulation of transcripts was specific to the respective infection stages. Two different modulations in expression patterns were reported by their team. At early-stage infection (1–4 days), downregulation of most of the transcripts such as Forkhead/HNF-3, digestive enzymes, peritrophic matrix protein, and enzymes involved in steroid metabolism was observed. However, few transcripts were upregulated, such as chymotrypsin-2, carboxypeptidase A, and trypsin-like serine protease. At a later stage of infection (6th day and afterward), the upregulation of most of the transcripts, such as detoxification-associated genes and lipids metabolism–associated genes, was reported. Among them, transcripts of p450 cytochrome, belonging to the detox functional category, were majorly upregulated on the 8th day of infection. The insect allergens proteins expressed in microvilli were also upregulated on 4 days and 6 days. The function of such proteins and their role during parasite infection are still unknown (Coutinho-Abreu et al. 2020). *Leishmania* modulates the majority of sand fly genes in the early stage of infection in order to survive in a hostile environment, with modest changes in gene expression being modulated by parasites in the anterior midgut during the later stages of infection.

4.14 Discussion and Conclusions

Leishmaniasis is one of the ignored tropical diseases caused by the protozoan parasite *Leishmania*. It acquires two alternative forms in its life cycle. One is the amastigote form, infective in the macrophages, and the other is the promastigote form, infective in the vector sand fly. Sand fly genera mainly involved in *Leishmania* infection are *Phlebotomus* and *Lutzomyia*. Parasites migrate and colonize the midgut, where they differentiate into infective forms, metacyclic promastigotes. Several other stages are known to lead to the mature infective form in the vector. To reach into infective form, the parasites have to face many host barriers, which otherwise would be detrimental to the parasites. Many studies are available regarding its infection and survival in the human host, but few are available for its infection and survival in sand flies. Even the immune response of sand flies against parasitic infection is not well known. This review attempts to delve into such factors, which positively or negatively affect the parasitic infection and its survival within the vector. As soon as the parasite enters, it interacts with the digestive enzymes of the vector. One such enzyme is trypsin; exposure of parasites to trypsin or trypsin-like enzymes has proven harmful to its growth. For a successful infection, the parasite downregulates the production of such enzymes. As it reaches the midgut, the attachment of the parasite with midgut epithelia governed by LPG becomes a crucial factor for further development. LPGs, during metacyclogenesis, undergo several structural changes that make the infective form susceptible to parasite attachment. Studies show that these attachments are species-specific and responsible for species-specific vector competence. Any changes in the sugar side chain of LPG would disrupt the attachment and result in the excretion of the parasite. Galectins act as a receptor of LPG, so its interaction with LPG is also necessary to maintain the infection. Furthermore, PM components like PpPer1 also play a role in the survival of the parasite. PpPer1 decreases the PM permeability causing the nutrients unavailability or longer exposure to enzymes and hence inhibiting the parasite escape from the ectoperitrophic membrane. There are factors produced by the vector itself used by the parasite for its survival. PpChit, chitinase, helps the parasite in membrane degradation and causes regurgitation of gut content whenever a sand fly takes a blood meal. Caspar is an enzyme known to suppress the Imd pathway, which is responsible for the immune response. The exact mechanism behind it is still a point of debate. Purines and nucleosides are sought to suppress the immune response. Studies show that adenine (purine) will induce metacyclogenesis in vector, whereas adenosine or AMP (nucleotide) present in salivary gland secretion will induce suppression by $A2_A R$- dependent mechanism as well as induce production of anti-inflammatory cytokines. The TGF-β is known to suppress the signal transduction pathway necessary for immune system activation. Silencing of TGF-β leads to the reduction in parasitic load. A similar role of TGF-β was known in mosquitoes, when infected with the *Plasmodium* parasite (Crampton and Luckhart 2001). *A. stephensi* shows the expression of the TGF- β gene when infected with *Plasmodium berghei*. Relish, a transcription factor in the Imd pathway, also affects the transmission of *Leishmania* parasites to the host. The Relish mutant harbors a large number of bacteria and parasites in its gut. The role of this factor was also well discussed in the transmission of malarial parasites, dengue virus, and the African trypanosomes to the *Anopheles gambiae*, *Aedes aegypti*, and *Glossina morsitans,* respectively (Hu and Aksoy 2006; Meister et al. 2005, 2009; Ramirez et al. 2012; Ramirez and Dimopoulos 2010; Xi et al. 2008). High glucose levels within the sand flies are known to reduce the immune response of the fly. It was also reported that hyperglycemic flies had enhanced survival rates and early development of metacyclic during the parasite infection. Various studies show that *Plasmodium* growth is also affected by glucose metabolism. Low sugar levels will lead to impaired growth and proliferation, whereas diabetic mice having elevated glucose levels show a higher level of parasite transmission (Ansell et al. 2002; Courtenay et al. 2017; Emami et al. 2017; Humeida et al. 2023; Pakpour et al. 2016). The sand fly releases certain factors that either cause infection or provide immunity to the other host. Mainly such factors are part of salivary gland secretions. Lundep, a salivary endonuclease and Hyaluronidase, which not only destroy the neutrophil extracellular traps (first line of defense) but also interferes with the intrinsic blood coagulation pathway by inhibiting the X11a factor (Chagas et al. 2014; Martin-Martin et al. 2018). It ultimately enhances the infection rate and survival of the parasites in mammals. These two enzymes have proven to be legitimate candidates for vaccine development. The mice were immunized with these two enzymes and showed immunity against cutaneous leishmaniasis. Researchers infected the C57BL/6 mice with the *Trypanosoma brucei brucei* and gave them diminazene aceturate treatment, followed by coinfection with *L. major* (Pereira et al. 2018). These mice showed a decrease in lesion size, which also led to a reduction in parasitic load. *T. brucei brucei* somehow activates T cells, which results in the secretion of proinflammatory cytokines such as IL-12, IFN-γ, and TNF-α. Survival of the parasites in such an environment is difficult, hence provides immunity against the parasite infection (Pereira et al. 2018). The complete eradication of the parasites can be done by using two approaches. First, by knockdown (Selvapandiyan et al. 2007, 2012b) or knockout (Selvapandiyan et al. 2012a; Avishek et al. 2016; Bhattacharya et al. 2016) of the genes essential for parasite survival or transmission. Second, by either upregulating or overexpressing the genes (Selvapandiyan et al. 2001), which are detrimental to the parasite's survival. This review compiles factors that can positively and negatively regulate parasite infection. A new eradication strategy is being used in which the mammalian host is provided with salivary proteins as an antigen, followed by the parasitic challenge. Such a host becomes immune to the parasitic infection, but this strategy is limited to the cutaneous leishmaniases. One such example is shown by a study where a recombinant lizard protozoan parasite, *Leishmania tarentolae* expressing cysteine proteinases A and B (CPA and CPB) with and without the salivary immunogen, PpSP15 of *P. papatasi* was used (Zahedifard et al. 2014). Susceptible BALB/c mice and resistant C57BL/6 infected with the *L. tarentolae* expressing CPA + CPB proteins along with PpSP15 DNA showed enhanced protective immune response against cutaneous leishmaniasis due to *L. major* (Zahedifard et al. 2014). PpSP15 of *P. papatasi* and the CpG as adjuvant were also sufficient to provide cell and humoral-mediated immunity in BALB/c mice against *L. major* infection (Katebi et al. 2015). Another member of the SP15 family, PpSP9 of *P. sergenti*, was also proven to have similar immunogenic properties providing cell-mediated immunity in BALB/c mice against the *L. tropica* infection (Gholami et al. 2019). Mammalian hosts also employ a range of factors that influence the pathogenesis (Selvapandiyan et al. 2023). However, the vector factors pass along with the parasites during the blood meal also influence/skew the host defense mechanism toward parasite survival. In conclusion, the survival of parasites in the vector that blood feed on human is one of the challenges (Selvapandiyan et al. 2019) that researchers study and attempt for a transmission blocking approach. A piece of deep knowledge about such factors opens up gates of advancement in this area and helps to overcome the challenges posed by the parasites.

Acknowledgments AS is supported by grants (No. 6/9-7(241)/2020-ECD-II; No. GIA/2/VBD/2021/ECD-II) from Indian Council of Medical Research, New Delhi, India.

References

Ansell J, Hamilton KA, Pinder M, Walraven GEL, Lindsay SW (2002) Short-range attractiveness of pregnant women to Anopheles gambiae mosquitoes. Trans R Soc Trop Med Hyg 96(2): 113–116. https://doi.org/10.1016/S0035-9203(02)90271-3

Anstead GM, Chandrasekar B, Zhao W, Yang J, Perez LE, Melby PC (2001) Malnutrition alters the innate immune response and increases early visceralization following *leishmania donovani* infection. Infect Immun 69(8):4709–4718. https://doi.org/10.1128/IAI.69.8.4709-4718.2001

Avishek K, Kaushal H, Gannavaram S, Dey R, Selvapandiyan A, Ramesh V, Negi NS, Dubey US, Nakhasi HL, Salotra P (2016) Gene deleted live attenuated leishmania vaccine candidates against visceral leishmaniasis elicit pro-inflammatory cytokines response in human PBMCs. Sci Rep 6(1):33059. https://doi.org/10.1038/srep33059

Bhattacharya P, Dey R, Dagur PK, Joshi AB, Ismail N, Gannavaram S, Debrabant A, Akue AD, KuKuruga MA, Selvapandiyan A, McCoy JP, Nakhasi HL (2016) Live attenuated leishmania donovani Centrin Knock out parasites generate non-inferior protective immune response in aged mice against visceral leishmaniasis. PLoS Negl Trop Dis 10(8):e0004963. https://doi.org/10.1371/journal.pntd.0004963

Boulanger N, Lowenberger C, Volf P, Ursic R, Sigutova L, Sabatier L, Svobodova M, Beverley SM, Späth G, Brun R, Pesson B, Bulet P (2004) Characterization of a defensin from the sand fly *Phlebotomus duboscqi* induced by challenge with bacteria or the protozoan parasite *leishmania major*. Infect Immun 72(12):7140–7146. https://doi.org/10.1128/IAI.72.12.7140-7146.2004

Carregaro V, Sá-Nunes A, Cunha TM, Grespan R, Oliveira CJF, Lima-Junior DS, Costa DL, Verri WA, Milanezi CM, Pham VM, Brand DD, Valenzuela JG, Silva JS, Ribeiro JMC, Cunha FQ (2011) Nucleosides from *Phlebotomus papatasi* salivary gland ameliorate murine collagen-induced arthritis by impairing dendritic cell functions. J Immunol 187(8):4347–4359. https://doi.org/10.4049/jimmunol.1003404

Carregaro V, Ribeiro JM, Valenzuela JG, Souza-Júnior DL, Costa DL, Oliveira CJF, Sacramento LA, Nascimento MSL, Milanezi CM, Cunha FQ, Silva JS (2015) Nucleosides present on phlebotomine saliva induce immunossuppression and promote the infection establishment. PLoS Negl Trop Dis 9(4):e0003600. https://doi.org/10.1371/journal.pntd.0003600

Chagas AC, Oliveira F, Debrabant A, Valenzuela JG, Ribeiro JMC, Calvo E (2014) Lundep, a sand fly salivary endonuclease increases leishmania parasite survival in neutrophils and inhibits XIIa contact activation in human plasma. PLoS Pathog 10(2):e1003923. https://doi.org/10.1371/journal.ppat.1003923

Colin Hughes R (2001) Galectins as modulators of cell adhesion. Biochimie 83(7):667–676. https://doi.org/10.1016/S0300-9084(01)01289-5

Courtenay O, Peters NC, Rogers ME, Bern C (2017) Combining epidemiology with basic biology of sand flies, parasites, and hosts to inform leishmaniasis transmission dynamics and control. PLoS Pathog 13(10):e1006571. https://doi.org/10.1371/journal.ppat.1006571

Coutinho-Abreu IV, Sharma NK, Robles-Murguia M, Ramalho-Ortigao M (2010) Targeting the midgut secreted PpChit1 reduces leishmania major development in its natural vector, the sand fly phlebotomus papatasi. PLoS Negl Trop Dis 4(11):e901. https://doi.org/10.1371/journal.pntd.0000901

Coutinho-Abreu IV, Sharma NK, Robles-Murguia M, Ramalho-Ortigao M (2013) Characterization of Phlebotomus papatasi peritrophins, and the role of PpPer1 in leishmania major survival in its natural vector. PLoS Negl Trop Dis 7(3):e2132. https://doi.org/10.1371/journal.pntd.0002132

Coutinho-Abreu IV, Serafim TD, Meneses C, Kamhawi S, Oliveira F, Valenzuela JG (2020) Leishmania infection induces a limited differential gene expression in the sand fly midgut. BMC Genomics 21(1):608. https://doi.org/10.1186/s12864-020-07025-8

Crampton A, Luckhart S (2001) The role of As60A, a TGF-β homolog, in Anopheles stephensi innate immunity and defense against plasmodium infection. Infect Genet Evol 1(2):131–141. https://doi.org/10.1016/S1567-1348(01)00017-X

Di-Blasi T, Telleria EL, Marques C, de Couto RM, da Silva-Neves M, Jancarova M, Volf P, Tempone AJ, Traub-Csekö YM (2019) Lutzomyia longipalpis TGF-β has a role in leishmania infantum chagasi survival in the vector. Front Cell Infect Microbiol 9:71. https://doi.org/10.3389/fcimb.2019.00071

Elnaiem DA, Ward RD, Young PE (1994) Development ofLeishmania chagasi (Kinetoplastida: Trypanosomatidae) in the second blood-meal of its vectorLutzomyia longipalpis (Diptera: Psychodidae). Parasitol Res 80(5):414–419. https://doi.org/10.1007/BF00932379

Emami SN, Lindberg BG, Hua S, Hill SR, Mozuraitis R, Lehmann P, Birgersson G, Borg-Karlson A-K, Ignell R, Faye I (2017) A key malaria metabolite modulates vector blood seeking, feeding, and susceptibility to infection. Science 355(6329):1076–1080. https://doi.org/10.1126/science.aah4563

Gholami E, Oliveira F, Taheri T, Seyed N, Gharibzadeh S, Gholami N, Mizbani A, Zali F, Habibzadeh S, Bakhadj DO, Meneses C, Kamyab-Hesari K, Sadeghipour A, Taslimi Y, Khadir F, Kamhawi S, Mazlomi MA, Valenzuela JG, Rafati S (2019) DNA plasmid coding for Phlebotomus sergenti salivary protein PsSP9, a member of the SP15 family of proteins, protects against leishmania tropica. PLoS Negl Trop Dis 13(1):e0007067. https://doi.org/10.1371/journal.pntd.0007067

Githure JI, Roberts CR, Lawyer PG, Anjili CO, Mebrahtu YB, Ngumbi PM, Koech DK, Odongo SO (1990) Development of leishmania major in Phlebotomus duboscqi and Sergentomyia schwetzi (Diptera: Psychodidae). Am J Trop Med Hyg 43(1):31–43. https://doi.org/10.4269/ajtmh.1990.43.31

Hendrickx S, Caljon G (2022) The effect of the sugar metabolism on leishmania infantum promastigotes inside the gut of Lutzomyia longipalpis: a sweet relationship? PLoS Negl Trop Dis 16(4):e0010293. https://doi.org/10.1371/journal.pntd.0010293

Hu C, Aksoy S (2006) Innate immune responses regulate trypanosome parasite infection of the tsetse fly glossina morsitans morsitans. Mol Microbiol 60(5):1194–1204. https://doi.org/10.1111/j.1365-2958.2006.05180.x

Humeida H, Pradel G, Stich A, Krawinkel M (2023) The effect of glucose and insulin on in vitro proliferation of plasmodium falciparum. J Diabetol 2(3):6. https://www.journalofdiabetology.org/article.asp?issn=2078-7685;year=2011;volume=2;issue=3;spage=6;epage=6;aulast=Humeida;type=0

Ibrahim MK, Barnes JL, Anstead GM, Jimenez F, Travi BL, Peniche AG, Osorio EY, Ahuja SS, Melby PC (2013) The malnutrition-related increase in early visceralization of leishmania donovani is associated with a reduced number of lymph node phagocytes and altered conduit system flow. PLoS Negl Trop Dis 7(8):e2329. https://doi.org/10.1371/journal.pntd.0002329

Kamhawi S (2006) Phlebotomine sand flies and leishmania parasites: friends or foes? Trends Parasitol 22(9):439–445. https://doi.org/10.1016/j.pt.2006.06.012

Kamhawi S, Ramalho-Ortigao M, van Pham M, Kumar S, Lawyer PG, Turco SJ, Barillas-Mury C, Sacks DL, Valenzuela JG (2004) A role for insect galectins in parasite survival. Cell 119(3): 329–341. https://doi.org/10.1016/j.cell.2004.10.009

Katebi A, Gholami E, Taheri T, Zahedifard F, Habibzadeh S, Taslimi Y, Shokri F, Papadopoulou B, Kamhawi S, Valenzuela JG, Rafati S (2015) Leishmania tarentolae secreting the sand fly salivary antigen PpSP15 confers protection against leishmania major infection in a susceptible BALB/c mice model. Mol Immunol 67(2):501–511. https://doi.org/10.1016/j.molimm.2015.08.001

Losada-Barragán M, Umaña-Pérez A, Durães J, Cuervo-Escobar S, Rodríguez-Vega A, Ribeiro-Gomes FL, Berbert LR, Morgado F, Porrozzi R, Mendes-da-Cruz DA, Aquino P, Carvalho PC, Savino W, Sánchez-Gómez M, Padrón G, Cuervo P (2019) Thymic microenvironment is modified by malnutrition and leishmania infantum infection. Front Cell Infect Microbiol 9: 252. https://doi.org/10.3389/fcimb.2019.00252

Louradour I, Ghosh K, Inbar E, Sacks DL (2019) CRISPR/Cas9 mutagenesis in Phlebotomus papatasi: the immune deficiency pathway impacts vector competence for leishmania major. MBio 10(4):e01941–e01919. https://doi.org/10.1128/mBio.01941-19

Mahoney AB, Sacks DL, Saraiva E, Modi G, Turco SJ (1999) Intra-species and stage-specific polymorphisms in lipophosphoglycan structure control *leishmania donovani* –sand fly interactions. Biochemistry 38(31):9813–9823. https://doi.org/10.1021/bi990741g

Martin-Martin I, Chagas AC, Guimaraes-Costa AB, Amo L, Oliveira F, Moore IN, DeSouza-Vieira TS, Sanchez EE, Suntravat M, Valenzuela JG, Ribeiro JMC, Calvo E (2018) Immunity to LuloHya and Lundep, the salivary spreading factors from Lutzomyia longipalpis, protects against leishmania major infection. PLoS Pathog 14(5):e1007006. https://doi.org/10.1371/journal.ppat.1007006

Meister S, Kanzok SM, Zheng X, Luna C, Li T-R, Hoa NT, Clayton JR, White KP, Kafatos FC, Christophides GK, Zheng L (2005) Immune signaling pathways regulating bacterial and malaria parasite infection of the mosquito *Anopheles gambiae*. Proc Natl Acad Sci 102(32): 11420–11425. https://doi.org/10.1073/pnas.0504950102

Meister S, Agianian B, Turlure F, Relógio A, Morlais I, Kafatos FC, Christophides GK (2009) Anopheles gambiae PGRPLC-mediated defense against bacteria modulates infections with malaria parasites. PLoS Pathog 5(8):e1000542. https://doi.org/10.1371/journal.ppat.1000542

Moraes CS, Aguiar-Martins K, Costa SG, Bates PA, Dillon RJ, Genta FA (2018) Second blood meal by female *Lutzomyia longipalpis* : enhancement by oviposition and its effects on digestion, longevity, and *leishmania* infection. Biomed Res Int 2018:1–10. https://doi.org/10.1155/2018/2472508

Novitskiy SV, Ryzhov S, Zaynagetdinov R, Goldstein AE, Huang Y, Tikhomirov OY, Blackburn MR, Biaggioni I, Carbone DP, Feoktistov I, Dikov MM (2008) Adenosine receptors in regulation of dendritic cell differentiation and function. Blood 112(5):1822–1831. https://doi.org/10.1182/blood-2008-02-136325

Nweze JA, Nweze EI, Onoja US (2020) Nutrition, malnutrition, and leishmaniasis. Nutrition 73: 110712. https://doi.org/10.1016/j.nut.2019.110712

Pace KE, Baum LG (2002) Insect galectins: roles in immunity and development. Glycoconj J 19(7–9):607–614. https://doi.org/10.1023/B:GLYC.0000014092.86763.2f

Pakpour N, Cheung KW, Luckhart S (2016) Enhanced transmission of malaria parasites to mosquitoes in a murine model of type 2 diabetes. Malar J 15(1):231. https://doi.org/10.1186/s12936-016-1277-7

Pereira L, Oliveira F, Townsend S, Metangmo S, Meneses C, Moore IN, Brodskyn CI, Valenzuela JG, Magez S, Kamhawi S (2018) Coinfection with Trypanosoma brucei confers protection against cutaneous leishmaniasis. Front Immunol 9:2855. https://doi.org/10.3389/fimmu.2018.02855

Pimenta PFP, Modi GB, Pereira ST, Shahabuddin M, Sacks DL (1997) A novel role for the peritrophic matrix in protecting *leishmania* from the hydrolytic activities of the sand fly midgut. Parasitology 115(4):359–369. https://doi.org/10.1017/S0031182097001510

Rabinovich GA, Baum LG, Tinari N, Paganelli R, Natoli C, Liu F-T, Iacobelli S (2002) Galectins and their ligands: amplifiers, silencers or tuners of the inflammatory response? Trends Immunol 23(6):313–320. https://doi.org/10.1016/S1471-4906(02)02232-9

Ramalho-Ortigão M, Jochim RC, Anderson JM, Lawyer PG, Pham V-M, Kamhawi S, Valenzuela JG (2007) Exploring the midgut transcriptome of Phlebotomus papatasi: comparative analysis of expression profiles of sugar-fed, blood-fed and leishmania major-infected sandflies. BMC Genomics 8(1):300. https://doi.org/10.1186/1471-2164-8-300

Ramirez JL, Dimopoulos G (2010) The toll immune signaling pathway control conserved anti-dengue defenses across diverse ae. Aegypti strains and against multiple dengue virus serotypes. Dev Comp Immunol 34(6):625–629. https://doi.org/10.1016/j.dci.2010.01.006

Ramirez JL, Souza-Neto J, Torres Cosme R, Rovira J, Ortiz A, Pascale JM, Dimopoulos G (2012) Reciprocal tripartite interactions between the Aedes aegypti midgut microbiota, innate immune system and dengue virus influences vector competence. PLoS Negl Trop Dis 6(3):e1561. https://doi.org/10.1371/journal.pntd.0001561

Rogers ME, Hajmová M, Joshi MB, Sadlova J, Dwyer DM, Volf P, Bates PA (2008) Leishmania chitinase facilitates colonization of sand fly vectors and enhances transmission to mice. Cell Microbiol 10(6):1363–1372. https://doi.org/10.1111/j.1462-5822.2008.01132.x

Sant'Anna MR, Diaz-Albiter H, Mubaraki M, Dillon RJ, Bates PA (2009) Inhibition of trypsin expression in Lutzomyia longipalpis using RNAi enhances the survival of leishmania. Parasit Vectors 2(1):62. https://doi.org/10.1186/1756-3305-2-62

Schlein Y, Jacobson RL (1998) Resistance of *Phlebotomus papatasi* to infection with *leishmania donovani* is modulated by components of the infective bloodmeal. Parasitology 117(5): 467–473. https://doi.org/10.1017/S0031182098003321

Schlein Y, Jacobson RL, Messer G (1992) Leishmania infections damage the feeding mechanism of the sandfly vector and implement parasite transmission by bite. Proc Natl Acad Sci 89(20): 9944–9948. https://doi.org/10.1073/pnas.89.20.9944

Selvapandiyan A, Duncan R, Debrabant A, Bertholet S, Sreenivas G, Negi NS, Salotra P, Nakhasi HL (2001) Expression of a mutant form of leishmania donovani centrin reduces the growth of the parasite. J Biol Chem 276(46):43253–43261. https://doi.org/10.1074/jbc.M106806200

Selvapandiyan A, Kumar P, Morris JC, Salisbury JL, Wang CC, Nakhasi HL (2007) Centrin1 is required for organelle segregation and cytokinesis in *Trypanosoma brucei*. Mol Biol Cell 18(9): 3290–3301. https://doi.org/10.1091/mbc.e07-01-0022

Selvapandiyan A, Dey R, Gannavaram S, Lakhal-Naouar I, Duncan R, Salotra P, Nakhasi HL (2012a) Immunity to visceral leishmaniasis using genetically defined live-attenuated parasites. J Trop Med 2012:1–12. https://doi.org/10.1155/2012/631460

Selvapandiyan A, Kumar P, Salisbury JL, Wang CC, Nakhasi HL (2012b) Role of centrins 2 and 3 in organelle segregation and cytokinesis in Trypanosoma brucei. PLoS One 7(9):e45288. https://doi.org/10.1371/journal.pone.0045288

Selvapandiyan A, Croft SL, Rijal S, Nakhasi HL, Ganguly NK (2019) Innovations for the elimination and control of visceral leishmaniasis. PLoS Negl Trop Dis 13(9):e0007616. https://doi.org/10.1371/journal.pntd.0007616

Selvapandiyan A, Puri N, Kumar P, Alam A, Ehtesham NZ, Griffin G, Hasnain SE (2023) Zooming in on common immune evasion mechanisms of pathogens in phagolysosomes: potential broad-spectrum therapeutic targets against infectious diseases. FEMS Microbiol Rev 47(1):fuac041. https://doi.org/10.1093/femsre/fuac041

Serafim TD, Malafaia G, Silva ME, Pedrosa ML, Rezende SA (2010) Immune response to leishmania (leishmania) chagasi infection is reduced in malnourished BALB/c mice. Mem Inst Oswaldo Cruz 105(6):811–817. https://doi.org/10.1590/S0074-02762010000600014

Serafim TD, Figueiredo AB, Costa PAC, Marques-da-Silva EA, Gonçalves R, de Moura SAL, Gontijo NF, da Silva SM, Michalick MSM, Meyer-Fernandes JR, de Carvalho RP, Uliana SRB,

Fietto JLR, Afonso LCC (2012) Leishmania metacyclogenesis is promoted in the absence of purines. PLoS Negl Trop Dis 6(9):e1833. https://doi.org/10.1371/journal.pntd.0001833
Soares RPP, Macedo ME, Ropert C, Gontijo NF, Almeida IC, Gazzinelli RT, Pimenta PFP, Turco SJ (2002) Leishmania chagasi: lipophosphoglycan characterization and binding to the midgut of the sand fly vector Lutzomyia longipalpis. Mol Biochem Parasitol 121(2):213–224. https://doi.org/10.1016/S0166-6851(02)00033-6
Stamper LW, Patrick RL, Fay MP, Lawyer PG, Elnaiem D-EA, Secundino N, Debrabant A, Sacks DL, Peters NC (2011) Infection parameters in the sand fly vector that predict transmission of leishmania major. PLoS Negl Trop Dis 5(8):e1288. https://doi.org/10.1371/journal.pntd.0001288
Telleria EL, Sant'Anna MRV, Ortigão-Farias JR, Pitaluga AN, Dillon VM, Bates PA, Traub-Csekö YM, Dillon RJ (2012) Caspar-like gene depletion reduces leishmania infection in sand fly host Lutzomyia longipalpis. J Biol Chem 287(16):12985–12993. https://doi.org/10.1074/jbc.M111.331561
Vivenes A, Oviedo M, Márquez JC, Montoya-Lerma J (2001) Effect of a second bloodmeal on the oesophagus colonization by leishmania mexicana complex in Lutzomyia evansi (Diptera: Psychodidae). Mem Inst Oswaldo Cruz 96(3):281–283. https://doi.org/10.1590/S0074-02762001000300001
Walters LL (1993) Leishmania differentiation in natural and unnatural sand fly hosts. J Eukaryot Microbiol 40(2):196–206. https://doi.org/10.1111/j.1550-7408.1993.tb04904.x
Walters LL, Chaplin GL, Modi GB, Tesh RB (1989) Ultrastructural biology of leishmania (Viannia) panamensis (=leishmania Braziliensis panamensis) in Lutzomyia gomezi (Diptera: psychodidae): a natural host-parasite association. Am J Trop Med Hyg 40(1):19–39. https://doi.org/10.4269/ajtmh.1989.40.19
Xi Z, Ramirez JL, Dimopoulos G (2008) The Aedes aegypti toll pathway controls dengue virus infection. PLoS Pathog 4(7):e1000098. https://doi.org/10.1371/journal.ppat.1000098
Yang R-Y, Liu F-T (2003) Galectins in cell growth and apoptosis. Cell Mol Life Sci 60(2):267–276. https://doi.org/10.1007/s000180300022
Young AR, Meeusen EN (2002) Galectins in parasite infection and allergic inflammation. Glycoconj J 19(7–9):601–606. https://doi.org/10.1023/B:GLYC.0000014091.00844.0a
Zahedifard F, Gholami E, Taheri T, Taslimi Y, Doustdari F, Seyed N, Torkashvand F, Meneses C, Papadopoulou B, Kamhawi S, Valenzuela JG, Rafati S (2014) Enhanced protective efficacy of nonpathogenic recombinant leishmania tarentolae expressing cysteine proteinases combined with a sand fly salivary antigen. PLoS Negl Trop Dis 8(3):e2751. https://doi.org/10.1371/journal.pntd.0002751

Leishmania Proteomics: Insight into Diagnostics and Vaccine Development

5

Dharmendra Kumar Maurya, Shyamali, Shyam Lal Mudavath, Shyam Sundar, and Om Prakash Singh

Abstract

Visceral leishmaniasis (VL), a life-threatening parasitic disease in the Indian subcontinent, is caused by the protozoan parasites - *Leishmania donovani* and transmitted by the bite of the *Phlebotomus* sand fly. These parasites have adopted many invasion mechanisms to survive the host's direct immune responses and established themselves in the host cells. Although efforts have been made to develop sensitive biomarkers for diagnosis and effective vaccines for VL treatment, we still need to achieve these goals with great success. Advances in various proteomic platforms have allowed us to identify large numbers of surface and cytosolic proteins from mixed stages of the Leishmania life cycle, which can be manipulated as a diagnostic marker or can be used as vaccine candidates. In this review, we highlight the proteins in amastigote and promastigote forms of Leishmania parasite and discuss their potential implications in future diagnostic and vaccine development.

Dharmendra Kumar Maurya and Shyamali contributed equally to this work.

D. K. Maurya · Shyamali · O. P. Singh (✉)
Department of Biochemistry, Institute of Science, Banaras Hindu University, Varanasi, Uttar Pradesh, India
e-mail: opsinghbiochem@bhu.ac.in

S. L. Mudavath
Infectious Disease Biology Laboratory, Chemical Biology Unit, Institute of Nano Science & Technology, Mohali, Punjab, India

S. Sundar
Department of Medicine, Institute of Medical Sciences, Banaras Hindu University, Varanasi, Uttar Pradesh, India

© The Author(s), under exclusive license to Springer Nature Singapore Pte Ltd. 2023

A. Selvapandiyan et al. (eds.), *Challenges and Solutions Against Visceral Leishmaniasis*, https://doi.org/10.1007/978-981-99-6999-9_5

Keywords

Leishmaniasis · Proteomics · Diagnostic markers · Vaccine candidate · Surface proteins · Cytosolic proteins

5.1 Background

Vector-borne diseases remain a major public health concern globally (thrive among impoverished populations, especially in tropical, subtropical, and Mediterranean regions), and an estimated 1.2 billion people are at risk (Engels and Zhou 2020). They include a range of chronic disabling infections, such as Chagas disease, Malaria, Yellow fever, human African trypanosomiasis (HAT), leishmaniasis, leprosy, lymphatic filariasis, but also more acute infections such as dengue, Chikungunia, Japanese Encephalitis, and Zika virus (Engels and Zhou 2020). Among these, parasitic diseases are responsible for an estimated 500,000 deaths each year. These parasites have adopted unique strategies to overcome and survive in extremely harsh and different environments between hosts (Kamhawi 2006). This includes the involvement of differential production of different proteins that form protective barriers against hostile forces (Pink et al. 2005). Leishmaniasis is one of the disorders in this group caused by the *Leishmania donovani* complex (Chappuis et al. 2007) and transmitted to humans through the bite of an infected female phlebotomine sand fly (Kushwaha et al. 2022). Once inside the human body, *Leishmania donovani* resides and proliferates inside the cells of the reticuloendothelial system (Singh et al. 2021). The parasites exist in two forms: the extracellular promastigote form that multiplies in the midgut of the fly and the intracellular amastigote form that infects phagocytic mononuclear cells of several human species (Sundar and Rai 2002). There are several forms of the disease, as defined by clinical syndromes, which are caused by various species of the parasite (Burza et al. 2018). The disease takes four main forms:

- *Visceral leishmaniasis (VL,* also known as *kala azar)*—the most serious form of the disease; fatal if untreated.
- *Cutaneous leishmaniasis (CL)*—the most common infection, causing one or more simple skin lesions, which heal after a few weeks or months, but leave unsightly scars.
- *Mucocutaneous leishmaniasis (MCL)*—begins with simple skin ulcers, which spread, causing hideous tissue destruction, particularly to the nose and mouth.
- *Diffuse cutaneous leishmaniasis (DCL)*—produces disseminated and chronic lesions, which resemble those of lepromatous leprosy and are the most difficult to treat.

The number of new cases is about 2 million (1.5 million of CL and 0.5 million of VL) annually. Apart from rising figures due to better reporting, there is also a clear and disturbing real increase in the number of people infected. Like many other tropical diseases, leishmaniases are related to economic development and man-made environmental changes, which increase exposure to the sandfly vector. VL is one of the most severe forms of leishmaniasis primarily caused by the Leishmania donovani complex, *including Leishmania donovani, Leishmania infantum, and Leishmania chagasi* (Burza et al. 2018). Classic VL, which is progressive and fatal if not treated, is an insidious, chronic disease that is characterized by irregular fever, anorexia, weight loss, cough, gross enlargement of the spleen and liver, mild anemia, and emaciation. This may be preceded by rigors and vomiting (Boelaert et al. 2009). VL is a potentially fatal human disease in the absence of treatment (Singh et al. 2016b; Sundar and Rai 2002). It is the second deadliest parasitic disease (after malaria), and around 200 million people are at the risk in the 47 affected countries, with an estimated annual incidence of approximately 200,000–400,000 cases (WHO n.d.). Epidemiological studies indicate that only one in 5–10 infected individuals develops clinical VL (Singh et al. 2014; Bucheton et al. 2002; Ho et al. 1982; Sacks et al. 1987), suggesting that the number of infected individuals and hence potential parasite reservoirs is far greater. VL is an anthroponotic disease in South East Asia and East Africa region and caused by *L. donovani*. However, in Central and South America and the Mediterranean region, it is zoonotic and caused by *L. infantum* and/or *L. chagasi* (Sundar and Singh 2018). About 90% of the all-annual cases of VL are from India, Sudan, Brazil, Ethiopia, Bangladesh, and Nepal (Singh et al. 2016a). In Eastern India (Bihar State), the situation is becoming more and more severe, with all districts of Bihar and regions of Jharkhand, West Bengal, and Uttar Pradesh facing the worst kala-azar epidemic since the late 1970s. On the order of 200,000–250,000 people are believed to have contracted the disease in 1992 (five times the official figures), and because of the rapid spread, an alarming situation exists (Singh and Sundar 2022). It has been estimated that 80% of the global burden of VL is present in the Indian continent (Singh et al. 2016a).

Recently, there has been an increase in the overlap of VL and human immunodeficiency virus (HIV) infection due to the spread of the acquired immunodeficiency syndrome (AIDS) pandemic. Leishmania/HIV coinfection is considered an "emerging disease," especially in southern Europe, where 25–70% of adult VL cases are related to HIV infection, and 1.5–9.5% of AIDS cases suffer from newly acquired or reactivated VL. Intravenous drug users have been identified as the main population at risk.

There has been increasing unresponsiveness to antimonials (occasionally associated with serious, not uncommonly fatal cardiotoxicity), especially in Bihar and the Terai regions of Nepal, where up to almost 60% of patients were found resistance to antimonial treatment (Sundar et al. 2000). Pentamidine (an alternative drug therapy) is extremely toxic and causing irreversible insulin-dependent diabetes mellitus and has also lost its initially high level of efficacy and is no longer recommended for VL treatment. Amphotericin B (highly effective but expensive) was rolled out in the year 1996 for the treatment of VL, but need of intravenous infusion and associated adverse reactions like myocarditis, hypokalemia, and nephrotoxicity were the major disadvantages for its use at remote public health centers (PHCs). Experiences with lipid formulations of amphotericin B (e.g., AmBisome or Abelcet) are excellent and highly been used as a rescue therapy because of its high cost. Overall, treatment options for VL are generally few, and this problem is exacerbated by a critical lack of appropriate diagnostic tools to guide treatment (Singh et al. 2016b). Due to this increasing resistance to standard therapy, there is a need for new safe, efficacious, low-cost therapies, diagnostics, and efficacious vaccines for the management of VL.

Recent technological developments have only made a limited contribution to Neglected Tropical Diseases (NTDs) in disease-affected countries, partly because of less profit and specific healthcare delivery challenges (Sundar et al. 2018). Therefore, a proper balance needs to be made between new biomedical technology and its translational application in healthcare systems in the most cost-effective manner. Furthermore, research needs to be expanded on the genomic level to develop markers that might be translated into simpler tools for tracking infections and treatment effectiveness (Blackwell et al. 2020; Garcia et al. 2021). Proteomics, along with developments in the blending of bioinformatics, has revolutionized the vaccinomics (i.e., characterization of the host response to immunization), offers an avenue for the identification of different developmental stages of *Leishmania* proteins (El Fakhry et al. 2002; Khan et al. 2020; Nugent et al. 2004; Thiel and Bruchhaus 2001) that may be responsible for parasite virulence and pathogenicity (de Araujo Soares et al. 2003). Some of these proteins can be used in diagnostic development and are potential targets based on the elucidation of the biological mechanism underlying drug resistance, pathogenesis, or involvement in differential expression (Drummelsmith et al. 2003). Despite plenty of studies on the identification of leishmanial proteins and their suitability as candidates for diagnosis and vaccine against human VL (Kushawaha et al. 2012; Murray et al. 2005; Routaray et al. 2022b; Singh et al. 2012), we are still struggling for highly effective vaccine or diagnostics (Brotherton et al. 2010). Also, the emergence of both resistant parasites (Croft et al. 2006; Sundar 2001) and vectors (Annual report, WHO 2015) offers a major hindrance to the current kala-azar control program. Hence, in the present situation, there is an urgent need to develop an effective vaccine against VL (Wei et al. 2021). In this review, we summarized the various known parasitic proteins and their clinical applications that can be considered and explored as potential molecules for developing an effective vaccine or diagnostics.

5.2 Structure and Biology of *Leishmania* Parasites

The Leishmania parasite is transmitted by female *Phlebotomine* sand flies as a flagellated, metacyclic promastigote, which is phagocytized by host macrophages and then differentiates into the nonflagellated, replicative amastigote(Stanley and Engwerda 2007). Thus, the parasite exists in two morphologically distinct life cycle stages (Fig. 5.1). The slender, flagellated promastigotes, ranging from 10 μm to

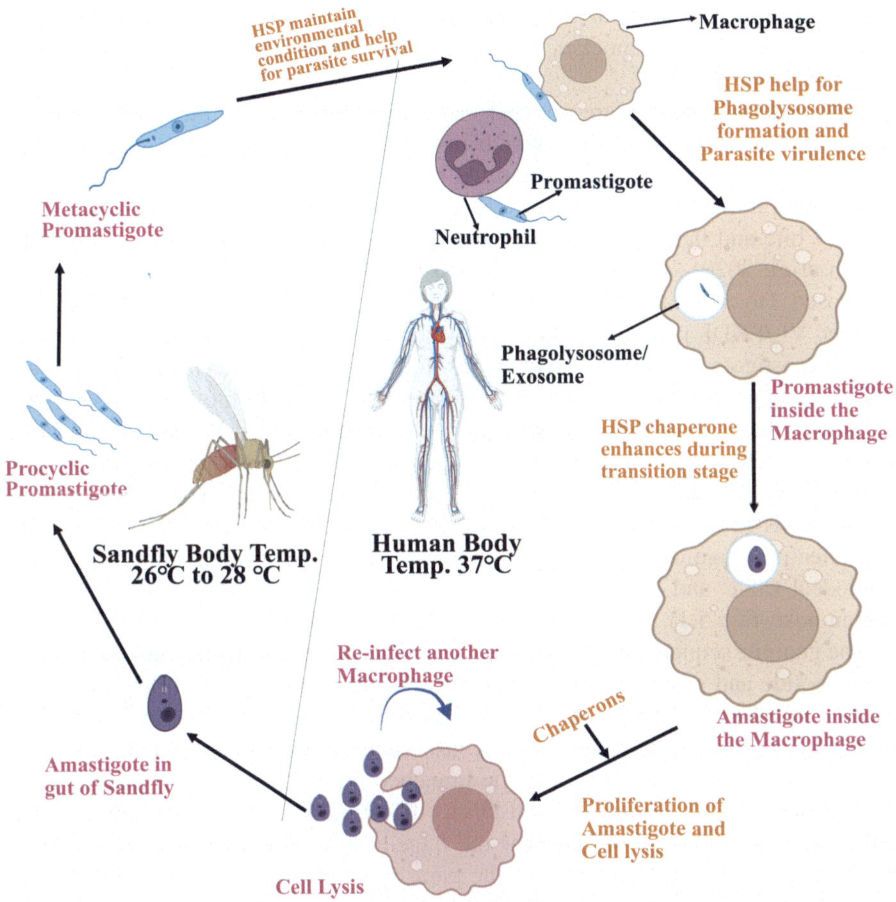

Fig. 5.1 *Leishmania donovani* life cycle among two different hosts: female Phlebotomine fly (sand fly) injects the infective stage promastigote (i.e., metacyclic promastigote) into a healthy host (human, rodents, and dog) during a blood meal. These promastigotes lodge inside the macrophage through the phagosome. Within the macrophage, the motile flagellated promastigote is converted to a nonmotile amastigote lacking flagella inside the phagolysosome. Amastigotes undergo growth and proliferation and infect a new set of macrophages for their further growth and proliferation. When the sand fly bites an infected individuals, the amastigotes circulating are picked up in a blood meal. Within the midgut of sand fly, amastigotes undergo growth and division to form procyclic promastigotes, which get differentiated into nondividing metacyclic promastigotes that are infectious to the healthy host

20 μm, proliferate in the alimentary tract of a female sand fly, *Phlebotomus argentipes*, at 26 °C (pH 7.4). The amastigote is a rounded intracellular parasite ranging from 3 μm to 7 μm in diameter with a protruding flagellum. Furthermore, amastigote numbers increase via binary fission and ultimately cause the bursting of the host cell, allowing the released parasites to infect other phagocytic cells(Bañuls et al. 2007), i.e., phagosomes at temperatures up to 39 °C (pH 5.5) (Sundar and Rai

2002). Importantly, in both developmental forms of the parasite, the flagellum emerges from a flagellar pocket, but it is almost wholly restricted in the amastigote (Rochette et al. 2009).

Promastigote parasites are injected into the human host via the bite of infected female phlebotomine sand flies, which have the parasites inside them. In humans, these promastigotes are quickly engulfed by phagocytic cells in which they differentiate and proliferate as amastigotes (Kaye and Scott 2011). The period between the sand fly bite and the appearance of symptoms is between 2 months and 6 months (Singh et al. 2020). This is quite a long period, and that is why VL is not easily diagnosed. However, sand flies get infected during the ingestion of blood meal from infected VL/PKDL patients. In the sand fly abdomen, these amastigotes differentiate into promastigotes via the different stages. Initially, amastigote changes into small, inactive procyclic with a tiny flagellar structure within 24–48 h of bloodmeal. After 2–3 days, these procyclics change into nectomonads, and that change into large slender, which assure the movement of parasites at the anterior midgut thoracic region for the formation of the colony (Kamhawi 2006). Nectomonads change into leptomonads within 4–5 days and undergo the second multiplication cycle in the fly. The infective stage of the promastigote is the metacyclic, present back side of the stomodeal valve, and they are very adapted for fortunate transmission to the mammalian host (Fig. 5.1).

The genome sequencing of a large number of *Leishmania* species and an increase in nucleotide and protein sequence databases accelerate the pace of diagnostic and vaccine development programs (Ashrafmansouri et al. 2022). The availability of such genome data revealed the overall conservation of gene order and chromosome structure. *Leishmania* contains a single-branched mitochondrion with a large DNA content (nearly 15% of the total cell DNA) that usually increases the length of the cell in the kinetoplast, a region near the flagellum basal body (Ivens et al. 2005; Rochette et al. 2008). Over 90% of the parasite's kinetoplast–DNA (k-DNA), are nonreplicative, and comprise 5000–10,000 catenated copies of the minicircle with 800–1200 bp each (Ivens and Blackwell 1999). The remaining 10% of k-DNA forms 25–50 copies of a maxicircle with about 20 kb each. In nonreplicating parasites, minicircles are covalently closed but topologically relaxed (Rauch et al. 1993). Maxicircles resemble eukaryotic mitochondrial DNA and encode some common rRNA such as subunits of cytochrome oxidase and NADH dehydrogenase (Shapiro and Englund 1995). Nuclei of both leishmanial stages are enveloped by the endoplasmic reticulum and divided by pores similar to eukaryotic cells. A centrally positioned prominent nucleolus 0.6–1.0 μm in diameter usually observed along with peripheral arrangement of nuclear chromatin (Vannier-Santos et al. 2002).

5.3 Proteomics of Life Cycle Stages of *Leishmania* Parasites

Proteomics study has become a principal tool for the basic biology of *Leishmania* parasites and host dialogues in the past few years. The proteomic study helps in understanding the mechanism of parasite development, the interaction between host and parasite, and effective vaccine development (Rashidi et al. 2018). Proteomic studies of any organism are complex because of the variability of proteins in biological systems. So far, proteomic studies in leishmaniasis highlighted many essential cytosolic and secretory proteins that play vital roles during *Leishmania* pathogenesis (El Fakhry et al. 2002; Zheng et al. 2020). The advancements in proteomic technologies have enabled researchers to better understand the role of these proteins in pathogenicity, associated drug resistance, virulence pathogenicity, and host–pathogen interactions, thereby defining novel therapeutic or vaccine targets (Zheng et al. 2020; Bag et al. 2014). The availability of the whole genome sequence of many *Leishmania* species provides a variable number of target identification, assessment in a biological pathway, and the essentiality of the gene in the mammalian stage of the parasite (Tasbihi et al. 2020; Zheng et al. 2020).

The identification of variability in different protein expression among *Leishmania* species (e.g., free and bound form) through proteomics approaches helped in the identification of these protein expression levels, posttranslational modifications (PTMs), interactions with other molecules, the structural and subcellular distribution of proteins in the parasite. Importantly, amastigote and promastigote parasites of *Leishmania* species consist of several proteins; among them, some proteins are species-specific and also parasites-specific by which they show transport, enzymatic and/or antigenic properties, thus can be used either as biomarkers or subunit vaccine (Alcolea et al. 2019; Zheng et al. 2020) (Fig. 5.2).

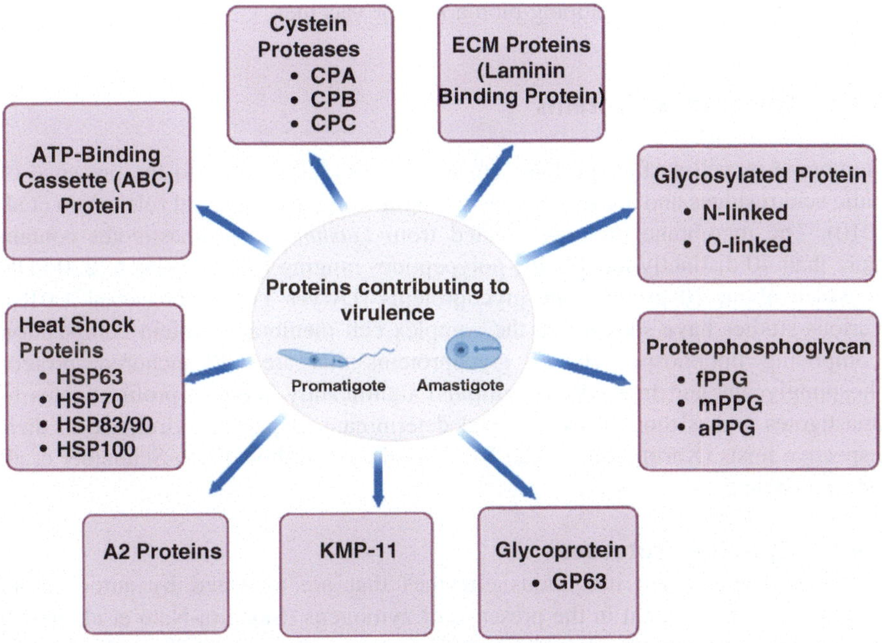

Fig. 5.2 Major stage-specific proteins in leishmania parasites

In a proteomic study of *L. donovani* promastigotes, approximately 2000 protein spots were identified, 31 of which were specific proteins, of which 65 proteins showed increased expression of amastigote differentiation in vitro, and four proteins revealed downregulated expression in amastigote differentiation (Alcolea et al. 2019). Based on this study, stage-specific proteins were recognized in promastigotes or amastigotes of *L. donovani*, and they can be separated into multiple proteins group with their function: cell cycle and proliferation, cytoskeleton, cell membrane, stress response (such as temperature, oxidative stress, pH), energy metabolism and phosphorylation; and amino acid metabolism (El Fakhry et al. 2002; Routaray et al. 2022a). Furthermore, amastigote-specific proteins have also been identified, such as amastin, gp63 isoform 7, several proteophosphoglycans, cysteine proteinase, heat shock proteins (HSPs)- 63, 70, 90, and 100, A2 protein family, an adenosine transporter, a cathepsin L-like protease, tryporedoxin peroxidase, and mitogen-activated protein kinases (MAPKs) (Bente et al. 2003). Walker et al. (2006), performed mass spectrometry and 2D gel electrophoresis and identified approximately 2000 proteins where more than 62 proteins are differentially expressed. Isocitrate dehydrogenase (IDH)-specific activity was approximately three times higher than that of promastigotes parasite in *L. infantum* (Walker et al. 2006). Several other researchers conducted proteomics experiments and found many specific proteins in different species of *Leishmania* that play a crucial role in the parasite growth, development, and host susceptibility.

In the following section, we will discuss some of the important proteins found in leishmania parasites, their crucial role in the parasite life cycle stage, and their clinical importance in developing biomarkers or vaccines.

5.3.1 Membrane Proteins

Membrane or cell surface proteins are negatively charged due to the presence of sialic acid residues and are responsible for significant physiological roles (Yao et al. 2010). The membrane proteins isolated from *Leishmania* promastigotes contain more than 40 distinctly identifiable polypeptides ranging from 12 kDa to 220 kDa, of which about 20 proteins are glycoproteins (Dwyer 1974; Ejazi et al. 2018). Various studies have shown that the complex cell membrane protein composition (comprising lipophospho-glycan, glycoproteins that are GPI-anchored, proteophosphoglycans and free GPI glycolipids) significantly alters in promastigotes to amastigotes and is thought to be novel determinant of parasite virulence in their respective hosts (Karim et al. 2022; McConville and Ralton 1997; Schneider et al. 1992). (Table 5.1).

5.3.1.1 Cysteine Proteases

Cysteine proteases are ubiquitous enzymes that are activated by autocatalytic processing of prodomain in the presence of zymogens (Siqueira-Neto et al. 2018). *Leishmania* demonstrates a broad range of cysteine proteases, which can be divided into three forms: CPA, CPB, and CPC. All three forms of cysteine proteases are a

Table 5.1 Proteins of *Leishmania donovani*: occurrence and possible biological role

Proteins	Promastigote	Amastigote	References
1. Membrane Proteins			
(a) gp 63 (glycoprotein zinc metallo protease)	Major cell surface glycoprotein found on entire surface, important role in multiplication, serves as a ligand for the macrophage receptor via complement component, protect against complement-mediated lysis, Proteolytically cleaves host macro-molecules	Expressed at lower level, found in the flagellar pocket	Brittingham et al. (1995, 1999), Button and McMaster (1988), Button et al. (1989), Funk et al. (1997), Olafson et al. (1990), Ramamoorthy et al. (1992), Schlagenhauf et al. (1998)
(b) LBP	Help parasites (promastigote) to glide through interstitial tissue, and behaves as an adhesin	Expressed at lower level, function is still unknown	Bandyopadhyay et al. (2002, 2003), Ghosh et al. (1996, 1999)
(c) KMP-11	Stabilizing molecule such LPG within membrane	Serves as a competitive inhibitor of nitric oxide synthase, allowing the parasite to proliferate intracellularly	Basu et al. (2007), Berberich et al. (1997), Jardim et al. (1995a, b), Roach et al. (1991)
(d) PPG (Proteophospho-glycans)	GPI anchored cell associated filamentous form termed m PPG. The gel-like matrix, formed by these interlocking filaments, traps the parasites in the sandfly's interior gut	Secrete their own non filamentous form termed a PPG, a PPG is believed to contribute to the formation of the parasitophorous vacuole; these participate in the maintenance of infection. Activate the complement system via the mannose binding pathway	Ilg (2000), Ilg et al. (1995, 1999a, b), Mukhopadhyay and Mandal (2006), Peters et al. (1997a, b)
(e) sAP (secreted acid phosphatase)	Secreted from flagella	Not reported	Ilg et al. (1991, 1994a, b, 1999a), Mukhopadhyay and Mandal (2006)
(f) ABC transporter	Drug resistance, ATP-dependent transport across membrane	Drug resistance, ATP-dependent transport across membrane	Araujo-Santos et al. (2005), Coelho et al. (2006), Henderson et al. (1992), Legare et al. (2001), Ouellette et al. (1994), Parodi-

(continued)

Table 5.1 (continued)

Proteins	Promastigote	Amastigote	References
			Talice et al. (2003), Singh (2006)
2. Cytosolic Proteins			
(a) Hsp 90 (also known as Hsp 83)	Feedback inhibitor of the cellular heat shock response	Not reported	Argaman et al. (1994), Hubel and Clos (1996), Shapira and Pinelli (1989), Wiesgigl and Clos (2001)
(b) Hsp 70	Carrier for antigenic peptide, acting as adjuvant	Target of Humoral and Cell immune response	Folgueira et al. (2007), Lee et al. (1988), MacFarlane et al. (1990)
(c) Hsp 100	Not reported	Function as an antagonist of amastigote to promastigote differentiation, promoter of full amastigote development	Clos et al. (2001), Hubel et al. (1997), Krobitsch et al. (1998), Krobitsch and Clos (1999)
(d) A2 protein	Not reported	Considered a virulence factor that is required for *L. donovani* survival in mammalian host, contributes to tropism for visceral infection	Carvalho et al. (2002), Charest and Matlashewski (1994), Matlashewski (2001), McCall and Matlashewski (2010), Zhang et al. (1996, 2003), Zhang and Matlashewski (1997)

ABC ATP-binding cassette proteins, *CPA* cysteine proteases A, *CPB* cysteine proteases B, *GP* glycosylated proteins, *HSPs* heat shock proteins, *IDH* isocitrate dehydrogenase, *KMP*-11 kinetoplastid membrane protein-11, *LBPR* laminin binding protein receptor, *Ld-mao1* maoc family dehydratase, *Ld-ppi1* peptidyl-prolyl cis-trans isomerase, *Ld-mad1* malate dehydrogenase, *Li-isd1* iron superoxide dismutase 1, *Li-txn1* tryparedoxin 1, *Li-ntf2* nuclear transport factor 2, *PPG* proteophosphoglycans, *SLD proteins* soluble *leishmania donovani* proteins, *TPI* triosephosphate isomerase

part of clan CA, family C1. Based on proteolytic mechanisms and divergent substrate binding differences, it is categorized into cathepsin L-like (CPA and CPB) and cathepsin B-like enzymes (CPC) (Souza et al. 1992). Cysteine proteases gene expression is responsible for the regulation of promastigotes and amastigotes stages, and it is significantly expressed in the mammalian amastigote stage rather than in the host or sand fly promastigote stage. However, CPC is expressed in procyclic promastigotes, while CPB1 and CPB2 are present more in the metacyclic promastigotes (Selzer et al. 1999). CPA knockout in *L. chagasi* and *L. infantum* decreased the infectivity during in vitro and in vivo conditions, but in *L. mexicana*, CPB knockout displayed impaired macrophage infectivity and delayed lesion progression. Several copies of CPB are necessary to significantly restore virulence or parasite multiplication. CPB alters infected macrophage Th1 immune responses, basically IFN-γ production in *L. mexicana* by degradation of the transcription factor NFκB and subsequent blocking of IL-12 production. *L. mexicana* cysteine proteases also degraded JNK and ERK MAP kinases, which are represented in Fig. 5.3. According to Omara et al., the cysteine protease gene is present on chromosome 10 in *L. donovani chagasi* and on chromosome 8 in *L. donovani donovani* genomes. This *Leishmania* species consist of two types of *L. donovani chagasi* cysteine protease gene (Ldccys); one is Ldccys1, and the other is Ldccys2 (Omara-Opyene and Gedamu 1997). The Ldccys1 genes are expressed abundantly in the amastigote stage compared to the promastigote stage of development, whereas the Ldccys2 gene exists in both the promastigote and amastigote phases of development.

5.3.1.2 Glycoprotein 63 (Gp63)

Gp63, also known as leishmanolysin, is the most widely studied protein, which is highly conserved among all Leishmania species. This zinc metalloprotease is expressed well both in promastigote and amastigote forms and is implicated in several mechanisms related to parasite virulence. Also, the proteinase activity of leishmanolysin results in increased resistance to complement-mediated lysis. All these make it an attractive vaccine candidate. Gp63 is expressed to a lower level in

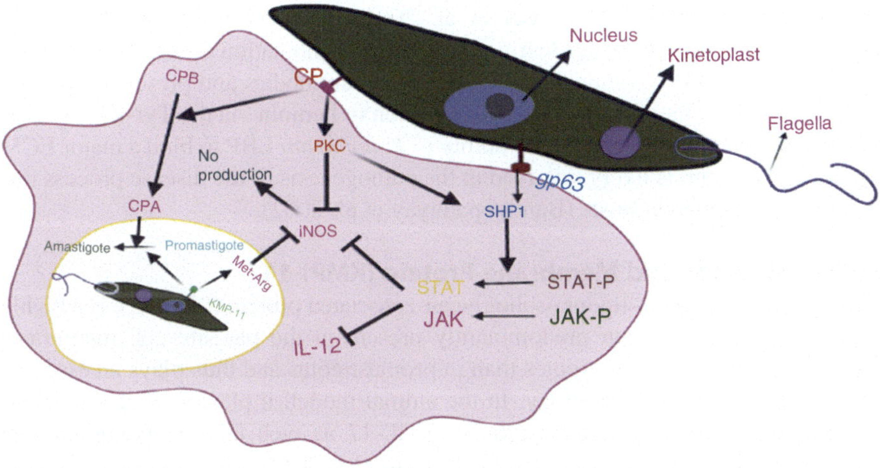

Fig. 5.3 Mechanism of differentiation from promastigote to amastigotes antigen presenting cell and involvement of cysteine protease. (1). Cysteine proteases (CP) activate protein Kinase-C (PKC), which inhibits iNOS, and this inhibition decreases NO (nitric oxide) production and prevents cell apoptosis; (2) Activation of protein tyrosine phosphatase 1 (SHP-1) is done by CP and gp63, which inhibit JAK-STAT pathway by dephosphorylating of JAK and STAT molecules and produced iNOS and IL12; (3) During the development of promastigote to amastigote CPA concentration increases.; (4) KMP-11 is involved in the activation of Met-Arg molecules, which block iNOS and reduce NO production while also preventing apoptosis. KMP-11 is also engaged in the development of the amastigote stage

amastigotes and is mainly found in the flagellar pocket region., However, in the promastigotes, it covers the entire surface (Button and McMaster 1988). Gp63 protein as a vaccine is highly protective and quite promising in the animal model (Bhowmick et al. 2008; Kaur et al. 2011), but the T-cell responses to Gp63 have been found variable in humans. Elfaki et al. (2012) used an algorithm, EpiMatrix to select some putative T-cell epitopes present in *L. donovani* GP63 in order to assess their in vitro immunogenicity. They found a significant decline in IL-10 levels in each peptide in comparison to unstimulated controls. Pooled peptides showed a moderate increase in IFN-γ levels in some volunteers, while individual peptides did not significantly differ from negative controls. Notably, gp63 has also been evaluated as a diagnostic and epidemiological marker for *L. donovani* infection in Ghana (Okong'o-Odera et al. 1993).

5.3.1.3 Laminin-Binding Protein

Laminin-Binding Protein (LBP), a major binding receptor protein, plays a crucial role in the parasite interaction with extracellular matrix (ECM) elements (Bandyopadhyay et al. 2003). Using immunofluorescence microscopy and flow cytometry, the surface location of the receptor was highlighted, as well as the detection of the 67 kDa protein in the biotinylated membrane fraction using an anti-LBP antiserum. Preliminary findings of the 67 kDa LBP may act as an adhesin protein, and the incubating parasites with laminin-coated microtiter wells in the presence of soluble receptor protein inhibited promastigote adhesion in a significant and specific way (Bandyopadhyay et al. 2001). In *L. donovani* promastigotes, laminin-binding receptor was identified by using a combination effect of Fluorescein isothiocyanate (FITC) conjugated goat anti-rabbit antibodies and present on numerous cells in which they involved in cell adhesion to laminin via the Tyr-Ile-Gly-Ser-Arg site on laminin's B1 chain. The ability of *L. donovani* LBP to bind a major ECM protein like laminin is likely involved in the pathogenesis of the disease process that exhibits in mammalian hosts (Bandyopadhyay et al. 2002).

5.3.1.4 Kinetoplastid Membrane Protein (KMP)-11

KMP-11, also known as lipophosphoglycan-associated protein (LPGAP), is a highly conserved 11 kDa protein predominantly present in the parasite cell membrane, expressed more in the amastigotes than in promastigotes and thus plays an essential role in the host–parasite interplay. In the animal model, it plays a vital role in the reversal of T-cell anergy and synthesis of IL-12 as well as the upregulation of inducible nitric oxide synthase (iNOS) (Jardim et al. 1991; Kemp et al. 1991) and peripheral blood lymphocyte (PBMCs) from the patients recovered from VL (Mendonca et al. 1991; Russo et al. 1992). Because of the high levels of KMP-11 transcription in axenic amastigotes, it contains a highly conserved T-cell epitope that aids in increasing T-cell-mediated immune responses. Basu et al. analyzed the entire episode of KMP–11 with clashing nonapeptides to decode the function of CD8+ T cells in infection defense and disease cure. Thirty peptides have been outlined that specifically boost interferon-γ secretion by human CD8+ T cells. However, four T-cell lines with distinct peptide specificities recognized Leishmania-infected autologous macrophages, displaying that KMP-11 is processed and presented via the MHC class I pathway of infected cells.

5.3.1.5 Glycosylated Proteins

Glycosylation is a process in which glycan macromolecules covalently conjugate with specific amino acid residues during co- and posttranslational modification and form glycoproteins and/or proteophosphoglycans (PPGs). The glycosylation process occurs in either the amino group or carboxylic group of amino-terminal and is called *N*-linked glycosylation and *O*-linked glycosylation, respectively, where the glycans are attached to asparagine (Asn) and serine (Ser)/threonine (Thr) amino acid side chains, respectively. N-linked glycosylation has been linked to pathogenicity, cytokine secretion, host–cell behavior (such as adhesion and invasion), host immune modulation, and protein structural integrity in *Leishmania* parasites (Olivier et al. 2012). The extremely O-glycosylated proteins play critical roles in the complement activation, serum inhibition, and opsonization of amastigote life stages. Another of the most studied *Leishmania N*-glycosylated proteins is gp63 (as discussed above), which is a 60–66 kDa zinc-metalloprotease that serves as a major surface antigen. According to the Uniprot database, gp63 protein shows varying numbers of *N*-linked glycosylation sites in *Leishmania* species, including *L. mexicana* (8 sites), *L. tropica* (7 sites), *L. guyanensis* (6 sites), *L. major* (3 sites), *L. amazonensis* (2 sites), *L. chagasi* (2 sites), and *L. donovani* (1 site), indicating a species-specific *N*-linked glycosylation pattern. In *L. major* and *L. donovani* $Man_6GlcNAc_2$ and $GlcMan_6GlcNAc_2$ *N*-linked oligosaccharides structures were reported in the promastigote life stage; however, glycosylation was not constant among the two species in the life stage of amastigote, with L. donovani missing N-linked glycans, demonstrating that protein N-linked glycosylation is species- and life stage–specific. Variations in glycosylated protein expression between different strains of *Leishmania* parasites may also determine parasite response.

5.3.1.6 Proteophosphoglycans (PPG)

The physiologically active phosphoglycan structures of lipophosphoglycan (LPG) have been found on numerous highly glycosylated leishmania proteins, which have been dubbed proteophosphoglycan (PPGs) (Ilg et al. 1999a). Different types of proteophosphoglycans (PPGs) are predominantly synthesized in the *Leishmania* promastigotes life stage, which includes filamentous (fPPG), membrane (mPPG), and secreted acid phosphatases (SAP), while nonfilamentous (aPPG) is secreted during amastigote life stage (Rogers et al. 2004). Most *Leishmania* species secreted promastigote secretory gel in the filamentous form called filamentous proteophosphoglycans (fPPG), which appeared thread-like when they are glycosylated with phosphoglycan during microscopic observation during condensation of fPPG form a gel-like compound known as promastigote secretory gel (Stierhof et al. 1999). Secundino et al. (2010) observed that *Leishmania* parasites LPG mutated and stuck to the surface of the host with the help of PPGs. PPG helps the parasite attach to macrophage receptors and appears to maintain the parasitophorous vacuole, increasing the parasite's chances of survival throughout infection (Secundino et al. 2010). Vaccination with purified aPPG leads to a weak B-cell response and no T-cell response (Aebischer et al. 1999). Finally, the important question to researchers is—if PPGs can supersede LPG as a possible target for leishmaniasis chemotherapeutic management.

5.3.1.7 ATP: Binding Cassette (ABC) Proteins

Of all the transmembrane proteins, ABC proteins are one of the largest transmembrane proteins found in all living organisms (Higgins 1992). It consists of four polypeptides, of which two are transmembrane domains (TMDs) and the other two nucleotide-binding domains (NBDs) (Legare et al. 2001). The ABC transporters are involved in different medical relevance, such as multidrug resistance (MDR) phenotyping and genetic disease. In total, 23 ABC transporter genes have been listed in the *Leishmania* genome project (Ivens et al. 2005), and three different classes of ABC transporters have been described (i.e., A, B, and C); two of them are linked with drug resistance and are homologous to the mammalian MDR-associated protein 1 (MRP1) and P-glycoprotein (Pgp) transporters. The third one is very similar to the mammalian ABCA family (Perez-Victoria et al. 2001). ABCA family is usually present in the plasma membrane and in the flagellar pocket. Research on ABC proteins, along with the analytical development of inhibitors that block these transporters, is a promising way to evade resistance to the drug in Leishmania.

5.3.2 Cytosolic Proteins

A large number of cytosolic proteins are required to regulate cell functions. These cytosolic proteins are involved in the transportation process, transcription, translation, surface presentation of molecules, protein folding, proteolysis, and metabolisms (Kumar et al. 2010). The interconversion of promastigotes to the amastigotes in *Leishmania* species required optimum temperature, pH adaptation against the host cytotoxicity. The amastigote-specific proteins identified in *L. donovani,* such as amastin, heat shock proteins (HSPs) 90 and 100, the A2 protein family, gp63, several proteophosphoglycans, an adenosine transporter, peroxidase, and mitogen-activated protein kinases (MAPKs) are cytosolic. Many proteins, such as ATP-dependent RNA helicase, UPB2, nucleoside diphosphate kinase b, histone H2B, fibrillarin, and DNA helicase, were all involved in nucleic acid metabolism.

5.3.2.1 Heat Shock Proteins

Heat shock protein (HSP) families are evolutionarily conserved cytosolic protein molecules, regulated at both transcriptional and translational levels and predominantly involved in the assembly, folding/refolding, and relocation of proteins across the cellular compartments (Folgueira and Requena 2007). Since HSPs are highly conserved throughout evolution, it is thought that recognizing epitopes shared by HSP from pathogens may provide the immune system with a universal signal for infection (Hunter et al. 1984). There are four different types of HSPs in leishmania: HSP 100, HSP 90, HSP 83, and HSP 70. HSP 100 proteins are expressed under heat stress. During the mammalian infective stage, HSP 90 keeps the cell cycle regulator proteins responsive, and HSP 83 acts as an immunostimulatory molecule. HSP 70 helps to fold newly synthesized proteins and refold misfolded and aggregated proteins (Fig. 5.1) (Hartl and Hayer-Hartl 2002). Hsp70, combined with Gp63, has enhanced immunogenicity and DNA vaccines against *L.donovani* in mice (Kaur et al. 2011; Sachdeva et al. 2009). Importantly, the role of HSP70 in response to antimony resistance has been illustrated and provides valuable target candidates to understand drug resistance for other antileishmanial drugs (Brochu et al. 2004).

5.3.2.2 A2 Protein

A2 protein is a cytosolic amastigote stage-specific protein family, identified in *L. donovani*, necessary for the survival of amastigotes in the visceral organs of mammalian host (Farahmand et al. 2008). At least seven members of the A2 gene family encode multiple copies of amastigote stage-specific proteins ranging from 45 kDa to 100 kDa and play a key role in inducing a strong Th1 immune response, thus conferring partial protection against natural infection.

Immunization of BALB/c mice with adenovirus expressing A2 (AdA2) resulted in a modest antibody response, despite the presence of significant quantities of IFN-producing CD4+ and CD8+ T cells specific for A2. Furthermore, vaccinated mice's A2-specific CD8+ T cells were capable of lysing sensitized target cells in vivo. They also found a link between A2-specific T cell responses and diminished parasitism in the liver and spleens of mice vaccinated with AdA2 and challenged with *L. (L.) chagasi*. Six L. major antigens were tested for putative CD8+ T cell activating 9-mer epitopes provided by HLA-A*0201. There was a specific response to LmsTI-1 and LPG-3-related peptides presented by HLA-A*0201. Specific response to LmsTI-1 and LPG-3-related peptides presented in HLA-A*0201 was demonstrated. Recently, Agallou et al. assessed eight peptides from four known antigenic *L. infantum* proteins, i.e., cysteine peptidase A (CPA), histone H1, KMP-11 and Leishmania eukaryotic initiation factor (LeIF), for their immunogenicity in BALB/c mice where they reported that CPA_p2, CPA_p3, LeIF_p3, and LeIF_p6 induced IFN-γ producing CD4+ T cells indicating a Th1-type response. In addition, CPA_p2, CPA_p3, and H1_p1 were found in CD8+ T cells producing cells.

5.4 Proteomics and Identification of Biomarkers for VL Diagnosis

Biomarkers have provided researchers with the improvised affirmation about the welfare and effectiveness of remedies from which more conclusive clinical data are accumulated. To screen and diagnose VL patients in the endemic area for leishmaniasis, efficient diagnostic techniques and novel biomarkers are needed. Antibodies in blood/serum are often used as markers of exposure/infection. If antibodies can reflect parasite persistence and/or host protection is uncertain. Mouse models have shown that neither B cells or antibodies are not involved in protection. Based on serological data from patients with a history of VL and endemic populations, antibody reactivity to Leishmania antigen over time is variable and probably poor at determining the longevity of immunity.

The rK39, an immunochromatographic test (ICT) that uses a product of the leishmania kinesin-like gene to detect circulating antibodies, was first described in 1993 (Sundar et al. 1998) and is the most widely used and important ICT for VL case confirmation. However, it exhibits lower sensitivity in East Africa when compared to the Indian subcontinent (Singh and Sundar 2015). Antibodies are produced against exposure to leishmania antigen proteins, and these specific antibodies produced in response to any diseased condition can serve as a virulence factor, and thus, they may serve as effective biomarkers. In a recent study of active VL patients assumed to be infected from pathogens *L. infantum* and *L. donovani,* some of the protein antigens are identified in their urine, which is further examined with ELISA assembled with monoclonal antibodies (mAbs); for that selection, generation, and purification of mAbs for respective leishmanial biomarkers Li-isd1(iron superoxide dismutase 1), Li-txn1 (tryparedoxin 1), Li-ntf2 (nuclear transport factor 2), Ld-mao1 (Maoc family dehydratase), Ld-ppi1(peptidyl-prolyl cis-trans isomerase), and Ld-mad1(malate dehydrogenase) were obtained, and these biomarkers were matched with their native protein's molecular weight using Western blot (Abeijon et al. 2020; Maia and Campino 2018). Marlais et al. (2018) analyzed IgG1 levels before and after the treatment and found high concentrations of IgG1 in relapsed cases while low IgG1 levels in cured patients using RDT (Sero K-SeT rapid diagnostic test) and ELISA. Kip et al. (2018) analyzed Neopterin, a macrophage activation marker, against VL. The authors identified pre and posttreatment concentrations of plasma neopterin in VL patients to detect early relapse. Adriansen et al. observed suppressed serum levels of macrophage activators (sCD40L and neopterin) in active VL with HIV patients while elicited levels in *Leishmania spp.* infected asymptomatic individuals.

A whole blood assay based on soluble leishmania antigen (SLA)-specific cytokine response was developed and is currently being used as a promising tool in immunological studies (Singh and Sundar 2014).

Urine proteomics offers an alternative to blood for detecting potential noninvasive biomarkers because urine is a blood filtrate containing protein components similar to those found in the blood (Singh and Sundar 2012). Changes in the urinary proteome can therefore be used to detect small circulating parasitic protein and peptide markers that can pass through glomeruli along with abnormalities within the kidney. Nowadays, antigen detection in the urine is increasingly being used to diagnose parasitic infections. With varied findings, a urinary antigen has recently been demonstrated in VL utilizing a latex agglutination test, the only commercially available urine test (KAtex) for the detection of VL. Low KAtex sensitivity (55.6–67%) has been observed in studies conducted in the Indian subcontinent and many African countries (El-Safi et al. 2003; Sundar et al. 2005). We have also analyzed and compared the total urinary proteome of different subjects by two-dimensional electrophoresis (2D-PAGE) but could not get specific parasitic protein in the patient's urine. Although we did not discover a clear biomarker protein for VL, our findings provided fundamental insights for the future development of noninvasive diagnostic tools (Singh and Sundar 2012).

5.5 Proteomics and Identification of Leishmania Antigens for Vaccine

Only a fraction of infected individuals develop clinical VL in an endemic area of VL, suggesting that many infected individuals develop natural immunity against the disease. Furthermore, most VL patients who have undergone successful drug treatment usually gain lifelong protection (Kaye et al. 2020). Experimental evidence, however, points to the fact that the persistence of parasites is required for protection and that a sterile cure leads to loss of protection. Information about long-term T-cell memory in humans is still minimal. Gaining insight into survival T-cell memory is important for understanding long-term herd immunity and population susceptibility to disease and providing important information regarding vaccines targeting CD4 T cells. (Kaye and Aebischer 2011). Importantly, developing effective vaccines has individual and public health implications. However, there is no effective human vaccine still available against any form of leishmaniasis. An effective vaccine is needed for the control of disease because there is a lack of adequate, affordable, minimally toxic drug therapies. As with any infectious disease, a prophylactic vaccine would be the most effective strategy to prevent the disease. Development of effective recombinant protein vaccines for leishmaniases is possible for the following reasons:

(a) Following treatment cure, most recovered patients develop life-long protection against disease.
(b) Antigenic variations in *Leishmania* are limited.
(c) In human, *Leishmania* parasites reside and proliferate in the macrophage, and there are lot of immunologic research studies on how to improve macrophage killing of harmful organisms.
(d) The disease in the mammalian host is linked with a single morphologic form, the amastigote.
(e) Most harmful types of *Leishmania* species share common key antigens.

Although most known vaccines include dead or attenuated parasites, short amino acid sequences (peptides) are more frequently used to elicit a protective immune response. Recent bioinformatic techniques use a variety of algorithms to predict epitopes, HLA-binding, transporter of antigen processing (TAP) affinity, proteasomal cleavage, and so on to investigate the usage of peptide epitopes with the highest chance of triggering protective immune responses. Synthetic peptide vaccines have several advantages over other vaccine types, including the lack of potentially infectious material, the ability to include multiple epitopes, the reduction of antigen amount and complexity, economic scale-up, and a lower risk of stimulating a response against self-antigens. Several proteins like Glycoprotein

63 (gp63), Kinetoplastid membrane protein 11 (kmp11), Amastigote virulence factor (A2), recombinant-antigens-Leishmun, HSP, MML-MPL, histones, PSA2, LACK, Lipophosphoglycan (LPG), and Cysteine proteinase both from promastigote and amastigote forms were evaluated and screened for development of peptide vaccine.

The first human leishmaniasis vaccine was developed using the recombinant fusion protein Leish-111f (L111f; LEISH-F1; comprising the L. major homologue of eukaryotic thiol–specific antioxidant (TSA), the L. major stress–inducible protein-1 (LmSTI1), and the *L. braziliensis* elongation and initiation factor (LeIF), in combination with MPL formulated in an oil-in-water emulsion (MPL-SE). In a recent small-scale clinical experiment in an *L. donovani* endemic location, Leish-F1-MPL-SE was shown to be safe and well tolerated in patients with and without prior VL exposure, and it generated robust antigen-specific T-cell responses. However, due to the high cost of producing the recombinant protein, this form of vaccination is relatively expensive and inadequate in terms of inducing both CD8+ and CD4+ T cells.

The proteomic technique paired with reverse vaccinology is also being used to find the new vaccine candidates and has so far been effective in identifying novel antigens (88). However, only a few research studies have utilized a wholly reversed method to create and evaluate putative vaccination antigens, beginning with genome, proteome, or transcriptome data (Kaye and Aebischer 2011).

5.6 Conclusion

Much laboratory work has been devoted to identifying differentially expressed Leishmania proteins and assessing their potential use as diagnostic tools or vaccines. With the advancement in biotechnological tools, proteomic researchers hope that their quest to find thousands of novel proteins will identify suitable candidates for either new therapeutic vaccines or developments of diagnostic. Precise identification and characterization of parasite proteins could allow more meaningful approaches to treatment regimens; however, there is a clear need for greater investment in research and development to move promising vaccine leads along the development pathway toward an effective and affordable antileishmanial vaccine. Importantly, until eradication of leishmaniasis (the reduction of transmission to zero worldwide) is achieved, efforts to eliminate *leishmania*sis will continue to require diagnostics/vaccine development strategies as reintroduction will remain possible. The development of next generation of diagnostic tests or vaccines should be affordable and easy to implement in remote, poor, and less developed rural areas where leishmaniasis is endemic.

Acknowledgment Authors would like to thank all research scholars of IDRL, Banaras Hindu University, for their valuable suggestions.

Funding The study was funded by the Extramural Program of the National Institute of Allergy and Infectious Diseases, National Institutes of Health (TMRC grant number U19AI074321), Science &

Engineering Research Board (SERB) [Grant No. SRG/2020/002558, MTR/2022/000980], UGC-BSR Research start-up Grant, and Institute of Eminence (IoE) grant of Banaras Hindu University. The funders had no role in design, decision to publish, or preparation of the report.

Disclosure Statement The authors report there are no competing interests to declare.

References

Abeijon C, Alves F, Monnerat S, Mbui J, Viana AG, Almeida RM, Bueno LL, Fujiwara RT, Campos-Neto A (2020) Urine-based antigen detection assay for diagnosis of visceral leishmaniasis using monoclonal antibodies specific for six protein biomarkers of Leishmania infantum/ Leishmania donovani. PLoS Negl Trop Dis 14:e0008246. https://doi.org/10.1371/journal.pntd.0008246

Aebischer T, Harbecke D, Ilg T (1999) Proteophosphoglycan, a major secreted product of intracellular Leishmania mexicana amastigotes, is a poor B-cell antigen and does not elicit a specific conventional CD4+ T-cell response. Infect Immun 67:5379–5385

Alcolea PJ, Alonso A, García-Tabares F, Mena MDC, Ciordia S, Larraga V (2019) An insight into the constitutive proteome throughout Leishmania donovani Promastigote growth and differentiation. Int Microbiol 22:143–154. https://doi.org/10.1007/s10123-018-00036-2

Araujo-Santos JM, Parodi-Talice A, Castanys S, Gamarro F (2005) The overexpression of an intracellular ABCA-like transporter alters phospholipid trafficking in Leishmania. Biochem Biophys Res Commun 330:349–355. https://doi.org/10.1016/j.bbrc.2005.02.176

Argaman M, Aly R, Shapira M (1994) Expression of heat shock protein 83 in Leishmania is regulated post-transcriptionally. Mol Biochem Parasitol 64:95–110

Ashrafmansouri M, Amiri-Dashatan N, Ahmadi N (2022) Identification of protein profile in metacyclic and amastigote-like stages of Leishmania tropica: a proteomic approach. AMB Express 12:142. https://doi.org/10.1186/s13568-022-01481-z

Bag AK, Saha S, Sundar S, Saha B, Chakrabarti A, Mandal C (2014) Comparative proteomics and glycoproteomics of plasma proteins in Indian visceral leishmaniasis. Proteome Sci 12:48. https://doi.org/10.1186/s12953-014-0048-z

Bandyopadhyay K, Karmakar S, Ghosh A, Das PK (2001) Role of 67 kDa cell surface laminin binding protein of Leishmania donovani in pathogenesis. J Biochem 130:141–148. https://doi.org/10.1093/oxfordjournals.jbchem.a002953

Bandyopadhyay K, Karmakar S, Ghosh A, Das PK (2002) High affinity binding between laminin and laminin binding protein of Leishmania is stimulated by zinc and may involve laminin zinc-finger like sequences. Eur J Biochem 269:1622–1629. https://doi.org/10.1046/j.1432-1327.2002.02793.x

Bandyopadhyay K, Karmakar S, Biswas A, Das PK (2003) Membrane orientation of laminin binding protein. Eur J Biochem 270:3806–3813. https://doi.org/10.1046/j.1432-1033.2003.03768.x

Bañuls A, Hide M, Prugnolle F (2007) *Leishmania* and the Leishmaniases: a parasite genetic update and advances in taxonomy, epidemiology and pathogenicity in humans. In: Baker RMJR, Rollinson D (eds) Advances in parasitology. Academic Press, pp 1–458

Basu R, Roy S, Walden P (2007) HLA class I-restricted T cell epitopes of the kinetoplastid membrane protein-11 presented by Leishmania donovani-infected human macrophages. J Infect Dis 195:1373–1380. https://doi.org/10.1086/513439

Bente M, Harder S, Wiesgigl M, Heukeshoven J, Gelhaus C, Krause E, Clos J, Bruchhaus I (2003) Developmentally induced changes of the proteome in the protozoan parasite Leishmania donovani. Proteomics 3:1811–1829. https://doi.org/10.1002/pmic.200300462

Berberich C, Requena JM, Alonso C (1997) Cloning of genes and expression and antigenicity analysis of the Leishmania infantum KMP-11 protein. Exp Parasitol 85:105–108. https://doi.org/10.1006/expr.1996.4120

Bhowmick S, Ravindran R, Ali N (2008) gp63 in stable cationic liposomes confers sustained vaccine immunity to susceptible BALB/c mice infected with Leishmania donovani. Infect Immun 76:1003–1015. https://doi.org/10.1128/IAI.00611-07

Blackwell JM, Fakiola M, Castellucci LC (2020) Human genetics of leishmania infections. Hum Genet 139:813–819. https://doi.org/10.1007/s00439-020-02130-w

Boelaert M, Meheus F, Sanchez A, Singh SP, Vanlerberghe V, Picado A, Meessen B, Sundar S (2009) The poorest of the poor: a poverty appraisal of households affected by visceral leishmaniasis in Bihar, India. Tropical Med Int Health 14:639–644. https://doi.org/10.1111/j.1365-3156.2009.02279.x

Brittingham A, Morrison CJ, McMaster WR, McGwire BS, Chang KP, Mosser DM (1995) Role of the Leishmania surface protease gp63 in complement fixation, cell adhesion, and resistance to complement-mediated lysis. J Immunol 155:3102–3111

Brittingham A, Chen G, McGwire BS, Chang KP, Mosser DM (1999) Interaction of Leishmania gp63 with cellular receptors for fibronectin. Infect Immun 67:4477–4484

Brochu C, Haimeur A, Ouellette M (2004) The heat shock protein HSP70 and heat shock cognate protein HSC70 contribute to antimony tolerance in the protozoan parasite leishmania. Cell Stress Chaperones 9:294–303. https://doi.org/10.1379/csc-15r1.1

Brotherton MC, Racine G, Foucher AL, Drummelsmith J, Papadopoulou B, Ouellette M (2010) Analysis of stage-specific expression of basic proteins in Leishmania infantum. J Proteome Res 9:3842–3853. https://doi.org/10.1021/pr100048m

Bucheton B, Kheir M, El-Safi S, Hammad A, Mergani A, Mary C, Abel L, Dessein A (2002) The interplay between environmental and host factors during an outbreak of visceral leishmaniasis in eastern Sudan. Microbes Infect 4:1449–1457. https://doi.org/10.1016/s1286-4579(02)00027-8

Burza S, Croft SL, Boelaert M (2018) Leishmaniasis. Lancet 392:951–970. https://doi.org/10.1016/S0140-6736(18)31204-2

Button LL, McMaster WR (1988) Molecular cloning of the major surface antigen of leishmania. J Exp Med 167:724–729. https://doi.org/10.1084/jem.167.2.724

Button LL, Russell DG, Klein HL, Medina-Acosta E, Karess RE, McMaster WR (1989) Genes encoding the major surface glycoprotein in Leishmania are tandemly linked at a single chromosomal locus and are constitutively transcribed. Mol Biochem Parasitol 32:271–283

Carvalho FA, Charest H, Tavares CA, Matlashewski G, Valente EP, Rabello A, Gazzinelli RT, Fernandes AP (2002) Diagnosis of American visceral leishmaniasis in humans and dogs using the recombinant Leishmania donovani A2 antigen. Diagn Microbiol Infect Dis 43:289–295

Chappuis F, Sundar S, Hailu A, Ghalib H, Rijal S, Peeling RW, Alvar J, Boelaert M (2007) Visceral leishmaniasis: what are the needs for diagnosis, treatment and control? Nat Rev Microbiol 5: 873–882. https://doi.org/10.1038/nrmicro1748

Charest H, Matlashewski G (1994) Developmental gene expression in Leishmania donovani: differential cloning and analysis of an amastigote-stage-specific gene. Mol Cell Biol 14:2975–2984. https://doi.org/10.1128/mcb.14.5.2975

Clos J, Klaholz L, Kroemer M, Krobitsch S, Lindquist S (2001) Heat shock protein 100 and the amastigote stage-specific A2 proteins of Leishmania donovani. Med Microbiol Immunol 190: 47–50

Coelho AC, Yamashiro-Kanashiro EH, Bastos SF, Mortara RA, Cotrim PC (2006) Intracellular location of the ABC transporter PRP1 related to pentamidine resistance in Leishmania major. Mol Biochem Parasitol 150:378–383. https://doi.org/10.1016/j.molbiopara.2006.08.013

Croft SL, Sundar S, Fairlamb AH (2006) Drug resistance in leishmaniasis. Clin Microbiol Rev 19: 111–126. 19/1/111 [pii]. https://doi.org/10.1128/CMR.19.1.111-126.2006

de Araujo Soares RM, dos Santos AL, Bonaldo MC, de Andrade AF, Alviano CS, Angluster J, Goldenberg S (2003) Leishmania (Leishmania) amazonensis: differential expression of proteinases and cell-surface polypeptides in avirulent and virulent promastigotes. Exp Parasitol 104:104–112

Drummelsmith J, Brochu V, Girard I, Messier N, Ouellette M (2003) Proteome mapping of the protozoan parasite Leishmania and application to the study of drug targets and resistance mechanisms. Mol Cell Proteomics 2:146–155. https://doi.org/10.1074/mcp.M200085-MCP200

Dwyer DM (1974) Lectin binding saccharides on a parasite protozoan. Science 184:471–474

Ejazi SA, Bhattacharyya A, Choudhury ST, Ghosh S, Sabur A, Pandey K, Das VNR, Das P, Rahaman M, Goswami RP, Ali N (2018) Immunoproteomic identification and characterization of Leishmania membrane proteins as non-invasive diagnostic candidates for clinical visceral Leishmaniasis. Sci Rep 8:12110. https://doi.org/10.1038/s41598-018-30546-y

El Fakhry Y, Ouellette M, Papadopoulou B (2002) A proteomic approach to identify developmentally regulated proteins in Leishmania infantum. Proteomics 2:1007–1017. https://doi.org/10.1002/1615-9861(200208)2:8<1007::AID-PROT1007>3.0.CO;2-G

Elfaki ME, Khalil EA, De Groot AS, Musa AM, Gutierrez A, Younis BM, Salih KA, El-Hassan AM (2012) Immunogenicity and immune modulatory effects of in silico predicted L. donovani candidate peptide vaccines. Hum Vaccin Immunother 8:1769–1774. https://doi.org/10.4161/hv.21881

El-Safi SH, Abdel-Haleem A, Hammad A, El-Basha I, Omer A, Kareem HG, Boelaert M, Chance M, Hommel M (2003) Field evaluation of latex agglutination test for detecting urinary antigens in visceral leishmaniasis in Sudan. East Mediterr Health J 9:844–855

Engels D, Zhou X-N (2020) Neglected tropical diseases: an effective global response to local poverty-related disease priorities. Infect Dis Poverty 9:10. https://doi.org/10.1186/s40249-020-0630-9

Farahmand M, Nahrevanian H, Assmar M, Mohebali M, Zarei Z (2008) Expression of A2 proteins in amastigotes of Leishmania infantum produced from canine isolates collected in the district of Meshkinshahr, in North-Western Iran. Ann Trop Med Parasitol 102:81–84. https://doi.org/10.1179/136485908X252160

Folgueira C, Requena JM (2007) A postgenomic view of the heat shock proteins in kinetoplastids. FEMS Microbiol Rev 31:359–377. https://doi.org/10.1111/j.1574-6976.2007.00069.x

Folgueira C, Canavate C, Chicharro C, Requena JM (2007) Genomic organization and expression of the HSP70 locus in new and Old World Leishmania species. Parasitology 134:369–377. https://doi.org/10.1017/S0031182006001570

Funk VA, Thomas-Oates JE, Kielland SL, Bates PA, Olafson RW (1997) A unique, terminally glucosylated oligosaccharide is a common feature on Leishmania cell surfaces. Mol Biochem Parasitol 84:33–48

Garcia GC, Carvalho A, Duarte MC, Silva M, Medeiros FAC, Coelho EAF, de Moura Franco DM, Gonçalves DU, de Oliveira Mendes TA, Menezes-Souza D (2021) Development of a chimeric protein based on a proteomic approach for the serological diagnosis of human tegumentary leishmaniasis. Appl Microbiol Biotechnol 105:6805–6817. https://doi.org/10.1007/s00253-021-11518-1

Ghosh A, Kole L, Bandyopadhyay K, Sarkar K, Das PK (1996) Evidence of a laminin binding protein on the surface of Leishmania donovani. Biochem Biophys Res Commun 226:101–106. https://doi.org/10.1006/bbrc.1996.1317

Ghosh A, Bandyopadhyay K, Kole L, Das PK (1999) Isolation of a laminin-binding protein from the protozoan parasite Leishmania donovani that may mediate cell adhesion. Biochem J 337 (Pt 3):551–558

Hartl FU, Hayer-Hartl M (2002) Molecular chaperones in the cytosol: from nascent chain to folded protein. Science 295(5561):1852–1858

Henderson DM, Sifri CD, Rodgers M, Wirth DF, Hendrickson N, Ullman B (1992) Multidrug resistance in Leishmania donovani is conferred by amplification of a gene homologous to the mammalian mdr1 gene. Mol Cell Biol 12:2855–2865. https://doi.org/10.1128/mcb.12.6.2855

Higgins CF (1992) ABC transporters: from microorganisms to man. Annu Rev Cell Biol 8:67–113

Ho M, Siongok T, Lyerly W, Smith D (1982) Prevalence and disease spectrum in a new focus of visceral leishmaniasis in Kenya. Trans R Soc Trop Med Hyg 76:741–746

Hubel A, Clos J (1996) The genomic organization of the HSP83 gene locus is conserved in three Leishmania species. Exp Parasitol 82:225–228. https://doi.org/10.1006/expr.1996.0029
Hubel A, Krobitsch S, Horauf A, Clos J (1997) Leishmania major Hsp100 is required chiefly in the mammalian stage of the parasite. Mol Cell Biol 17:5987–5995. https://doi.org/10.1128/mcb.17.10.5987
Hunter KW, Cook CL, Hayunga EG (1984) Leishmanial differentiation in vitro: induction of heat shock proteins. Biochem Biophys Res Commun 125:755–760. https://doi.org/10.1016/0006-291x(84)90603-x
Ilg T (2000) Proteophosphoglycans of Leishmania. Parasitol Today 16:489–497
Ilg T, Stierhof YD, Etges R, Adrian M, Harbecke D, Overath P (1991) Secreted acid phosphatase of Leishmania mexicana: a filamentous phosphoglycoprotein polymer. Proc Natl Acad Sci U S A 88:8774–8778. https://doi.org/10.1073/pnas.88.19.8774
Ilg T, Overath P, Ferguson MA, Rutherford T, Campbell DG, McConville MJ (1994a) O- and N-glycosylation of the Leishmania mexicana-secreted acid phosphatase. Characterization of a new class of phosphoserine-linked glycans. J Biol Chem 269:24073–24081
Ilg T, Stierhof YD, Wiese M, McConville MJ, Overath P (1994b) Characterization of phosphoglycan-containing secretory products of Leishmania. Parasitology 108(Suppl):S63–S71
Ilg T, Stierhof YD, McConville MJ, Overath P (1995) Purification, partial characterization and immunolocalization of a proteophosphoglycan secreted by Leishmania mexicana amastigotes. Eur J Cell Biol 66:205–215
Ilg T, Handman E, Stierhof YD (1999a) Proteophosphoglycans from Leishmania promastigotes and amastigotes. Biochem Soc Trans 27:518–525. https://doi.org/10.1042/bst0270518
Ilg T, Montgomery J, Stierhof YD, Handman E (1999b) Molecular cloning and characterization of a novel repeat-containing Leishmania major gene, ppg1, that encodes a membrane-associated form of proteophosphoglycan with a putative glycosylphosphatidylinositol anchor. J Biol Chem 274:31410–31420. https://doi.org/10.1074/jbc.274.44.31410
Ivens AC, Blackwell JM (1999) The Leishmania genome comes of age. Parasitol Today 15:225–231
Ivens AC, Peacock CS, Worthey EA, Murphy L, Aggarwal G, Berriman M, Sisk E, Rajandream MA, Adlem E, Aert R et al (2005) The genome of the kinetoplastid parasite, Leishmania major. Science 309:436–442. https://doi.org/10.1126/science.1112680
Jardim A, Tolson DL, Turco SJ, Pearson TW, Olafson RW (1991) The Leishmania donovani lipophosphoglycan T lymphocyte-reactive component is a tightly associated protein complex. J Immunol 147:3538–3544
Jardim A, Funk V, Caprioli RM, Olafson RW (1995a) Isolation and structural characterization of the Leishmania donovani kinetoplastid membrane protein-11, a major immunoreactive membrane glycoprotein. Biochem J 305(Pt 1):307–313. https://doi.org/10.1042/bj3050307
Jardim A, Hanson S, Ullman B, McCubbin WD, Kay CM, Olafson RW (1995b) Cloning and structure-function analysis of the Leishmania donovani kinetoplastid membrane protein-11. Biochem J 305(Pt 1):315–320. https://doi.org/10.1042/bj3050315
Kala-Azar elimination programme: report of a WHO consultation of partners, Geneva, Switzerland, 10–11 February 2015. https://www.who.int/publications/i/item/9789241509497
Kamhawi S (2006) Phlebotomine sand flies and Leishmania parasites: friends or foes? Trends Parasitol 22:439–445. https://doi.org/10.1016/j.pt.2006.06.012
Karim M, Singh G, Thakur S, Rana A, Rub A, Akhter Y (2022) Evaluating complete surface-associated and secretory proteome of Leishmania donovani for discovering novel vaccines and diagnostic targets. Arch Microbiol 204:604. https://doi.org/10.1007/s00203-022-03219-8
Kaur T, Sobti RC, Kaur S (2011) Cocktail of gp63 and Hsp70 induces protection against Leishmania donovani in BALB/c mice. Parasite Immunol 33:95–103. https://doi.org/10.1111/j.1365-3024.2010.01253.x
Kaye PM, Aebischer T (2011) Visceral leishmaniasis: immunology and prospects for a vaccine. Clin Microbiol Infect 17:1462–1470. https://doi.org/10.1111/j.1469-0691.2011.03610.x
Kaye P, Scott P (2011) Leishmaniasis: complexity at the host-pathogen interface. Nat Rev Microbiol 9:604–615. https://doi.org/10.1038/nrmicro2608

Kaye PM, Cruz I, Picado A, Van Bocxlaer K, Croft SL (2020) Leishmaniasis immunopathology-impact on design and use of vaccines, diagnostics and drugs. Semin Immunopathol 42:247–264. https://doi.org/10.1007/s00281-020-00788-y

Kemp M, Theander TG, Handman E, Hey AS, Kurtzhals JA, Hviid L, Sorensen AL, Were JO, Koech DK, Kharazmi A (1991) Activation of human T lymphocytes by Leishmania lipophosphoglycan. Scand J Immunol 33:219–224

Khan MAA, Ami JQ, Faisal K, Chowdhury R, Ghosh P, Hossain F, Abd El Wahed A, Mondal D (2020) An immunoinformatic approach driven by experimental proteomics: in silico design of a subunit candidate vaccine targeting secretory proteins of Leishmania donovani amastigotes. Parasit Vectors 13:196. https://doi.org/10.1186/s13071-020-04064-8

Kip AE, Wasunna M, Alves F, Schellens JHM, Beijnen JH, Musa AM, Khalil EAG, Dorlo TPC (2018) Macrophage activation marker Neopterin: a candidate biomarker for treatment response and relapse in visceral Leishmaniasis. Front Cell Infect Microbiol 8:181. https://doi.org/10.3389/fcimb.2018.00181

Krobitsch S, Clos J (1999) A novel role for 100 kD heat shock proteins in the parasite Leishmania donovani. Cell Stress Chaperones 4:191–198. https://doi.org/10.1379/1466-1268(1999)004<0191:anrfkh>2.3.co;2

Krobitsch S, Brandau S, Hoyer C, Schmetz C, Hubel A, Clos J (1998) Leishmania donovani heat shock protein 100. Characterization and function in amastigote stage differentiation. J Biol Chem 273:6488–6494. https://doi.org/10.1074/jbc.273.11.6488

Kumar A, Sisodia B, Misra P, Sundar S, Shasany AK, Dube A (2010) Proteome mapping of overexpressed membrane-enriched and cytosolic proteins in sodium antimony gluconate (SAG) resistant clinical isolate of Leishmania donovani. Br J Clin Pharmacol 70:609–617. https://doi.org/10.1111/j.1365-2125.2010.03716.x

Kushawaha PK, Gupta R, Tripathi CD, Khare P, Jaiswal AK, Sundar S, Dube A (2012) Leishmania donovani triose phosphate isomerase: a potential vaccine target against visceral leishmaniasis. PLoS One 7:e45766. https://doi.org/10.1371/journal.pone.0045766

Kushwaha AK, Scorza BM, Singh OP, Rowton E, Lawyer P, Sundar S, Petersen CA (2022) Domestic mammals as reservoirs for Leishmania donovani on the Indian subcontinent: possibility and consequences on elimination. Transbound Emerg Dis 69:268–277. https://doi.org/10.1111/tbed.14061

Lee MG, Atkinson BL, Giannini SH, Van der Ploeg LH (1988) Structure and expression of the hsp 70 gene family of Leishmania major. Nucleic Acids Res 16:9567–9585. https://doi.org/10.1093/nar/16.20.9567

Legare D, Cayer S, Singh AK, Richard D, Papadopoulou B, Ouellette M (2001) ABC proteins of Leishmania. J Bioenerg Biomembr 33:469–474

MacFarlane J, Blaxter ML, Bishop RP, Miles MA, Kelly JM (1990) Identification and characterisation of a Leishmania donovani antigen belonging to the 70-kDa heat-shock protein family. Eur J Biochem 190:377–384. https://doi.org/10.1111/j.1432-1033.1990.tb15586.x

Maia C, Campino L (2018) Biomarkers associated with Leishmania infantum exposure, infection, and disease in dogs. Front Cell Infect Microbiol 8:302. https://doi.org/10.3389/fcimb.2018.00302

Marlais T, Bhattacharyya T, Singh OP, Mertens P, Gilleman Q, Thunissen C, Hinckel BCB, Pearson C, Gardner BL, Airs S et al (2018) Visceral Leishmaniasis IgG1 rapid monitoring of cure vs. relapse, and potential for diagnosis of post Kala-Azar dermal Leishmaniasis. Front Cell Infect Microbiol 8:427. https://doi.org/10.3389/fcimb.2018.00427

Matlashewski G (2001) Leishmania infection and virulence. Med Microbiol Immunol 190:37–42

McCall LI, Matlashewski G (2010) Localization and induction of the A2 virulence factor in Leishmania: evidence that A2 is a stress response protein. Mol Microbiol 77:518–530. MMI7229 [pii]. https://doi.org/10.1111/j.1365-2958.2010.07229.x

McConville MJ, Ralton JE (1997) Developmentally regulated changes in the cell surface architecture of Leishmania parasites. Behring Inst Mitt 99:34–43

Mendonca SC, Russell DG, Coutinho SG (1991) Analysis of the human T cell responsiveness to purified antigens of Leishmania: lipophosphoglycan (LPG) and glycoprotein 63 (gp 63). Clin Exp Immunol 83:472–478. https://doi.org/10.1111/j.1365-2249.1991.tb05663.x

Mukhopadhyay S, Mandal C (2006) Glycobiology of Leishmania donovani. Indian J Med Res 123: 203–220

Murray HW, Berman JD, Davies CR, Saravia NG (2005) Advances in leishmaniasis. Lancet 366: 1561–1577. https://doi.org/10.1016/S0140-6736(05)67629-5

Nugent PG, Karsani SA, Wait R, Tempero J, Smith DF (2004) Proteomic analysis of Leishmania mexicana differentiation. Mol Biochem Parasitol 136:51–62. https://doi.org/10.1016/j.molbiopara.2004.02.009

Okong'o-Odera EA, Kurtzhals JA, Hey AS, Kharazmi A (1993) Measurement of serum antibodies against native Leishmania gp63 distinguishes between ongoing and previous L. donovani infection. APMIS 101:642–646

Olafson RW, Thomas JR, Ferguson MA, Dwek RA, Chaudhuri M, Chang KP, Rademacher TW (1990) Structures of the N-linked oligosaccharides of Gp63, the major surface glycoprotein, from Leishmania mexicana amazonensis. J Biol Chem 265:12240–12247

Olivier M, Atayde VD, Isnard A, Hassani K, Shio MT (2012) Leishmania virulence factors: focus on the metalloprotease GP63. Microbes Infect 14:1377–1389. https://doi.org/10.1016/j.micinf.2012.05.014

Omara-Opyene AL, Gedamu L (1997) Molecular cloning, characterization and overexpression of two distinct cysteine protease cDNAs from Leishmania donovani chagasi. Mol Biochem Parasitol 90:247–267

Ouellette M, Legare D, Papadopoulou B (1994) Microbial multidrug-resistance ABC transporters. Trends Microbiol 2:407–411

Parodi-Talice A, Araujo JM, Torres C, Perez-Victoria JM, Gamarro F, Castanys S (2003) The overexpression of a new ABC transporter in Leishmania is related to phospholipid trafficking and reduced infectivity. Biochim Biophys Acta 1612:195–207. https://doi.org/10.1016/s0005-2736(03)00131-7

Perez-Victoria JM, Parodi-Talice A, Torres C, Gamarro F, Castanys S (2001) ABC transporters in the protozoan parasite Leishmania. Int Microbiol 4:159–166. https://doi.org/10.1007/s10123-001-0031-2

Peters C, Kawakami M, Kaul M, Ilg T, Overath P, Aebischer T (1997a) Secreted proteophosphoglycan of Leishmania mexicana amastigotes activates complement by triggering the mannan binding lectin pathway. Eur J Immunol 27:2666–2672. https://doi.org/10.1002/eji.1830271028

Peters C, Stierhof YD, Ilg T (1997b) Proteophosphoglycan secreted by Leishmania mexicana amastigotes causes vacuole formation in macrophages. Infect Immun 65:783–786

Pink R, Hudson A, Mouries MA, Bending M (2005) Opportunities and challenges in antiparasitic drug discovery. Nat Rev Drug Discov 4(9):727–740

Ramamoorthy R, Donelson JE, Paetz KE, Maybodi M, Roberts SC, Wilson ME (1992) Three distinct RNAs for the surface protease gp63 are differentially expressed during development of Leishmania donovani chagasi promastigotes to an infectious form. J Biol Chem 267:1888–1895

Rashidi S, Kalantar K, Hatam G (2018) Using proteomics as a powerful tool to develop a vaccine against Mediterranean visceral leishmaniasis. J Parasit Dis 42:162–170. https://doi.org/10.1007/s12639-018-0986-y

Rauch CA, Perez-Morga D, Cozzaelli NR, Englund PT (1993) The absence of supercoiling in kinetoplast DNA minicircle. EMBO J 12:403–411

Roach TI, Kiderlen AF, Blackwell JM (1991) Role of inorganic nitrogen oxides and tumor necrosis factor alpha in killing Leishmania donovani amastigotes in gamma interferon-lipopolysaccharide-activated macrophages from Lshs and Lshr congenic mouse strains. Infect Immun 59:3935–3944

Rochette A, Raymond F, Ubeda JM, Smith M, Messier N, Boisvert S, Rigault P, Corbeil J, Ouellette M, Papadopoulou B (2008) Genome-wide gene expression profiling analysis of Leishmania major and Leishmania infantum developmental stages reveals substantial differences between the two species. BMC Genomics 9:255. https://doi.org/10.1186/1471-2164-9-255

Rochette A, Raymond F, Corbeil J, Ouellette M, Papadopoulou B (2009) Whole-genome comparative RNA expression profiling of axenic and intracellular amastigote forms of Leishmania infantum. Mol Biochem Parasitol 165:32–47. https://doi.org/10.1016/j.molbiopara.2008.12.012

Rogers ME, Ilg T, Nikolaev AV, Ferguson MA, Bates PA (2004) Transmission of cutaneous leishmaniasis by sand flies is enhanced by regurgitation of fPPG. Nature 430:463–467. https://doi.org/10.1038/nature02675

Routaray CB, Choudhari V, Prakash D, Patil R, Jagtap S, Bai S, Kulkarni MJ, Kuchipudi SV, Pai K (2022a) Quantitative proteomic analysis reveals differential modulation of crucial stage specific proteins during promastigote to amastigote differentiation in Leishmania donovani. Journal of Proteins and Proteomics 13:17–27. https://doi.org/10.1007/s42485-021-00080-z

Routaray CB, Kumar A, Sundar S, Sathe G, Pawar H, Pai K (2022b) An in-depth proteomic map of Leishmania donovani isolate from post Kala-azar dermal Leishmaniasis (PKDL) patient. Acta Parasitol 67:687–696. https://doi.org/10.1007/s11686-021-00511-3

Russo DM, Turco SJ, Burns JM Jr, Reed SG (1992) Stimulation of human T lymphocytes by Leishmania lipophosphoglycan-associated proteins. J Immunol 148:202–207

Sachdeva R, Banerjea AC, Malla N, Dubey ML (2009) Immunogenicity and efficacy of single antigen Gp63, polytope and polytopeHSP70 DNA vaccines against visceral Leishmaniasis in experimental mouse model. PLoS One 4:e7880. https://doi.org/10.1371/journal.pone.0007880

Sacks D, Lal S, Shrivastava S, Blackwell J, Neva F (1987) An analysis of T cell responsiveness in Indian kala-azar. J Immunol 138:908–913

Schlagenhauf E, Etges R, Metcalf P (1998) The crystal structure of the Leishmania major surface proteinase leishmanolysin (gp63). Structure 6:1035–1046

Schneider P, Bordier C, Etges R (1992) Membrane proteins and enzymes of Leishmania. Subcell Biochem 18:39–72

Secundino N, Kimblin N, Peters NC, Lawyer P, Capul AA, Beverley SM, Turco SJ, Sacks D (2010) Proteophosphoglycan confers resistance of Leishmania major to midgut digestive enzymes induced by blood feeding in vector sand flies. Cell Microbiol 12:906–918. https://doi.org/10.1111/j.1462-5822.2010.01439.x

Selzer PM, Pingel S, Hsieh I, Ugele B, Chan VJ, Engel JC, Bogyo M, Russell DG, Sakanari JA, McKerrow JH (1999) Cysteine protease inhibitors as chemotherapy: lessons from a parasite target. Proc Natl Acad Sci U S A 96:11015–11022. https://doi.org/10.1073/pnas.96.20.11015

Shapira M, Pinelli E (1989) Heat-shock protein 83 of Leishmania mexicana amazonensis is an abundant cytoplasmic protein with a tandemly repeated genomic arrangement. Eur J Biochem 185:231–236. https://doi.org/10.1111/j.1432-1033.1989.tb15107.x

Shapiro TA, Englund PT (1995) The structure and replication of kinetoplast DNA. Annu Rev Microbiol 49:117–143

Singh N (2006) Drug resistance mechanisms in clinical isolates of Leishmania donovani. Indian J Med Res 123:411–422

Singh OP, Sundar S (2012) Analysis of Total urine proteins: towards a non-invasive approach for diagnosis of visceral Leishmaniasis. J Mol Biomark Daign 3:131. https://doi.org/10.4172/2155-9929.1000131

Singh OP, Sundar S (2014) Whole blood assay and visceral leishmaniasis: challenges and promises. Immunobiology 219:323–328. https://doi.org/10.1016/j.imbio.2014.01.005

Singh OP, Sundar S (2015) Developments in diagnosis of visceral Leishmaniasis in the elimination era. J Parasitol Res 2015:239469. https://doi.org/10.1155/2015/239469

Singh OP, Sundar S (2022) Visceral leishmaniasis elimination in India: progress and the road ahead. Expert Rev Anti-Infect Ther 20:1381–1388. https://doi.org/10.1080/14787210.2022.2126352

Singh, O.P., Stober, C.B., Singh, A.K., Blackwell, J.M., and Sundar, S. (2012). Cytokine responses to novel antigens in an Indian population living in an area endemic for of *visceral leishmaniasis* PLoS Negl trop dis *in press*

Singh OP, Hasker E, Sacks D, Boelaert M, Sundar S (2014) Asymptomatic Leishmania infection: a new challenge for Leishmania control. Clin Infect Dis 58:1424–1429. https://doi.org/10.1093/cid/ciu102

Singh OP, Hasker E, Boelaert M, Sundar S (2016a) Elimination of visceral leishmaniasis on the Indian subcontinent. Lancet Infect Dis 16:e304–e309. https://doi.org/10.1016/s1473-3099(16)30140-2

Singh OP, Singh B, Chakravarty J, Sundar S (2016b) Current challenges in treatment options for visceral leishmaniasis in India: a public health perspective. Infect Dis Poverty 5:19. https://doi.org/10.1186/s40249-016-0112-2

Singh OP, Hasker E, Boelaert M, Sacks D, Sundar S (2020) Xenodiagnosis to address key questions in visceral leishmaniasis control and elimination. PLoS Negl Trop Dis 14:e0008363. https://doi.org/10.1371/journal.pntd.0008363

Singh OP, Syn G, Nylen S, Engwerda C, Sacks D, Wilson ME, Kumar R, Chakravarty J, Sundar S, Blackwell JM, Fakiola M (2021) Anti-Interleukin-10 unleashes transcriptional response to Leishmanial antigens in visceral Leishmaniasis patients. J Infect Dis 223:517–521. https://doi.org/10.1093/infdis/jiaa381

Siqueira-Neto JL, Debnath A, McCall LI, Bernatchez JA, Ndao M, Reed SL, Rosenthal PJ (2018) Cysteine proteases in protozoan parasites. PLoS Negl Trop Dis 12:e0006512. https://doi.org/10.1371/journal.pntd.0006512

Souza AE, Waugh S, Coombs GH, Mottram JC (1992) Characterization of a multi-copy gene for a major stage-specific cysteine proteinase of Leishmania mexicana. FEBS Lett 311:124–127. https://doi.org/10.1016/0014-5793(92)81382-v

Stanley A, Engwerda C (2007) Balancing immunity and pathology in visceral leishmaniasis. Immunol Cell Biol 85:138–147. https://doi.org/10.1038/sj.icb.7100011

Stierhof YD, Bates PA, Jacobson RL, Rogers ME, Schlein Y, Handman E, Ilg T (1999) Filamentous proteophosphoglycan secreted by Leishmania promastigotes forms gel-like three-dimensional networks that obstruct the digestive tract of infected sandfly vectors. Eur J Cell Biol 78:675–689. https://doi.org/10.1016/S0171-9335(99)80036-3

Sundar S, More DK, Singh MK, Singh VP, Sharma S, Makharia A, Kumar PC, Murray HW. (2000) Failure of pentavalent antimony in visceral leishmaniasis in India: report from the center of the Indian epidemic. Clin Infect Dis. 2000 Oct;31(4):1104–7. https://doi.org/10.1086/318121.

Sundar S (2001) Drug resistance in Indian visceral leishmaniasis. Tropical Med Int Health 6:849–854

Sundar S, Rai M (2002) Laboratory diagnosis of visceral leishmaniasis. Clin Diagn Lab Immunol 9:951–958. https://doi.org/10.1128/cdli.9.5.951-958.2002

Sundar S, Singh OP (2018) Molecular diagnosis of visceral Leishmaniasis. Mol Diagn Ther 22:443–457. https://doi.org/10.1007/s40291-018-0343-y

Sundar S, Reed SG, Singh VP, Kumar PC, Murray HW (1998) Rapid accurate field diagnosis of Indian visceral leishmaniasis. Lancet 351:563–565. https://doi.org/10.1016/S0140-6736(97)04350-X

Sundar S, Agrawal S, Pai K, Chance M, Hommel M (2005) Detection of leishmanial antigen in the urine of patients with visceral leishmaniasis by a latex agglutination test. Am J Trop Med Hyg 73:269–271

Sundar S, Singh OP, Chakravarty J (2018) Visceral leishmaniasis elimination targets in India, strategies for preventing resurgence. Expert Rev Anti-Infect Ther 16:805–812. https://doi.org/10.1080/14787210.2018.1532790

Tasbihi M, Shekari F, Hajjaran H, Khanmohammadi M, Hadighi R (2020) Comparative mitochondrial proteomics of Leishmania tropica clinical isolates resistant and sensitive to meglumine antimoniate. Parasitol Res 119:1857–1871. https://doi.org/10.1007/s00436-020-06671-x

Thiel M, Bruchhaus I (2001) Comparative proteome analysis of Leishmania donovani at different stages of transformation from promastigotes to amastigotes. Med Microbiol Immunol 190:33–36

Vannier-Santos MA, Martiny A, de Souza W (2002) Cell biology of Leishmania spp.: invading and evading. Curr Pharm Des 8:297–318. https://doi.org/10.2174/1381612023396230

Walker J, Vasquez JJ, Gomez MA, Drummelsmith J, Burchmore R, Girard I, Ouellette M (2006) Identification of developmentally-regulated proteins in Leishmania panamensis by proteome

profiling of promastigotes and axenic amastigotes. Mol Biochem Parasitol 147:64–73. https://doi.org/10.1016/j.molbiopara.2006.01.008
Wei FR, Gao CH, Wang JY, Yang YT, Shi F, Zheng B (2021) Label-free quantitative proteomic analysis of three strains of Viscerotropic Leishmania isolated from patients with different epidemiological types of visceral Leishmaniasis in China. Acta Parasitol 66:1366–1386. https://doi.org/10.1007/s11686-021-00387-3
WHO. (n.d.). https://www.who.int/health-topics/neglected-tropical-diseases#tab=tab_1
Wiesgigl M, Clos J (2001) Heat shock protein 90 homeostasis controls stage differentiation in Leishmania donovani. Mol Biol Cell 12:3307–3316. https://doi.org/10.1091/mbc.12.11.3307
Yao C, Li Y, Donelson JE, Wilson ME (2010) Proteomic examination of Leishmania chagasi plasma membrane proteins: contrast between avirulent and virulent (metacyclic) parasite forms. Proteomics Clin Appl 4:4–16. https://doi.org/10.1002/prca.200900050
Zhang WW, Matlashewski G (1997) Loss of virulence in Leishmania donovani deficient in an amastigote-specific protein, A2. Proc Natl Acad Sci U S A 94:8807–8811. https://doi.org/10.1073/pnas.94.16.8807
Zhang WW, Charest H, Ghedin E, Matlashewski G (1996) Identification and overexpression of the A2 amastigote-specific protein in Leishmania donovani. Mol Biochem Parasitol 78:79–90
Zhang WW, Mendez S, Ghosh A, Myler P, Ivens A, Clos J, Sacks DL, Matlashewski G (2003) Comparison of the A2 gene locus in Leishmania donovani and Leishmania major and its control over cutaneous infection. J Biol Chem 278:35508–35515. https://doi.org/10.1074/jbc.M305030200
Zheng Z, Chen J, Ma G, Satoskar AR, Li J (2020) Integrative genomic, proteomic and phenotypic studies of Leishmania donovani strains revealed genetic features associated with virulence and antimony-resistance. Parasit Vectors 13:510. https://doi.org/10.1186/s13071-020-04397-4

Post Kala-Azar Dermal Leishmaniasis: Diagnosis and Treatment

6

Mudsser Azam, V. Ramesh, Poonam Salotra, and Ruchi Singh

Abstract

Post kala-azar dermal leishmaniasis (PKDL) cases play a vital role in maintaining the parasite reservoir during interepidemic periods of visceral leishmaniasis. Timely diagnosis and treatment of PKDL cases are paramount in achieving sustained kala-azar elimination. Diagnosis of PKDL remains a major challenge due to sparse parasite loads, especially in macular cases. A substantial number of PKDL cases are misdiagnosed at primary health centers. Conventional methods of parasite detection such as stained smears, histopathology, and culture have low sensitivity and require invasive procedures. Due to the persistence of antileishmanial antibodies after VL, serological tests such as DAT, ELISA, and rK-39 are inconclusive for diagnosis of PKDL. PCR, real-time PCR, and isothermal assays, among others, have overcome these limitations being highly sensitive and specific nucleic acid–based tests. For an accurate and rapid diagnosis of PKDL, loop-mediated isothermal amplification (LAMP) and recombinase polymerase amplification (RPA) have been used. Such assays can be used to assess the efficacy of new antileishmanial drugs and to assess the effectiveness of cures. The standard treatment for PKDL still needs improvement. The oral drug miltefosine was introduced as a first-line treatment due to increased resistance to antimonials. With miltefosine monotherapy, PKDL relapses have increased, indicating the need for more robust regimens based on other drugs and combinations. We will discuss various approaches to diagnosing and treating this disease.

M. Azam · P. Salotra · R. Singh (✉)
ICMR-National Institute of Pathology, Safdarjung Hospital Campus, New Delhi, India
e-mail: ruchisingh.nip@gov.in

V. Ramesh
Department of Dermatology, ESIC Hospital, Faridabad, India

© The Author(s), under exclusive license to Springer Nature Singapore Pte Ltd. 2023

A. Selvapandiyan et al. (eds.), *Challenges and Solutions Against Visceral Leishmaniasis*, https://doi.org/10.1007/978-981-99-6999-9_6

Keywords

Leishmaniasis · Post kala-azar dermal leishmaniasis · *Leishmania donovani* · Diagnostics · Treatment

6.1 Introduction

Leishmaniasis is a chronic parasitic human disease caused by the protozoa *Leishmania.* Around 54 different *Leishmania* species have been identified; among those, around 21 species are known to cause the disease. More than 90 sand fly species (*Phlebotomus* spp) transmit *Leishmania*, and about 70 animal species, including humans, are hosts and reservoirs of the parasite (WHO 2023). Around 0.7–1 million new cases of leishmaniasis are annually reported, ranking it the second most prevalent pathogenic disease. Risk factors for leishmaniasis include poor socioeconomic conditions, malnutrition, movement of nonimmune individuals into disease-endemic areas, and environmental change (WHO 2023).

Leishmaniasis has a complex of clinical pleomorphism in the form of cutaneous leishmaniasis (CL), mucocutaneous leishmaniasis (MCL), and visceral leishmaniasis (VL). VL is the fatal form if left untreated and is characterized by the clinical manifestations of prolonged fever, hepatosplenomegaly, anemia, and weight loss. Post kala-azar dermal leishmaniasis (PKDL) is the dermotropic sequel of VL and manifests with a range of clinical spectrum from monomorphic macular form predominantly presenting with hypopigmented lesions to infiltrated plaques and even to the most aggravated and chronic nodular form (Mukhopadhyay et al. 2014). In addition, patients may develop a combination of lesions as macules and papulonodules concurrently describing the polymorphic form (Ramesh et al. 2007; Zijlstra et al. 2003). The intracellular and extracellular *Leishmania donovani* bodies (LDBs) are mainly present in the superficial epidermis, having low parasite count with a poor detection rate in macular and papular lesions, while abundant in nodular lesions (Beena et al. 2003; Islam and Basher 2017).

The number of PKDL incidences and the time of onset vary between countries, and accordingly, PKDL has two major geographical zones comprising East Africa (Sudan) and South Asia (India, Bangladesh, and Nepal). PKDL develops in 5–15% of apparently cured VL patients after 6 months to up to 5 years of incidence of VL in the Indian subcontinent (ISC) (Desjeux and Ramesh 2011; Ganguly et al. 2010; WHO 2023). In Sudan, the disease occurs as papular or nodular lesions during or immediately after apparent treatment of VL in 50–60% of cases (Zijlstra et al. 2003; Dixit et al. 2020). Up to 20% of PKDL cases in ISC and 15% in Sudan do not show a prior history of kala-azar infection (Hossain and Jamil 2011; Ramesh et al. 2015a). Demarcating the two regional forms, lymphocytes, macrophages, and plasma cells infiltrate to the granuloma of Indian PKDL, while plasma cells are virtually absent in Sudanese PKDL (Ramesh et al. 2007; Zijlstra et al. 2003). Sudanese PKDL patients have been described to suffer from peripheral nerve involvement, lymphadenopathy, mucosal lesions with destructive complications in the oronasal region, and ocular lesions, but these lesions appear rare in Indian PKDL patients (Khandpur et al. 2004; Zijlstra et al. 2003).

In addition to the classical presentation, the disease may have unusual presentations such as erythrodermic, fibroid type, and plaque. The disease severity is graded as mild, if lesions are very few, moderate, when having generalized and easily visible lesions, and severe, when the body is densely covered with lesions. In Sudan, PKDL lesions generally start on the face as papules and gradually increase in size to form nodules or plaques spreading to other parts of body. The combined maculopapular rash is common, while sometimes the micropapular rashes develop, resembling measles (WHO 2012). Though ulceration is often seen in Sudanese PKDL, lesions rarely ulcerate, and nodules may coalesce to form large plaques in Indian PKDL. In addition, treatment is always necessary for Indian PKDL, while Sudanese PKDL is usually self-healing (Zijlstra et al. 2003). Para-PKDL, a condition of concurrent VL and PKDL, is seen in approximately 16% of cases in Sudan, but seldom seen in Indian PKDL (Maruf et al. 2023; Zijlstra et al. 2000).

6.2 The Interplay of Factors Leading to PKDL

Host factors like immunosuppression, genetic susceptibility, reactivation of residual parasites, or reinfection in a viscerally immune person may trigger the development of PKDL (Mukhopadhyay et al. 2014). The persistence of parasites in the host is thought to be influenced by interactions between host and parasite factors, but the actual mechanisms are still unknown. Few cases have been reported in which UV-induced skin damage contributed to PKDL. Genetic polymorphism at well-defined loci (LdP13 and β-tubulin) in the *L. donovani* isolated from patients with PKDL and VL has been identified in the parasite causing the visceral and dermal manifestations (Dey and Singh 2007; Sreenivas et al. 2004b). Moreover, different *Leishmania* species and sand fly vectors complicate the pathogenicity of the disease and its epidemiology in different regions.

6.3 Need for Diagnosis of PKDL

PKDL burden is poorly investigated, and unclear pathogenesis makes the disease a challenge for researchers and clinicians. Patients have the reluctance to report early or seek medical treatment, and it has been reported that 25% of the reporting patients seek medical treatment majorly for cosmetic reasons (Singh et al. 2012). The hypomelanotic dermal lesions of PKDL are the parasite reservoirs and help in transmission (anthroponotic), especially during interepidemic periods of VL (Islam and Basher 2017; Mondal et al. 2019). Due to poor diagnosis tools, similarities with leprosy, and patients' reluctance or noncompliance to therapy, macular PKDL goes untreated more often than VL, resulting in a greater cumulative transmission potential (Trindade et al. 2015). As per the WHO's road map for neglected tropical diseases 2021–2030, about 75% of the VL endemic countries will validate the elimination of disease (defined as <1% case fatality due to primary VL) by 2025, and the number will increase to 85% by 2030 (WHO 2020). Hence, identifying and treating PKDL are essential to achieving and maintaining VL elimination (Le Rutte et al. 2019; Mondal et al. 2019).

6.4 Challenges in Clinical Investigation

The diagnosis of PKDL is based on clinical examination, the previous episode of kala-azar, the morphology of the dermal lesions, the patient's response to a given treatment, and the exclusion of other differential manifestations (Ganguly et al. 2010). In PKDL, hypopigmented macules, erythematous plaques, and papulonodular lesions may appear singly or in combination. Hypopigmented macular forms resemble vitiligo and are difficult to diagnose due to low parasite density (Ramesh et al. 2008). In its nodular form, the disease is often confused with a variety of dermatological conditions, including Hansen's disease, secondary syphilis, and sarcoidosis (el Hassan et al. 1992; Salotra and Singh 2006; Zijlstra et al. 2017). In addition, the PKDL presentation is similar to leprosy, where the coendemicity of PKDL and leprosy as chronic infectious dermatosis in the same geographical location makes the diagnosis more challenging.

The typical clinical presentation of leishmaniasis is generally well known to clinicians. Absences of VL history (15–20% cases of PKDL) potentially due to subclinical infection complicate the PKDL suspicion. Moreover, atypical presentations of PKDL present another challenge for clinicians to diagnose PKDL and require the inclusivity of differential diagnosis. Leishmaniasis may be unusual in terms of viscerotropic species causing dermotropic leishmaniasis or vice versa, rare morphological presentation, and unusual sites and numbers of lesions. Atrophic, xanthomatous, verrucous, papillomatous, hypertrophic, fibroid, mucosal, and extensive tumorous forms of PKDL go undetected even in endemic regions (Yadav et al. 2023). PKDL nodules involving mucosa in the oral and upper respiratory tract leading to ulcerations have been documented (Ramesh et al. 2007; Zijlstra et al. 2003). Hypopigmented macules coalescing to form patches can be misdiagnosed as pityriasis versicolor. A PKDL patient with polymorphic skin lesions involving mucosa developed erythematous papular and nodular lesions (Salam et al. 2013). PKDL with *L. infantum* can appear as immune reconstitution inflammatory syndrome (IRIS) or develop mucocutaneous lesions that resemble Kaposi's sarcoma (Yadav et al. 2023).

Eye infections associated with Leishmaniasis are rare, and the parasite demonstration is challenging (el Hassan 1998). There are several forms of ocular leishmaniasis, including post kala-azar conjunctivitis, blepharitis, and post kala-azar anterior uveitis (Mignot et al. 2021; Yadav et al. 2023). The complex combination of factors like individual immunity, parasite virulence, and drug response may explain the unusual forms of leishmaniasis, but a clear understanding of those factors is needed.

6.5 Challenges in Histopathology-Based Diagnosis

After clinical observation, PKDL is traditionally demonstrated by microscopic examination of Leishman–Donovan bodies (LDB) amastigotes in tissue or slit-skin smears, and it is considered the gold standard because of its high specificity. Slides stained with hematoxylin and eosin (H&E) are used to demonstrate LDBs in skin specimens. However, the positivity rate for the detection of LDBs varies depending on the clinical lesions. In the case of macular PKDL, the epidermal layer is healthy with the low number of LDBs, giving the histopathology positivity rate of 7–33%, papular 36–69%, and nodular PKDL shows atrophied epidermis with abundant LDBs having identification in 67–100% cases (Mondal et al. 2010; Ramesh and Mukherjee 1995; Salotra and Singh 2006; Zijlstra 2016). The application of monoclonal antibodies specific to *Leishmania* in immunohistochemistry (IHC) has increased the detection of LDBs by 80% (compared to 50% localization in H&E staining). The demonstration of LDB in macrophages is pathognomonic; however, the procedure's invasiveness, the need for an experienced clinician, poor sensitivity involving prolonged searches, patient resistance to skin biopsy, and inadequacy of field application limit its use for diagnostics.

6.6 Challenges in Serological Diagnosis

Serological assays developed for the diagnosis of VL also find application for the diagnosis of PKDL. The serological methods like direct agglutination test (DAT), enzyme-linked immunosorbent assay (ELISA), indirect fluorescent antibody test (IFAT), and immunoblotting have high sensitivity and overcome the limitations of microscopy-based diagnosis of PKDL.

El Harith and colleagues used DAT for the diagnosis of PKDL using antigens prepared from indigenous isolates in freeze-dried form or in suspension that can be applied to whole blood, plasma, or serum under field and laboratory conditions (El Harith et al. 1996). DAT for the diagnosis of PKDL showed sensitivity in the range of 94–100% and specificity between 40% and 100% (Adams et al. 2013). In Nepal, application of DAT for the diagnosis of PKDL showed 100% sensitivity and specificity (Das et al. 2007). DAT antigen developed using promastigote of an indigenous Indian *L. donovani* isolate could correctly detect 100% polymorphic cases and 95.4% macular PKDL cases, whereas detection of PKDL with DAT based on axenic amastigote antigen was 100% sensitive and specific (Singh et al. 2005). However, the requirement of preparing serial dilutions of blood samples and an incubation period of approximately 18 h and antigen variability limit the wide use of DAT to diagnose PKDL.

For the diagnosis of VL and PKDL, the two antigens rK39 (from *L. chagasi*) and rKE16 (from *L. donovani*) are commercially available in different formats, including ELISA, latex agglutination tests, and lateral flow rapid tests. In Sudan, recombinant *L. donovani* gene B protein GBP (rGBP) in ELISA demonstrated a higher sensitivity of 93% for PKDL and 92% for VL (Jensen et al. 1999). While glucose-regulated protein 78 (GRP78) had reactivity among 89% of VL, 78% of PKDL and 85% of CL patients (Jensen et al. 2002). Several other genes like rGBP, ORF F, rH2a, rH2b, rLACK, rgp63, rP20, rPSA, and purified LPG have been cloned and exploited in serological methods for the diagnosis of VL and PKDL (Dixit et al. 2020). Recently, by engineering a kinesin antigen with an optimized kinesin structure (repeats) and sequence, an improved kinesin antigen-based assay (rKLi8.3) with better diagnostic performance for VL was developed and is under evaluation [(Improved diagnosis and therapy of visceral Leishmaniasis—LOEWE-Zentrum DRUID (loewe-druid.de) accessed on 12.03.2023].

The *Leishmania* parasites, with a vast repertoire of carbohydrate moieties, elicit a specific and strong humoral response in patients. Antileishmanial IgG, IgM, IgA, and IgG subclass (IgG1, IgG2, and IgG3) antibodies are increased in PKDL patients as compared to cured VL. The lack or low expression of IgE and IgG4 has been observed in patients with PKDL, as opposed to those with active VL (Saha et al. 2005). In comparison to the parasite (promastigote/amastigotes) antigen or rK39 antigen-based ELISA, IgG-based ELISA demonstrated higher sensitivity (100%) and specificity (96.7%) for the diagnosis of PKDL (Saha et al. 2005). VL Sero K-SeT low-cost RDT detects an increased anti-*Leishmania* IgG1 among posttreatment VL relapse or PKDL cases (Marlais et al. 2018). Anti-gal antibodies, when used in ELISA in Indian patients, proved successful in serological diagnosis of PKDL and differentiating cases of past cured VL infections (Datta et al. 2019).

Western blot assay is one of the most sensitive and specific serological methods that can distinguish PKDL from other parasitic infections based on their antigenic profiles and IgG reactivity. The antibodies generated in the sera of patients with VL and PKDL are intrinsically different. A promastigote membrane-specific polypeptide (31 kDa) demonstrated the potential to differentiate PKDL from past and ongoing VL infection by inducing antibody formation in the sera of PKDL patients compared to sera of VL patients, both before and after treatment (Saha et al. 2005). Antibodies were also generated against two other antigens (110 kDa and 65 kDa) only among patients with PKDL, not in patients with leprosy or vitiligo (Salotra et al. 1999).

Patients with VL are diagnosed using the Leishmanin skin test (LST), which measures delayed-type hypersensitivity reactions in response to intradermal injections of leishmanin antigen (killed *L. donovani* parasite) in their forearms. Induration in response to hypersensitivity is measured (in mm) around 48–72 h, and a reaction of $\geq$5 mm is considered positive. Among the Sudanese PKDL patients, LST demonstrated higher positivity in patients with PKDL only as compared to PKDL patients with concomitant VL (Zijlstra et al. 2000). However, the LST being a skin test encounters limitations like a lack of reference standards, performance scale, and poor sensitivity and specificity of antigens from crude extracts of parasites.

The most widely used rK39 RDT is the rapid and sensitive test for the detection of both VL and PKDL. The anti-K39 IgG circulating in the Indian subcontinent can be detected among 95–100% of patients with kala-azar infection. Recombinant antigen (rK39) has shown higher sensitivity and specificity than the crude *Leishmania* antigen for the PKDL diagnosis (Salotra et al. 2002). PKDL-suspected cases are routinely tested with the strip, regardless of a previous history of VL (Ganguly et al. 2015). The assay can be performed with serum samples, less-invasive slit aspirates, and patient-friendly, minimally invasive fingerstick-obtained blood samples for the diagnosis of PKDL without affecting the test performance (Verma et al. 2013).

In further advancements, urine ELISA is a noninvasive test for the diagnosis of PKDL with sensitivity and specificity of 97.94% and 100% (Ejazi et al. 2016). A dipstick test using recombinant glycoprotein 63 (rGP63) and recombinant cysteine protease A (rCPA) yielded 100% sensitivity and specificity for PKDL. Despite improved sensitivity and specificity, the available consortium of serological tests fails to differentiate between VL (active or cured) and PKDL infection. Circulation of antileishmanial antibodies for an extended period post-VL cure limits the application of serological assessments for the diagnosis of PKDL.

6.7 Challenges in Molecular Diagnosis

Molecular methods harnessing the nucleic acid (DNA/RNA) detection of the specific parasite are the most promising techniques for the clinical diagnosis of PKDL that overcome the drawbacks of histopathological and serological methods. Nucleic acid detection–based molecular methods can actively differentiate the cases of PKDL from that of past VL. PKDL cases with a low parasite abundance, especially macular and maculopapular PKDL, are challenging to diagnose, but molecular techniques (such as PCR, RFLP, nested PCR, and qPCR) are effective in confirming the diagnosis (Mondal et al. 2010). Several species-specific targets like ribosomal RNA, small subunit (SSU) rRNA, mini-exon-derived RNA (med RNA), internal transcribed spacer (ITS) regions, and microsatellite DNAs, β-tubulin, and gp63 gene locus have been exploited for the PCR-based amplification for PKDL diagnosis (Dixit et al. 2020). Multicopy PCR targets (like maxi- and minicircles of kinetoplast DNA (kDNA)) raise the sensitivity detecting even very low levels of parasitemia up to 10 fg of parasite DNA (equivalent to one parasite) when million-fold human DNA is present in the test reaction (Salotra et al. 2001). Combining the primary and nested PCR improves the positivity rate of PKDL diagnosis with higher sensitivity and reliability (Nasreen et al. 2012; Sreenivas et al. 2004a). In addition, nucleic acid detection methods can be used to detect HIV-*Leishmania* coinfections and detect parasites in suspected cases prior to the appearance of any clinical symptoms or signs (Singh et al. 2011).

PCR using clinical specimens like skin biopsy of lesions (positivity rate 93.8%), slit aspirates (82%), or lymph node aspirates (83%) have been used for PKDL diagnosis with rapidity (Osman et al. 1998; Salotra et al. 2001; Tavares et al. 2003). The detection of *Leishmania* DNA in peripheral blood buffy coats was helpful in diagnosing 40–75% of clinically suspected PKDL cases, especially those with hypopigmented maculae (Mondal et al. 2010).

Real-time PCR (RT-PCR) or q-PCR has the edge over conventional PCR for its high sensitivity, quantification, automation, potential for high throughput, and rapid detection capabilities (Galluzzi et al. 2018). RT-PCR prevents cross-contamination as it does not require the opening of tubes for post-PCR analysis. RT-PCR is commonly used in referral laboratories for epidemiological and diagnostic studies of leishmaniasis and for assessing parasite burdens in both symptomatic and asymptomatic carriers. RT-PCR targeting conserved regions (18S rRNA region and ITS) facilitates species typing using high-resolution melt (HRM) analysis (Schulz et al. 2003).

The qPCR showed a high sensitivity in DNA extracted from archived skin biopsies (85%) and active macular cases with multiple skin samples (91%) (Ghosh et al. 2018; Hossain et al. 2017). More recently, the utility of less-invasive blood samples for the qPCR-based diagnosis of PKDL, particularly for the macular variant, was established (Dixit et al. 2021b). Patients having higher parasite load are at a higher risk of relapse (Ramesh et al. 2015b; Verma et al. 2013). A thorough study of parasite kinetics by qPCR in PKDL patients shows a resurgence of the parasite in a subset of patients post 6 months of treatment with miltefosine and liposomal amphotericin-B (LAmB), indicating treatment inadequacy (Moulik et al. 2018).

Despite significant progress in PCR technology, molecular assays have yet to be adopted in routine clinical settings, especially in underdeveloped nations. The major hitch for routine application of PCR-based techniques is the high cost of equipment (thermocyclers) and reagents and the technical complexity of reaction setup requiring well-equipped laboratories and skilled personnel. In addition, utmost care is required to overcome the limitations including appropriate and sufficient positive and negative controls, and the risk of contamination due to postreaction analysis in standard PCR. A portable diagnostic tool that allows the detection of parasites, especially in field conditions, is necessary to identify and treat leishmaniasis early (Martins et al. 2020).

Several isothermal amplification–based techniques can be used as a substitute for expensive molecular methods, including nucleic acid sequence–based amplification, strand displacement amplification, loop-mediated isothermal amplification (LAMP), and helicase-dependent amplification. These techniques offer a rapid, sensitive, and low-cost alternative for detecting and characterizing pathogens in clinical settings. Furthermore, these methods are easy to use, require minimal technical expertise, and are ideal for use in resource-limited settings, where access to traditional detection methods may be limited. The recombinase polymerase amplification (RPA) assay with a mobile suitcase laboratory approach endorses its application in field conditions for the diagnosis of PKDL (Mondal et al. 2016). Loop-mediated isothermal amplification (LAMP) harnesses the ability of *Bst* polymerase to amplify the DNA in isothermal conditions by its unique strand displacement activity. This technique does not require any sophisticated instrument or complicated analysis and is not constrained by the inhibitory elements present in clinical samples (Mori et al. 2013). By analyzing direct clinical samples and eliminating the need to extract DNA from clinical samples, the assay turns out to be faster and more cost-effective. A VL LAMP based on direct-blood lysis has been reported to have a sensitivity of 93.06% (Dixit et al. 2018). The direct sample lysis method using slit aspirate assay has also been developed and validated as an alternative, less-invasive method for PKDL with remarkable sensitivity of 96%. However, Loopamp™ Leishmania Detection Kit (Eiken Chemical Co. Ltd., Japan), when used for the diagnosis of PKDL in skin samples, using kit-based DNA extract as a template showed a sensitivity of 72%, which declined in comparable reactions, where DNA was extracted using boil-spin method (Ghosh et al. 2022). Chowdhury et al. also showed that kit-based DNA extraction (Qiagen) has higher efficacy than the boil and spin method when used in RPA, qPCR, and LAMP (Chowdhury et al. 2020). The kDNA-based LAMP assay, developed by us, detects *Leishmania* species in PKDL with the greatest sensitivity for *L. donovani* (1 fg), followed by *L. tropica* (1 pg) and *L. major* (100 pg) (Verma et al. 2017). Our other closed-tube SYBR greenI–based LAMP assay diagnosed VL and PKDL with high sensitivity (97%) and specificity (100%) (Dixit et al. 2018). The assay has been developed in a kit format finding its applicability in field conditions for mass screening of *Leishmania* infection in endemic regions.

A real-time fluorimetry loop-mediated isothermal amplification (RealAmp) test was devised and demonstrated a 100% detection rate for the PKDL. Using RealAmp as a test of cure for PKDL, RealAmp detected parasites in two posttreatment cases reporting relapse on follow-up. These results suggest that RealAmp can be employed to monitor posttreatment relapse and can be a powerful tool to monitor and evaluate the efficacy of treatment regimens. The sensitivity of the assay using direct sample lysis of slit aspirates was 89% (Dixit et al. 2021a).

Multiplex LAMP assays using a real-time fluorometer revealed differential detection of *Mycobacterium leprae* and *L. donovani* with a 15-min mean amplification time and sensitivity of 1 fg and 100 fg, respectively (Joshi et al. 2021). In addition, a rapid diagnosis by kDNA minicircle sequence–based speciation of *L. donovani* causing PKDL was established (Selvapandiyan et al. 2008).

6.8 Nanotechnology for *Leishmania* Detection

Nanoparticles (NPs) with excellent physicochemical, structural, and optical profiles that vary with their size/shape resulting in different absorption bands, and electrical, magnetic, and catalytic properties have aided in the development of sensors for the diagnosis of leishmaniasis (Diouani et al. 2019; Farshchi et al. 2020; Martins et al. 2020; Nazari-Vanani et al. 2018). An RDT containing a recombinant chimeric antigen (rCAg) capturing antileishmanial antibodies and protein A labeled with gold NPs demonstrated excellent sensitivity and specificity with long-term storage stability (Anfossi et al. 2018). An electrochemical immunosensor containing crude *Leishmania* antigen functionalized with gold NPs recognized the specific epitope of anti-*Leishmania* antibodies present in the VL patient serum with 100% specificity (Martins et al. 2020).

An ultrasensitive genosensor discriminating the *Leishmania* genome from mismatched sequences was developed with a thiolated DNA probe immobilized on gold nanoparticles (Nazari-Vanani et al. 2018). Another cost-effective paper-based genosensor electrodeposited with cysteamine functionalized gold NPs was also developed for *Leishmania* DNA detection (Farshchi et al. 2020). As a further demonstration of nanosensors' applicability, magnetic particles were used as capture probes for *Leishmania* surface proteins and DNA, and cadmium selenide quantum dots (CdSe QDs) were used as detection probes. Both probes were 100% sensitive and specific for the samples tested (Andreadou et al. 2016).

Currently, there are no reliable biomarkers for predicting the response to antileishmanial therapy. Skin lesions are evaluated clinically for size reductions or color changes, both of which are subjective. A handheld three-dimensional optical scanning device has been developed for evaluating skin lesions objectively using height maps for changes in size and color over time. As a result, PKDL lesions can be quantified with high accuracy and minimal discomfort (Zijlstra et al. 2020).

6.9 Treatment of PKDL

Treating PKDL patients is vital to VL control programs as they are regarded as the major reservoir in anthroponotic transmission settings. Compared to VL standard treatment regimens, PKDL treatment regimens are 3–4 times longer for all the available drugs. Most PKDL patients in Sudan self-heal, and only severe and chronic cases are treated, whereas treatment is always necessary for Indian forms of PKDL.

Since the 1950s, pentavalent antimonials (sodium stibogluconate SSG, generic sodium antimony gluconate, and meglumine antimoniate) have remained the standard drugs. In India, SSG was extensively used at a dose of 20 mg/kg/day for 120 days, with cure rates of 64–92% until the last decade (Salotra et al. 2013; Zijlstra et al. 2003). However, a high rate of clinical nonresponsiveness to pentavalent antimonials has led to their abandonment in Bihar state, India, though they are widely used in other endemic areas (Croft and Olliaro 2011).

The frontline drugs for VL, a polyene antibiotic Amphotericin B and oral drug miltefosine, are effective for treating PKDL patients. Based on recommendations from WHO, Table 6.1 summarizes the drugs and treatment regimens applicable in two geographical regions for PKDL. After initial studies proved the safety and efficacy of miltefosine for PKDL treatment, the drug became India's first-line treatment for the disease.

In Africa, where other drugs are limited, SSG is still widely used despite toxicity, lot-to-lot variations, and sometimes the hospitalization requirement (Croft and Olliaro 2011). In Sudanese PKDL patients with severe lesions, SSG at 20 mg/kg/day for 30 days is the best treatment, which can be prolonged to 2–3 months if needed (Zijlstra et al. 2003). An infusion of 2.5 mg/kg/day of liposomal amphotericin B for 20 days has been shown to be effective, with an 83% cure rate with no side effects in Sudan (Musa et al. 2005). PKDL treatment using alum-precipitated autoclaved *L. major* vaccine and Bacille Calmette-Guerin (BCG) and SSG cured 87% of Sudanese patients by day 60 (Musa et al. 2008). Several antifungal drugs with variable cure rates, including allopurinol, ketoconazole, itraconazole, terbinafine, and antibacterial rifampicin, have been used in India and Sudan for the treatment of PKDL (Musa et al. 2013; Ramesh et al. 2010).

Table 6.1 WHO-recommended treatment for PKDL (WHO 2010)

Bangladesh, India and Nepal		
Drug	Dosage and duration	Remarks
Pentavalent antimonials (sodium stibogluconate SSG, generic sodium antimony gluconate, and meglumine antimoniate)	20 mg/kg/day for 120 days or more intramuscular/ intravenous (4X of VL)	Not recommended for PKDL therapy in India due to widespread antimony resistance
Miltefosine (*MIL*)	100 mg/day for 12 weeks100 mg/day for 12 weeks for patients weighing >25 kg; 50 mg/day for 12 weeks for patients weighing <25 kg. (3X)	Oral, first line therapy in India; Due to increasing reporting of ophthalmic complications in the last few years, the drug should be used with caution
Amphotericin-B (*AmB*)	1 mg/kg/day 60–80 doses over 4 months (4X)	Hospitalization required
Liposomal amphotericin-B	5 mg/kg, infusion 2 times per week for 3 weeks for a total dose of 30 mg/kg (3X)	Expensive
East Africa		
Pentavalent antimonials	20 mg/kg/day for 30–60 days intramuscular/ intravenous	
Liposomal amphotericin-B	2.5 mg/kg, for 20 days infusion	
Miltefosine	100 mg/day for 30 days for in patients co-infected with HIV and PKDL)	
Combination treatment of antimonial and paromomycin	Pentavalent antimonial (20 mg/kg/day intramuscular/ intravenous) for 17–60 days plus paromomycin (11 mg intramuscularly) for 17 days	

PKDL treatment is plagued by long treatment schedules, antimony resistance, increased relapse rates from miltefosine treatment, side effects, and poor patient compliance. Therefore, systematic studies on larger sets of patients to evaluate drugs for the treatment of PKDL are urgently needed to achieve improved, effective, and shorter treatments.

6.10 Newer Treatment Regimes

Over the last decade, varying doses of miltefosine and miltefosine combined with other injectable drugs have been explored to establish new effective and shortened treatment regimens for PKDL patients on the Indian subcontinent. Several studies evaluating monotherapy and combination therapy are presented in Table 6.2.

Other than these studies, systematic studies on a larger set of PKDL patients have recently been proclaimed by DNDi. These include (1) Completed open-label,

Table 6.2 Treatment approaches for PKDL in the Indian subcontinent

Drug	Studies	Treatment regimens and doses	Efficacy	Side effects	Issues to consider
Miltefosine (MIL)	Ramesh et al. (2011)	50 mg thrice daily for 60 days or twice daily for 90 days	96%; no relapses at 1 year follow up	Anorexia, nausea, vomiting and occasionally diarrhoea occurred beyond 1–2 weeks of therapy Elevated SGOT (AST) and SGPT (ALT) aming 4.1% patients	• Increasing relapse rate to miltefosine therapy on follwoup beoynd 12 months • Miltefosine tolerance increased >twofold in parasites isolated from patients after relapse • Macular PKDL patients variant demonstrated slow repigmentation process sometimes a year posttreatment
	Sundar et al. (2013)	2.5 mg/kg/day for 8 weeks or 12 weeks utilizing 50 mg capsules; two capsules for patients > = 25 kg or one capsule <25 kg	93%		
	Ramesh et al. (2015b)	Regimen I—50 mg twice daily for 90 days Regimen II—50 mg thrice for 60 days	85%; Relapse rate 15% at 18 months follow up Regimen II 31% relapse rate		
	Ramesh et al. (2020)	100 mg/day for 90 days; $n = 16$	100%; 4 of 16 patients (25%) reported relapse (3 patients relapsed at 18 and 1 relapsed at 6 months		
	Pandey et al. (2021)	2.5 mg/kg or 100 mg/day for 12 weeks	87% of the cases cured and 13% relapsed		

Amphotericin-B deoxycholate (AmB)	Thakur et al. (1997)	1 mg/kg/day by intra venous infusion in 5% dextrose on days 1–20, 21–40 & 61–80	100%; no relapse in 12 months follow up	Mild; signs of infusion-related toxicity (shivering and rigor) occurred in all patients	
	Rabi Das et al. (2017)	Group-A—0.5 mg/kg daily for 20 infusions for 3 courses at an interval of 15 days. Group-B—1 mg/kg on alternate days, 20 infusions for 3 courses at an interval of 15 days	Group-A—92% Group-B—88% 1 relapsed in each study groups	Nephrotoxicity in both groups. Serious adverse events observed in 52% and 64% patients in group A and group B respectively	
Liposomal amphotericin B (LAmB)	Moulik et al. (2018)	5 mg/kg body weight intravenous twice weekly for 3 weeks	Lesions persisted in 35% of cases indicating cure rate to be 65%. A resurgence of parasites on 6 months follow up in 35% cases		
	Pandey et al. (2021)	30 mg/kg, six doses each 5 mg/kg, biweekly for 3 weeks	25% patients relapsed		
Paromomycin	Sundar et al. (2014)	11 mg/kg intramuscular injection for 45 days	37.5%	45.2% complained of pain at Injection site	
Combination therapy					
Amphotericin B (AmB) + Miltefosine (MIL)	Ramesh et al. (2014)	Infusion of 50 mg AmB for 20 days plus 50 mg oral MIL 3 times a day for 40 days	100%; follow up for 2 years with no reports of relapse	–	

(continued)

Table 6.2 (continued)

Drug	Studies	Treatment regimens and doses	Efficacy	Side effects	Issues to consider
Liposomal Amphotericin B (LAmB) + Miltefosine (MIL)	Ramesh et al. (2020)	Three intravenous injections of LAmB (5 mg/kg body weight) were given on day 1, 8, and 15, and MIL capsules were administered after the first dose of LAmB and continued daily until The end of 45 days, $n = 16$	100%; no reports of relapse at 18 months' follow-up		
Miltefosine (MIL) + Paromomycin (PMM)	Pandey et al. (2017)	Intramuscular injection of Group-A—PMM 11 mg/kg plus Miltefosine capsule orally (2.5 mg/kg/day) daily for 10 days for 2 courses with a gap of 15 days; Group-B—3 courses of the same combination	83.33%; Relapse rate—16.67%	Pain at injection site, increased hepatic enzymes, eosinophilia and vomiting among 80% of patients	

randomized clinical trial of two regimens-First- Ambisome® monotherapy regimen (5 × 4 mg/kg IV, twice per week, a total dose of 20 mg/kg) Total duration of 15 days (Ambisome injections at D1, D4, D8, D11, D15) and a second combination of Ambisome® (5 × 4 mg/kg IV, twice per week, total dose of 20 mg/kg) with miltefosine allometric dose orally for 3 weeks to assess their safety and efficacy for treatment of PKDL patients in India and Bangladesh and (2) In East Africa, the ongoing DND*i*-IEND clinical trial will assess the safety and efficacy of two combination treatments based on miltefosine that is not yet approved for clinical use in Africa. One is a combination of injectable paromomycin with the oral drug miltefosine, and the other is a combination of intravenous liposomal amphotericin B with miltefosine (https://dndi.org/press-releases/2018/new-hope-pkdl-patients-africa-clinical-trial/).

To meet the need for newer treatment for persistent PKDL, a phase IIa safety, and immunogenicity, a trial of ChAd63-KH vaccine was conducted in 16 adults and eight adolescent patients using a single intramuscular vaccination (Younis et al. 2021). In patients with PKDL, ChAd63-KH vaccine showed minimal adverse events and induced potent innate and cellular immunity. In addition, the clinical improvement rate was over 90% in 7/23 patients, and partial improvement was seen in 5/23. The ChAd63-KH vaccine is currently in the midst of a randomized controlled trial to establish vaccine-related clinical responses and its clinical utility (Younis et al. 2021).

In any chronic disease, there is always scope to increase the duration of treatment till a complete cure is achieved. This is greatly limited by the intolerable side effects. For example, vomiting episodes during miltefosine therapy and the recent occurrence of ocular damage do not augur well for the continuance of the drug. Combination therapy still revolves around juggling with the available two or three drugs. The current need is to provide better antileishmanial drugs to introduce robust regimens.

References

Adams ER, Versteeg I, Leeflang MM (2013) Systematic review into diagnostics for post-kala-azar dermal leishmaniasis (PKDL). J Trop Med 2013:150746. https://doi.org/10.1155/2013/150746

Andreadou M, Liandris E, Gazouli M, Mataragka A, Tachtsidis I, Goutas N, Vlachodimitropoulos D, Ikonomopoulos J (2016) Detection of Leishmania-specific DNA and surface antigens using a combination of functionalized magnetic beads and cadmium selenite quantum dots. J Microbiol Methods 123:62–67. https://doi.org/10.1016/j.mimet.2015.11.019

Anfossi L, Di Nardo F, Profiti M, Nogarol C, Cavalera S, Baggiani C, Giovannoli C, Spano G, Ferroglio E, Mignone W, Rosati S (2018) A versatile and sensitive lateral flow immunoassay for the rapid diagnosis of visceral leishmaniasis. Anal Bioanal Chem 410(17):4123–4134. https://doi.org/10.1007/s00216-018-1067-x

Beena KR, Ramesh V, Mukherjee A (2003) Identification of parasite antigen, correlation of parasite density and inflammation in skin lesions of post kala-azar dermal leishmaniasis. J Cutan Pathol 30(10):616–620. https://doi.org/10.1034/j.1600-0560.2003.00125.x

Chowdhury R, Ghosh P, Khan MAA, Hossain F, Faisal K, Nath R, Baker J, Wahed AAE, Maruf S, Nath P, Ghosh D, Masud-Ur-Rashid M, Utba Bin Rashid M, Duthie MS, Mondal D (2020) Evaluation of rapid extraction methods coupled with a recombinase polymerase amplification

assay for point-of-need diagnosis of post-kala-azar dermal Leishmaniasis. Trop Med Infect Dis 5(2):95. https://doi.org/10.3390/tropicalmed5020095
Croft SL, Olliaro P (2011) Leishmaniasis chemotherapy--challenges and opportunities. Clin Microbiol Infect 17(10):1478–1483. https://doi.org/10.1111/j.1469-0691.2011.03630.x
Das ML, Deb M, Karki BM, Sarif M, Khanal B, Bhattacharya SK, Agrawal S, Koirala S (2007) Use of rK39 for diagnosis of post kala-azar dermal leishmaniasis in Nepal. Southeast Asian J Trop Med Public Health 38(4):619–625
Datta S, Ghosh M, Sarkar S, Saha B, Mukhopadhyay S (2019) Evaluation of anti-gal enzyme-linked immunosorbent assay for the diagnosis of Indian post-kala-azar dermal leishmaniasis. Indian J Dermatol Venereol Leprol 85(6):578–589. https://doi.org/10.4103/ijdvl.IJDVL_485_18
Desjeux P, Ramesh V (2011) Post-kala-azar dermal Leishmaniasis: facing the challenge of eliminating kala-azar from South Asia. In: Jha TK, Noiri E (eds) Kala azar in South Asia: current status and challenges ahead. Springer, Netherlands, pp 111–124. https://doi.org/10.1007/978-94-007-0277-6_11
Dey A, Singh S (2007) Genetic heterogeneity among visceral and post-kala-azar dermal leishmaniasis strains from eastern India. Infect Genet Evol 7(2):219–222. https://doi.org/10.1016/j.meegid.2006.09.001
Diouani MF, Ouerghi O, Belgacem K, Sayhi M, Ionescu R, Laouini D (2019) Casein-conjugated gold nanoparticles for amperometric detection of Leishmania infantum. Biosensors 9(2):68. https://doi.org/10.3390/bios9020068
Dixit KK, Verma S, Singh OP, Singh D, Singh AP, Gupta R, Negi NS, Das P, Sundar S, Singh R, Salotra P (2018) Validation of SYBR green I based closed tube loop mediated isothermal amplification (LAMP) assay and simplified direct-blood-lysis (DBL)-lamp assay for diagnosis of visceral leishmaniasis (VL). PLoS Negl Trop Dis 12(11):e0006922. https://doi.org/10.1371/journal.pntd.0006922
Dixit KK, Singh R, Salotra P (2020) Advancement in molecular diagnosis of post kala-azar dermal leishmaniasis. Indian J Dermatol 65(6):465–472. https://doi.org/10.4103/ijd.IJD_311_19
Dixit KK, Ramesh V, Gupta R, Negi NS, Singh R, Salotra P (2021a) Real-time fluorimetry loop-mediated isothermal amplification for diagnosis of leishmaniasis and as a tool for assessment of cure for post-kala-azar dermal leishmaniasis. Am J Trop Med Hyg 104(6):2097–2107. https://doi.org/10.4269/ajtmh.20-1057
Dixit KK, Ramesh V, Upadhyay S, Singh AK, Singh OP, Sundar S, Singh R, Salotra P (2021b) Utility of blood as the clinical specimen for the molecular diagnosis of post-kala-azar dermal leishmaniasis. J Clin Microbiol 59(9):e0013221. https://doi.org/10.1128/JCM.00132-21
Ejazi SA, Bhattacharya P, Bakhteyar MA, Mumtaz AA, Pandey K, Das VN, Das P, Rahaman M, Goswami RP, Ali N (2016) Noninvasive diagnosis of visceral leishmaniasis: development and evaluation of two urine-based immunoassays for detection of leishmania donovani infection in India. PLoS Negl Trop Dis 10(10):e0005035. https://doi.org/10.1371/journal.pntd.0005035
El Harith A, Chowdhury S, Al-Masum A, Semião-Santos S, Das PK, Akhter S, Vetter JC, Haq I (1996) Reactivity of various leishmanial antigens in a direct agglutination test and their value in differentiating post-kala azar dermal leishmaniasis from leprosy and other skin conditions. J Med Microbiol 44(2):141–146. https://doi.org/10.1099/00222615-44-2-141
el Hassan AM, Ghalib HW, Zijlstra EE, Eltoum IA, Satti M, Ali MS, Ali HM (1992) Post kala-azar dermal leishmaniasis in the Sudan: clinical features, pathology and treatment. Trans R Soc Trop Med Hyg 86(3):245–248. https://doi.org/10.1016/0035-9203(92)90294-m
el Hassan AM, Khalil EA, el Sheikh EA, Zijlstra EE, Osman A, Ibrahim ME (1998) Post kala-azar ocular leishmaniasis. Trans. R. Soc. Trop. Med. Hyg. 92:177–179. https://doi.org/10.1016/s0035-9203(98)90736-2
Farshchi F, Saadati A, Hasanzadeh M (2020) Optimized DNA-based biosensor for monitoring Leishmania infantum in human plasma samples using biomacromolecular interaction: a novel platform for infectious disease diagnosis. Anal Methods 12(39):4759–4768

Galluzzi L, Ceccarelli M, Diotallevi A, Menotta M, Magnani M (2018) Real-time PCR applications for diagnosis of leishmaniasis. Parasit Vectors 11(1):273. https://doi.org/10.1186/s13071-018-2859-8

Ganguly S, Das NK, Barbhuiya JN, Chatterjee M (2010) Post-kala-azar dermal leishmaniasis--an overview. Int J Dermatol 49(8):921–931. https://doi.org/10.1111/j.1365-4632.2010.04558.x

Ganguly S, Saha P, Chatterjee M, Roy S, Ghosh TK, Guha SK, Kundu PK, Bera DK, Basu N, Maji AK (2015) PKDL--a silent parasite pool for transmission of leishmaniasis in kala-azar endemic areas of Malda district, West Bengal, India. *PLoS Negl Trop Dis* 9(10):e0004138. https://doi.org/10.1371/journal.pntd.0004138

Ghosh P, Hasnain MG, Hossain F, Khan MAA, Chowdhury R, Faisal K, Mural MA, Baker J, Nath R, Ghosh D, Maruf S, Shomik MS, Haque R, Matlashewski G, Hamano S, Duthie MS, Mondal D (2018) Evaluation of real-time PCR for diagnosis of post-kala-azar dermal leishmaniasis in endemic foci of Bangladesh. *Open Forum Infect Dis* 5(10):ofy234. https://doi.org/10.1093/ofid/ofy234

Ghosh P, Chowdhury R, Maruf S, Picado A, Hossain F, Owen SI, Nath R, Baker J, Hasnain MG, Shomik MS, Ghosh D, Rashid M, Rashid MU, Sagar SK, Rahat MA, Basher A, Nath P, Edwards T, Andrews JR et al (2022) Gauging the skin resident Leishmania parasites through a loop mediated isothermal amplification (LAMP) assay in post-kala-azar dermal leishmaniasis. Sci Rep 12(1):18069. https://doi.org/10.1038/s41598-022-21497-6

Hossain M, Jamil KM (2011) Pathology and mechanism of disease in kala-azar and post-kala-azar dermal leishmaniasis. In: Kala azar in South Asia: current status and challenges ahead, pp 11–14

Hossain F, Ghosh P, Khan MAA, Duthie MS, Vallur AC, Picone A, Howard RF, Reed SG, Mondal D (2017) Real-time PCR in detection and quantitation of leishmania donovani for the diagnosis of visceral leishmaniasis patients and the monitoring of their response to treatment. PLoS One 12(9):e0185606. https://doi.org/10.1371/journal.pone.0185606

Islam QT, Basher A (2017) Atypical presentation of post-kala-azar dermal leishmaniasis. Lancet Infect Dis 17(11):1218. https://doi.org/10.1016/S1473-3099(17)30583-2

Jensen AT, Gasim S, Moller T, Ismail A, Gaafar A, Kemp M, el Hassan AM, Kharazmi A, Alce TM, Smith DF, Theander TG (1999) Serodiagnosis of Leishmania donovani infections: assessment of enzyme-linked immunosorbent assays using recombinant L. donovani gene B protein (GBP) and a peptide sequence of L. donovani GBP. Trans R Soc Trop Med Hyg 93(2):157–160. https://doi.org/10.1016/s0035-9203(99)90291-2

Jensen AT, Ismail A, Gaafar A, El Hassan AM, Theander TG (2002) Humoral and cellular immune responses to glucose regulated protein 78 -- a novel Leishmania donovani antigen. Tropical Med Int Health 7(5):471–476. https://doi.org/10.1046/j.1365-3156.2002.00880.x

Joshi S, Dixit KK, Sharma V, Ramesh V, Singh R, Salotra P (2021) Rapid multiplex loop-mediated isothermal amplification (m-LAMP) assay for differential diagnosis of leprosy and post-kala-azar dermal leishmaniasis. Am J Trop Med Hyg 104(6):2085–2090. https://doi.org/10.4269/ajtmh.19-0313

Khandpur S, Ramam M, Sharma VK, Salotra P, Singh MK, Malhotra A (2004) Nerve involvement in Indian post kala-azar dermal leishmaniasis. Acta Derm Venereol 84(3):245–246. https://doi.org/10.1080/00015550410024607

Le Rutte EA, Zijlstra EE, de Vlas SJ (2019) Post-kala-azar dermal leishmaniasis as a reservoir for visceral leishmaniasis transmission. Trends Parasitol 35(8):590–592. https://doi.org/10.1016/j.pt.2019.06.007

Marlais T, Bhattacharyya T, Singh OP, Mertens P, Gilleman Q, Thunissen C, Hinckel BCB, Pearson C, Gardner BL, Airs S, de la Roche M, Hayes K, Hafezi H, Falconar AK, Eisa O, Saad A, Khanal B, Bhattarai NR, Rijal S et al (2018) Visceral leishmaniasis IgG1 rapid monitoring of cure vs. relapse, and potential for diagnosis of post kala-azar dermal leishmaniasis. Front Cell Infect Microbiol 8:427. https://doi.org/10.3389/fcimb.2018.00427

Martins BR, Barbosa YO, Andrade CMR, Pereira LQ, Simão GF, de Oliveira CJ, Correia D, Oliveira RTS, da Silva MV, Silva ACA, Dantas NO, Rodrigues V, Muñoz RAA, Alves-Balvedi RP (2020) Development of an electrochemical Immunosensor for specific detection of visceral leishmaniasis using gold-modified screen-printed carbon electrodes. Biosensors 10(8):81. https://www.mdpi.com/2079-6374/10/8/81

Maruf S, Sagar SK, Rashid MMU, Nath P, Islam MS, Ghosh P, Rashid MU, Mondal D, Abd El Wahed A, Basher A (2023) Revisiting the diagnosis and treatment of para kala-azar dermal leishmaniasis in the endemic foci of Bangladesh. PLoS One 18(1):e0280747. https://doi.org/10.1371/journal.pone.0280747

Mignot G, Bhattacharya Y, Reddy A (2021) Ocular leishmaniasis - a systematic review. Indian J Ophthalmol 69(5):1052–1060. https://doi.org/10.4103/ijo.IJO_2232_20

Mondal D, Nasrin KN, Huda MM, Kabir M, Hossain MS, Kroeger A, Thomas T, Haque R (2010) Enhanced case detection and improved diagnosis of PKDL in a kala-azar-endemic area of Bangladesh. PLoS Negl Trop Dis 4(10):e832. https://doi.org/10.1371/journal.pntd.0000832

Mondal D, Ghosh P, Khan MA, Hossain F, Böhlken-Fascher S, Matlashewski G, Kroeger A, Olliaro P, Abd El Wahed A (2016) Mobile suitcase laboratory for rapid detection of Leishmania donovani using recombinase polymerase amplification assay. Parasit Vectors 9(1):281. https://doi.org/10.1186/s13071-016-1572-8

Mondal D, Bern C, Ghosh D, Rashid M, Molina R, Chowdhury R, Nath R, Ghosh P, Chapman LAC, Alim A, Bilbe G, Alvar J (2019) Quantifying the infectiousness of post-kala-azar dermal leishmaniasis toward sand flies. Clin Infect Dis 69(2):251–258. https://doi.org/10.1093/cid/ciy891

Mori Y, Kanda H, Notomi T (2013) Loop-mediated isothermal amplification (LAMP): recent progress in research and development. J Infect Chemother 19(3):404–411. https://doi.org/10.1007/s10156-013-0590-0

Moulik S, Chaudhuri SJ, Sardar B, Ghosh M, Saha B, Das NK, Chatterjee M (2018) Monitoring of parasite kinetics in Indian post-kala-azar dermal leishmaniasis. Clin Infect Dis 66(3):404–410. https://doi.org/10.1093/cid/cix808

Mukhopadhyay D, Dalton JE, Kaye PM, Chatterjee M (2014) Post kala-azar dermal leishmaniasis: an unresolved mystery. Trends Parasitol 30(2):65–74. https://doi.org/10.1016/j.pt.2013.12.004

Musa AM, Khalil EA, Mahgoub FA, Hamad S, Elkadaru AM, El Hassan AM (2005) Efficacy of liposomal amphotericin B (AmBisome) in the treatment of persistent post-kala-azar dermal leishmaniasis (PKDL). Ann Trop Med Parasitol 99(6):563–569. https://doi.org/10.1179/136485905x514127

Musa AM, Khalil EA, Mahgoub FA, Elgawi SH, Modabber F, Elkadaru AE, Aboud MH, Noazin S, Ghalib HW, El-Hassan AM (2008) Immunochemotherapy of persistent post-kala-azar dermal leishmaniasis: a novel approach to treatment. Trans R Soc Trop Med Hyg 102(1):58–63. https://doi.org/10.1016/j.trstmh.2007.08.006

Musa AM, Khalil EA, Younis BM, Elfaki ME, Elamin MY, Adam AO, Mohamed HA, Dafalla MM, Abuzaid AA, El-Hassan AM (2013) Treatment-based strategy for the management of post-kala-azar dermal leishmaniasis patients in the Sudan. J Trop Med 2013:708391. https://doi.org/10.1155/2013/708391

Nasreen SA, Hossain MA, Paul SK, Mahmud MC, Ahmed S, Ghosh S, Kobayashi N (2012) PCR-based detection of Leishmania DNA in skin samples of post kala-azar dermal leishmaniasis patients from an endemic area of Bangladesh. Jpn J Infect Dis 65(4):315–317. https://doi.org/10.7883/yoken.65.315

Nazari-Vanani R, Sattarahmady N, Yadegari H, Delshadi N, Hatam G, Heli H (2018) Electrochemical quantitation of Leishmania infantum based on detection of its kDNA genome and transduction of non-spherical gold nanoparticles. Anal Chim Acta 1041:40–49

Osman OF, Oskam L, Kroon NC, Schoone GJ, Khalil ET, El-Hassan AM, Zijlstra EE, Kager PA (1998) Use of PCR for diagnosis of post-kala-azar dermal leishmaniasis. J Clin Microbiol 36(6):1621–1624. https://doi.org/10.1128/JCM.36.6.1621-1624.1998

Pandey K, Pal B, Das VNR, Murti K, Lal CS, Verma N, Bimal S, Ali V, Verma RB, Topno RK, Siddiqi NA, Das P (2017) Safety and efficacy of a combination of paromomycin and miltefosine for two vs. three courses in patients with post-kala-azar dermal leishmaniasis: an observational pilot study. Br J Dermatol 177(2):557–559. https://doi.org/10.1111/bjd.15119

Pandey K, Pal B, Siddiqui NA, Lal CS, Ali V, Bimal S, Kumar A, Verma N, Das VNR, Singh SK, Topno RK, Das P (2021) A randomized, open-label study to evaluate the efficacy and safety of liposomal amphotericin B (AmBisome) versus miltefosine in patients with post-kala-azar dermal leishmaniasis. Indian J Dermatol Venereol Leprol 87(1):34–41. https://doi.org/10.25259/ijdvl_410_19

Rabi Das VN, Siddiqui NA, Pal B, Lal CS, Verma N, Kumar A, Verma RB, Kumar D, Das P, Pandey K (2017) To evaluate efficacy and safety of amphotericin B in two different doses in the treatment of post kala-azar dermal leishmaniasis (PKDL). PLoS One 12(3):e0174497. https://doi.org/10.1371/journal.pone.0174497

Ramesh V, Mukherjee A (1995) Post-kala-azar dermal leishmaniasis. Int J Dermatol 34(2):85–91. https://doi.org/10.1111/j.1365-4362.1995.tb03584.x

Ramesh V, Singh R, Salotra P (2007) Short communication: post-kala-azar dermal leishmaniasis--an appraisal. Tropical Med Int Health 12(7):848–851. https://doi.org/10.1111/j.1365-3156.2007.01854.x

Ramesh V, Ramam M, Singh R, Salotra P (2008) Hypopigmented post-kala-azar dermal leishmaniasis. Int J Dermatol 47(4):414–416. https://doi.org/10.1111/j.1365-4632.2008.03621.x

Ramesh V, Kumar J, Kumar D, Salotra P (2010) A retrospective study of intravenous sodium stibogluconate alone and in combinations with allopurinol, rifampicin, and an immunomodulator in the treatment of Indian post-kala-azar dermal leishmaniasis. Indian J Dermatol Venereol Leprol 76(2):138–144. https://doi.org/10.4103/0378-6323.60553

Ramesh V, Katara GK, Verma S, Salotra P (2011) Miltefosine as an effective choice in the treatment of post-kala-azar dermal leishmaniasis. Br J Dermatol 165(2):411–414. https://doi.org/10.1111/j.1365-2133.2011.10402.x

Ramesh V, Avishek K, Sharma V, Salotra P (2014) Combination therapy with amphotericin-B and miltefosine for post-kala-azar dermal leishmaniasis: a preliminary report. Acta Derm Venereol 94(2):242–243. https://doi.org/10.2340/00015555-1582

Ramesh V, Kaushal H, Mishra AK, Singh R, Salotra P (2015a) Clinico-epidemiological analysis of post kala-azar dermal leishmaniasis (PKDL) cases in India over last two decades: a hospital based retrospective study. BMC Public Health 15:1092. https://doi.org/10.1186/s12889-015-2424-8

Ramesh V, Singh R, Avishek K, Verma A, Deep DK, Verma S, Salotra P (2015b) Decline in clinical efficacy of oral miltefosine in treatment of post kala-azar dermal leishmaniasis (PKDL) in India. PLoS Negl Trop Dis 9(10):e0004093. https://doi.org/10.1371/journal.pntd.0004093

Ramesh V, Dixit KK, Sharma N, Singh R, Salotra P (2020) Assessing the efficacy and safety of liposomal amphotericin B and miltefosine in combination for treatment of post kala-azar dermal Leishmaniasis. J Infect Dis 221(4):608–617. https://doi.org/10.1093/infdis/jiz486

Saha S, Mazumdar T, Anam K, Ravindran R, Bairagi B, Saha B, Goswami R, Pramanik N, Guha SK, Kar S, Banerjee D, Ali N (2005) Leishmania promastigote membrane antigen-based enzyme-linked immunosorbent assay and immunoblotting for differential diagnosis of Indian post-kala-azar dermal leishmaniasis. J Clin Microbiol 43(3):1269–1277. https://doi.org/10.1128/jcm.43.3.1269-1277.2005

Salam MA, Siddiqui MA, Nabi SG, Bhaskar KR, Mondal D (2013) Post-kala-azar dermal leishmaniasis with mucosal involvement: an unusual case presentation including successful treatment with miltefosine. J Health Popul Nutr 31(2):294–297. https://doi.org/10.3329/jhpn.v31i2.16395

Salotra P, Singh R (2006) Challenges in the diagnosis of post kala-azar dermal leishmaniasis. Indian J Med Res 123(3):295–310. https://www.ncbi.nlm.nih.gov/pubmed/16778312

Salotra P, Raina A, Ramesh V (1999) Western blot analysis of humoral immune response to Leishmania donovani antigens in patients with post-kala-azar dermal leishmaniasis. Trans R Soc Trop Med Hyg 93(1):98–101. https://doi.org/10.1016/s0035-9203(99)90197-9

Salotra P, Sreenivas G, Pogue GP, Lee N, Nakhasi HL, Ramesh V, Negi NS (2001) Development of a species-specific PCR assay for detection of Leishmania donovani in clinical samples from patients with kala-azar and post-kala-azar dermal leishmaniasis. J Clin Microbiol 39(3): 849–854. https://doi.org/10.1128/JCM.39.3.849-854.2001

Salotra P, Sreenivas G, Nasim AA, Subba Raju BV, Ramesh V (2002) Evaluation of enzyme-linked immunosorbent assay for diagnosis of post-kala-azar dermal leishmaniasis with crude or recombinant k39 antigen. Clin Diagn Lab Immunol 9(2):370–373. https://doi.org/10.1128/cdli.9.2.370-373.2002

Salotra P, Singh R, Seifert K (2013) Visceral leishmaniasis – current treatments and needs. In: Trypanosomatid Diseases. Wiley, pp 1–15. https://doi.org/10.1002/9783527670383.ch1

Schulz A, Mellenthin K, Schönian G, Fleischer B, Drosten C (2003) Detection, differentiation, and quantitation of pathogenic leishmania organisms by a fluorescence resonance energy transfer-based real-time PCR assay. J Clin Microbiol 41(4):1529–1535. https://doi.org/10.1128/jcm.41.4.1529-1535.2003

Selvapandiyan A, Duncan R, Mendez J, Kumar R, Salotra P, Cardo LJ, Nakhasi HL (2008) A Leishmania minicircle DNA footprint assay for sensitive detection and rapid speciation of clinical isolates. Transfusion 48(9):1787–1798. https://doi.org/10.1111/j.1537-2995.2008.01798.x

Singh R, Subba Raju BV, Jain RK, Salotra P (2005) Potential of direct agglutination test based on promastigote and amastigote antigens for serodiagnosis of post-kala-azar dermal leishmaniasis. Clin Diagn Lab Immunol 12(10):1191–1194. https://doi.org/10.1128/CDLI.12.10.1191-1194.2005

Singh S, Sharma U, Mishra J (2011) Post-kala-azar dermal leishmaniasis: recent developments. Int J Dermatol 50(9):1099–1108. https://doi.org/10.1111/j.1365-4632.2011.04925.x

Singh RP, Picado A, Alam S, Hasker E, Singh SP, Ostyn B, Chappuis F, Sundar S, Boelaert M (2012) Post-kala-azar dermal leishmaniasis in visceral leishmaniasis-endemic communities in Bihar, India. *Trop Med Int Health* 17(11):1345–1348. https://doi.org/10.1111/j.1365-3156.2012.03067.x

Sreenivas G, Ansari NA, Kataria J, Salotra P (2004a) Nested PCR assay for detection of Leishmania donovani in slit aspirates from post-kala-azar dermal leishmaniasis lesions. J Clin Microbiol 42(4):1777–1778. https://doi.org/10.1128/JCM.42.4.1777-1778.2004

Sreenivas G, Raju BV, Singh R, Selvapandiyan A, Duncan R, Sarkar D, Nakhasi HL, Salotra P (2004b) DNA polymorphism assay distinguishes isolates of Leishmania donovani that cause kala-azar from those that cause post-kala-azar dermal leishmaniasis in humans. J Clin Microbiol 42(4):1739–1741. https://doi.org/10.1128/JCM.42.4.1739-1741.2004

Sundar S, Sinha P, Jha TK, Chakravarty J, Rai M, Kumar N, Pandey K, Narain MK, Verma N, Das VN, Das P, Berman J, Arana B (2013) Oral miltefosine for Indian post-kala-azar dermal leishmaniasis: a randomised trial. Tropical Med Int Health 18(1):96–100. https://doi.org/10.1111/tmi.12015

Sundar S, Singh A, Tiwari A, Shukla S, Chakravarty J, Rai M (2014) Efficacy and safety of paromomycin in treatment of post-kala-azar dermal leishmaniasis. ISRN Parasitol 2014: 548010. https://doi.org/10.1155/2014/548010

Tavares CA, Fernandes AP, Melo MN (2003) Molecular diagnosis of leishmaniasis. Expert Rev Mol Diagn 3(5):657–667. https://doi.org/10.1586/14737159.3.5.657

Thakur CP, Narain S, Kumar N, Hassan SM, Jha DK, Kumar A (1997) Amphotericin B is superior to sodium antimony gluconate in the treatment of Indian post-kala-azar dermal leishmaniasis. Ann Trop Med Parasitol 91(6):611–616. https://doi.org/10.1080/00034989760707

Trindade MAB, Silva LLDC, Braz LMA, Amato VS, Naafs B, Sotto MN (2015) Post-kala-azar dermal leishmaniasis and leprosy: case report and literature review. BMC Infect Dis 15(1):543. https://doi.org/10.1186/s12879-015-1260-x

Verma S, Bhandari V, Avishek K, Ramesh V, Salotra P (2013) Reliable diagnosis of post-kala-azar dermal leishmaniasis (PKDL) using slit aspirate specimen to avoid invasive sampling procedures. Tropical Med Int Health 18(3):268–275. https://doi.org/10.1111/tmi.12047

Verma S, Singh R, Sharma V, Bumb RA, Negi NS, Ramesh V, Salotra P (2017) Development of a rapid loop-mediated isothermal amplification assay for diagnosis and assessment of cure of Leishmania infection. BMC Infect Dis 17(1):223. https://doi.org/10.1186/s12879-017-2318-8

WHO. (2010). Control of the leishmaniases: WHO TRS N°949, report of a meeting of the WHO expert committee on the control of Leishmaniases, Geneva, 22–26 march 2010

WHO. (2012). The post Kala-azar dermal leishmaniasis (PKDL) atlas: a manual for health workers. E. E. Z. A. J. Alvar, Ed. WHO. https://www.who.int/publications/i/item/9789241504102

WHO (2020) Ending the neglect to attain the sustainable development goals: a road map for neglected tropical diseases 2021–2030. World Health Organization, Geneva

WHO. (2023). Factsheet-leishmaniasis. https://www.who.int/news-room/fact-sheets/detail/leishmaniasis

Yadav P, Azam M, Ramesh V, Singh R (2023) Unusual Observations in Leishmaniasis-An Overview. *Pathogens* 12(2):297. https://doi.org/10.3390/pathogens12020297

Younis BM, Osman M, Khalil EAG, Santoro F, Furini S, Wiggins R, Keding A, Carraro M, Musa AEA, Abdarahaman MAA, Mandefield L, Bland M, Aebischer T, Gabe R, Layton AM, Lacey CJN, Kaye PM, Musa AM (2021) Safety and immunogenicity of ChAd63-KH vaccine in post-kala-azar dermal leishmaniasis patients in Sudan. Mol Ther 29(7):2366–2377. https://doi.org/10.1016/j.ymthe.2021.03.020

Zijlstra EE (2016) The immunology of post-kala-azar dermal leishmaniasis (PKDL). Parasit Vectors 9(1):464. https://doi.org/10.1186/s13071-016-1721-0

Zijlstra EE, Khalil EA, Kager PA, El-Hassan AM (2000) Post-kala-azar dermal leishmaniasis in the Sudan: clinical presentation and differential diagnosis. Br J Dermatol 143(1):136–143. https://doi.org/10.1046/j.1365-2133.2000.03603.x

Zijlstra EE, Musa AM, Khalil EA, El-Hassan IM, El-Hassan AM (2003) Post-kala-azar dermal leishmaniasis. Lancet Infect Dis 3(2):87–98. https://doi.org/10.1016/s1473-3099(03)00517-6

Zijlstra EE, Alves F, Rijal S, Arana B, Alvar J (2017) Post-kala-azar dermal leishmaniasis in the Indian subcontinent: a threat to the South-East Asia region kala-azar elimination programme. PLoS Negl Trop Dis 11(11):e0005877. https://doi.org/10.1371/journal.pntd.0005877

Zijlstra EE, Liberton N, Musa AM, Te Slaa S, Wolff J (2020) Three-dimensional optical scanning in post-kala-azar dermal leishmaniasis (PKDL). Am J Trop Med Hyg 102(2):286–288. https://doi.org/10.4269/ajtmh.19-0370

Emergence of Novel *Leishmania* Genetic Variants: A New Challenge to the Ongoing Leishmaniasis Elimination Program in the Indian Subcontinent

7

Yogesh Chauhan, Priyanka Madaan, and Manju Jain

Abstract

Leishmaniasis is a disease complex caused by the protozoan parasite, *Leishmania* with clinical manifestations ranging from cutaneous leishmaniasis (CL) with skin lesions to mucocutaneous leishmaniasis (MCL) involving the mucous membrane to systemic visceral leishmaniasis (VL). The disease outcome is mediated by the parasite determinants along with the host immunity elicited by the infecting species involved. *L. donovani* or *L. infantum* is responsible for VL; the *L. major*, *L. mexicana,* or *L. amazonensis* causes CL, and the subgenus *L. viannia* causes CL and MCL as per the classical association of specific parasite species with distinct clinical outcomes. This association between the infecting *Leishmania* species and clinical outcome is more recently changing with the emergence of newer parasite variants in known and newer endemic regions, such that parasite causing visceral disease is able to cause cutaneous disease and vice versa. Examples of such atypical leishmaniasis forms are on an increase with reports of atypical CL caused by *L. donovani* genetic variants from the Indian subcontinent. The phenomenon points toward the possible emergence of novel parasite variants and calls for molecular surveillance of the heterogeneous parasite isolates from the known and newer endemic sites to understand the changing disease landscape. The increased recognition of atypical leishmaniasis as an outcome of parasite variants has major implications for leishmaniasis control and elimination. Identifying molecular correlates of parasite isolates associated with the disease diversification, prediction of transmission patterns of region-specific parasite variants can provide baseline data of relevance for

Y. Chauhan · P. Madaan · M. Jain (✉)
Department of Biochemistry, Central University of Punjab, Bathinda, Punjab, India
e-mail: manju.jain@cup.edu.in

© The Author(s), under exclusive license to Springer Nature Singapore Pte Ltd. 2023

A. Selvapandiyan et al. (eds.), *Challenges and Solutions Against Visceral Leishmaniasis*, https://doi.org/10.1007/978-981-99-6999-9_7

better control programs to counteract the novel parasite variants posing newer challenges to disease elimination.

Keywords

Visceral leishmaniasis · Cutaneous leishmaniasis · Leishmania donovani · Atypical leishmaniasis · Genetic variant

7.1 Introduction

Leishmaniasis is one of the infectious tropical diseases on priority for control and elimination. It is a disease complex caused by a protozoan parasite of the genus *Leishmania*. More than 20 species of *Leishmania* are responsible for different clinical manifestations, majorly prominent in developing regions in approximately 98 countries and territories in the tropics, subtropics, and Southern Europe (Alvar et al. 2012; World Health Organization 2022; Centers for Disease Control and Prevention 2020). *L. major, L. tropica and L. aethiopica* cause cutaneous leishmaniasis (CL) in the Old World, whereas *L. mexicana* and *L. braziliensis* complex cause CL in the New World. Members of subgenus *Viannia*: *L.* (*V.*) *braziliensis, L.* (*V.*) *panamensis* and *L.* (*V.*) *guyanensis* are responsible for mucocutaneous leishmaniaisis (MCL). *L. donovani* and *L. infantum/chagasi* cause the fatal visceral leishmaniasis (VL) (Centers for Disease Control and Prevention 2020). Approximately 0.6–1 million new CL cases and approximately 50,000–90,000 new VL cases are reported annually. The population involved in outdoor activities, immunocompromised individuals with immune defects and coinfections, and those living in conditions of malnutrition and poverty are more prone to leishmaniasis (World Health Organization 2022; Centers for Disease Control and Prevention 2020; Picado et al. 2014; Thakur et al. 2018) Infected female phlebotomine sand flies inject parasite promastigotes via blood meal in the mammalian host that transform into nonflagellated amastigote form and multiply in host macrophage cells. The infected host further transmits the parasite to uninfected sand flies as a part of the digenetic life cycle back to the promastigote stage. The parasite starts by infecting the dermis and can be confined to the skin region or extent to the mucosa of nasopharyngeal region or may pass through the lymphatic and vascular systems and infect other phagocytic cells of reticuloendothelial system to invade the bone marrow, liver, spleen, and lymph nodes depending on the parasite species/strains involved and host immunemodulation.

The timelines for the leishmaniasis control and elimination programs are extended at the national and global levels due to lack of awareness, poor health service infrastructure, underreported cases, asymptomatic and post kala-azar dermal leishmaniasis (PKDL) cases as parasite reservoirs along with the emergence of drug-resistant parasite variants (Hirve et al. 2016; Olliaro et al. 2017; World Health Organization. Regional Office for South-East, A. 2005). Recently, a newer phenomenon with the emergence of novel parasite variants in known and newer endemic regions, unlike the classical species-specific disease phenotype, is witnessed. In such endemic foci with atypical leishmaniasis, unusual disease manifestations are seen such that *L. donovani* complex classically known to cause visceral disease can cause cutaneous disease, and *L. tropica* complex known to cause cutaneous disease can cause viscerotropic outcome (Thakur et al. 2018; Nawaratna et al. 2009). The situation warrants a better understanding of the genetic variation and relatedness of the parasite isolates from classical and atypical leishmaniasis cases to gauge the origin, transmission, and disease outcome in newer regions endemic for leishmaniasis and/or known endemic foci with continued disease prevalence (Kumar Bhat et al. 2017; Raina et al. 2016; Ready 2010; Siriwardana et al. 2017; Thakur et al. 2020; Kariyawasam et al. 2017). The current disease landscape is discussed in the chapter in line with classical and atypical manifestations of leishmaniasis.

7.2 Classical Leishmaniasis

The classical disease complex comprises different clinical manifestations caused in a species-specific manner such that clinical-parasitological features are often used to differentiate the infecting parasite species. The visceral, cutaneous, and mucocutaneous infections with the known causative parasite species/subspecies and the associated pathophysiology are discussed under this section. *L. donovani* and *L. infantum/chagasi* are the main etiologic agents that cause VL, whereas CL is mainly caused by *L. major* and *L. tropica*.

7.3 Visceral Leishmaniasis

VL is the systemic disease caused by *L. donovani* and *L. infantum/chagasi* in the old world and new world, respectively. Countries including Brazil, China, Ethiopia, Kenya, India, Sudan, and South Sudan contribute to more than 90% of cases (World Health Organization 2022). *L. chagasi* was considered as a separate species in the New World, but a study conducted by Lukes, et al. identified it as a subpopulation of *L. infantum* (Lukes et al. 2007). Zoonotic VL involving transmission from animals to vector to human has been reported in *L. infantum* endemic areas, namely Latin America, Europe, and North Africa, and anthroponotic VL with transmission from human to vector to human in *L. donovani* endemic areas is reported in East Africa and Indian subcontinent (Chappuis et al. 2007). In Indian subcontinent, northeastern India, western and central districts of Bangladesh, the southeastern Terai region of Nepal, and more recently, Bhutan and Sri Lanka comprise the endemic regions (Thakur et al. 2018; World Health Organization. Regional Office for South-East, A. 2005).

Interestingly, parasite isolates from different endemic regions exhibit discrete genetic lineages and variations often associated with new phenotypes possibly associated with increased dissemination and persistence of the disease parasite despite ongoing elimination programs. Recent developments in whole-genome sequencing (WGS) have helped to decipher the genetic makeup of parasite isolates in relation to differential disease phenotypes, the emergence of drug resistance, and in defining the population structure circulating in the known and newer endemic regions (Downing et al. 2011; Gupta et al. 2015; Imamura et al. 2016; Zackay et al. 2018). In addition, newer endemic regions are reported with further dissemination of the disease (Nawaratna et al. 2009; Kumar Bhat et al. 2017; Raina et al. 2016; Ready 2010; Siriwardana et al. 2017; Thakur et al. 2020; Kariyawasam et al. 2017).

7.4 Cutaneous and Mucocutaneous Leishmaniasis

CL and MCL are restricted to infection of the skin and mucous membrane, respectively. *L. major, L. tropica, and L. aethiopica* cause CL in the Old World, whereas *L. amazonensis, L. mexicana,* and *L. braziliensis* cause CL in the New World. Subgenus *L. viannia* either remains in the skin or penetrates into the mucocutaneous layer to cause CL and MCL respectively. Approximately 0.6–1 million new CL cases are reported every year. Americas, Middle East, Central Asia, and the Mediterranean basin contribute approximately 95% CL cases (World Health Organization 2022).

CL can exhibit a gradation of disease severity involving differences in number and nature of lesions, load and persistence of parasite, localized versus diffuse lesions, nonhealing to healing phenotypes with varying timelines (Scorza et al. 2017). Similar to VL, an increasing number of CL cases have been identified in North African countries with increasing geographical distribution of disease in Morocco, Libya, and Tunisia that reported with more than thousands of CL cases every year (Aoun and Bouratbine 2014). In the Indian subcontinent, CL caused by *L. tropica* is reported in dry and hot pockets of Rajasthan, with Bikaner as the major disease region (Kumar et al. 2007). Interestingly, a high number of CL patients are documented from hilly regions in Jammu and Kashmir Valley (Kaul et al. 2016; Wani et al. 2015). In Nepal, sporadic CL cases have been referred, with *L. major* as the causative agent in one such case (Kumar et al. 2008). Also, case reports of CL in immigrants from the Middle East are documented in Bangladesh (Basher et al. 2017).

L. braziliensis is responsible for most cases of MCL, while *L.* (*V.*) *panamensis* and *L.* (*V.*) *guyanensis* may also cause MCL. Brazil, Peru, Ethiopia, and Bolivia contribute to more than 90% of MCL cases in the world. The disease form gets manifested as ulcerative mucosal lesions involving the mucosa of nose, mouth, pharynx, and larynx of the affected regions (World Health Organization 2022; Pinart et al. 2020).

7.5 Atypical Leishmaniasis

The classical landscape of species-specific disease phenotype is assuming new dimensions with the demonstration of genetic variants of classically viscerotropic and/or dermotropic parasite species causing atypical disease manifestations. In this context, dermotropic species have been detected as causing agents of VL manifestation, while parasite species known to cause visceral disease have been identified from CL and MCL-like patients. The phenomenon implies the emergence and circulation of parasite variants as the potential basis of atypical disease manifestations. Parasite isolates from specific pockets of atypical VL and CL are beginning to be explored for potential genetic variability and to identify genes associated with differential tropism and virulence in relation to atypical VL and CL disease phenotypes.

7.6 Atypical Visceral Leishmaniasis (AVL)

Visceral manifestations have been reported to be caused by *L. tropica*, *L. major,* and *L. amazonensis* classically known as the causative agents of CL. Such AVL cases although rare have been identified and demonstrated to have parasite dissemination to the visceral organs with a heterogeneous disease phenotype often overlapping with a cutaneous disease-like features (Thakur et al. 2018). AVL by *L. amazonensis* is reported with characteristic hepatosplenomegaly, hepatocellular injury, anemia, and leukopenia and generalized adenopathy (Aleixo et al. 2006). *L. tropica*–mediated AVL has been reported in Iran as scattered cases with diffuse CL-like manifestation alongside in one such case (Alborzi et al. 2006, 2008). In one such instance, *L. tropica* clinical isolate from a patient with viscerotropic disease manifestation was shown to exhibit genetic variation compared to the reference *L. tropica* (Sarkari et al. 2016). In India, a fraction of parasite isolates from the northeast VL belt classical VL have been typed as *L. tropica* with genetic dissimilarities from those isolated from typical *L. tropica*–mediated CL patients (Khanra et al. 2012; Krayter et al. 2014). VL causing *L. tropica*, isolated from a patient in Israel, was found to be closest to *L. tropica* on the basis of serotyping such that the mobilities of the enzymes pyruvate kinase, glucose phosphate isomerase, glucose-6-phosphate dehydrogenase, proline iminopeptidase, aspartate aminotransferase, and alanine aminotransferase were similar to *L. donovani* (Oren et al. 1991), while VL causing drug-resistant *L. tropica* isolates have been reported from Kenyan patients (Mebrahtu et al. 1989). *L. tropica* has also been reported to cause PKDL cases, where patients may or may not be responsive to sodium antimony gluconate (Sacks et al. 1995).

L. amazonensis has the capacity to cause MCL, CL, and VL, with genetic heterogeneity possibly associated with the different disease manifestations (de Oliveira et al. 2007). Additionally, PKDL has also been reported due to *L. amazonensis,* in Bahia, Brazil (Barral et al. 1991). *L. major* is one of the etiologic agents of CL and has been documented to subsequently exhibit visceral dissemination, in a patient in Bushehr Province of Southern Iran (Karamian et al. 2007).

7.7 Atypical Cutaneous Leishmaniasis (ACL)

Cutaneous manifestation in atypical cases infected with *L. donovani* and *L. infantum/chagasi* classically known to cause VL is an ongoing phenomenon in different parts of the world. The atypical *L. infantum* CL has been reported in parts of the Mediterranean basin, North Africa, and Europe, as well as countries in South America with autochthonous origin often associated with detection of parasite variants (Aoun and Bouratbine 2014; Convit et al. 2005; del Giudice et al. 1998; Rioux et al. 1980). Similar to *L. infantum*–mediated CL, distinct zymodemes and genetic variants of *L. donovani* have been demonstrated to cause cutaneous manifestations in countries like Sudan and Kenya.

L. donavani variants with dermotropic capabilities have also been reported in the Indian Subcontinent with the increasing number of cases in disease foci now recognized as endemic for ACL in Sri Lanka and India. High incidence of patients with cutaneous lesions lacking VL-like symptoms is reported in Sri Lanka, such that the clinical isolates from skin lesions are identified as *L. donovani* MON-37 zymodeme, genotypically related, but distinct from the VL causing *L. donovani* MON-2 in India (Karunaweera et al. 2003). In India, specific pockets in the states of Kerala and Himachal Pradesh have been demonstrated to be endemic for the atypical CL. The forest tribal settlements, namely Kani tribe and Agasthyamala Biosphere Reserve forest along the Western Ghats, comprise the endemic region in Kerala (Kumar et al. 2015). The clinical isolates from CL cases in the region have been identified as *L. donovani* variants, similar to *L. donovani* isolates from Sri Lankan ACL cases from atypical CL cases. Himachal Pradesh, the hilly state in northwest India, is emerging as a new disease state in India. The state with the increasing number of ACL cases along the Sutlej river belt is now considered an endemic focus (Thakur et al. 2020; Sharma et al. 2005, 2009). A comprehensive molecular analysis of clinical isolates from the region identifies them as CL causing *L. donovani* genetic variants different from the VL causing *L. donovani* in India but related to CL causing *L. donovani* isolates from Kerala and Sri Lanka (Thakur et al. 2020).

The atypical VL and CL disease etiology needs to be understood in terms of parasite determinants in line with increasing evidence of genetic heterogeneity of parasite isolates from such cases. Studies on clinical isolates from distinct atypical VL and CL endemic regions have unraveled specific genetic differences, based on distinct gene sequences and polymorphisms defining parasite variants compared to standard species-specific parasite genotypes associated with classical VL and/or CL disease phenotype. The emergence and circulation of novel genetic variants can possibly explain the emergence of atypical leishmaniasis and thus warrant more studies on genetic analysis of the clinical isolates from different disease foci for an insight into unusual disease biology (Fig. 7.1).

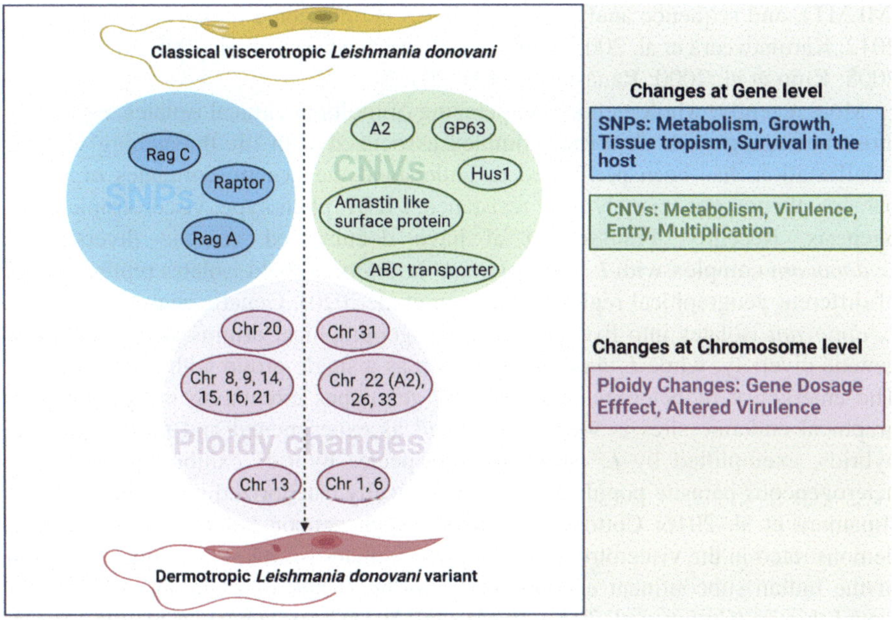

Fig. 7.1 Genetic basis of *Leishmania donovani*–mediated atypical cutaneous leishmaniasis. (created with BioRender.com)

7.8 Insight into Genetic Analysis of Parasite to Understand Atypical VL and Atypical CL

The rationale behind the association of distinct infecting *Leishmania* species and the clinical outcome is a question still under exploration for the leishmaniasis disease complex. More recently, the significance of parasite determinants in driving the biology of disease outcome is being explored, with the demonstration of novel parasite genotypes in relation to distinct disease phenotypes (McCall et al. 2013; Zhang and Matlashewski 2010). Knowledge on the genetic basis of classical viscerotropic versus dermotropic disease manifestation by specific parasite species is far from sufficient. Plasticity of *Leishmania* genome exhibited in laboratory-grown cultures and clinical isolates due to environmental selection pressures as per geographical locations associated with different vector species, host populations, and varied disease outcomes is a hindrance to deciphering genotype–phenotype correlates (Downing et al. 2011; Zackay et al. 2018; McCall et al. 2013; Franssen et al. 2020; Sinha et al. 2018). Nonetheless, exploration on the molecular mechanism of differential tissue tropism and parasite species/subspecies pathogenic capabilities to cause visceral, cutaneous, and mucocutaneous disease is underway. Parasite isolates from VL and CL cases have been characterized based on isoenzyme electrophoretic mobility, PCR-RFLP patterns, Multilocus microsatellite typing

(MLMT), and sequence analysis of specific gene/intergenic regions. (Khanra et al. 2012; Karunaweera et al. 2003; Kumar et al. 2015; Sharma et al. 2005; Elamin et al. 2008; Knio et al. 2000; Ranasinghe et al. 2012).

More recently, whole-genome sequencing of multiple clinical isolates, especially those belonging to *L. donovani* complex associated with life-threatening visceral manifestation, has been performed to understand the genetic correlates of varied parasite phenotypes, namely drug resistance, PKDL phenotype, viscerotropic pathogenesis. Recently, Franssen et al have documented genome diversity of *L. donovani* complex with *L. donovani* and *L. infantum* field isolates representative of different geographical regions (Franssen et al. 2020). Genetic analysis clustered *L. donovani* isolates into five groups as per geographical origins albeit with huge genetic diversity, while *L. infantum* came up as a single group with little diversity. The emergence of heterogeneous parasite genotypes circulating in specific geographical endemic sites is well documented as experimental and natural parasite hybrids, exemplified by *L. donovani* intraspecies hybrids exhibiting genetically heterogeneous parasite population phylogenetically categorized into distinct clades (Imamura et al. 2016; Cotton et al. 2020). Such genetic heterogeneity has been demonstrated in the viscerotropic *L. donovani* isolates from key VL endemic region of the Indian subcontinent encompassing northeast belt of India and VL isolates from Ethiopia (Cotton et al. 2020; Zhang et al. 2014; Lypaczewski et al. 2018, 2022; Lypaczewski and Matlashewski 2021).

Apart from understanding the parasite genotypes associated with classical VL and CL, it is imperative to understand the phenomenon of atypical leishmaniasis in terms of the genetic makeup of the causative parasite. Clinical isolates from atypical VL and CL cases provide useful model organisms to decipher the genetic basis of visceral versus cutaneous pathogenicity. Thus, the molecular survey of parasite isolates from atypical VL and CL endemic foci can help to identify the genetic basis of differential tropism and virulence. Multiple studies on this aspect have unraveled specific parasite genes, single-nucleotide polymorphisms (SNPs), gene amplification and/or deletion, copy number variations along with changes in chromosome ploidy associated with atypical disease outcome (Thakur et al. 2020; Kariyawasam et al. 2017; Zhang et al. 2014; Lypaczewski et al. 2018, 2021, 2022; Lypaczewski and Matlashewski 2021).

The atypical VL cases caused by *L. tropica*, although fewer in number, were detected among the predominant classical VL cases in the northeastern India based on species-specific PCR-RFLP (Khanra et al. 2012). In line with this initial finding, multilocus microsatellite typing (MLMT) of Indian *L. tropica* isolates causing cutaneous leishmaniasis in Bikaner and atypical visceral cases in Bihar from Indian Subcontinent exhibited genetic relatedness albeit with discrete genetic differences (Krayter et al. 2014). The genetic differences in the *L. tropica* subpopulations could potentially explain the basis of altered disease outcomes by *L. tropica* variants (Krayter et al. 2014). Similarly, intraspecies genetic variation based on sequence analysis of N-Acetylglucosamine-1-Phosphate Transferase (nagt) gene identified *L. tropica* as the etiologic agent of viscerotropic manifestations in a VL endemic area in southwestern Iran (Sarkari et al. 2016).

Much of the documented incidences of atypical leishmaniasis comprise atypical cutaneous manifestation caused by *L. donovani* variants with most of them coming from the Indian Subcontinent. Countries with ACL include Nepal, Sri Lanka, and India with endemic foci in the states of Kerala and Himachal Pradesh (Thakur et al. 2020; Kumar et al. 2015; Parajuli et al. 2020). Initial detection of *L. donovani* from indigenous cases of CL from these regions was based on species-specific genic and intergenic sequence comparisons of Heat Shock Protein 70 (HSP70), 6-phosphogluconate dehydrogenase (6PGDH), Internal Transcribed spacer (ITS, ITS1), minicircle kDNA, and RFLP patterns (Thakur et al. 2020; Khanra et al. 2012; Kumar et al. 2015; Parajuli et al. 2020; Tabbabi et al. 2020). The CL causing *L. donovani* isolates from Himachal Pradesh were similar to isolates from CL cases in Kerala and Sri Lanka based on partial sequence comparison of 6PGDH with aspartic acid at 326 position and differed from the classical VL causing *L. donovani* isolates from India and Bangladesh with asparagine at 326 position (Thakur et al. 2020).

With detection and confirmation of atypical CL caused by *L. donovani* in these regions, it is imperative to understand the parasite determinants that drive the disease phenotype toward visceral disease versus cutaneous manifestation. In an attempt toward this end, in vitro experiments wherein *L. donovani* genes, namely A2 gene cluster and Ld28.0340, a cytosolic protein of unknown function are known to enhance survival of transgenic *L. major* in the viscera of infected mice (McCall et al. 2013; Zhang and Matlashewski 2010). In multiple studies toward exploring the genetic basis of classical visceral and cutaneous disease phenotype, a relatively small number of genes have been shown to be differentially present in viscerotropic versus dermotropic species with most of these lacking functional annotation (Smith et al. 2007).

More recent comparative genome studies on clinical isolates from classical versus atypical disease cases, namely *L. donovani* isolates from classical VL cases and those from atypical CL cases in India and Sri Lanka, have provided substantial insight into the genetics of disease outcome. Much of the findings imply the role of species-specific parasite genes, gene-specific single nucleotide polymorphisms, and copy number variations in dictating disease phenotypes (Tables 7.1 and 7.2). A decreased copy number of the virulence A2 gene cluster and a mutation in the Rag C GTPase have been identified as determinants of attenuated CL-causing *L. donovani* strain in Sri Lanka (Zhang et al. 2014; Lypaczewski et al. 2021, 2022). A2 gene family plays an important role in visceralization of the parasite (Zhang et al. 2003). A2 gene products are stress-inducible factors, required for visceral infection and survival in host's visceral organs. Two opposing A2 gene clusters are present on chromosome 22, while absent in *L. major*. Transgenic *L. major* expressing A2 genes, when injected in mice, resulted in increased survival in the spleen (Zhang et al. 2003). An altered copy number and transcript abundance of the A2 gene cluster have been proved to be important for visceral pathogenesis. Experimentally increased expression of A2 gene, under high temperature, in CL causing MON-37 zymodeme isolated from Sri Lanka exhibits increased survival in the visceral organs (Zhang et al. 2014). The functional characterization studies mark A2 proteins as

Table 7.1 Genetic variations (SNPs) at gene level in atypical cutaneous leishmaniasis

Sr. No.	Gene (Gene ID)	Chr	Gene variant	Function of gene/ gene product	Endemic site of geographical isolation	Functional validation	References
1.	*Rag C* (ras like GTPase C) gene (LdBPK_366140.1)	36	R231C	Component of mTOR pathway, helps in survival in the host visceral organ (VO)	Sri Lanka	Validated	Zhang et al. (2014)
2.	DnaJ family protein, putative, highly conserved (LdBPK_252290.1)	25	S160G	Role in protein translation, folding, translocationand degradation	Sri Lanka	Validated	Zhang et al. (2014)
3.	Phosphoinositide phosphatase (LdBPK_220120.1)	22	G226D	Cell signaling	Sri Lanka	Validated	Zhang et al. (2014)
4.	Rab GTPase family 2/Rab2 (LdBPK_322160.1)	32	Q163H	Involved in membrane trafficking	Sri Lanka	Validated	Zhang et al. (2014)
5.	3-oxo-5-alphasteroid 4-dehydrogenase (LdBPK_171200)	17	P147Q	Synthesis of dihydrotestosterone in higher eukaryotes	Sri Lanka	Validated	Zhang et al. (2014)
6.	Raptor N-terminal CASPase like domain containing protein (*LdBPK_250620* LdCL_25001140)	25	Ala969Glu Replacement of alanine by glutamine at 696 position	Cell signaling	Sri Lanka	Not validated	Lypaczewski et al. (2018)
7.	Sucrose hydrolase-like protein (LdCL_23001750)	23	INS:446Glu Insertion of glutamic acid at 446 position	Carbohydrate metabolism	Sri Lanka	Not validated	Lypaczewski et al. (2018)
8.	ATP-binding cassette protein subfamily A (*LdBPK_270830* LdCL_27001490)	27	Ala1493Thr Alanine replaced by threonine at 1493 position	Transport	Sri Lanka	Not validated	Lypaczewski et al. (2018)

9.	Kinesin K39 (*LdBPK_250620* LdCL_14001770)	14	Glu1034Asp Glutamic acid replaced by aspartic acid at 1034 position	Parasite motility	Sri Lanka	Not validated	Lypaczewski et al. (2018)
10.	long-chain-fatty-acid-CoA ligase, putative (LdCL_010009900)	1	T127149C (Val18Ala)	Fatty acid metabolism	India	Not validated	Lypaczewski et al. (2022)
11.	Potassium channel subunit-like protein (LdCL_010013500)	1	AGCACCG265108A (Pro767_Ala768del)	Ion transport	India	Not validated	Lypaczewski et al. (2022)
12.	phosphoglycan beta 1,3 galactosyltransferase 4 (LdCL_020005000)	2	A5979C (Leu108Arg)	Carbohydrate metabolism	India	Not validated	Lypaczewski et al. (2022)
13.	quinone oxidoreductase, putative (LdCL_030010600)	3	T214597C (Val198Ala)	Energy metabolism	India	Not validated	Lypaczewski et al. (2022)
14.	phosphatidylinositol kinase related protein, putative (LdCL_020006200)	2	T47087C (Ser4134Gly)	Cell Signaling	India	Not validated	Lypaczewski et al. (2022)
15.	phosphoglycan beta 1,3 galactosyltransferase (LdCL_020006800)	2	T87617C (Asn569Ser)	Carbohydrate metabolism	India	Not validated	Lypaczewski et al. (2022)
16.	Mitochondrial protein 81 (LdCL_020009200)	2	T184767G (Pro192Thr)	RNA editing	India	Not validated	Lypaczewski et al. (2022)
17.	Dehydrogenase/oxidoreductase-like protein (LdCL_020012500)	2	A354547G (Asn35Asp)	Oxidation-reduction	India	Not validated	Lypaczewski et al. (2022)
18.	Serine/threonine-protein kinase, putative (LdCL_030012800)	3	G294063A (Ala2846Val)	Signaling	India	Not validated	Lypaczewski et al. (2022)
19.	Adenosine monophosphate deaminase, putative (LdCL_040008200)	4	C95836A (His1464Asn)	Nucleotide metabolism	India	Not validated	Lypaczewski et al. (2022)

(continued)

Table 7.1 (continued)

Sr. No.	Gene (Gene ID)	Chr	Gene variant	Function of gene/gene product	Endemic site of geographical isolation	Functional validation	References
20.	Nucleotidyltransferase domain containing protein, putative (LdCL_040010400)	4	G194511GGCTGCT (Ala2150_Ala2151dup)	Nucleotide modifications	India	Not validated	Lypaczewski et al. (2022)
21.	Tyrosine phospatase-like protein (LdCL_040014000)	4	A336138G (Ser200Pro)	Signaling	India	Not validated	Lypaczewski et al. (2022)
22.	Trypanothione reductase (LdCL_050008500)	5	G105073T (Arg250Leu)	Oxidation-reduction balance	India	Not validated	Lypaczewski et al. (2022)
23.	Kinetoplast-associated protein-like protein (LdCL_050010400)	5	T174721C (Val575Ala)	Motility	India	Not validated	Lypaczewski et al. (2022)
24.	ATP-dependent RNA helicase, putative (LdCL_050011000)	5	G196107T (Leu384Phe)	DNA repair	India	Not validated	Lypaczewski et al. (2022)
25.	Paraflagellar rod protein, putative (LdCL_050014300)	5	G344161A (Gly769Ser)	Motility	India	Not validated	Lypaczewski et al. (2022)
26.	Metallo-peptidase, Clan M-, Family M49 (LdCL_050014700)	5	G360427T (Gln266His)	Lysosomal and autophagy	India	Not validated	Lypaczewski et al. (2022)
27.	Adenylate cyclase regulatory protein-like protein (LdCL_050021400)	5	C471639T (Arg549Cys)	Nucleotide metabolism	India	Not validated	Lypaczewski et al. (2022)
28.	Dihydrofolate reductase-thymidylate synthase (LdCL_060014000)	6	G379736A (Gly132Asp)	Folate metabolism	India	Not validated	Lypaczewski et al. (2022)

29.	Pteridine transporter, putative (LdCL_060018500)	6	T506736C (His285A)	Pteridine transport	India	Not validated	Lypaczewski et al. (2022)
30.	Paraflagellar rod protein, putative (LdCL_070008200)	7	G137543A (Ala724Val)	Motility	India	Not validated	Lypaczewski et al. (2022)
31.	Dynein heavy chain, N-terminal region 2, putative (LdCL_070010300)	7	A228028G (His3990Arg)	Motility	India	Not validated	Lypaczewski et al. (2022)
32.	DNAJ-domain transmembrane-like protein (LdCL_070013600)	7	C372774CT (Ala101fs)	Stress response	India	Not validated	Lypaczewski et al. (2022)
33.	Amidinotransferase, putative (LdCL_080017400)	8	A507115G (Lys168Glu)	Polyamine metabolism	India	Not validated	Lypaczewski et al. (2022)
34.	Phospholipid: diacylglycerol acyltransferase, putative (LdCL_090017600)	9	G421459A (Ala36Val)	Phospholipid metabolism	India	Not validated	Lypaczewski et al. (2022)
35.	ABC transporter, putative (LdCL_090021500)	9	T536741C (Ile19Val)	Transport	India	Not validated	Lypaczewski et al. (2022)
36.	Phosphate-Repressible Phosphate Permease-like protein (LdCL_100021600)	10	G590903A (Glu276Lys), G590915A (Asp280Asn)	Transport	India	Not validated	Lypaczewski et al. (2022)
37.	Phosphate-Repressible Phosphate Permease-like protein (LdCL_100005200)	10	C12125T (Asp280Asn)	Transport	India	Not validated	Lypaczewski et al. (2022)
38.	Pteridine transporter, putative (LdCL_100009000)	10	G163291A (Val362Ile)	Pteridine transport	India	Not validated	Lypaczewski et al. (2022)
39.	GP63, leishmanolysin (LdCL_100010300)	10	G225451T (Gly184Val), A225454G (Asp185Gly), C254762G (Thr146Ser)	Parasite migration, attachment, internalization and survival in host macrophages	India	Not validated	Lypaczewski et al. (2022)

(continued)

Table 7.1 (continued)

Sr. No.	Gene (Gene ID)	Chr	Gene variant	Function of gene/ gene product	Endemic site of geographical isolation	Functional validation	References
40.	GP63, leishmanolysin (LdCL_100011800)	10	A270962T (Ile413Leu), T270977TA (Phe418fs), A270983T (Met420Leu)	Parasite migration, attachment, internalization and survival in host macrophages	India	Not validated	Lypaczewski et al. (2022)
41.	Translation initiation factor eIF-2B subunit beta, putative (LdCL_100017200)	10	C477757T (Gly227Asp)	Translation	India	Not validated	Lypaczewski et al. (2022)
42.	Inositol-1,4,5-trisphosphate (IP3) 5-phosphatase, putative (LdCL_110015900)	11	C393542T (Ala133Val)	Cell signaling	India	Not validated	Lypaczewski et al. (2022)
43.	Dolichyl-P-Man:GDP-Man5GlcNAc2-PP-dolichyl alpha-1,2-mannosyltransferase, putative (LdCL_120006700)	12	G49025A (Arg274Gln), C50210T (Ala669Val)	Sterol metabolism	India	Not validated	Lypaczewski et al. (2022)
44.	Mitogen-activated protein kinase kinase 2 (LdCL_130008700)	13	C123417T (Gly409Arg)	Cell signaling	India	Not validated	Lypaczewski et al. (2022)
45.	Amastin surface glycoprotein,putative (LdCL_140010300)	14	T176318C (ASN162SER)	Virulence	India	Not validated	Lypaczewski et al. (2022)
46.	Kinesin K39, putative (LdCL_140017600)	14	A505325G (Ser141Gly), A505338G (Asn145Ser), G507308T	Motility	India	Not validated	Lypaczewski et al. (2022)

			(Ala802Ser), G507314A (Ala804Thr), G507922C (Glu1006Asp)				
47.	Kinesin K39, putative (LdCL_140017800)	14	A533766G (Thr1110Ala), C534774T (Arg1446Cys), A535171G (Asp1578Gly)	Motility	India	Not validated	Lypaczewski et al. (2022)
48.	NADP oxidoreductase coenzyme F420-dependent, putative (LdCL_140019500)	14	G591605T (Cys388Phe)	Oxidation, reduction	India	Not validated	Lypaczewski et al. (2022)
49.	Calmodulin-like protein (LdCL_150015400)	15	T424607 TGCAGCC (Ala13_Ala14dup)	Cell Signaling	India	Not validated	Lypaczewski et al. (2022)
50.	Nucleoside transporter 1, putative (LdCL_150017800)	15	A504699G (Ser43Pro)	Nucleoside transport	India	Not validated	Lypaczewski et al. (2022)
51.	Carbamoyl-phosphate synthase, putative (LdCL_160011000)	16	T207412G (Lys633Thr)	Carbamoyl-phosphate	India	Not validated	Lypaczewski et al. (2022)
52.	3′5’-cyclic nucleotide phosphodiesterase, putative (LdCL_170016500)	17	G456233A (Val334Met), G456744A (Gly504Glu)	Nucleic acid metabolism	India	Not validated	Lypaczewski et al. (2022)
53.	L-galactonolactone oxidase, putative (LdCL_170020900)	17	CGCT640082T (Leu182del)	Carbohydrate metabolism	India	Not validated	Lypaczewski et al. (2022)
54.	Citrate synthase, putative (LdCL_180012000)	17	G285628T (Glu420Asp)	Amino acid metabolism	India	Not validated	Lypaczewski et al. (2022)

(continued)

Table 7.1 (continued)

Sr. No.	Gene (Gene ID)	Chr	Gene variant	Function of gene/ gene product	Endemic site of geographical isolation	Functional validation	References
55.	Adenylate cyclase regulatory protein-like protein (LdCL_190022700)	19	C704055T (Arg549Cys)	Cell signaling	India	Not validated	Lypaczewski et al. (2022)
56.	Phosphoglycerate kinase C, glycosomal (LdCL_200006100)	20	C37873T (Val80Met)	Carbohydrate metabolism	India	Not validated	Lypaczewski et al. (2022)
57.	GMP synthase (glutamine-hydrolysing) (LdCL_220006100)	22	C46920G (Glu417Asp)	Nucleic acid metabolism	India	Not validated	Lypaczewski et al. (2022)
58.	Pentamidine resistance protein 1 (LdCL_230007300)	23	G77126A (Glu1545Lys)	Drug resistance	India	Not validated	Lypaczewski et al. (2022)
59.	Pentamidine resistance protein 1 (LdCL_230007800)	23	G93956A (Glu1554Lys)	Drug resistance	India	Not validated	Lypaczewski et al. (2022)
60.	Pteridine reductase 1 (LdCL_230008500)	23	G119537A (Ala255Thr)	Pteridine reduction	India	Not validated	Lypaczewski et al. (2022)
61.	Galactose oxidase, central domain containing protein, putative (LdCL_230013300)	23	T247404C (Glu2100Gly)	Carbohydrate metabolism	India	Not validated	Lypaczewski et al. (2022)
62.	UDP-galactose transporter, putative (LdCL_240008500)	24	G119651T (Ala358Ser)	Nucleotide-sugan metabolism	India	Not validated	Lypaczewski et al. (2022)
63.	Thioredoxin/Thioredoxin-like, putative (LdCL_250012500)	25	G253662A (Pro19Leu)	Oxidation-reduction balance	India	Not validated	Lypaczewski et al. (2022)
64.	Galactose oxidase, central domain containing protein, putative (LdCL_250016900)	25	G436025A (Ala1291Val)	Carbohydrate metabolism	India	Not validated	Lypaczewski et al. (2022)

65.	ATP synthase subunit beta, mitochondrial, putative (LdCL_250017700)	25	T460235C (Lys234Arg)	Energy metabolism	India	Not validated	Lypaczewski et al. (2022)
66.	Type II (glutathione peroxidase-like) tryparedoxin peroxidase (LdCL_260013400)	26	A238091G (Phe165Ser)	Oxidation-reduction balance	India	Not validated	Lypaczewski et al. (2022)
67.	Intraflagellar transport protein 88, putative (LdCL_270016600)	27	C483715T (Val729Ile)	Transport	India	Not validated	Lypaczewski et al. (2022)
68.	Phosphatidylinositol 3- and 4-kinase, putative (LdCL_270027900)	27	G936481A (Ala288Val)	Cell signaling	India	Not validated	Lypaczewski et al. (2022)
69.	Acyl-CoA oxidase, putative (LdCL_270029500)	27	G990021A (Pro571Ser)	Fatty acid metabolism	India	Not validated	Lypaczewski et al. (2022)
70.	RNA polymerase I (LdCL_280013000)	28	C284167A (Ser295Arg)	Transcription	India	Not validated	Lypaczewski et al. (2022)
71.	Galactose oxidase, central domain/ Kelch motif containing protein, putative (LdCL_280028100)	28	C840793A (Ala358Asp)	Carbohydrate metabolism	India	Not validated	Lypaczewski et al. (2022)
72.	Mannosyl oligosaccharide glucosidase, putative (LdCL_280028600)	28	C855180A (Ala361Glu)	Carbohydrate metabolism	India	Not validated	Lypaczewski et al. (2022)
73.	Flagellar Member 6 (LdCL_290016400)	29	G404179A (Ala380Thr)	Motility	India	Not validated	Lypaczewski et al. (2022)
74.	Asparagine synthase, putative (LdCL_290021500)	29	CAGG689132C (Pro443del)	Amino acid metabolism	India	Not validated	Lypaczewski et al. (2022)
75.	5-Methyltetrahydropteroyltriglutamate—homocysteine S-methyltransferase (LdCL_310005000)	31	C4044T (Glu433Lys)	Not known	India	Not validated	Lypaczewski et al. (2022)

(continued)

Table 7.1 (continued)

Sr. No.	Gene (Gene ID)	Chr	Gene variant	Function of gene/ gene product	Endemic site of geographical isolation	Functional validation	References
76.	Aquaglyceroporin 1 (LdCL_310005100)	31	G7750A (Arg228Cys)	Transport	India	Not validated	Lypaczewski et al. (2022)
77.	Calpain-like protein, putative (LdCL_310009300)	31	G127746A (Pro670Ser)	Virulence	India	Not validated	Lypaczewski et al. (2022)
78.	Calpain-like cysteine peptidase, putative (LdCL_310009500)	31	C141379T (Gly208Asp), T141620C (Thr128Ala)	Virulence	India	Not validated	Lypaczewski et al. (2022)
79.	Calpain-like cysteine peptidase, putative (LdCL_310010200)	31	C174928T (Cys866Tyr)	Virulence	India	Not validated	Lypaczewski et al. (2022)
80.	Pentamidine resistance protein 1 (LdCL_310021500)	31	G683347A (Pro1128Ser)	Drug resistance	India	Not validated	Lypaczewski et al. (2022)
81.	Amino acid permease (LdCL_310025400)	31	G876950C (Thr347Ser), T876951C (Thr347Ala), G877052A (Ala313Val)	Transport	India	Not validated	Lypaczewski et al. (2022)
82.	Amino acid permease (LdCL_310025500)	31	G884673T (Arg17Ser)	Transport	India	Not validated	Lypaczewski et al. (2022)
83.	Amino acid permease (LdCL_310025800)	31	C893579A (Ala526Ser), C893975G (Val394Leu)	Transport	India	Not validated	Lypaczewski et al. (2022)

84.	Mannosyltransferase-like protein (LdCL_310026900)	31	G961354A (His334Tyr)	Glycosylation	India	Not validated	Lypaczewski et al. (2022)
85.	Mannosyltransferase-like protein (LdCL_310027000)	31	A964013G (Tyr938His), G965825A (His334Tyr)	Glycosylation	India	Not validated	Lypaczewski et al. (2022)
86.	Squalene synthase, putative (LdCL_310038800)	31	G1428805A (Ser7Leu)	Lipid metabolism	India	Not validated	Lypaczewski et al. (2022)
87.	Phosphoglycan beta 1,3 galactosyltransferase 4 (LdCL_310041600)	31	G1520387A (Cys61Tyr)	Lipophosphoglycan metabolism	India	Not validated	Lypaczewski et al. (2022)
88.	Serine hydrolase (FSH1)/50S ribosome-binding GTPase/ GTPase putative (LdCL_320039100)	32	G1246285A (Ala200Thr)	Not known	India	Not validated	Lypaczewski et al. (2022)
89.	Flagellar Member 7 (LdCL_320039400)	32	C1257896T (Ala341Val) G1260295T (Val1141Phe) T1260296C (Val1141Ala) G1260304A (Val1144Met) G1261987T (Val1705Phe) G1261996A (Val1708Met)	Motility	India	Not validated	Lypaczewski et al. (2022)
90.	3-hydroxyisobutyryl-coenzyme a hydrolase-like protein (LdCL_320044100)	31	G1449228C (Gly81Ala)	Amino acid metabolism	India	Not validated	Lypaczewski et al. (2022)

(continued)

Table 7.1 (continued)

Sr. No.	Gene (Gene ID)	Chr	Gene variant	Function of gene/ gene product	Endemic site of geographical isolation	Functional validation	References
91.	3-hydroxyisobutyryl-coenzyme a hydrolase-like protein (LdCL_320044200)	31	G1453712C (Gly81Ala)	Amino acid metabolism	India	Not validated	Lypaczewski et al. (2022)
92.	Ras-like small GTPase, putative (LdCL_330032400)	33	A1022160C (Ser3Arg)	Cell signaling, survival in host visceral organs	India	Not validated	Lypaczewski et al. (2022)
93.	MRB1-associated protein (LdCL_330035800)	33	A1165645G (Gln84Arg)	RNA processing	India	Not validated	Lypaczewski et al. (2022)
94.	Tyrosyl-DNA phosphodiesterase 1 (LdCL_330039400)	33	C1405672G (Leu465Val)	Nucleic acid metabolism	India	Not validated	Lypaczewski et al. (2022)
95.	46 kD virulence factor (LdCL_330039500)	33	C1409404T (Thr102Met)	Virulence	India	Not validated	Lypaczewski et al. (2022)
96.	Flagellar Member 8 (LdCL_330039800)	33	T1418181C (Tyr2085His)	Motility	India	Not validated	Lypaczewski et al. (2022)
97.	Flagellar Member 8 (LdCL_330039900)	33	T1419731C (Tyr295His), G1420985A (Ala713Thr), G1426261T (Gln2471His)	Motility	India	Not validated	Lypaczewski et al. (2022)
98.	Phosphoglycan beta 1,2 arabinosyltransferase, (SCA like) (LdCL_340010900)	34	G228246T (Ala804Ser)	Transport	India	Not validated	Lypaczewski et al. (2022)
99.	Elongation factor 1-beta (LdCL_340014200)	34	TGAG384471T (Glu94del)	Translation	India	Not validated	Lypaczewski et al. (2022)

100.	Kelch motif/Galactose oxidase, central domain containing protein, putative (LdCL_340027900)	34	G941094A (Ala1105Val)	Carbohydrate metabolism	India	Not validated	Lypaczewski et al. (2022)
101.	Phosphoglycan beta 1,3 galactosyltransferase 4 (LdCL_350005000)	35	C5560T (Gly236Arg), T5563C (Lys235Glu), A5565C (Leu234Arg), C5567G (Arg233Ser)	Transport	India	Not validated	Lypaczewski et al. (2022)
102.	U5 snRNA-associated splicing factor (LdCL_350045100)	35	G1499624T (Lys2327Asn)	RNA splicing	India	Not validated	Lypaczewski et al. (2022)
103.	phosphatidylserine decarboxylase, putative (LdCL_350051700)	35	C1764266T (Pro264Ser)	Phospholipid biosynthesis	India	Not validated	Lypaczewski et al. (2022)
104.	glucose transporter 2 (LdCL_360072900)	36	C2443786T (Glu6Lys),	Carbohydrate transport	India	Not validated	Lypaczewski et al. (2022)
105.	glucose transporter 2 (LdCL_360073000)	36	C2447346T (Glu6Lys)	Carbohydrate transport	India	Not validated	Lypaczewski et al. (2022)
106.	glucose transporter 2 (LdCL_360073100)	36	C2450908T (Glu6Lys),	Carbohydrate transport	India	Not validated	Lypaczewski et al. (2022)
107.	glucose transporter 2 (LdCL_360073200)	36	C2454468T (Glu6Lys)	Carbohydrate transport	India	Not validated	Lypaczewski et al. (2022)

CL cutaneous leishmaniasis, *VL* visceral leishmaniasis, *SNP* single nucleotide polymorphism, *Non-Syn SNP* nonsynonymous single-nucleotide polymorphism, *CNV* copy number variation, *INS* insertion, *F/S* frameshift, *Chr* chromosome, *Rab2* Rab GTPase family 2, *Rag C* Ras-like GTPase C, *GP63* glycoprotein 63

Table 7.2 Genetic variations (CNV) at gene level in atypical cutaneous leishmaniasis

Sr No.	Gene (Gene ID)	Chr	Gene variant	Function of gene/gene product	Endemic site of geographical isolation	Functional validation done/not done	References
1.	A2 gene (LdBPK_220670.1)	22	Decreased copy number	Stress inducible factors, required for visceral infection and survival in host's visceral organs	Sri Lanka	Validated	Zhang et al. (2014) Put all rows for 57 together followed with rows from ref. Samarasinghe et al. (2018)
2.	ABC transporter-like protein (LdBPK_111220.1)	11	Decreased copy number (VL/CL Ratio: 2.94)	Transport	Sri Lanka	Not validated	Lypaczewski and Matlashewski (2021)
3.	Phosphoglycerate kinase B, cytosolic (LdBPK_200120.1)	20	Decreased copy number (VL/CL Ratio: 2.53)	Generate ATP in cytosol of Trypanosomatids	Sri Lanka	Not validated	Lypaczewski and Matlashewski (2021)
4.	ABC-thiol transporter (MRPA) (LdBPK_230290.1)	23	Increased copy number (VL/CL Ratio: 0.46)	Transport	Sri Lanka	Not validated	Lypaczewski and Matlashewski (2021)
5.	ABC-thiol transporter (MRPA) H-region (LdBPK_230250.1)	23	Increased copy number (VL/CL Ratio: 0.42)	Transport	Sri Lanka	Not validated	Lypaczewski and Matlashewski (2021)
6.	Terbinafine resistance locus protein, YIP1 (LdBPK_230280.1)	23	Increased copy number (VL/CL Ratio: 0.43)	Terbinafine resistance	Sri Lanka	Not validated	Lypaczewski and Matlashewski (2021)

7.	Amastin-Like Surface Protein, Putative (Fragment) (LdBPK_341720)	34	Decreased copy number (VL/CL Ratio: 2.10)	Act as vital transporters and signal transducers for Intracellular survival and adaptation in host	Sri Lanka	Not validated	Samarasinghe et al. (2018)
8.	Gp63, Leishmanolysin (LdBPK_100520)	10	Decreased copy number (VL/CL Ratio: 4.698)	Silent entry into macrophage and parasite multiplication in side host	Sri Lanka	Not validated	Samarasinghe et al. (2018)
9.	Gp63, Leishmanolysin (Fragment) (LdBPK_100510)	10	Decreased copy number (VL/CL Ratio: 2.55)	Parasite migration, attachment, internalization and survival in host macrophages	Sri Lanka	Not validated	Samarasinghe et al. (2018)
10.	Histone H3 (LdBPK_100920)	10	Decreased copy number (VL/CL Ratio: 3.55)	DNA metabolism	Sri Lanka	Not validated	Samarasinghe et al. (2018)
11.	Histone H4 (LdBPK_210020)	25	Decreased copy number (VL/CL Ratio: 3.80)	DNA metabolism	Sri Lanka	Not validated	Samarasinghe et al. (2018)
12.	D-Lactate Dehydrogenase-Like Protein (LdBPK_271940)	27	Decreased copy number (VL/CL Ratio: 7.96)	involved in energy metabolism	Sri Lanka	Not validated	Samarasinghe et al. (2018)
13.	Branched-Chain Amino Acid Aminotransferase, putative (fragment) (LdBPK_271950)	27	Decreased copy number (VL/CL Ratio: 6.84)	Amino acid metabolism	Sri Lanka	Not validated	Samarasinghe et al. (2018)
14.	Checkpoint Protein Hus1, Putative (LdBPK_230330)	23	Decreased copy number (VL/CL Ratio: 3.85)	Help the parasite to cope with replicative stress	Sri Lanka	Not validated	Samarasinghe et al. (2018)

(continued)

Table 7.2 (continued)

Sr No.	Gene (Gene ID)	Chr	Gene variant	Function of gene/gene product	Endemic site of geographical isolation	Functional validation done/not done	References
15.	Abc-Thiol Transporter (Fragment) (LdBPK_230290*)	23	Decreased copy number (VL/CL Ratio: 2.49)	Transport	Sri Lnaka	Not validated	Samarasinghe et al. (2018)

factors that help the parasite to adapt to visceral heat and oxidative stress conditions (McCall and Matlashewski 2012). In line with the significance of the A2 gene product in visceral infection, a decrease in the A2 gene copy number is associated with dermotropic Sri Lankan ACL isolates (Zhang et al. 2014). The second functionally characterized mutation is represented by a single-nucleotide polymorphism in the RagC (small ras-like GTPase) gene involved in the pathway leading to the activation of the target of the mammalian rapamycin (mTOR) protein, involved in mTOR signaling pathway that affects cell stress and proliferation (Lypaczewski et al. 2021; Jewell and Guan 2013). Out of the three TOR genes (TOR 1–3) identified in the *Leishmania* genome, TOR 1 and TOR 2 have been shown to be required for survival in the promastigote stage (Madeira da Silva and Beverley 2010). With a highly conserved Arginine at 231 position in Rag C protein encoded by Rag C gene in *Trypanosomes* and *Leishmania*, CL causing *L. donovani* isolate from Sri Lanka exhibits SNP change from arginine to cysteine at position 231 (Zhang et al. 2014). CL causing *L. donovani* isolates with mutated *Rag C* gene (R231C) when transfected with a plasmid expressing wildtype *RagC* gene resulted in increased survival in the visceral organs (Zhang et al. 2014). Also, VL causing *L. donovani* 1S2D strain, when edited to express mutated *RagC* gene (arginine replaced by cysteine at 231 position: R231C, exhibited reduced virulence (Lypaczewski et al. 2021). Apart from this, with an improved sequencing annotation and comparative genomic studies of VL causing standard *L. donovani* isolate and CL causing *L. donovani* from Sri Lanka, novel genes with alterations were detected (Lypaczewski et al. 2018). Among the novel genes with genetic variation, Raptor and RagA genes were the important genes involved in the TOR pathway indicating the significance of the pathway in *L. donovani* visceral outcome as RagC R231C had a demonstrable role in atypical CL pathology. RagA is shown to be essential for promastigote survival based on RagA disruption experiments.

Additionally, a filtered set of potentially high-impact SNPs in the conserved gene sequences detected through genome comparison of cutaneous, visceral, and gain-of-function *L. donovani* strain (retrieved after multiple passaging of the attenuated CL *L. donovani* through mice) that can potentially define altered parasite virulence need to be further characterized in future studies. Some of the genes with SNPs of significance that may help to further define the genetic basis of Sri Lankan CL are enlisted in Table 7.1. Many of the sequence variations fall in regions that correspond to hypothetical proteins with unknown functions (Zhang et al. 2014; Lypaczewski et al. 2018). Chromosome aneuploidy changes associated with copy number variations of specific genes and SNPs in protein-coding genomic regions with a potential role in determining disease phenotype have also been documented, albeit with heterogeneous findings emphasizing the circulation of multiple genetic variants of ACL *L. donovani* in Sri Lanka (Samarasinghe et al. 2018).

Genome-level analysis of a representative ACL *L. donovani* isolate from Himachal Pradesh, India, in comparison to Sri Lankan CL isolates and standard VL isolate further adds to the diverse genetic variations associated with CL causing *L. donovani* isolates. Polymorphic genome-wide changes unique to *L. donovani* CL isolate from HP gave a list of around 1704 SNPs with coding change effects. A major fraction of these SNPs mapped to genes with yet unknown function. The nature of polymorphic genes with known or putative protein product implies potential alterations in metabolic pathways, oxidative stress machinery, parasite growth and motility, drug resistance, and virulence factors to count on (Lypaczewski et al. 2022).

Leishmania genome is characterized by chromosome aneuploidy often associated with gene copy number variations (Table 7.3). *L. donovani* population from the Indian Subcontinent circulating in Nepal and North-east Indian VL belt exhibits extensive variation in chromosome copy number (Imamura et al. 2016). A similar chromosome number variation is documented for Ethiopian *L. donovani* isolates with each exhibiting a unique karyotype (Kariyawasam et al. 2017; Zackay et al. 2018). Chromosome somy variation is also documented in CL causing *L. donovani* variants compared to VL causing isolates from Sri Lanka and Himachal Pradesh in India (Zhang et al. 2014; Samarasinghe et al. 2018). Thus, the occurrence of chromosome aneuploidy is associated with hybrid genomes albeit with no tight correlation with the gene transcript and/or protein levels. Studies on differential expression of transcript, protein, and metabolite profile of *L. donovani* cutaneous and visceral isolates from Sri Lanka and India (HP) need to be carried out to understand the biology of differential virulence and tropism (McCall et al. 2015).

Origin of hybrid ACL *L. donovani* strains circulating in Sri Lanka and HP in India reflects different backgrounds. A single HP cutaneous isolate has been genetically analyzed to be an intraspecies *L. donovani* hybrid between parents originating from "Yeti" ISC1 clade of *L. donovani* representing a recently emerging divergent subpopulation within the ISC parasite, in the highlands of Nepal (Lypaczewski et al. 2022). In contrast, the *L. donovani* genetic variants capable of cutaneous disease in Sri Lanka are deciphered to emerge from Ethiopia as an interspecies hybrid between *L. donovani-L. tropica* and *L. donovani-L. major*.

Thus, integration of heterogeneous genetic variations as intraspecies and interspecies hybrids explains the basis of atypical disease outcome, albeit genotype to atypical phenotype correlates await further exploration to understand the molecular basis of altered parasite tropism. It is important to recognize the emerging phenomenon of atypical cutaneous leishmaniasis in light of genetic makeup of parasite strains circulating in known and the newer endemic foci, specially observed in the Indian Subcontinent. The phenomenon holds the potential of reversal of cutaneous disease to fatal visceral disease via the emergence of newer *L. donovani* variants posing newer challenges in the path of the ongoing VL elimination program. Thus, a continuous program focused on molecular surveillance of parasite isolates from different endemic regions is required to access the dynamics of the current disease epidemiology encompassing atypical manifestation. The situation warrants an urgent call from the policymakers to implement a continuous and comprehensive epidemiological survey with multicenter studies on genome-wide typing of region-specific parasite variants in relation to disease prevalence involving surveillance bodies, medical staff, and researchers as a part of leishmaniasis elimination and maintenance program.

Table 7.3 Genetic variation at chromosome level in atypical cutaneous leishmaniasis

Sr. No.	Chr	Chromosome in VL causing *L. donovani*	Chromosome variant in CL causing *L. donovani*	Nature of change	Endemic site of geographical isolation	References
1.	20	Trisomic	Disomic	Aneuploidy	Sri Lanka	Zhang et al. (2014)
2.	13	Trisomic	Disomic/Trisomic	Aneuploidy	Sri Lanka, India	Zhang et al. (2014) and Lypaczewski et al. (2022)
3.	31	Trisomic/Tetrasomic	Tetrasomic/Pentasomic	Aneuploidy	India, Sri Lanka, Nepal	Zhang et al. (2014), Lypaczewski et al. (2022) and Samarasinghe et al. (2018)
4.	1	Disomic	Tetrasomic	Aneuploidy	India	Lypaczewski et al. (2022)
5.	6	Disomic	Tetrasomic	Aneuploidy	India	Lypaczewski et al. (2022)
6.	8	Disomic	Trisomic	Aneuploidy	India	Lypaczewski et al. (2022)
7.	9	Disomic	Trisomic	Aneuploidy	India	Lypaczewski et al. (2022)
8.	14	Disomic	Trisomic	Aneuploidy	India	Lypaczewski et al. (2022)
9.	15	Disomic	Trisomic	Aneuploidy	India	Lypaczewski et al. (2022)
10.	16	Disomic	Trisomic	Aneuploidy	India	Lypaczewski et al. (2022)
11.	21	Disomic	Trisomic	Aneuploidy	India	Lypaczewski et al. (2022)
12.	33	Disomic	Trisomic	Aneuploidy	India	Lypaczewski et al. (2022)
13.	22	Disomic	Trisomic	Aneuploidy	Sri Lanka	Samarasinghe et al. (2018)
14.	26	Disomic	Trisomic	Aneuploidy	Sri Lanka	Samarasinghe et al. (2018)

Acknowldegments We acknowledge Dr. Angamuthu Selvapandiyan, Associate Professor, Department of Molecular Medicine, Jamia Hamdard, New Delhi, for putting forth the idea and critically proofreading the chapter.

References

Alborzi A, Rasouli M, Shamsizadeh A (2006) Leishmania tropica-isolated patient with visceral leishmaniasis in southern Iran. Am J Trop Med Hyg 74(2):306–307

Alborzi A et al (2008) Isolation of Leishmania tropica from a patient with visceral leishmaniasis and disseminated cutaneous leishmaniasis, southern Iran. Am J Trop Med Hyg 79(3):435–437

Aleixo JA et al (2006) Atypical American visceral leishmaniasis caused by disseminated Leishmania amazonensis infection presenting with hepatitis and adenopathy. Trans R Soc Trop Med Hyg 100(1):79–82

Alvar J et al (2012) Leishmaniasis worldwide and global estimates of its incidence. PLoS One 7(5): e35671

Aoun K, Bouratbine A (2014) Cutaneous leishmaniasis in North Africa: a review. Parasite 21:14

Barral A et al (1991) Leishmaniasis in Bahia, Brazil: evidence that Leishmania amazonensis produces a wide spectrum of clinical disease. Am J Trop Med Hyg 44(5):536–546

Basher A et al (2017) Cutaneous leishmaniasis in immigrant workers returning to Bangladesh - An emerging problem. Travel Med Infect Dis 19:62–63

Centers for Disease Control and Prevention, *Leishmaniasis*. 2020. [2020 February 18; education and information about leishmaniasis including fact sheets, information for special groups, prevention and control, epidemiology, diagnosis and treatment]. https://www.cdc.gov/parasites/leishmaniasis/

Chappuis F et al (2007) Visceral leishmaniasis: what are the needs for diagnosis, treatment and control? Nat Rev Microbiol 5(11):873–882

Convit J et al (2005) Atypical cutaneous leishmaniasis in Central America: possible interaction between infectious and environmental elements. Trans R Soc Trop Med Hyg 99(1):13–17

Cotton JA et al (2020) Genomic analysis of natural intra-specific hybrids among Ethiopian isolates of Leishmania donovani. PLoS Negl Trop Dis 14(4):e0007143

de Oliveira JP et al (2007) Genetic diversity of Leishmania amazonensis strains isolated in northeastern Brazil as revealed by DNA sequencing, PCR-based analyses and molecular karyotyping. Kinetoplastid Biol Dis 6:5

del Giudice P et al (1998) Cutaneous leishmaniasis due to Leishmania infantum. Case reports and literature review. Arch Dermatol 134(2):193–198

Downing T et al (2011) Whole genome sequencing of multiple Leishmania donovani clinical isolates provides insights into population structure and mechanisms of drug resistance. Genome Res 21(12):2143–2156

Elamin EM et al (2008) Identification of Leishmania donovani as a cause of cutaneous leishmaniasis in Sudan. Trans R Soc Trop Med Hyg 102(1):54–57

Franssen SU et al (2020) Global genome diversity of the Leishmania donovani complex. elife 9: e51243

Gupta AK et al (2015) De Novo whole-genome sequence and annotation of a Leishmania strain isolated from a case of post-kala-azar dermal leishmaniasis. Genome Announc 3(4):e00809–e00815

Hirve S et al (2016) Transmission dynamics of visceral leishmaniasis in the Indian subcontinent - a systematic literature review. PLoS Negl Trop Dis 10(8):e0004896

Imamura H et al (2016) Evolutionary genomics of epidemic visceral leishmaniasis in the Indian subcontinent. elife 5:e12613

Jewell JL, Guan KL (2013) Nutrient signaling to mTOR and cell growth. Trends Biochem Sci 38(5):233–242

Karamian M et al (2007) Leishmania major infection in a patient with visceral leishmaniasis: treatment with Amphotericin B. Parasitol Res 101(5):1431–1434

Kariyawasam UL et al (2017) Genetic diversity of Leishmania donovani that causes cutaneous leishmaniasis in Sri Lanka: a cross sectional study with regional comparisons. BMC Infect Dis 17(1):791

Karunaweera ND et al (2003) Sri Lankan cutaneous leishmaniasis is caused by Leishmania donovani zymodeme MON-37. Trans R Soc Trop Med Hyg 97(4):380–381

Kaul N et al (2016) A new focus of cutaneous leishmaniasis in Jammu division of Jammu and Kashmir State, *India*. Indian J Dermatol Venereol Leprol 82(2):145–150

Khanra S et al (2012) RFLPs of ITS, ITS1 and hsp70 amplicons and sequencing of ITS1 of recent clinical isolates of kala-azar from India and Bangladesh confirms the association of L. tropica with the disease. Acta Trop 124(3):229–234

Knio KN et al (2000) Isoenzyme characterization of Leishmania isolates from Lebanon and Syria. Am J Trop Med Hyg 63(1–2):43–47

Krayter L et al (2014) Multilocus microsatellite typing reveals a genetic relationship but, also, genetic differences between Indian strains of Leishmania tropica causing cutaneous leishmaniasis and those causing visceral leishmaniasis. Parasit Vectors 7:123

Kumar Bhat N et al (2017) Changing epidemiology: a new focus of kala-azar at high-altitude Garhwal region of North India. J Trop Pediatr 63(2):104–108

Kumar R et al (2007) Cutaneous leishmaniasis caused by Leishmania tropica in Bikaner, India: parasite identification and characterization using molecular and immunologic tools. Am J Trop Med Hyg 76(5):896–901

Kumar R et al (2008) Cutaneous leishmaniasis in Nepal: Leishmania major as a cause. Trans R Soc Trop Med Hyg 102(2):202–203

Kumar NP et al (2015) Cutaneous leishmaniasis caused by Leishmania donovani in the tribal population of the Agasthyamala Biosphere Reserve forest, Western Ghats, Kerala, India. J Med Microbiol 64(Pt 2):157–163

Lukes J et al (2007) Evolutionary and geographical history of the Leishmania donovani complex with a revision of current taxonomy. Proc Natl Acad Sci U S A 104(22):9375–9380

Lypaczewski P, Matlashewski G (2021) Leishmania donovani hybridisation and introgression in nature: a comparative genomic investigation. The Lancet Microbe 2:e250–e258

Lypaczewski P et al (2018) A complete Leishmania donovani reference genome identifies novel genetic variations associated with virulence. Sci Rep 8(1):16549

Lypaczewski P, Zhang WW, Matlashewski G (2021) Evidence that a naturally occurring single nucleotide polymorphism in the RagC gene of Leishmania donovani contributes to reduced virulence. PLoS Negl Trop Dis 15(2):e0009079

Lypaczewski P et al (2022) An intraspecies Leishmania donovani hybrid from the Indian subcontinent is associated with an atypical phenotype of cutaneous disease. iScience 25(2):103802

Madeira da Silva L, Beverley SM (2010) Expansion of the target of rapamycin (TOR) kinase family and function in Leishmania shows that TOR3 is required for acidocalcisome biogenesis and animal infectivity. Proc Natl Acad Sci U S A 107(26):11965–11970

McCall LI, Matlashewski G (2012) Involvement of the Leishmania donovani virulence factor A2 in protection against heat and oxidative stress. Exp Parasitol 132(2):109–115

McCall LI, Zhang WW, Matlashewski G (2013) Determinants for the development of visceral leishmaniasis disease. PLoS Pathog 9(1):e1003053

McCall LI et al (2015) Adaptation of Leishmania donovani to cutaneous and visceral environments: in vivo selection and proteomic analysis. J Proteome Res 14(2):1033–1059

Mebrahtu Y et al (1989) Visceral leishmaniasis unresponsive to pentostam caused by Leishmania tropica in Kenya. Am J Trop Med Hyg 41(3):289–294

Nawaratna SS, Weilgama DJ, Rajapaksha K (2009) Cutaneous leishmaniasis in Sri Lanka: a study of possible animal reservoirs. Int J Infect Dis 13(4):513–517

Olliaro PL et al (2017) Investments in research and surveillance are needed to go beyond elimination and stop transmission of Leishmania in the Indian subcontinent. PLoS Negl Trop Dis 11(1): e0005190

Oren R et al (1991) Visceral leishmaniasis: a difficult diagnosis and unusual causative agent. J Infect Dis 164(4):746–749

Parajuli N et al (2020) Cutaneous leishmaniasis of lip and role of polymerase chain reaction: a case report. JNMA J Nepal Med Assoc 58(227):494–496

Picado A et al (2014) Risk factors for visceral leishmaniasis and asymptomatic Leishmania donovani infection in India and Nepal. PLoS One 9(1):e87641

Pinart M et al (2020) Interventions for American cutaneous and mucocutaneous leishmaniasis. Cochrane Database Syst Rev 8(8):Cd004834

Raina S et al (2016) Expansion of visceral leishmaniasis to northwest sub-Himalayan region of India: a case series. J Vector Borne Dis 53(2):188–191

Ranasinghe S et al (2012) Leishmania donovani zymodeme MON-37 isolated from an autochthonous visceral leishmaniasis patient in Sri Lanka. Pathog Glob Health 106(7):421–424

Ready PD (2010) Leishmaniasis emergence in Europe. Euro Surveill 15(10):19505

Rioux JA et al (1980) Leishmania infantum Nicolle, 1908, the agent of the autochthonous oriental sore. Apropos of the biochemical identification of 2 strains isolated in the eastern pyrenees. C R Seances Acad Sci D 291(8):701–703

Sacks DL et al (1995) Indian kala-azar caused by Leishmania tropica. Lancet 345(8955):959–961

Samarasinghe SR et al (2018) Genomic insights into virulence mechanisms of Leishmania donovani: evidence from an atypical strain. BMC Genomics 19(1):843

Sarkari B et al (2016) molecular evaluation of a case of visceral leishmaniasis due to Leishmania tropica in Southwestern Iran. Iran J Parasitol 11(1):126–130

Scorza BM, Carvalho EM, Wilson ME (2017) Cutaneous manifestations of human and murine leishmaniasis. Int J Mol Sci 18(6):1296

Sharma NL et al (2005) Localized cutaneous leishmaniasis due to Leishmania donovani and Leishmania tropica: preliminary findings of the study of 161 new cases from a new endemic focus in himachal pradesh, India. Am J Trop Med Hyg 72(6):819–824

Sharma NL et al (2009) The sandflies of the Satluj river valley, Himachal Pradesh (India): some possible vectors of the parasite causing human cutaneous and visceral leishmaniases in this endemic focus. J Vector Borne Dis 46(2):136–140

Sinha R et al (2018) Genome plasticity in cultured Leishmania donovani: comparison of early and late passages. Front Microbiol 9:1279–1279

Siriwardana H et al (2017) Emergence of visceral leishmaniasis in Sri Lanka: a newly established health threat. Pathog Glob Health 111(6):317–326

Smith DF, Peacock CS, Cruz AK (2007) Comparative genomics: from genotype to disease phenotype in the leishmaniases. Int J Parasitol 37(11):1173–1186

Tabbabi A et al (2020) Nuclear and kinetoplast DNA analyses reveal genetically complex Leishmania strains with hybrid and mito-nuclear discordance in Peru. PLoS Negl Trop Dis 14(10): e0008797

Thakur L et al (2018) Atypical leishmaniasis: A global perspective with emphasis on the Indian subcontinent. PLoS Negl Trop Dis 12(9):e0006659

Thakur L et al (2020) Leishmania donovani infection with atypical cutaneous manifestations, Himachal Pradesh, India, 2014-2018. Emerg Infect Dis 26(8):1864–1869

Wani GM, Ahmad SM, Khursheed B (2015) Clinical study of cutaneous leishmaniasis in the Kashmir valley. Indian Dermatol Online J 6(6):387–392

World Health Organization. Leishmaniasis 2022. https://www.who.int/news-room/fact-sheets/detail/leishmaniasis.

World Health Organization. Regional Office for South-East, A (2005) Regional strategic framework for elimination of kala azar from the South-East Asia region (2005–2015). WHO Regional Office for South-East Asia, New Delhi

Zackay A et al (2018) Genome wide comparison of Ethiopian Leishmania donovani strains reveals differences potentially related to parasite survival. PLoS Genet 14(1):e1007133
Zhang WW, Matlashewski G (2010) Screening Leishmania donovani-specific genes required for visceral infection. Mol Microbiol 77(2):505–517
Zhang WW et al (2003) Comparison of the A2 gene locus in Leishmania donovani and Leishmania major and its control over cutaneous infection. J Biol Chem 278(37):35508–35515
Zhang WW et al (2014) Genetic analysis of Leishmania donovani tropism using a naturally attenuated cutaneous strain. PLoS Pathog 10(7):e1004244

Atypical *Leishmania donovani* Infections in Sri Lanka: Challenges for Control and Elimination

8

Nadira D. Karunaweera and Rajika Dewasurendra

Abstract

Leishmaniasis is a neglected tropical disease, which is transmitted via infective bites of the female phlebotomine sand flies. There are three main forms of the disease, cutaneous leishmaniasis, mucocutaneous leishmaniasis, and visceral leishmaniasis. The causative agent is the protozoan parasite belonging to the genus *Leishmania*. *Leishmania donovani* is one of over 20 species of *Leishmania* that causes visceral leishmaniasis. Interestingly, the MON37 zymodeme of this species is almost exclusively responsible for causing cutaneous leishmaniasis in Sri Lanka. The disease presentation of this atypical form of leishmaniasis is nonitching and painless skin lesions. Genetic studies on the local parasite have shown that it has a unique genetic makeup though closely related to the other *L. donovani* species, particularly those in the region. The first case of leishmaniasis was documented in the early 1990s, and it is an endemic disease in Sri Lanka with cases reported from all parts of the island. A steady escalation of numbers, especially in recent years, indicates the importance of the implementation of a focused control program. Lack of awareness of the disease among the general public and even some medical and paramedical staff, limitations in diagnosis, treatment failure, the appearance of other *Leishmania* species, limited knowledge in vector bionomics, absence of a vector control program, and likely involvement of animal reservoirs in disease transmission may be listed as major challenges associated with infection control. This chapter discusses the challenges faced in controlling leishmaniasis in Sri Lanka and possible strategies to overcome them.

N. D. Karunaweera (✉) · R. Dewasurendra
Department of Parasitology, Faculty of Medicine, University of Colombo, Colombo, Sri Lanka
e-mail: nadira@parasit.cmb.ac.lk; raika@parasit.cmb.ac.lk

© The Author(s), under exclusive license to Springer Nature Singapore Pte Ltd. 2023

A. Selvapandiyan et al. (eds.), *Challenges and Solutions Against Visceral Leishmaniasis*, https://doi.org/10.1007/978-981-99-6999-9_8

Keywords

Leishmaniasis · Vector-borne diseases · Neglected tropical diseases · Parasitic infections · Skin lesions · Vector control · Disease surveillance · Drug resistance

8.1 Leishmaniasis: Overview/Causative Agent in Sri Lanka

Leishmaniasis is a neglected tropical disease (NTD) that is found mainly in the tropics, subtropics, and some parts of Southern Europe. There are three types of leishmaniasis, namely visceral leishmaniasis (VL), cutaneous leishmaniasis (CL), and mucocutaneous leishmaniasis (MCL). VL, which affects the internal organs (spleen, liver, bone marrow), is considered the most dangerous and is potentially fatal (95% of the cases are fatal) if untreated. VL is mostly prevalent in Brazil, East Africa, and the Indian subcontinent and an estimated 50,000–90,000 new cases are reported every year. CL is considered the most common type of leishmaniasis. It usually causes lesions on the exposed skin, which may appear as a papule that could develop into a nodule and ultimately into an ulcer. CL is mainly seen in the Americas, the Mediterranean basin, the Middle East, and Central Asia. An estimated 600,000–1,000,000 new CL cases occur worldwide annually. MCL, on the other hand, affects the mucous membranes, i.e., nose, mouth, and throat, and is prevalent mainly in Brazil, Ethiopia, Peru, and Bolivia (WHO 2023).

Leishmaniasis is caused by a protozoan parasite belonging to the genus *Leishmania*. The type of leishmaniasis prevalent is largely dependent on the causative species of *Leishmania*. Out of the 20 *Leishmania* species, which is known to cause this disease, VL is mainly caused by *L. donovani* and *L. infantum* in the old world, and *L. chagasi* in the new world; while *L. major* and *L. tropica* are responsible for CL in the old world, and *L. mexicana, L. amazonensis*, and *L. braziliensis* are the major species of *Leishmania* responsible for CL in the new world (Table 8.1).

Leishmaniasis was considered an imported disease in Sri Lanka before the 1990s. It was reported from only a few individuals returning from foreign countries where leishmaniasis was prevalent, especially the Middle East (Naotunne et al. 1990). As per widely published information, the first case of locally acquired CL in Sri Lanka was reported in 1992 (Athukorale et al. 1992), although the actual time of the introduction of leishmaniasis remains debatable. There are a few sporadic case reports available since its introduction, especially from the Southern and Northern areas. However, a full-blown outbreak was started in 2001 (Siriwardena et al. 2003), and the case numbers have increased rapidly, since then with patients reported from all 25 administrative districts of the island (Karunaweera et al. 2020; Siriwardana et al. 2019; Kariyawasam et al. 2015). CL is thus an endemic disease in Sri Lanka, and the causative agent is a strain of *L. donovani*, i.e., *L. donovani*—zymodeme MON37 (Karunaweera et al. 2003). Intraspecies variations are apparent in *L. donovani* with possible genetic hybridization, which may explain the phenotypic variations seen (Kariyawasam et al. 2017; Samarasinghe et al. 2018; Lypaczewski et al. 2021).

Table 8.1 Species of *Leishmania* that cause leishmaniasis, their geographical distribution, vertebrate, and insect vectors involved (Espinosa et al. 2018; Akhoundi et al. 2016)

Leishmania species	Vertebrate host	Insect vector	Geographical distribution
Cutaneous leishmaniasis			
L. aethiopica	Humans, hyraxes	*Phlebotomus* spp.	Ethiopia, Kenya
L. tropica minor	Humans, dogs, rodents	*Phlebotomus* spp.	Mediterranean
L. tropica major	Humans, dogs, rodents	*Phlebotomus* spp.	Mediterranean
L. peruviana	Humans, dogs	*Lutzomyia* spp.	Peru
L. mexicana mexicana	Humans, rodents	*Lutzomyia* spp.	Central America, Mexico
L. mexicana amazonensis	Humans, rodents	*Lutzomyia* spp.	South America
L. mexicana pifanoi	Humans, rodents	*Lutzomyia* spp.	Venezuela
L. braziliensis	Humans, rodents	*Lutzomyia* spp.	Mexico, Brazil
Visceral leishmaniasis			
L. donovani donovani	Humans, dogs, foxes	*Phlebotomus* spp.	Mediterranean, South America, India
L. donovani infantum	Humans, dogs	*Phlebotomus* spp.	Mediterranean
L. donovani chagasi	Humans, cats, foxes	*Lutzomyia* spp.	South America

8.2 Clinical Features of Leishmaniasis Reported in Sri Lanka

Even though the *L. donovani* MON37 strain is closely related to the known *L. donovani* visceralizing strains, the phenotypes or the resultant clinical outcomes of the two genotypes are quite distinct. The clinical presentation of CL caused by *L. donovani* MON37 comprises typically nonitching and painless skin lesions. They can be small and inconspicuous papules, nodules of varying sizes, or ulcers usually affecting the exposed body parts. Occasionally, skin lesions could take the form of dermal plaques, erythematous ulcerative patches, or lesions affecting the mucosal tissues (Karunaweera et al. 2020) (Fig. 8.1a–d). A few cases of visceral disease also have been reported caused by the same causative agent as CL (Refai et al. 2016; Ranasinghe et al. 2012). While one such patient has had comorbidities including primary immune suppression, others, as far as it is known, have not had any evidence of coexisting immune-compromised status (Refai et al. 2016).

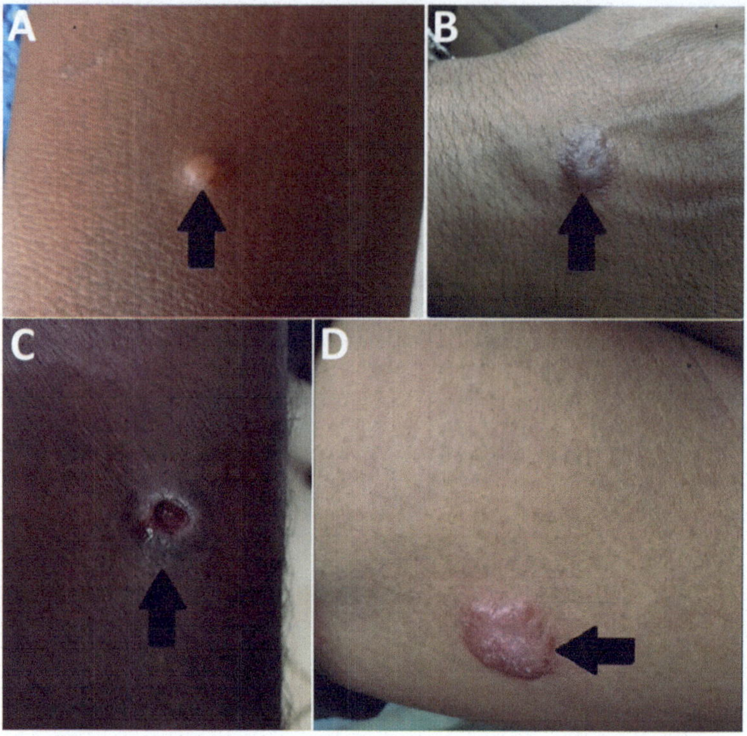

Fig. 8.1 Types of skin lesions. (**a**)—papule, (**b**)—Nodule, (**c**)—Ulcer, (**d**)—Plaque (extracted with permission from Karunaweera et al. (2020))

The skin lesions of *L. donovani* CL are generally single with the majority being <2 cm in diameter. Furthermore, none of the systemic features of VL infections caused by *L. donovani* (i.e., fever, loss of appetite, loss of weight, splenomegaly anemia, and hepatomegaly) are seen in association with local *L. donovani* CL infections either during or following local infections (Kariyawasam et al. 2018; Siriwardana et al. 2021). Therefore, the local *L. donovani* is considered as an attenuated strain of the species, which is essentially dermotropic.

8.3 Genetic Characterization of the Causative Agent of CL in Sri Lanka: *L. donovani* (MON 37)

The *L. donovani* MON37 strain is responsible for causing both CL and VL in Sri Lanka. It is suggested that alterations in the genome (mutations or amplifications) could lead to these differences in its virulence (Zhang et al. 2014).

The local strain shows a unique genetic makeup compared to *L. donovani* strains from other geographic areas, i.e., Nepal and India in the phylogenetic analysis (Kariyawasam et al. 2017; Thakur et al. 2018). Genetic characterization of the

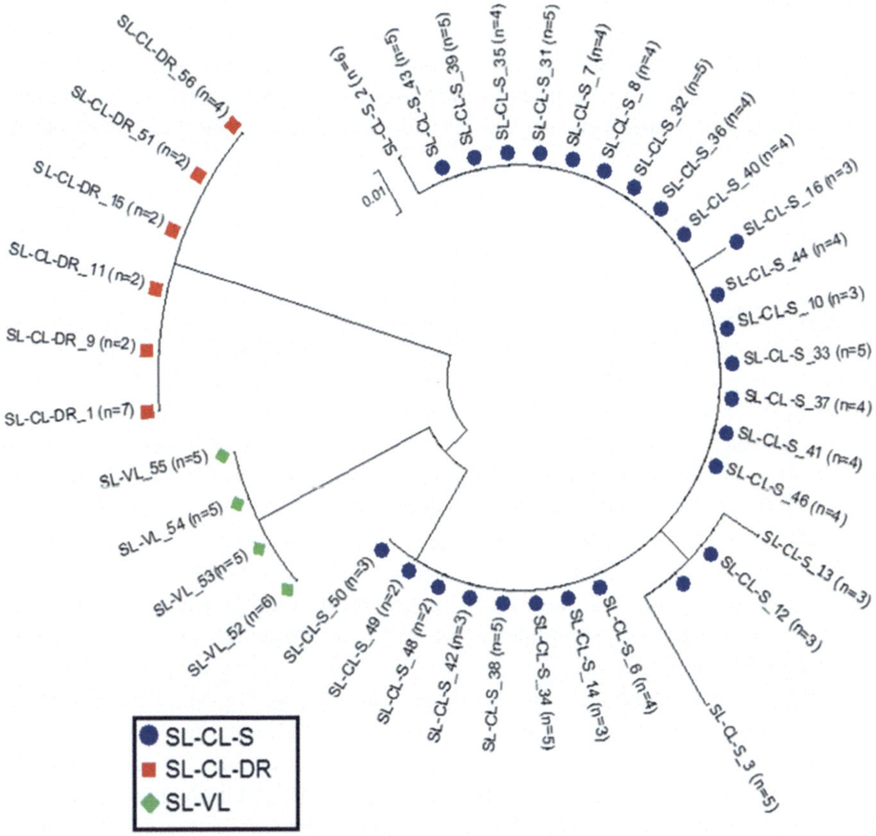

Fig. 8.2 Phylogenetic tree of kDNA minicircle sequences of *Leishmania* parasite isolates from Sri Lanka (Extracted with permission from Kariyawasam et al. (2017)). SL-CL-S: Parasite isolates from skin lesions of laboratory-confirmed CL patients whose lesions healed following intralesional antimony treatment. SL-CL-DR: Parasite isolates from skin lesions of laboratory-confirmed CL patients who failed to respond to antimony. SL-VL: Parasite isolates from VL patients

local *L. donovani* MON37 parasite strain has been performed using various methods. In the kDNA footprint assay (Selvapandiyan et al. 2008), the Sri Lankan isolates formed a cluster separated from clusters of isolates from the region, namely Nepalian and Indian strains, indicate a significant genetic differentiation from isolates from other geographic regions. Phylogenetic analysis indicated that the Sri Lankan *L. donovani* isolates were clearly different from other *Leishmania* species, which cause CL, but better linked with VL-causing parasite strains (Kariyawasam et al. 2017). Further, the Sri Lankan parasite isolates formed distinct genetic clusters according to clinical phenotypes, namely parasite isolates from patients who responded to antimony therapy versus those with poor responsiveness to antimony indicating intraspecies variability within the local *L. donovani* parasite population (see Fig. 8.2) (Kariyawasam et al. 2017).

Similar findings were also apparent in whole-genome sequence analyses. Copy number variation of 248 genes differed more than twice between the two phenotypic groups. Furthermore, the chromosome and single-nucleotide polymorphisms (SNPs) profiles showed that the two groups formed two distinct clusters. Therefore, genetic structure and differentiations appear to influence the responses of *Leishmania* parasites to frequently used antileishmanial drugs, namely sodium stibogluconate (Samarasinghe et al. 2018).

Further characterization of *L. donovani* isolates from Sri Lanka also indicated that the studied isolates fall into three major groups:

Group 1—Isolates more closely related to isolates from India.

Group 2—Isolates closely related to isolates from Africa.

Group 3—Isolates that are not related to either India or Africa.

The three groups differed in heterozygosity as well as in the SNP profiles indicating the extensive genetic diversity of the local parasite. Exorbitant heterozygosity in isolates not related to Indian or African origin (i.e., Group 3) suggested hybridization of non–*L. donovani* species (e.g., *L. major*, *L. tropica*) with the local parasite strain (Lypaczewski and Matlashewski 2021).

It has been also discovered that there are 91 alterations in the genome of *L. donovani* MON37 strain, which causes CL, compared to the typical *L. donovani* that causes VL, and mutations in RagC (R321C) and Raptor genes that may have major contributions in reducing the virulence of the local strain (Lypaczewski et al. 2021).

Since CL is an established disease in Sri Lanka, it is important to implement surveillance programs to enable effective control and elimination of this disease, where diagnosis, treatment, and vector control play a major role. However, delayed diagnosis, variable treatment-seeking behavioral patterns, and high exposure to sand fly bites in endemic areas have contributed to the increasing spread of the disease (Traoré et al. 2016). The situation is compounded by the absence of a vaccine (Srivastava et al. 2016).

8.4 Strategies to Overcome Challenges Faced Against Control and Elimination

8.4.1 Knowledge Gap

Even though CL is an endemic disease that has prevailed for over two decades, a recent community survey revealed that the knowledge related to the disease, especially regarding the vector, preventive methods, and treatment, is poor even in areas with a high burden of the disease (Dewasurendra et al. 2022). Public awareness of a disease in a local community where active transmission exists, covering areas such as how the infection is acquired and spread, resultant symptoms and control methods, is important for effective planning and implementation of prevention as well as control of the disease. Public awareness of basic information about the disease (i.e., knowledge of the correct disease name as "leishmaniasis" or identifying the disease as a fly-induced skin disease, the symptoms, whether the disease is curable or not, and the name of the vector) appeared satisfactory (~70% of the study population was knowledgeable of these details, as indicated through an island-wide survey). However, details that are relevant for disease containment such as vector-breeding places, vector-biting times, and information on how to prevent vector bites have not reached the communities. Over 90% of the study population was ignorant regarding the peak biting times of the sand fly and vector control methods, which is of concern. Less than 2% were aware of the breeding places of the vector. Such deficiencies would need to be addressed in the future to enable better disease control, particularly through the adoption of personal protective methods that in turn will enhance any future organized control efforts that will be made by the Ministry of Health.

Ignorance of the need to seek medical advice for apparently nonthreatening skin lesions tends to cause delays in diagnosis and initiation of treatment. This in turn will perpetuate further infection spread. Therefore, awareness of the general public to ensure early treatment-seeking behavior and alertness of healthcare personnel in the area are also important factors that would enable prompt management that will arrest the spread.

These knowledge gaps exposed (Dewasurendra et al. 2022) would facilitate the planning of a future disease awareness program as part of a national effort in infection control.

Although a considerable amount of scientific research has been conducted by various universities and research centers across the island, the knowledge gained has failed to reach the general public, despite the scientific publications produced and information disseminated through scientific fora that target the scientific community. This highlights the need for better public outreach programs in the country and the need for more effective science communication.

The communication channels that are widely available and readily accessible to the community (i.e., television, radio, mass media, posters, newspapers, and social media) should be utilized to convey key health-related messages to enhance public awareness. Similarly, conducting seminars, workshops, and awareness programs for the public as well as targeted populations (e.g., school children, social or healthcare workers) could help in decreasing this knowledge gap. Focus Group Discussions (FGDs) are primarily a research methodology, used to generate research data by means of discussions centered on a specific research topic within a small group of participants. Instead of interviewer and questionnaire-based data collection, the information is gathered based on the exchange of views and ideas within the focus group (Wong 2008). Besides data gathering, FGDs may be used to increase awareness as well as diminish misconceptions on the topic of interest among community members. Further, the information gathered also will give an idea of the deficiencies in public health programs that are specific to that community such as inadequate vector control methods in the area and limited access to diagnostic methods, which can be considered by the health officials in prioritization of the issues when implementing control and/or surveillance programs.

8.4.2 Early Case Detection and Patient Management

Early diagnosis and effective treatment of CL are key for preventing disease spread and infection control in a community. In Sri Lanka, the widely used confirmatory diagnostic test for CL is the direct microscopic examination of lesion samples.

While such laboratory diagnostic test is available in main state-sector hospitals, university laboratories and selected private-sector laboratories, peripheral health centers lack such expertise. Upon clinical suspicion of CL, the clinicians (usually dermatologists) direct the patients to obtain confirmatory diagnostic tests. Unfortunately, CL can mimic many dermatoses, which may lead to misdiagnosis (Gurel et al. 2020). Therefore, clinical suspicion alone isn't considered reliable to initiate treatment.

Establishing a standard case definition may help in minimizing the misdiagnosis since disease symptoms alone or in combination are not specific enough to differentiate CL from other similar infections (Gurel et al. 2020). The consensus view of clinicians to enable a case definition for leishmaniasis in the local setting has been initiated, which would help health professionals in more accurate clinical diagnosis of the disease in instances where laboratory confirmation is not possible (Ministry of Health Sri Lanka 2019).

CL microscopy is a simple technique, which, however, requires technical expertise. A smear stained with Giemsa's stain would be made using a slit-skin or lesion aspirate taken from the lesion. The stained smears are examined under a light microscope, to identify the presence of amastigotes (Hawash et al. 2018). Visualization of parasites under the microscope could be done with or without prior in vitro culture of parasites in a suitable culture media (M199, RPMI1640). Such additional procedures become particularly useful in cases where parasites are scant and microscopy is negative in the first instance (Ihalamulla et al. 2002, 2008). Although diagnosis done using light microscopy is considered relatively cost-effective, it requires well-trained personnel both to obtain the lesion sample and for the detection of parasites by microscopy. Lack of such trained technical persons in many laboratories in the island, especially in the rural areas, may delay the confirmatory diagnosis of CL. In this scenario, when a suspected case is presented in a peripheral healthcare unit, the patient is usually referred to a laboratory, where such facilities are available. This may cause delays in diagnosis, particularly if the patient has difficulties in reaching such centers, with the resultant persistence of parasites that will continue transmission of leishmaniasis in the population that will remain unrecorded in the healthcare system.

Identification of the *Leishmania* parasites in patient material using genetic tools, e.g., PCR is available only in a few laboratories on the island since it requires sophisticated machines and reagents that are relatively costly (Kothalawala and Karunaweera 2016; Deepachandi et al. 2019).

In-house ELISA methods also have been developed to identify the presence of antibodies specific for antigens of *L. donovani* (Piyasiri et al. 2022; Deepachandi et al. 2020). The availability of an optimized in-house ELISA assay for IgG antibodies against *L. donovani* soluble antigen with 98% sensitivity and 90.3%

specificity coupled with the observation of accelerated reduction of antibody levels over time offers an acceptable method, to be used particularly in monitoring the treatment response (Piyasiri et al. 2022).

Almost all these diagnostic procedures, however, require either trained personnel or well-equipped laboratory facilities. Sometimes, multiple testing methods need to be carried out to confirm the disease. Therefore, the need for a simpler and more accurate method of diagnosis, preferably in the form of a bedside diagnostic (or rapid diagnostic test), is evident. Although commercial diagnostic kits for rapid diagnosis (RDT kits) are available for *L. donovani*-induced leishmaniasis, the sensitivity of such assays in the local setting is very low (e.g., rK39 RDT, *InBios*) (Kariyawasam et al. 2018). The lack of efficacy and sensitivity of existing RDTs against *L. donovani*-induced CL could be due to the differences in antigens used in some assays and the species-specific nature of immunogenic molecules may account for such observations. Therefore, the introduction of an easy-to-use, simple, and inexpensive diagnostic method (i.e., RDT), which is sensitive to the local parasite strain, may improve case detection in the community.

Further studies on serology-based ELISA and/or molecular diagnosis (LAMP), which may lead to the development of RDTs, could enable the eventual availability of a point-of-care diagnosis with wider access.

Passive case detection is a method used in leishmaniasis for screening of cases in healthcare facilities/laboratories, while active case detection is done by screening for leishmaniasis cases through home visits (house-to-house searches, index case approach, incentive-based approach) (Das et al. 2014). Although passive case detection is routinely performed in the country, active case detection does not take place, unless it is executed for research purposes (Kariyawasam et al. 2015). Performing active case detection programs, at least in areas with a high burden of leishmaniasis, including the southern and north-central provinces, could help in early case detection leading to more effective and efficient management of cases, which will reduce reservoirs of infection, the disease transmission, and disease burden.

8.4.3 Delayed Response, Treatment Failure, and Possible Drug Resistance

Control of CL requires the early diagnosis of the disease as well as rapid and effective treatment (Silva et al. 2021a). A successful treatment method would accelerate the healing of the lesion, reduce the risk of further complications, and minimize the possibility of further transmission. The effectiveness of the treatment method will depend primarily on the parasite drug susceptibility, and other factors like intrinsic host factors, i.e., interplay between the host immune system and the parasites (Regli et al. 2018).

In Sri Lanka, the standard treatment carried out for *L. donovani*-induced CL is intralesional injection of sodium stibogluconate (IL-SSG) given at weekly intervals for several weeks. Cryotherapy is also used in some centers, particularly if SSG isn't available and such a mode of treatment involves application of liquid nitrogen to the site of the lesion (Karunaweera et al. 2013). Treatment failure to SSG remains a major concern in Sri Lanka with the such burden being as high as over 75% in more recent studies (Refai et al. 2016; Silva et al. 2021a). Treatment failure could be due to a multitude of factors including environmental factors, host factors (e.g., nutritional status of the host, host immune responses), drug-related factors, and most importantly drug resistance of the parasite.

The emergence of drug resistance can be due to many reasons, which are almost always difficult to overcome. Such known factors include a reduction of drug concentration within the parasite caused either by the decrease in drug uptake or inhibition of drug activation or inactivation of active drugs (Capela et al. 2019). The above can ensue through genetic alterations within the parasites due to the use of the same kind of treatment over a long period of time. The best strategy to avoid the emergence of drug resistance is by minimizing drug pressure, which could be achieved through the reduction of the transmission rate of an infection, which, in turn, will minimize the use of a specific drug (De Koning 2017). Thus, methods such as vector control, improving diagnostic methods and improving the healthcare system for satisfactory treatment access, play an important role indirectly in reducing the emergence of drug resistance.

A more recent discovery of *L. tropica* infections from CL in Sri Lanka has complicated the picture even more (Silva 2021). This may at least partly explain the antimony nonresponsiveness of a larger proportion of local patients (Silva et al. 2021a). Furthermore, the mixed picture of SNP profiles also may point toward the contribution of *L. tropica* in possible hybridization events, as suggested earlier (Lypaczewski et al. 2021; Lypaczewski and Matlashewski 2021).

Furthermore, ensuring that appropriate treatment in line with the national and regional policies is initiated following confirmation of diagnosis would help minimize the misuse of drugs, which in turn would help in preventing drug resistance. It is equally important to educate the communities to seek treatment from qualified medical personnel at the earliest following the emergence of symptoms.

8.4.4 Vector Control

The incriminated vector of *L. donovani* in Sri Lanka is a dipteran fly (sand fly) *Phelobotomus argentipes* (Gajapathy et al. 2013). Although both male and female sand flies feed on plant material, the female requires a blood meal for the development of eggs (Malaviya et al. 2014).

Leishmaniasis is associated with the social and economic status of communities, which is often reflected by the housing conditions. Earthen houses, with dark and damp interiors, are considered popular breeding places for sand flies, making the inhabitants more exposed to the disease (Calderon-Anyosa et al. 2018). Thus, for an effective vector control program to fully succeed, it would require addressing the problems associated with poverty in these communities, which are multifactorial and complex in nature. Therefore, a multisectoral approach involving many parties including the government, nongovernment organizations, and health authorities, similar to what was in place that led to the eventual elimination of malaria in Sri Lanka (Carter and Karunaweera 2020), will be required in order to yield optimal benefits of a control program for leishmaniasis.

Reduction of man–vector contact can be achieved through vector control using chemicals and/or environmental management and personal protection (Claborn 2010). Vector control by insecticide residual spraying (IRS) of houses and cattle sheds is the main strategy adopted by leishmaniasis control programs across the world. IRS using insecticides such as dichlorodiphenyltrichloroethane (DDT) and malathion was carried out island-wide, especially in the dry zone by the antimalaria campaign, Ministry of Health, Sri Lanka, as a strategy for malaria control in mid-1940s and early 1990s. Similarly, in 2004, insecticide-treated nets (ITNs) and long-lasting impregnated nets (LLINs) were distributed in order to reduce malaria incidence. Although not aimed at sand fly control, these methods may have played a positive role in reducing the sand fly densities in these areas preventing sand fly bites of net users. With the elimination of malaria in 2016, vector control strategies, especially IRS, were relaxed. Currently, there are no formal or regular programs for IRS in Sri Lanka that target any vector-borne disease in the island. This, in fact, is considered as a likely reason for the flare-up of leishmaniasis cases in 2001 (Karunaweera et al. 2014).

Knowledge regarding the type of the vector as well as the vector population density of the area and its bionomics is essential to plan a successful program for the control of a vector-borne disease (Malaviya et al. 2014). Owing to the difficulties of identifying the sand fly due to its minute nature (CVBD 2022) and poor information flow on the insect concerned, public awareness of the leishmaniasis vector remains poor in Sri Lanka (Dewasurendra et al., 2022). Although vector prevalence and details on biological properties, particularly those related to insecticide susceptibility, are available in Sri Lanka (Pathirage et al. 2020; Surendran et al. 2007), further research studies on detailed vector bionomics should be done to obtain data to formulate a successful vector control program.

Avoiding sand fly bites by avoiding times and places of sand fly activity also helps in reducing transmission. Therefore, findings of studies related to vector behavior (Tonelli et al. 2021; Hassaballa et al. 2021) should be effectively disseminated to the public to enable optimal benefits.

The primary method of vector control used is indoor residual spraying (IRS) (Garlapati et al. 2021). Introduction of IRS for the control of sand fly populations, at least in disease hotspots with close monitoring (Deb et al. 2021; Chowdhury et al. 2017), may reduce the escalating numbers of infections as recorded more recently (Karunaweera et al. 2020). Therefore, planning and implementation of a leishmaniasis control program are a timely necessity in the country. Furthermore, more scientific research should be carried out to reveal detailed information on insecticide susceptibility patterns and the impact of the insecticides on the environment prior to the implementation of a successful IRS program.

Proper management of sand fly habitats and breeding places is equally important in controlling vector populations. Damp agricultural areas, garbage pits, and dark and damp resting places, e.g., rodent burrows, cattle sheds, and houses made of wattle and daub, are ideal sand fly breeding and resting places (Bennis et al. 2015). Maintenance of a clean environment, the targeted program of IRS for sand fly control in these specific places could help in vector control. Although environmental management measures could result in reducing sand fly populations and thereby minimize human–vector contact, alterations to the environment should be preceded by a careful and thorough investigation of local ecological systems and the possible impacts of such interventions on the environment.

Distribution of LLINs could be beneficial to prevent sand fly bites especially, during nighttime. Although the habit of usage of LLINs, as well as regular bed nets, is high in the country, the types of nets usually used are not very effective for the prevention of sand fly bites due to the larger pore size designed for the prevention of mosquito entry. Thus, national programs and/or organized community level programs to promote the availability and the use of suitable bed nets may need to be implemented possibly with the aid of nongovernmental organizations, health sector administrators, and research groups in the future.

8.4.5 Animal Reservoirs

Leishmania sp. is considered a heteroxenous parasite that can survive within several different hosts, namely a range of mammals including rodents, wild canines, as well as some domestic animals (i.e., dogs) acting as reservoir hosts (Wijerathna et al. 2017). Since the transmission cycle of *Leishmania* sp. can be maintained entirely within the reservoir host without any involvement of the human host (who is considered as an accidental host for most species), the reservoir host plays an important role in *Leishmania* life cycle (Centers for Disease Control and Prevention 2013). The parasite *L. donovani*, which is responsible for CL in Sri Lanka, is the causative agent of VL in other geographical areas, e.g., India, Somalia, and Nepal (Cloots et al. 2020; Aalto et al. 2020; Joji et al. 2020). New World VL is a zoonotic disease, with the involvement of domestic dogs (*Canis familiaris*) and wild fox species (*Dusicyon vetulus* and *Cerdocyon thous*). A similar situation exists in Mediterranean countries, where VL transmission involves reservoir hosts (Elnaiem et al. 2001). However, records on potential reservoir hosts in Sri Lanka are scarce. Initial evidence of the presence of possible animal reservoirs in Sri Lanka has been accrued through the detection of antibodies against the *Leishmania* antigen rK39 in dogs (Rosypal et al. 2010) and observation of parasite stages (amastigotes) in Giemsa-stained smears made from lesion material of dogs (Nawaratna et al. 2009). In a recent study carried out, it was evident that the seropositivity of domestic dogs for antileishmania antibodies is as high as 13.33% (Samaranayake et al., unpublished data). Therefore, it could be suspected that local leishmania parasites may involve zoonotic reservoir host(s). Such an assumption is in line with the initial observation of the current outbreak in the north-central province in association with an army camp located adjacent to a jungle (Karunaweera 2009).

Cats and dogs are usually reared as pets in Sri Lankan households, and the presence of stray dogs and cats is a common sight, particularly in rural areas. Furthermore, small mammals, i.e., bandicoots, mongoose, and house rats are very common in households and neighborhood areas. Thus, it could indeed be possible that they actively contribute to the transmission of leishmaniasis.

Although these small mammals can be regarded as potential reservoir hosts, there is yet no backup evidence through scientific research to support such assumptions. This might be considered a major drawback in the planning of an effective control or surveillance program. If warranted, such a program could involve mass screening of pets, and if possible stray animals within village/city limits that could help in identifying possible reservoirs. In case such evidence is revealed in the future, additional measures will be required for disease containment. However, the approach to minimizing the effect of such animal reservoirs in disease transmission would be different according to the type of animal dealt with, which may include the following: Keeping the animal sheds clean, spraying insecticides inside the sheds, inserting insecticide-impregnated collars to pets and reared animals (especially mammals, cattle, goats), using insect repellents that will reduce the exposure of these animals to the sand flies. Various control tools especially to control canine leishmaniasis (of pets) have been developed, i.e., insecticides, vaccines, and chemotherapy (Otranto et al. 2013); but these methods are difficult to practice for wild and farm animals when they are reared at larger scale. Increasing awareness on the role of reared animals or pets in disease transmission might help in reducing the risk of these animals spreading the disease. When dealing with wild animals, the involvement with relevant government authorities, i.e., the Department of Wildlife and the national zoological gardens would be essential for successful control of the disease. Further to these, the use of technology (e.g., drone pictures) to locate burrows of rodents and other small mammals could help in destroying the burrow systems by plowing or by applying suitable chemicals, which can minimize the involvement of reservoir hosts in disease transmission (Akhavan et al. 2014).

8.4.6 Lack of a Planned Disease Control Program and Disease Surveillance

The first case of CL in Sri Lanka was identified in the early 1990s thus considered a newly established disease in Sri Lanka, and its spread was considered sparse at that time. The number of cases began to increase only after the year 2001 followed by a drastic increase after 2018. Interestingly, a detailed investigation of molecular signatures of the causative agent *L. donovani*, however, revealed a longer existence of this local strain of the parasite (Karunaweera 2009). Furthermore, though CL in Sri Lanka is caused by *L. donovani*, which is recognized widely as a visceralizing species, there is no scientific evidence thus far as to its visceralization following natural transmission (Kariyawasam et al. 2018). Due to the status quo that prevails, neither the Sri Lankan health authorities nor the international policymakers see CL as an urgent or priority health threat in Sri Lanka (Karunaweera et al. 2020). Other than routine disease diagnosis and treatment, active interventions are not implemented either targeting the high transmission hotspots or as a general nationwide program to control and/or eliminate CL from Sri Lanka (Karunaweera et al. 2021).

A successful control program will reduce the disease burden and may lead to the eventual elimination of the disease from a defined geographic area (Ejov and Dagne 2014). Identification, definition, and review of goals and objectives as well as setting up clear definitions for planning, implementation, sustenance, and evaluation are essential attributes of a control program. Forming committees comprising decision-makers, policymakers, and senior staff of specialized health services, operating at the national level would be essential to provide technical and operational guidance in a satisfactory manner. Good program management inclusive of case detection/management, disease surveillance, control of reservoir host, if any, integrated vector control, environment management and personal protection, epidemic preparedness/response, and health education should be implemented. However, in order to ensure optimal outcomes through such efforts, it would be essential to establish effective partnerships and collaborations with funding agencies as well as international health institutions and research organizations (Ejov and Dagne 2014).

One of the major advantages Sri Lanka has in implementing a successful leishmaniasis control program is the well-established health network and already established malaria/dengue control programs, which operate at Grama Niladhari Division (GN Division) level, which is the smallest grassroots-level administrative health unit within the country. Furthermore, the existence of already trained staff for implementing control strategies also can be considered as an added advantage with past success of malaria elimination providing further confidence.

Sri Lanka currently lacks a reliable and sensitive surveillance program. Systematic collection, analysis, and interpretation of disease-related data are essential for a successful surveillance program, and currently, this is done only at a limited level. A proper mechanism should be initiated to make use of the valuable data generated through leishmaniasis research, in planning and implementing control/surveillance programs and later to monitor and evaluate the established programs.

8.4.7 Use of Modern Tools in Surveillance, Education, and Disease Management

Localized or national outbreaks of leishmaniasis as predicted through modeling (Karunaweera et al. 2021) a possibility with the sand fly vector and the parasite distributed throughout the country. Therefore, a reliable mechanism should be in place to predict outbreaks and rapidly respond to such a situation. Criteria for epidemic alerts should be established by the healthcare administrative bodies and facilities should be provided accordingly.

Further studies in closing gaps in leishmaniasis research should be encouraged, which may include the development of rapid diagnostics and patient and vector surveillance tools, and the results generated should be used in developing focused control strategies.

Training and education of healthcare personnel (i.e., physicians, entomologists, parasitologists, veterinarians) on risk factors, diagnosis, treatment, prevention, control methods as well as training laboratory staff in microscopy and molecular methods of parasite identification can be equally helpful for a successful control program.

A review of existing treatment protocols seems timely in the backdrop of the development of high rates of treatment failures. This will pave the way for the use of newer methods of treatment of CL, such as radio-frequency heat therapy or thermotherapy, which has been tested and proven to be efficient and cost-effective in the local setting for both uncomplicated CL and CL that has failed antimony treatment (Refai et al. 2016; Silva et al. 2021b).

Leishmaniasis is a disease with a wide variety of causative parasite species, reservoir hosts as well as vectors involved in disease transmission. Globally, leishmaniasis is among the top ten neglected tropical diseases with a high disease burden and mortality (Pan America Health Organization 2017). Although the disease is not yet considered a greater threat to public health in Sri Lanka, the possibility of it developing into a serious health hazard cannot be disregarded. Therefore, it is important to initiate a carefully planned disease control and surveillance program paying attention to diagnosis, patient management, and vector control. It is equally important to focus on increasing public awareness of the disease, which will help the control efforts imposed by the ministry of health.

Acknowledgment The authors like to thank the researchers of the Parasitic Diseases Research Unit (PDRU), Faculty of Medicine, University of Colombo, Sri Lanka, whose research findings on leishmaniasis in Sri Lanka have contributed to compiling this book chapter. Professor Nilakshi Samaranayake, of Department of Parasitology, Faculty of Medicine, University of Colombo, is greatly appreciated for reviewing and proof reading. The authors like to acknowledge the financial contributions for leishmania research at the PDRU made through the TMRC/NIH grant (No: U01AI136033). The content is solely the responsibility of the authors and does not necessarily represent the official views of the National Institutes of Health, USA.

References

Aalto MK, Sunyoto T, Ahmed M, Yusuf A, Mohamed AA, Van Der AG et al (2020) Visceral leishmaniasis, Northern Somalia, 2013–2019. Emerg Infect Dis 26(1):153–154

Akhavan AA, Veysi A, Arandian MH, Vatandoost H, Yaghoobi-Ershadi MR, Hosseini M et al (2014) Field evaluation of phostoxin and zinc phosphide for the control of zoonotic cutaneous leishmaniasis in a hyperendemic area, Central Iran. J Vector Borne Dis 51:307–312

Akhoundi M, Kuhls K, Cannet A, Votýpka J, Marty P, Delaunay P et al (2016) A historical overview of the classification, evolution, and dispersion of Leishmania parasites and sandflies. PLoS Negl Trop Dis 10:1–40

Athukorale DNN, Seneviratne JKK, Ihalamulla RLL, Premaratne UNN (1992) Locally acquired cutaneous leishmaniasis in Sri Lanka. J Trop Med Hygeine 95:432–433

Bennis I, De Brouwere V, Ameur B, El Idrissi Laamrani A, Chichaoui S, Hamid S et al (2015) Control of cutaneous leishmaniasis caused by Leishmania major in South-Eastern Morocco. Trop Med Int Heal 20:1297–1305

Calderon-Anyosa R, Galvez-Petzoldt C, Garcia PJ, Carcamo CP (2018) Housing characteristics and leishmaniasis: a systematic review. Am J Trop Med Hyg. 99:1547–1554

Capela R, Moreira R, Lopes F (2019) An overview of drug resistance in protozoal diseases. Int J Mol Sci 20:5748

Carter R, Karunaweera ND (2020) The role of improved housing and living environments in malaria control and elimination. Malar J 19:1–6. https://doi.org/10.1186/s12936-020-03450-y

Centers for Disease Control and Prevention. Leishmaniasis. 2013. http://www.cdc.gov/parasites/leishmaniasis.

Chowdhury R, Faria S, Huda MM, Chowdhury V, Maheswary NP, Mondal D et al (2017) Control of phlebotomus argentipes (diptera: psychodidae) sand fly in Bangladesh: a cluster randomized controlled trial. PLoS Negl Trop Dis 11:e0005890. https://pubmed.ncbi.nlm.nih.gov/28873425/

Claborn D (2010) The biology and control of leishmaniasis vectors. J Glob Infect Dis 2:127

Cloots K, Uranw S, Ostyn B, Bhattarai NR, Le Rutte E, Khanal B et al (2020) Impact of the visceral leishmaniasis elimination initiative on Leishmania donovani transmission in Nepal: a 10-year repeat survey. Lancet Glob Heal. The Author(s). Published by Elsevier Ltd. This is an open access article under the CC BY-NC-ND 4.0 license 8:e237–e243. https://doi.org/10.1016/S2214-109X(19)30536-4

CVBD (2022) Sand flies morphology|Vectors. CVBD. https://cvbd.elanco.com/vectors/sand-flies/morphology

Das AK, Harries AD, Hinderaker SG, Zachariah R, Ahmed B, Shah GN et al (2014) Active and passive case detection strategies for the control of leishmaniasis in Bangladesh. Public Heal Action 4:15–21

De Koning HP (2017) Drug resistance in protozoan parasites. Emerg Top Life Sci 1:627–632

Deb R, Singh RP, Mishra PK, Hitchins L, Reid E, Barwa AM et al (2021) Impact of IRS: four-years of entomological surveillance of the Indian visceral leishmaniases elimination programme. PLoS Negl Trop Dis 15:e0009101. https://pubmed.ncbi.nlm.nih.gov/34370731/

Deepachandi B, Weerasinghe S, Soysa P, Karunaweera N, Siriwardana Y (2019) A highly sensitive modified nested PCR to enhance case detection in leishmaniasis. BMC Infect Dis 3:1–10

Deepachandi B, Weerasinghe S, Gunathilake H, Andrahennadi TP, Wickramanayake MN, Siri S et al (2020) Prevalidation of an ELISA for detection of a new clinical entity: Leishmania donovani -induced cutaneous leishmaniasis. Int J Anal Chem 2020:9289651

Dewasurendra R, Silva H, Samaranayake N, Manamperi N, Silva D, Karunanayake P et al (2022) Assessment of knowledge and perceptions on leishmaniasis : an island-wide study in Sri Lanka. PLoS Negl Trop Dis 16(10):e0010821. https://doi.org/10.1371/journal.pntd.0010821

Ejov M, Dagne D (2014) Strategic framework for leishmaniasis control in the WHO European region 2014-2020. World Health Organanization, pp 1–17

Elnaiem DA, Hassan MM, Maingon R, Nureldin GH, Mekawi AM, Miles M et al (2001) The egyptian mongoose, Herpestes ichneumon, is a possible reservoir host of visceral leishmaniasis in eastern Sudan. Parasitology 122:531–536

Espinosa OA, Serrano MG, Camargo EP, Teixeira MMG, Shaw JJ (2018) An appraisal of the taxonomy and nomenclature of trypanosomatids presently classified as Leishmania and Endotrypanum. Parasitology 145:430–442

Gajapathy K, Peiris LB, Goodacre SL, Silva A, Jude PJ, Surendran SN (2013) Molecular identification of potential leishmaniasis vector species within the Phlebotomus (Euphlebotomus) argentipes species complex in Sri Lanka. Parasites and Vectors 6:1–9

Garlapati R, Iniguez E, Serafim TD, Mishra PK, Rooj B, Sinha B et al (2021) Towards a sustainable vector-control strategy in the post kala-azar elimination era. Front Cell Infect Microbiol 11:1–11

Gurel MS, Tekin B, Uzun S (2020) Cutaneous leishmaniasis: a great imitator. Clin Dermatol 38: 140–151

Hassaballa IB, Sole CL, Cheseto X, Torto B, Tchouassi DP (2021) Afrotropical sand fly-host plant relationships in a leishmaniasis endemic area, Kenya. PLoS Negl Trop Dis 15:1–18. Public Library of Science; /pmc/articles/PMC7895382/

Hawash YA, Ismail KA, Abdel-Wahab MM, Khalifa M (2018) Diagnosis, treatment and clinical features of cutaneous leishmaniasis in Saudi Arabia. Korean J Parasitol 56:229–236

Ihalamulla RL, Siriwardana HVYD, Karunaweera ND (2002) First successful in-vitro cultivation of autochthonous Leishmania species in Sri Lanka. Ceylon Med J 47:58

Ihalamulla RL, Siriwardana HV, Karunaweer ND (2008) Efficacy of RPMI 1640 and M199 media in the isolation of Leishmania from cutaneous lesions. Ann Trop Med Parasitol 102:173–175

Joji A, Rajesh KR, Reena KJ (2020) Visceral leishmaniasis caused by Leishmania donovani zymodeme MON-37, Western Ghats, India. Emerg Infect Dis 26:1956–1958

Kariyawasam KKGDUL, Edirisuriya CS, Senerath U, Hensmen D, Siriwardana HVYD, Karunaweera ND (2015) Characterisation of cutaneous leishmaniasis in Matara district, southern Sri Lanka: evidence for case clustering. Pathog Glob Health. Taylor & Francis 109:336–343. https://doi.org/10.1179/2047773215Y.0000000032

Kariyawasam K, Selvapandiyan A, Rai K, Wani TH, Ahuja K, Beg MA et al (2017) Genetic diversity of Leishmania donovani that causes cutaneous leishmaniasis in Sri Lanka : a cross sectional study with regional comparisons. BMC Infect Dis 17:791

Kariyawasam KKGDUL, Selvapandiyan A, Siriwardana HVYD, Dube A, Karunanayake P, Senanayake SASC et al (2018) Dermotropic Leishmania donovani in Sri Lanka: visceralizing potential in clinical and preclinical studies. Parasitology 145:443–452

Karunaweera ND (2009) Leishmania donovani causing cutaneous leishmaniasis in Sri Lanka: a wolf in sheep's clothing? Trends Parasitol 25:458–463

Karunaweera ND, Pratlong F, Siriwardane HVYD, Ihalamulla RL, Dedet JP (2003) Sri Lankan cutaneous leishmaniasis is caused by Leishmania donovani zymodeme MON-37. Trans R Soc Trop Med Hyg 97:380–381

Karunaweera N, Siriwardana HVYD, Karunanayake PH (2013) Management of leishmaniasis. Sri Lanka Prescr 21:1–5

Karunaweera ND, Galappaththy GN, Wirth DF (2014) On the road to eliminate malaria in Sri Lanka: lessons from history, challenges, gaps in knowledge and research needs. Malar J 13:59. http://www.malariajournal.com/content/13/1/59

Karunaweera N, Ginige S, Senanayake S, Silva H, Manamperi N, Samaranayake N et al (2020) Spatial epidemiologic trends and hotspots of leishmaniasis, Sri Lanka, 2001-2018. Emerg Infect Dis 26:1–10

Karunaweera ND, Senanayake S, Ginige S, Silva H, Manamperi N, Samaranayake N et al (2021) Spatiotemporal distribution of cutaneous leishmaniasis in Sri Lanka and future case burden estimates. PLoS Negl Trop Dis 15:1–16. https://doi.org/10.1371/journal.pntd.0009346

Kothalawala HS, Karunaweera ND (2016) Loop-mediated isothermal amplification assay as a sensitive diagnostic tool for Leishmania donovani infections in Sri Lanka. Ceylon Med J 61: 68–70

Lypaczewski P, Matlashewski G (2021) Leishmania donovani hybridisation and introgression in nature: a comparative genomic investigation. The Lancet Microbe. The Author(s). published by Elsevier Ltd. This is an open access article under the CC BY-NC-ND 4.0 license 2:e250–e258. https://doi.org/10.1016/S2666-5247(21)00028-8

Lypaczewski P, Zhang WW, Matlashewski G (2021) Evidence that a naturally occurring single nucleotide polymorphism in the ragc gene of Leishmania donovani contributes to reduced virulence. PLoS Negl Trop Dis 15:1–21. https://doi.org/10.1371/journal.pntd.0009079

Malaviya P, Hasker E, Picado A, Mishra M, Van Geertruyden J, Das ML et al (2014) Exposrue to Phlebotomus argentipes (Diptera, Psychodidae, Phlebotominae) sand flies in rural areas of Bihar, India: the role of housing conditions. PLoS One 9:e106771

Ministry of Health Sri Lanka. Guidelines on prevention and control of leishmaniasis. Colombo; 2019

Naotunne TD, Rajakulendran S, Abeywickreme W, Kulasiri CD, Perera J, Premaratne UN et al (1990) Cutaneous leishmaniasis in Sri Lanka. An imported disease linked to the middle east and African employment boom. Trop Geogr Med 42:72–74

Nawaratna SSK, Weilgama DJ, Rajapaksha K (2009) Cutaneous leishmaniasis in Sri Lanka: a study of possible animal reservoirs. Int J Infect Dis 13:513–517

Otranto D, Dantas-Torres F, de Capariis D, Paola GD, Tarallo VD, Latrofa MS et al (2013) Prevention of canine leishmaniosis in a hyper-endemic area using a combination of 10% imidacloprid/4.5%flumethrin. PLoS One 8:e56374

Pan America Health Organization (2017) Leishmaniasis key data. Neglected Infect Dis 1. https://www.paho.org/hq/dmdocuments/2017/2017-cha-leishmaniasis-factsheet-work.pdf

Pathirage DRK, Karunaratne S, Senanayake SC, Karunaweera ND (2020) Insecticide susceptibility of the sand fly leishmaniasis vector Phlebotomus argentipes in Sri Lanka. Parasites and Vectors 13:246

Piyasiri SB, Samaranayake TN, Silva H, Manamperi NH, Karunaweera ND (2022) ELISA-based evaluation of antibody response to Leishmania in a region endemic for cutaneous leishmaniasis. Parasite Immunol 2022:e1294

Ranasinghe S, Zhang W-W, Wickremasinghe R, Abeygunasekera P, Chandrasekharan V, Athauda S et al (2012) Leishmania donovani zymodeme MON-37 isolated from an autochthonous visceral leishmaniasis patient in Sri Lanka. Pathog Glob Health 106:421–424

Refai F, Madarasingha N, Fernandopulle R, Karunaweera N (2016) Nonresponsiveness to standard treatment in cutaneous leishmaniasis: a case series from Sri Lanka. Trop Parasitol 6:155. http://www.tropicalparasitology.org/text.asp?2016/6/2/155/190835

Regli IB, Fernández OL, Martínez-Salazar B, Gómez MA, Saravia NG, Tacchini-Cottier F (2018) Resistance of Leishmania (Viannia) panamensis to meglumine antimoniate or miltefosine modulates neutrophil effector functions. Front Immunol 9:1–11

Rosypal AC, Tripp S, Kinlaw C, Hailemariam S, Lindsay RRRDS, Rajapakse R et al (2010) Surveillance for antibodies to Leishmania spp. in dogs from Sri Lanka. J Parasitol 96:230–231

Samarasinghe SR, Samaranayake N, Kariyawasam UL, Siriwardana YD, Imamura H, Karunaweera ND (2018) Genomic insights into virulence mechanisms of Leishmania donovani: evidence from an atypical strain. BMC Genomics 19(1):843

Selvapandiyan A, Duncan R, Mendez J, Kumar R, Salotra P, Cardo LJ et al (2008) A Leishmania minicircle DNA footprint assay for sensitive detection and rapid speciation of clinical isolates. Transfusion 48:1788–1798

Silva H (2021) Parasite determinants that influence the treatment outcome in cutaneous lesihmaniasis in Sri Lanka. University of Colombo, Sri Lanka

Silva H, Liyanage A, Deerasinghe T, Chandrasekara V, Chellappan K, Karunaweera ND (2021a) Treatment failure to sodium stibogluconate in cutaneous leishmaniasis: a challenge to infection control and disease elimination. PLoS One 16:1–16. https://doi.org/10.1371/journal.pone.0259009

Silva H, Liyanage A, Deerasinghe T, Sumanasena B, Munidasa D, de Silva H et al (2021b) Therapeutic response to thermotherapy in cutaneous leishmaniasis treatment failures for sodium stibogluconate: a randomized controlled proof of principle clinical trial. Am J Trop Med Hyg 104:945–950

Siriwardana Y, Zhou G, Deepachandi B, Akarawita J, Wickremarathne C, Warnasuriya W et al (2019) Trends in recently emerged Leishmania donovani induced cutaneous leishmaniasis, Sri Lanka, for the first 13 years. Biomed Res Int 2019:4093603

Siriwardana Y, Deepachandi B, Weerasinghe S, Karunaweera N, Udagedara C, Warnasuriya W et al (2021) First evidence from Sri Lanka for subphenotypic diversity within L. donovani -induced classical cutaneous leishmaniasis. Biomed Res Int 2021:3537968

Siriwardena HV, Udagedara CU, Karunaweera ND (2003) Clinical features, risk factors and efficacy of cryotherapy in cutaneous leishmaniasis in Sri Lanka. Ceylon Med J 48:10–12

Srivastava S, Shankar P, Mishra J, Singh S (2016) Possibilities and challenges for developing a successful vaccine for leishmaniasis. Parasites and Vectors. Parasites & Vectors 9:1–15. https://doi.org/10.1186/s13071-016-1553-y

Surendran SN, Kajatheepan A, Ramasamy R (2007) Socio-environmental factors and sandfly prevalence in Delft Island, Sri Lanka: implications for leishmaniasis vector control. J Vector Borne Dis 44:65–68

Thakur L, Singh KK, Shanker V, Negi A, Jain A, Matlashewski G et al (2018) Atypical leishmaniasis: a global perspective with emphasis on the Indian subcontinent. PLoS Negl Trop Dis 12:1–17

Tonelli GB, Binder C, Margonari C, Filho JDA (2021) Sand fly behavior: much more than weak-flying. Mem Inst Oswaldo Cruz 116:e210230. /pmc/articles/PMC8672444/

Traoré B, Oliveira F, Faye O, Dicko A, Coulibaly CA, Sissoko IM et al (2016) Prevalence of cutaneous leishmaniasis in districts of high and low endemicity in Mali. PLoS Negl Trop Dis 10: 1–12

WHO. 2023. Leishmaniasis. https://www.who.int/health-topics/leishmaniasis#tab=tab_1

Wijerathna T, Gunathilaka N, Gunawardana K, Rodrigo W (2017) Potential challenges of controlling leishmaniasis in Sri Lanka at a disease outbreak. Biomed Res Int 2017:6931497

Wong L (2008) Focus group discussion: a tool for health and medical research. Singapore Med J 49: 256. http://smj.sma.org.sg/4903/4903me1.pdf

Zhang WW, Ramasamy G, McCall LI, Haydock A, Ranasinghe S, Abeygunasekara P et al (2014) Genetic analysis of Leishmania donovani tropism using a naturally attenuated cutaneous strain. PLoS Pathog 10:e1004244

Critical Roles of Micro-RNAs in the Pathogenesis and Immunoregulation of *Leishmania* Infection

9

Chaitenya Verma, Ryan H. Huston, Abigail R. Wharton, Rebecca Fultz, Samer Zidan, Greta Volpedo, and Abhay R. Satoskar

Abstract

The genus *Leishmania* was first described in 1903 with the highly pathogenic species *L. donovani*, but several other pathogenic species have been reported after that. These species, endemic in more than 98 countries, are associated with morbidity, mortality, and economic losses. *Leishmania* is transmitted to the mammalian host by the bite of infected female phlebotomine sandflies, leading to a wide range of symptoms from skin lesions to fatal leishmaniasis, depending on the species. These protozoan parasites strategically alter the host's immunity to survive and replicate within the host. Over the past decade, micro-RNAs (miRNAs) have been reported to regulate both acquired and innate immunity in the host, especially during infection and disease states. miRNAs, noncoding single-stranded RNA molecules, regulate gene expression processes at the post-transcriptional level. Recent evidence has shown that miRNAs also play a critical role in immune regulation during leishmaniasis. This chapter will discuss how miRNAs regulate different biological processes during cell growth and development, proliferation, inflammation, cell-cell interaction, and immune response modulation during infection with *Leishmania* parasites. We will also highlight the potential diagnostic and prognostic uses of miRNAs against leishmaniasis.

C. Verma · A. R. Wharton · R. Fultz · S. Zidan
Department of Pathology, Wexner Medical Center, The Ohio State University, Columbus, OH, USA

R. H. Huston · G. Volpedo · A. R. Satoskar (✉)
Department of Pathology, Wexner Medical Center, The Ohio State University, Columbus, OH, USA

Department of Microbiology, The Ohio State University, Columbus, OH, USA
e-mail: abhay.satoskar@osumc.edu

© The Author(s), under exclusive license to Springer Nature Singapore Pte Ltd. 2023

A. Selvapandiyan et al. (eds.), *Challenges and Solutions Against Visceral Leishmaniasis*, https://doi.org/10.1007/978-981-99-6999-9_9

Keywords

Leishmania · Leishmaniasis · miRNA · Macrophages · Dendritic cells · T cells · B cells · Therapeutic targets

Abbreviations

AGO	Argonaute protein
AID	Activation-induced cytidine deaminase
AmB	Amphotericin B
APCs	Antigen-presenting cells
ATG-5	Autophagy related-5
BIC	B-cell-integrated cluster
BMDC	Bone marrow dendritic cells
BMDM	Bone-marrow-derived macrophage
BNIP-3	BCL2/adenovirus E1B 19 kDa protein-interacting protein-3
CL	Cutaneous leishmaniasis
CLL	Chronic lymphocytic leukemia
CLP	Common lymphoid progenitor
CSR	Class-switch recombinase
CTS-E	Cathepsin-E
CVL	Canine visceral leishmania
DC	Dendritic cells
DC-SIGN	Dendritic-cell-specific ICAM-3-grabbing nonintegrin
ER	Endoplasmic reticulum
GZMB	Granzyme B
HIF	Hypoxia-inducible factor
IFN	Interferon
Ig	Immunoglobulin
IGF	Insulin-like growth factor
IgH	Immunoglobulin heavy chain
IgL	Immunoglobulin light chain
IL	Interleukin
LRV-1	Leishmania RNA virus 1
MCL	Mucocutaneous leishmaniasis
MDDC	Monocyte-derived dendritic cells
METTL3	Methyle-transferase-like-3 protein
MHC	Major histocompatibility complex
MIL	Miltefosine
mTOR	Mechanistic target of rapamycin
NF-kB	Nuclear factor kappa beta
NK	Natural killer
NKG7	Natural killer cell granule protein 7

NLRP	Nucleotide-binding domain leucine-rich repeat and pyrin domain-containing receptor
NOS-2	Nitric oxide synthase 2
Oct	Octamer-binding transcription factor
p-Akt	Phospho-protein kinase B
PBMCs	Peripheral blood mononuclear cells
PDL	Program death ligand
p-IkB-α	Phospho-inhibitor of kappa B alpha
PKDL	*Post-kala-azar-dermal leishmaniasis*
POU family	Pit-Oct-Unc
p-PLCγ	Phospholipase C gamma
Pre-BCR	Pre-B cell receptor
RISC	RNA-induced silencing complex
SHIP	Src homology-2 domain-containing inositol 5-phosphate
SLC	Surrogate light chain
SOCS	Suppressor of cytokine signaling
TAP1	ER antigen peptide transporter 1
TBX	T-box transcription factor
TCR	T-cell receptor
TGF	Transforming growth factor
Th	T helper
TLR	Toll-like receptor
TNF	Tumor necrosis factor
T-reg	Regulatory T cells
VL	Visceral leishmaniasis
XBP-1	X-box-binding protein-1

9.1 Introduction: Immunology of Leishmaniasis

Leishmania parasite heterogenicity complicates the immunopathology of leishmaniasis. The parasite can be internalized by a range of phagocytic cells, such as neutrophils, dendritic cells, monocytes, and macrophages. *Leishmania* parasites prevent the apoptotic process in infected host cells in order to persist for longer periods of time. Promastigotes differentiate into amastigotes inside the host cells and divide rapidly by binary fission until they lyse the cells. Released amastigotes infect nearby cells and disseminate to the bone marrow, spleen, liver, and lymph nodes in the case of visceral leishmaniasis (VL), to the skin in the case of cutaneous leishmaniasis (CL), and to the mucosa in the case of mucocutaneous leishmaniasis (MCL) (Pareyn et al. 2020; Rodrigues et al. 2016). Infection with *Leishmania* species is demonstrated by the production of anti-*Leishmania* antibodies, such as IgG, and subclasses of IgG are predominately elevated in CL and VL cases (Ravindran et al. 2004; Ozbılge et al. 2006).

Leishmania parasites rely on evasion and the suppression of both innate and adaptive branches of the immune system to maintain and perpetuate infection. *Leishmania* parasites have been shown to promote an anti-inflammatory M2 phenotype in macrophages, creating a parasite-permissive environment. *L. donovani* and *L. infantum* also impair dendritic cell activation, maturation, and antigen presentation. An extensive review of the determinants of innate immunity in VL can be found in Volpedo et al. (2021). Furthermore, cell-mediated immunity is crucial for controlling VL. In particular, T-helper (Th) 1 cells, characterized by the production of interferon (IFN)-γ, are important for granuloma formation and the activation of macrophages to engage in antiparasitic activities. While the activity of Th2 cells is mostly detrimental in CL, a beneficial role has been demonstrated for these cells in VL (Stäger et al. 2003; McFarlane et al. 2011). Finally, B cells and humoral immunity play a detrimental role in clinical and preclinical VL (Kumar and Nylén 2012).

Overall, *Leishmania* parasites rely on biological, mechanical, and genetic mechanisms in the host to confer susceptibility. Recent studies have found several micro-RNAs (miRNAs or miRs) to be differentially expressed in the host cells during *Leishmania* infection and to be able to modulate host immunity. Therefore, we aim to assess the impact of miRNAs on host defense mechanisms and their regulation in different cell types.

9.2 Overview of miRNAs: Biogenesis and Function

Micro-RNAs (miRNAs) are a type of small, nonprotein encoding RNAs that regulate protein levels through both transcriptional and posttranscriptional mechanisms via the complementary base interactions of miRNAs to a target DNA or mRNA sequence (Saliminejad et al. 2019; Pu et al. 2019). These diverse miRNAs are present in numerous species and have developed a diverse array of regulatory mechanisms for their control, pointing to their important evolutionary origins and the deleterious effects of miRNA malfunctions.

miRNAs are produced by the cell from the intronic regions of the genome and are involved in multiple processing steps before maturing and interacting with their intended target; furthermore, mature miRNAs are ~22 nucleotides in length (Pu et al. 2019). As shown in Fig. 9.1, most miRNAs originate from transcription, producing hairpin primary (pri)-miRNAs from the genomic introns via RNA polymerase II and transcription factors bound to the DNA upstream of the miRNA start site (Georgakilas et al. 2018; Wang et al. 2021; Piriyapongsa et al. 2011; Vishnoi and Rani 2017). After transcription, pri-miRNAs are produced. These are short hairpin structures with long 5′ and 3′ tails that are not complementary (Yeo and Chong 2011). These pri-miRNAs are then recognized inside the nucleus by the RNA-binding protein DGCR8 through methylated adenosine groups contributed by the mammalian methyle-transferase-like-3 protein (METTL3) to provide greater specificity for DGCR8 to pri-miRNAs (Alarcón et al. 2015). After DGCR8 association with the methylated pri-miRNA, the RNase known as Drosha then recognizes

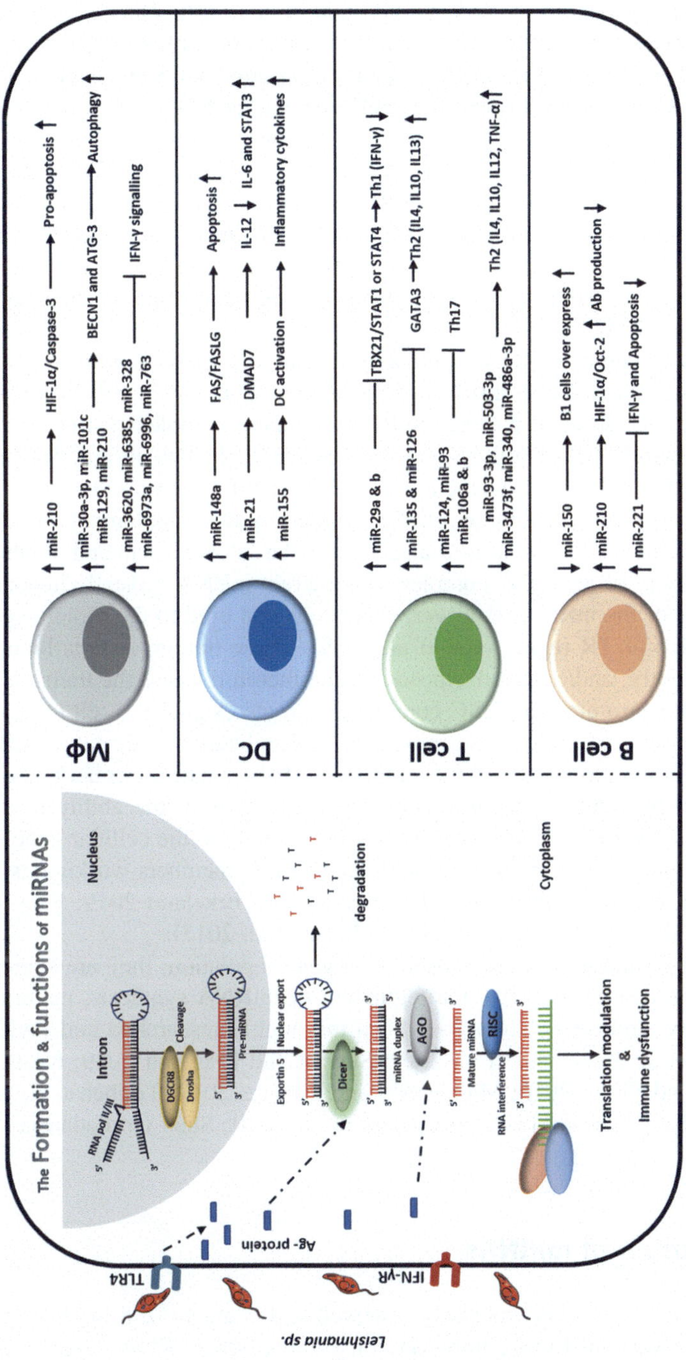

Fig. 9.1 *miRNA biogenesis and targeted signaling pathways in different types of cells.* The left panel shows a schematic of the formation of miRNAs from transcription in the nucleus to their final form in the cytoplasm. The right panels show the major miRNAs identified in macrophages, dendritic cells, T cells, and B cells and their main functions

where the helix hairpin gives way to single-stranded RNA and cuts the pri-miRNA at that site, as recently confirmed by Cryo-EM (Jin et al. 2020). The cut pri-miRNA product, now termed "pre-miRNA," remains in a hairpin formation but now with much shorter 3′ and 5′ tails. Pre-miRNAs are then exported out from the nucleus into the cytoplasm. Cells further process pre-miRNAs via an RNase III protein called Dicer, which breaks the single-strand hairpin structure into two distinct miRNAs (Vishnoi and Rani 2017; Liu et al. 2018). Among the diversity of miRNAs, there are additional pathways to generate mature miRNAs, such as by mRNA degradation and splicing or by noncanonical mechanisms that allow for maturation independent of Drosha and/or Dicer (Yeo and Chong 2011). However, these outlier mechanisms, which are discussed further by Yeo and Chong, may be specific to each miRNA (Yeo and Chong 2011).

Functionally, mature miRNAs can associate in the cytosol with Argonaute (AGO) proteins, which shield miRNAs from degradation, aid in cell localization, and initiate the formation of a larger miRNA silencing complex (Pu et al. 2019; Kingston and Bartel 2019; Carthew and Sontheimer 2009; Fabian and Sonenberg 2012). The miRNA-induced silencing complex (RISC) recruits various other proteins before binding to the target mRNA, guided by the complementarity of the miRNA to the target mRNA (Carthew and Sontheimer 2009; Fabian and Sonenberg 2012). The binding of the RISC complex to the target mRNA typically happens at the 3' UTR site of the mRNA; however, exceptions can bind to the protein-coding regions and the 5' UTR region (Cai et al. 2009). These binding events block the translocations, entry, and/or exit of ribosomes and therefore cease the translation of the target mRNA into protein. RISC complexes can also signal the cell to degrade the mRNA transcript through the decapping and destabilization of the target mRNA, further inhibiting the production of the target gene (Fabian and Sonenberg 2012). A minority of miRNAs have also been shown recently to have the ability to act as transcriptional activators and silencers by binding directly to the cellular genome at promoter or enhancer sequences or through AGO/RISC members working as transcription factor scaffolds (Pu et al. 2019; Stavast and Erkeland 2019; Kim et al. 2008; Place et al. 2008; Huang et al. 2012; Matsui et al. 2013).

Due to the prevailing effect of miRNAs on gene regulation, they are subject to regulation in the cell on all steps: the regulation of miRNA synthesis, maturation, target specificity, and degradation have been thoroughly researched and reviewed (Piriyapongsa et al. 2011; Cai et al. 2009; Stavast and Erkeland 2019; Krol et al. 2010a; Siomi and Siomi 2010; Michlewski and Cáceres 2019; Treiber et al. 2019; Krol et al. 2010b; Han et al. 2020; Cazalla et al. 2010; Sheu-Gruttadauria et al. 2019).

9.3 Evolution of miRNAs

Small RNAs have been experimentally observed and characterized in all levels of eukaryotes and various forms; however, only those small RNAs created from two-step cleavage of hairpin precursors are genuine miRNAs (Hombach and Kretz

2016; Kim 2005). The enzymes canonically required for creating miRNAs are highly conserved in eukaryotes, not just for their function in miRNAs but also for their other functions, such as RNase III enzymes (Johanson et al. 2013), which led to the independent evolution of miRNAs in five separate eukaryotic clades, as characterized to date (Song et al. 2019; Tarver et al. 2012). The evolutionary origin of miRNA sequences in the genome may have historically resulted from transposable elements (Cho 2018; Cui et al. 2017), viral insertion fragments (Ying et al. 2008), partially inverted gene duplication events (Cui et al. 2017; Gramzow et al. 2020), or regular RNA splicing events in the cell (Carthew and Sontheimer 2009). Despite sharing primary eukaryotic RNA and DNA processing mechanisms and recent studies confirming miRNAs in single-celled green and brown algae (Tarver et al. 2012), *Leishmania*-origin small RNAs have not yet been demonstrated to fall under the miRNA category (Atayde et al. 2013); therefore, all miRNAs discussed in the remainder of this chapter are host-produced miRNAs.

9.4 Role of miRNAs in Modulating Immune Mechanism in Leishmaniasis

miRNAs play an essential role in the survival and functioning of various immune cells. The characteristics of these miRNAs make them potential candidates for managing the immune system during infection and disease. Moreover, a variety of miRNAs is expressed from different cell types, depending upon pathogen, pathological site, and responding cells. A complete list of miRNAs known to date to be involved in leishmaniasis can be found in Table 9.1, divided by host cell type and *Leishmania* species.

9.4.1 Expression and Function of miRNAs in *Leishmania*-Infected Macrophages

Macrophages play different roles during *Leishmania* infections. Their main functions are to present parasite antigens to T cells and engage in parasitic killing at the healing stage of the disease (Mauël 1990). Monocytes/macrophages are considered the canonical hosts for *Leishmania*; they are recruited at the bite site with the help of chemokines such as MCP-1, MIG, IP-10, and MIP-1 to play an essential role during the initial counter with the parasite (Pacheco-Fernandez et al. 2021). Mukherjee et al. showed that antimony-resistant *L. donovani* parasites exploit miRNA (miR)-466i to deactivate MyD88 pathways and create a disproportional interleukin (IL)-10/IL-12 ratio during the initial hour of infection (Mukherjee et al. 2015), suggesting a role for miR-466i to be exploited as a therapeutic target. Evidence has revealed that host macrophages differentially express miR-22, miR-133b, miR-155, and miR-210. These levels are associated with host apoptosis processes (Cheng et al. 2005; Lemaire et al. 2013). miR-210 is known to be upregulated under hypoxic conditions and regulated through hypoxia-inducing

Table 9.1 *miRNA profiling in leishmaniasis*. This table summarized the known preclinically investigated miRNA pathways and their cellular targets based on the host cell type and *Leishmania* species involved

Cell types	Causative species	miRNAs	Cellular pathways	References
Macrophage (Mφ)	*L. major*	miR-22, miR-133b, and miR-155	Associated with the host apoptosis process	Cheng et al. (2005) and Lemaire et al. (2013)
		miR-210, miR-23, miR-24, miR-26a, miR-291, and miR-107	Modulate apoptosis through HIF-1α-mediated proapoptotic caspase-3 regulation	Lemaire et al. (2013)
		miR-24-3p	Inhibits proapoptotic genes (caspases 3 and 7)	Lasjerdi et al. (2020)
		miR-155 inhibitor and miR-15a analog	Elevate apoptosis level by reducing parasite burden	Gholamrezaei et al. (2020)
		miR-101c, miR-129, and miR-210	Induce autophagy by subverting phospho-mTOR and upregulating ATG-5, BNIP-3, and cathepsin-E	Frank et al. (2015)
		miR-146a-3p, miR-146a-5p, miR-122, miR-155-3p, miR-188, miR-9-3p, miR-9-5p, and miR-147	Regulate inflammatory response	Nimsarkar et al. (2020)
		miR-340 and miR-27a	Promote IL-10 and TGF-b1 production	Hamidi et al. (2021)
	L. donovani	Inhibition of miR-210	Activates NF-kB-p50 and p60-inducing pro-inflammatory cytokines	Kumar et al. (2018)
		miR-155, miR-335, miR-143, miR-221, miR-93	Negative regulation of apoptosis	Kumar et al. (2020a)
		miR-146, miR-106, miR-324,	Downregulation of IFN-γ signaling	

(continued)

Table 9.1 (continued)

Cell types	Causative species	miRNAs	Cellular pathways	References
		miR-221, miR-9, and miR-115		
		miR-3620, miR-6385, miR-6973a, miR-6996, miR-328, miR-8113, miR-3473f, miR-763, miR-6540, and miR-1264	Inhibition of apoptosis, phagocytosis, and T-cell differentiation	Tiwari et al. (2017)
		miR-30a-3p	Negatively regulates BECN1 and ATG-3-mediated autophagy	Singh et al. (2016), Su et al. (2015), and Ishikawa et al. (1987)
		Inhibition of miR-21	Promotes M2-like macrophage phenotype	Varikuti et al. (2021a)
	L. major & *L. donovani*	miR-210	Affects apoptosis through HIF-1α-mediated proapoptotic caspase-3 regulation	Lemaire et al. (2013) and Kumar et al. (2018)
		miR-146a and miR-122	Regulate IL-6-mediated inflammatory response	Bruscia et al. (2011) and Nakamura et al. (2015)
	L. infantum & *L. Viannia*	miR-346	Targets immune function of MHC class-1 gene, IFN-induced gene, and TAP1	Diotallevi et al. (2018) and Solov'eva and Volkova (1990)
	L. amazonensis	miR-294 and miR-721	Induce negative regulation of NOS-2 and NO production by depleting NOS2 mRNA	Muxel et al. (2017) and Fernandes et al. (2019)
		Inhibition of miR-181c-5p, miR-294-3p, miR-30e-5p, and miR-303d-3p	Increase NOS2, TNF, MCP-1/CCL2, and RANTES/CCL5 expression in the host	Fernandes et al. (2019)
	L. viannia			

(continued)

Table 9.1 (continued)

Cell types	Causative species	miRNAs	Cellular pathways	References
		Inhibition of miR-548d-3p	Increases MCP-1/ CCL2, RANTES/ CCL5, and IP-10/ CXCL10 production	Souza et al. (2021)
	Various disease models	Knockdown of miR-155	Induces M2 phenotypes in humans	Martinez-Nunez et al. (2011)
Dendritic cell (DC)	*L. braziliensis*	miR-193b and miR-671	Regulate CD40- and TNFR-mediated response	Nunes et al. (2018)
	L. infantum	miR-148a	Targets FAS and FASLG	Melo et al. (2019)
		miR-150	Targets TNF-α and DNAM-1	Bragato et al. (2018)
		miR-192	Target CD80 and FASL	
		miR-194		
		miR-371	Targets IL-6 and TNF-α	
		miR-21	Targets FAS, FASL, TNF-α, CD40L, NFKB, IFN-γ, CD69, and CCR7	
	L. major & L. donovani	Let-7a and let-7b miRNA	Targets inflammatory genes and possible feedback inhibitors	Geraci et al. (2015) and Staedel and Darfeuille (2013)
		miR-103	Interferon regulators control insulin-like growth factor (IGF)-1 expression	Reis et al. (2013) and Liao and Lönnerdal (2010)
	L. donovani	miR-511	Positive regulator of TLR-4 and CD80	Tserel et al. (2011)
		miR-21	Enhances expression of DMAD7 gene, reduces production of IL-12, and triggers IL-6 and STAT3 responses	Geraci et al. (2015), Bhagat et al. (2013), Gannavaram et al. (2019) and Dos Santos et al. (2016)

(continued)

Table 9.1 (continued)

Cell types	Causative species	miRNAs	Cellular pathways	References
		miR-146b-5p	Targets TRAF6 associated with TGF-β signaling and interferes with DC-SIGN	Geraci et al. (2015), Curtale et al. (2013), and Liao et al. (2013)
	L. major	miR-155	Increases DC activation to induce inflammatory response	Du et al. (2014)
T cell	*L. donovani*	miR-29a and miR-29b	Suppress Th1-mediated response	Pandey et al. (2016)
		miR-135 and miR 126	Regulate GATA-3 transcriptional factor altering Th2 responses	
		miR-124, miR-93, miR-106a, and miR-106b	Deterioration of Th17 immune mechanism	
		miR-744	Suppresses TGF-β and decreases T-reg maturation and differentiation	
		miR-69,994-5p	Suppresses NLRP3 function	(Kumar et al. 2020b)
		miR-7093-3p miR-5128	Inhibit IL12rb1 gene function corresponding to dysregulated production of IL-1B and IFN-γ	
		miR-690, miR-5128, miR-8096, and miR-8094	Control genes affecting TCR signaling pathways, Th1 cytokine production, and helper T-cell differentiation	
		miR-7a-1-3p, miR-574-5p, miR-7235-5p, and miR-7673-5p	Modulates T-cell differentiation, TCR signaling pathway, and TLR4 signaling pathways	

(continued)

Table 9.1 (continued)

Cell types	Causative species	miRNAs	Cellular pathways	References
		Downregulation of miR-93-3p, miR-503-3p, miR-3473f, miR-340-5p, and miR-486a-3p	Affect T-cell tolerance, macrophage activation, cytokine (IL-4, IL-10, IL-12, TNF) production, and Th2 proliferation	
		miR-155	Promotes CD4+ Th1 response and IFN-γ production through SOCS-1 and SHIP-1	O'Connell et al. (2010) and Baumjohann and Ansel (2013)
	L. infantum Canine VL	miR-148a	FAS and FASLG, for the regulation of apoptosis in CD4+ and CD8+ T cells	Melo et al. (2019)
		miR-21	IL-12 production and T-bet/GATA-3 regulation	
	L. major	Downmodulation of miR-10a	Controls IFN-γ production	Kelada et al. (2013)
		miR-182	Associated with IL-4 and regulates Th2-associated T-reg cells	
	L. (viannia) braziliensis	miR-361-3p miR-140-3p	Mediates NK cells and CD8+ T-cell recruitment and induces TNF	Lago et al. (2018), Ribeiro-de-Jesus et al. (1998), Bacellar et al. (2002) and Da-Cruz et al. (2002)
	L. guyanesis	miR499a	Upregulates Th17 expression and IL-8 cytokines	de Mesquita et al. (2021)
B cell	*L. infantum*	miR-150	Encourages differentiation and overexpression of B1 cells	Bragato et al. (2018), Xiao et al. (2007), and Zhou et al. (2007)
	L. donovani	miR-122	Elevates serum cholesterol to reduce liver parasite load	Ghosh et al. (2013), Mok et al. (2013), and Kumar et al. (2018)

(continued)

Table 9.1 (continued)

Cell types	Causative species	miRNAs	Cellular pathways	References
		miR-210	Upregulating HIF-1α or Oct-2 and altering antibody production	Mok et al. (2013) and Kumar et al. (2018)
		miR-221	Downregulates IFN-γ signaling and modulates the apoptosis process	Kumar et al. (2020a)
		Loss of miR-17~92	Upregulation of proapoptotic protein Bim	Koralov et al. (2008)
	L. guyanensis	miR-155	TLR3-dependent p-Akt signaling induces high parasitic burden and lesion swelling	Eren et al. (2016)
	Various disease models	miR-181a	Regulates B-cell linage differentiation	Lwin et al. (2010) and Chen et al. (2004)
		miR-181b	Impairs CSR through AID	de Yébenes et al. (2008)

factors (HIFs), which mediates caspase-3-dependent apoptotic activity in malignant conditions (Fig. 9.1) (Dang and Myers 2015). Similarly, *L. major*- and *L. donovani*-infected macrophages increase miR-210, miR-23, miR-24, miR-26a, miR-291, and miR-107, resulting in the modulation of apoptosis through HIF-1α-mediated proapoptotic caspase-3 regulation (Lemaire et al. 2013; Kumar et al. 2018). Also, a microarray analysis unveiled the anti-apoptotic activity of miR-24-3p in *L. major*-infected RAW 264.7 macrophages, compared to uninfected macrophages. Furthermore, real-time PCR validated that the upregulation of miR-24-3p can hinder the expression of caspases 3- and 7-mediated proapoptotic genes (Lasjerdi et al. 2020). Moreover, the inhibition of miR-210 in *L. donovani*-infected macrophages activates NF-kB-p50 and p60, which significantly induce pro-inflammatory cytokines such as tumor necrosis factor (TNF)-α and IL-12, and reduces anti-inflammatory IL-10 levels (Kumar et al. 2018). Similarly to miR-466i, miR-210 can be exploited as a therapeutic target for *L. donovani* infection. Furthermore, using miR-155 inhibitors or analogs of miR-15a increases apoptosis and reduces parasite burden in *L. major*-infected macrophages (Gholamrezaei et al. 2020). *Kumar* et al. also found that miR-155, miR-335, miR-143, miR-221, miR-93, and let7c are associated with the negative regulation of apoptosis in macrophages infected with clinical isolates of *L. donovani* and post-kala-azar-dermal leishmaniasis (PKDL)-causing species. In silico analysis also demonstrates dampened IFN-γ signaling due to the upregulation of miR-146. miR-106, miR-324, miR-221, miR-9, and miR-115 facilitate disease progression, while macrophages were infected with PKDL isolates (Kumar et al. 2020a).

In contrast, *Singh* et al. elucidate the role of miR-30a-3p in modulating autophagy, followed by *L. donovani* infection, in both THP-1 and human-monocyte-derived macrophages (Singh et al. 2016). Those infected macrophages displayed enhanced levels of miR-30a-3p, which negatively regulated beclin-1- and ATG-3-mediated autophagy (Fig. 9.1) (Singh et al. 2016; Kumar et al. 2018; Su et al. 2015; Ishikawa et al. 1987). Frank et al. also described the role of micro-RNAs, such as miR-101c, miR-129, and miR-210, in the autophagic digestion of *L. major* by host macrophages (Frank et al. 2015). They found that autophagy was induced in *L. major*-infected bone-marrow-derived macrophages (BMDMs) by subverting mTOR (mechanistic target of rapamycin) phosphorylation and upregulating proteins ATG-5 (autophagy related-5), BNIP-3 (BCL2/adenovirus E1B 19 kDa protein-interacting protein-3), and CTS-E (cathepsin-E). In the late phase of infection, they found lower miR-101c and miR-129-5p expression. However, miR-210 gets elevated along with the LC3B-II/LC3B-I ratio, which helps increase the autophagic flux in host macrophages (Frank et al. 2015).

A recent study by *Tiwari* et al. characterized several micro-RNAs in *L. donovani*-infected macrophages and found many macrophage regulators, including miR-3620, miR-6385, miR-6973a, miR-6996, miR-328, miR-8113, miR-3473f, miR-763, miR-6540, and miR-1264. Gene prediction and biological interaction analysis revealed the involvement of these miRNAs in apoptosis inhibition, phagocytosis, T-cell differentiation, and IFN-γ signaling (Fig. 9.1) (Tiwari et al. 2017). Another in vitro study on THP-1 macrophages infected with *L. major* identified miR-146a-3p, miR-146a-5p, miR-122, miR-155-3p, miR-188, miR-9-3p, miR-9-5p, and miR-147 as key regulators in generating an inflammatory response (Nimsarkar et al. 2020). Furthermore, *L. donovani*-infected BMDMs showed the upregulation of miR-146a-5p, which mediates macrophage polarization (Das et al. 2021). The upregulation of miR-146a and miR-122 regulates IL-6-mediated inflammatory response by altering TLR4 signaling in macrophages (Bruscia et al. 2011; Nakamura et al. 2015). In contrast, miR-155 and miR-188 were acknowledged as regulators of macrophage plasticity (Curtale et al. 2019) and anti-inflammatory activity (Zhang et al. 2018), respectively. mir-155 targets C/EBP-β (a signature marker for M2 macrophage), which in turn dysregulates arginase-1 and modulates Akt-dependent M1/M2 macrophage polarization (Arranz et al. 2012). The knockdown of miR-155 induces M2 phenotypes in humans by modulating IL-13Rα1 and deteriorating STAT6 activation (Martinez-Nunez et al. 2011). Likewise, an M2 phenotype was induced in the macrophages of *L. donovani*-infected C57bl/6 mice by inhibiting miR-21 expression (Varikuti et al. 2021a). These results highlight the importance of researching the mechanisms of action of miR-155 and miR-21 for the development of novel immunomodulatory agents against VL.

L. infantum- and *L. viannia*-infected U937/THP-1 macrophages displayed the upregulation of miR-346 (Diotallevi et al. 2018). Plausibly, associated transcription factor X-box-binding protein-1 (XBP-1) is essential for ER (endoplasmic reticulum) stress related to miR-346 induction. miR-346 targets immune function governed by the major histocompatibility complex (MHC) class-1 gene, IFN-induced gene, and ER antigen peptide transporter 1 (TAP1) (Diotallevi et al. 2018; Bartoszewski et al. 2011; Solov'eva and Volkova 1990). Other miRNAs, such as miR-294 and miR-721, were upregulated in bone-marrow-derived macrophages infected with *L. amazonensis* (Muxel et al. 2017). Furthermore, the enhancement of the polyamine pathway in *L. amazonensis*-infected macrophages increases L-arginine uptake by increasing CAT2B and CAT1 (L-arginine transporter) expression. Simultaneously, miR-294 and miR-721 induced the negative regulation of nitric oxide synthase-2 (NOS-2) and NO production by depleting NOS2 mRNA. In reverse, the inhibition of miR-181c-5p, miR-294-3p, miR-30e-5p, and miR-303d-3p increases NOS2, TNF, MCP-1/CCL2, and RANTES/CCL5 expression in the host and controls *L. amazonensis* infectivity in BMDM (Fernandes et al. 2019). Similarly, the inhibition of miR-548d-3p declined *L. (viannia) braziliensis* growth in macrophages by increasing MCP-1/CCL2, RANTES/CCL5, and IP-10/CXCL10 production (Souza et al. 2021). Furthermore, *L. major* parasites alter IL-10 and transforming growth factor (TGF)-β1 mechanisms to escape the immune system. Downregulated miR-340 and miR-27a results in increased levels of IL-10 and TGF-β1 in *L. major*-infected macrophage (Hamidi et al. 2021).

9.4.2 Role of miRNAs in *Leishmania*-Infected Dendritic Cells

During infection by leishmaniasis, dendritic cells (DCs) are essential in protecting organisms by bridging the innate and adaptive immune systems. DCs release IL-12, which is vital for activating Th1 differentiation and promoting resistance to leishmaniasis (Kautz-Neu et al. 2012). In particular, the pro-inflammatory mechanisms that are influenced by DCs control disease outcomes from the infection (Von Stebut 2007). It has been well established in previous miRNA profiling studies that miRNAs play a role in leishmaniasis by regulating immune responses, including those of DCs (Geraci et al. 2015; Singh et al. 2016). A study by Chaussabel et al. regarding the effects of *L. major* and *L. donovani* gene expression in human monocyte-derived DCs (MDDC) found similarities in response to the two parasite species within-host gene expression patterns. Among IL-12-associated genes, there is an extensive range of divergent effects on gene expression profiles in MDDCs that suggest species-specific mechanisms in DC responses (Chaussabel et al. 2003; Cummings et al. 2010; Kedzierski and Evans 2014). It has become apparent that different immune mechanisms control *L. major* and *L. donovani* infections, which extends to miRNAs.

In a study performed by Geraci et al., differences between *L. major*-infected DCs and *L. donovani*-infected DCs were determined. miRNAs (95 were identified) were suppressed in *L. major* infection and were upregulated in *L. donovani* infection (45 were identified) (Geraci et al. 2015). The study also found that DCs infected by *L. donovani* had a higher expression of miRNAs than cells infected by *L. major*. There are various distinct examples of this higher expression of mature miRNAs.

DCs infected with *L. donovani* had an upregulation of let-7a and let-7b miRNA, which were downregulated in *L. major* (Geraci et al. 2015). The let-7 miRNA family assists in targeting genes responsible for inflammation during bacterial infections and that are possible feedback inhibitors (Staedel and Darfeuille 2013). Similarly, miR-103 was upregulated in *L. donovani* and suppressed in *L. major*. miR-103 expression may be due to insulin-like growth factor (IGF)-1 expression controlled by interferon regulators (Liao and Lönnerdal 2010; Reis et al. 2013). Another study also found that miR-511 had high upregulation in DCs that were infected with *L. donovani* only. This miRNA acts as a positive regulator of TLR-4 and CD80, important for the activation of the immune response against Leishmania parasites (Tserel et al. 2011). The differential expression of these various miRNAs could be related to the species-specific mechanisms of infection (Geraci et al. 2015). Further studies to identify miRNA profiles in infection caused by different *Leishmania* species are necessary to develop immunomodulatory and therapeutic agents tailored to specific infections.

In *L. donovani* infection, miR-21 has been linked with the TGF-β signaling pathways via the expression of the SMAD7 gene (Geraci et al. 2015; Bhagat et al. 2013) and, specifically for DCs, the reduced production of IL-12 (Gannavaram et al. 2019). A study performed by our group established the importance of miR-21 in *L. donovani* infection due to the miRNA's distinctive expressions that promote infection through cytokine responses and the polarization of infected cells. The study results found that within bone-marrow-derived DCs (BMDCs), the suppression of miR-21 led to increased IL-12 levels and induced IFN-γ-associated Th1 immunity. This illustrates that increased miR-21 expression caused by leishmaniasis suppresses IL-12 production, decreasing Th1 immune responses and leading to worsened infection (Varikuti et al. 2021a). Furthermore, the increased miR-21 expression associated with *L. donovani* triggers IL-6- and STAT3-mediated responses, which correlate with the susceptibility and severity of the VL infection (Dos Santos et al. 2016). Taken together, these results show that miR-21 is an inhibitor of protective immune responses in VL and highlight its role as an immunotherapeutic target.

Within the TGF-β signaling pathway, miR-146b-5p plays an essential role during *L. donovani* infection (Geraci et al. 2015). miR-146b-5p is known to target TRAF6, a gene transcript associated with TGF-β signaling (Curtale et al. 2013). While this may seem counterintuitive, miR-146b-5p is part of a negative feedback loop controlling TGF-β signaling in infected DCs (Liao et al. 2013). Also, within *L. donovani* infection, miR146a-5p interferes with DC-SIGN (dendritic-cell-specific ICAM-3-grabbing nonintegrin), which plays a significant role in promastigote attachment and entry in DCs during leishmaniasis (Das et al. 2021; Martinez-Nunez et al. 2009).

In general, miR-155 assists in the regulation of immune response and cell function and increases DC activation to induce inflammatory responses (Du et al. 2014), and in *L. major*-infected DCs, miR-155 has a vital role (Fig. 9.1). Within CL infection, both the Th1 and Th2 immune responses determine the outcome of the disease and are influenced by the upregulation of miR155 (Varikuti et al. 2021b).

Also, the suppression of miR-155 in *L. major* infection diminished the Th2 immune response and increased the Th1 response, which is responsible for controlling the infection. MiR-155 mutant DCs infected with L. major showed NO suppression, lower IL-12 levels, and enhanced IFN-γ production from activated T cells in comparison to wild-type DCs (Varikuti et al. 2021b). These results highlight the role of miR-155 in CL susceptibility in *L. major* infections. By suppressing miR-155 function, the Th1 immune response is increased and Th2 response is decreased. Various studies point to miR-155 influencing immunity in *L. major* infections, as opposed to *L. donovani* (Geraci et al. 2015; Varikuti et al. 2021b).

The innate immune response against *L. braziliensis* includes crosstalk between DCs and DC maturation (Geiger et al. 2016; Nunes et al. 2018). A study performed by Vasco et al. found that miR-193b and miR-671 regulated inflammation associated with this innate immune response. miR-193b was found to be negatively correlated with its target genes, which means it could be playing a role in augmenting inflammatory response to *L. braziliensis* (Nunes et al. 2018). On the other hand, miR 671 had a positive correlation with the TNF receptor (TNFR), which is known to regulate inflammation. This suggests that miR-671 might be enhancing the expression of TNFR (Nunes et al. 2018). In contrast, with *L. braziliensis*, crosstalk between DCs in *L. infantum*-infected dogs seems to be controlled by miR-148a (Fig. 9.1) (Melo et al. 2019). The crosstalk between DCs and natural killer cells is possibly regulated by miR-150 (targets TNF-α and DNAM-1), miR-194, miR-192 (targets CD80 and FASL), miR-371 (targets IL-6 and TNF-α), and miR-21 (targets FAS, FASL, TNF-α, CD40L, NFKB, IFN-γ, CD69, and CCR7) (Fig. 9.1) (Bragato et al. 2018). This once again demonstrates the species-specific controls of leishmaniasis.

9.4.3 miRNAs as Regulators of T-Cell Subsets During *Leishmania* Infection

Leishmanial infection and its presentation by antigen-presenting cells (APCs) induce variable pro-inflammatory responses directly related to the causative *Leishmania* species. Studies in experimental mice infected with *L. donovani* or *L. infantum* presented with antiparasitic IFN-γ-producing CD4+ T cells (Bunn et al. 2014; Solano-Gallego et al. 2016). This presentation of CD4+ T cells is essential to show resistance against infection. However, many patients infected with parasites fail to generate potent cell-mediated immune responses against parasitic antigen, the primary cause of disease progression. Although elevated levels of TNF-α, IL-12, and IFN-γ in VL patients show protective Th1 differentiation of naïve CD4+ T cells, concurrently, the parasites take advantage of an immunosuppressive mechanism to manage their proliferation and survival (Schmid-Hempel 2009). This T-cell proliferation and differentiation, dependent on micro-RNA-mediated gene regulation, leads to fine-tuning for developing Th1, Th2, Th17, and T-reg cells to uphold cellular homeostasis.

A computational study by *Pandey* et al. demonstrates that miR-29a and miR-29b have an affinity to the T-box transcription factor (TBX)-21, which may suppress T-helper (Th)-1-mediated responses during VL (Pandey et al. 2016). On the other hand, miR-135 and miR-126 are associated with the GATA-3 transcriptional factor and alter Th2 responses. Also, a deteriorated Th17 immune mechanism is associated with miR-124, miR-93, miR-106a, and miR-106b, resulting in *L. donovani* parasite survival and replication in macrophages (Fig. 9.1). Furthermore, decreased T-reg maturation and differentiation through the suppression of transforming growth factor (TGF)-β is linked to miR-744 (Pandey et al. 2016). Taken together, these results highlight the importance of miRNA profiling to develop immunomodulatory agents targeting adaptive immunity in VL.

Kelada et al. reported that miR-10a and miR-182 are predominately associated with regulatory T (T-reg) cells to tune Th1 and Th2 responses. The downmodulation of miR-10a in Th1-associated T-reg cells controls IFN-γ production, but in contrast, miR-182 is associated with IL-4 and regulates Th2-associated T-reg cells. Together, these miRNAs modulate the CD4+FoxP3+ cells' microenvironment and shape T-reg's stability and suppressor function (Kelada et al. 2013). Dogs suffering from canine visceral leishmaniasis (CVL) caused by *L. infantum* parasites also showed enhanced expression of miR-148a, miR-21, miR-7, and miR615 and the downregulation of miR-150, miR-125a, and miR-125b. The identified genes involved with miR-21 associated with IL-12 production and T-bet/GATA-3 regulation and miR148a are FAS and FAS-L, which have been suggested to cause the death of CD8+ and CD4+ T cells in CVL-infected dogs (Melo et al. 2019).

miR155 also plays an important role in T-cell differentiation by promoting a CD4+ Th1 response and IFN-γ production through the suppressor of cytokine signaling (SOCS)-1 and Src homology-2 domain-containing inositol 5-phosphate (SHIP)-1 (O'Connell et al. 2010; Baumjohann and Ansel 2013). miR155-knock out mice infected with VL showed high levels of program death ligand (PDL)-1 expressing Ly6C hi inflammatory monocytes (Varikuti et al. 2019). At the same time, the miR155-knock out CL model infected with *L. major* demonstrates a lack of miR155 promoting Th2 response by inhibiting DC function (Varikuti et al. 2021b). Another patient with CL infected with *L. (viannia) braziliensis* displayed a high expression of miR-361-3p and miR-140-3p at the skin lesion site, compared to normal skin (Lago et al. 2018). The genes associated with miR-361-3p (TNF and GZMB; granzyme B) and miR-140-3p (NKG7; natural killer cell granule protein 7) were also overexpressed and mediated natural killer (NK) cells and CD8+ T-cell recruitment at the infection site. Additionally, it increases necrosis, cell adhesion, and cytotoxicity while inducing the pro-inflammatory cytokines from TNF family (Ribeiro-de-Jesus et al. 1998; Bacellar et al. 2002; Da-Cruz et al. 2002). In parentheses, granzyme-B promotes inflammation by activating IL-1α, IL-1β, and IL-18 (Omoto et al. 2010; Santos et al. 2013). miR499a is associated with susceptibility to *L. guyanensis*-mediated CL infection by upregulating Th17 expression and IL-8 cytokines during infection (de Mesquita et al. 2021). IL-8 cytokines expressed the variability of cells like neutrophils, granulocytes, and lymphocytes, including

T cells. They regulate T-cell activation and increase cytotoxic CD8+ T cells at the disease site (Wechsler et al. 1994; Hess et al. 2004).

Recently, Kumar et al. identified several micro-RNAs in *L. donovani*-infected mice that affect CD4+-mediated cell functions (Kumar et al. 2020b). For example, the upregulation of miR-69994-5p suppresses nucleotide-binding domain leucine-rich repeat and pyrin domain-containing receptor (NLRP)-3 function. Additionally, miR-7093-3p and miR-5128 inhibit IL12rb1 gene function, which is correspondingly observed with dysregulated production of IL-1β and IFN-γ. Simultaneously, the upregulation of miR-690, miR-5128, miR-8096, and miR-8094 controls several genes, such as *CD38*, *IL-12β*, and *CD247*, respectively, which affect several T-cell receptor (TCR) signaling pathways, Th1 cytokine productions, and helper T-cell differentiation. Similarly, the upregulation of miR-71a-1-3p, miR-574-5p, miR-7235-5p, and miR7673-5p modulates T-cell differentiation, TCR signaling pathway, and toll-like receptor (TLR)-4 signaling pathways. On the other hand, the downregulation of miR-8096, miR-93-3p, miR-503-3p, miR-3473f, miR-340-5p, and miR-486a-3p affects T-cell tolerance, macrophage activation, cytokine (IL-4, IL-10, IL-12, TNF) production, and Th2 proliferation (Kumar et al. 2020b).

9.4.4 miRNAs Alter B-Cell Development and Functions During Leishmaniasis

B cells originate from hematopoietic stem cells, which develop in the bone marrow. In particular, B cells develop from common lymphoid progenitor (CLP) cells, identified as $lin^{(-)}$ $c\text{-}kit^{Low}$ $Sca\text{-}1^{Low}$ $IL7R^{(+)}$ (Kondo et al. 1997). The first B-lineage-restricted cells, called pro-B cells, initiate gene rearrangement at immunoglobulin (Ig) loci. The process of rearrangement and recombination of variable (v), joining (J), and diverse (D) gene segments, called V(D)J recombination, is mediated by RAG-1 and two recombinase enzymes and grows as pro-B cells. Later, immunoglobulin heavy chain (IgH) and light chain (IgL) assemble with surrogate light chain (SLC) and form the pre-B cell receptor (pre-BCR) or immature B cells. Several miRNAs have been reported to play a vital role in B-cell development independently. In 2004, miR-181a was identified in B lymphoid cells of the bone marrow for the first time (Chen et al. 2004); thus, the overexpression of miR-181a promotes an increase in B-cell lineages in mice. On the other hand, *Bragato* et al. found a twofold downregulation of miR-150 in VL-infected dogs that negatively correlated with *Leishmania* parasite load in the peripheral blood (Bragato et al. 2018). MiR-150 mutants, or knockout mice, displayed increased expression of cell growth and differentiation gene c-Myb, which promoted the differentiation of a high number of B1 cells (Fig. 9.1) (Xiao et al. 2007; Zhou et al. 2007). Eventually, these overdifferentiated B1 cells produced all immunoglobulin classes superfluously, causing a hypergammaglobulinemic condition in VL patients (Bragato et al. 2018). However, *Ghosh* et al. discovered that a glycoprotein from *L. donovani* parasites targets host Dicer1 (RNase-III enzyme) and cleaves it to downregulate miR-122. The result was an increased parasitic burden in mouse livers by altering the posttranscriptional regulation of host mRNA/miRNA interactions (Ghosh et al. 2013). Studies show that conditional deletion of the Dicer gene among B-cell progenitors blocks the differentiation of pro-B cells and upregulates miR-17~92, including proapoptotic Bim molecules (a BCL-2 family protein), and regulates apoptosis during *Leishmania* infection (Koralov et al. 2008).

Another subset of miRNAs, including miR-29a, miRNA-181a, and miRNA-181b, also regulates B-cell function during various disease conditions. miR-29a alters the expression of the TCL-1, MCL10, and CDK6 genes and becomes upregulated in indolent B-cell chronic lymphocytic leukemia (CLL), compared to healthy B cells (Pekarsky et al. 2006). miRNA-181a modulates proapoptotic activity through the BIM protein, simultaneously regulating B-cell lineage differentiation (Lwin et al. 2010; Chen et al. 2004). In contrast, miRNA 181b impairs class-switch recombinase (CSR) through activation-induced cytidine deaminase (AID) (de Yébenes et al. 2008). Although the role of these miRNAs has not yet been demonstrated in leishmaniasis, future studies are needed to investigate their function in B-cell differentiation during VL.

Micro-RNA-155 (miR-155), a remarkably well-illustrated miRNA, has been known as a regulator of assorted immune cells, such as monocytes, macrophages, T cells, including B cells (Rothchild et al. 2016; Vigorito et al. 2013; Alivernini et al. 2017). First of all, Costinean et al. developed a genetically engineered mouse (Eμ-miR-155) to overexpress miR-155 in B cells (Costinean et al. 2006). Eμ-miR-155 transgenic mice produce high levels of TNF-α against LPS/D-galactosamine-induced septic shock (Tili et al. 2007). This accumulation of miR-155 leads to oncogenesis, which causes B-cell lymphoma (Eis et al. 2005). On the other hand, miR-155 deficiency in mice affects B-cell generation and maturation and produces competent antigen-specific antibodies for efficient immune responses (Thai et al. 2007; Rodriguez et al. 2007). Eren et al. also observed that mice infected with *L. guyanensis* harboring *Leishmania* RNA virus 1 (LRV-1) simultaneously showed elevated levels of miR-155 and high parasitic burden and lesion swelling in comparison to WT mice (Eren et al. 2016). miR-155 is positioned in the third exon of the noncoding region of the B-cell-integrated cluster (BIC) gene (Tam 2001). Remarkably, the upregulation of miR-155 is dependent on TLR-3/TRIF signaling (Yutiao and Kinzie 1975; O'Connell et al. 2008). Furthermore, this TLR-3 activation promotes parasitic persistence by enhancing macrophage survival partially through miR-155-dependent Akt activation (Eren et al. 2016).

miR-210 is another micro-RNA that can dysregulate B-cell function. It upregulates hypoxia-inducible factor (HIF)1-alpha or octamer-binding transcription factor (Oct)-2 and alters antibody production (Mok et al. 2013). *L. donovani* activates HIF-1α and miR-210 by downmodulating nuclear factor kappa beta (NF-kB)-mediated pro-inflammatory response for their survival within the host cells (Kumar et al. 2018). In addition, *L. infantum* was observed to elevate miR-210 expression in canine peripheral blood mononuclear cells (PBMCs) (Soares et al. 2021). Conversely, the Oct-2 transcriptional factor, which is a member of the POU (Pit-Oct-Unc) family of transcriptional factors and is expressed on B cells (Latchman 1996), alters miR-210 and its associated CD36 expression, resulting in the modulation of immune globulin (Ig)-M production in *L. major*-infected mice (Fig. 9.1) (Corcoran et al. 2002).

miR-221 also plays a substantial role in B-cell development in the bone marrow (Knoll et al. 2013). They promote Ig-E-mediated activation of mast cells. miR-221-expressing cells encourage cellular degranulation by the unbalanced function of phospholipase C gamma (p-PLCγ)-1, phospho-protein kinase B (p-Akt), phospho-inhibitor of kappa B alpha (p-IkB-α), NF-kB, and calcium $(Ca)^{2+}$ signaling pathways (Xu et al. 2016). Kumar et al. found that a higher expression of hs-miR-221 in VL and PKDL patients downregulates IFN-γ signaling and modulates apoptosis, ultimately aiding disease progression (Fig. 9.1) (Kumar et al. 2020a).

9.5 Discussion: The Role of miRNAs as Biomarkers and Therapeutic Targets

Identifying and characterizing altered miRNA profiles induced by *Leishmania* infection in the host can provide new insight regarding disease pathogenesis and novel drug targets. Several studies, both in vitro and in vivo, have shown a differential expression of miRNAs in healthy compared to diseased/infected cells, tissue, serum, and other fluids. These miRNAs have substantial potential to become prognostic and diagnostic biomarkers for the efficacy of vaccination and treatment strategies against *Leishmania* parasites. Despite prevailing standard protocols for the diagnosis and treatment of parasitic diseases, miRNA-based identification of disease and monitoring of disease progression can bring more reliability to the current clinical practice. Further studies are required to identify specific miRNAs and elucidate their role in regulating pathological and physiological processes. Now, researchers are exploring signature miRNAs that characterize the disease state and validate them as biomarkers with appropriate sensitivity and specificity. Identified miRNAs regulate host immune cells such as macrophages and DCs at the molecular level and affect their functional activity by modulating critical immune pathways, such as apoptosis and autophagy. Similarly, miRNAs induce several signaling molecules and secrete cytokines, which determine cellular fate during infection. Using/developing specific inhibitors or promoters of identified miRNAs associated with *Leishmania* parasitic infection may accelerate or assist the current treatment strategy to control the disease. On the other hand, the analogs of those miRNAs could be used as host-directed immunotherapies to uphold immune response against parasitic infection.

In conclusion, miRNAs have great potential for the development of a novel class of therapeutics or diagnostic tools. Therefore, the biological information relative to miRNAs is expected to increase and be translated to the field of parasitology and other infectious diseases in the coming years.

Conflict of Interest No conflict of interest was disclosed by the authors.

References

Alarcón CR, Lee H, Goodarzi H, Halberg N, Tavazoie SF (2015) N6-methyladenosine marks primary microRNAs for processing. Nature 519(7544):482–485. https://doi.org/10.1038/nature14281

Alivernini S, Gremese E, McSharry C, Tolusso B, Ferraccioli G, McInnes IB, Kurowska-Stolarska M (2017) MicroRNA-155-at the critical Interface of Innate and adaptive immunity in arthritis. Front Immunol 8:1932. https://doi.org/10.3389/fimmu.2017.01932

Arranz A, Doxaki C, Vergadi E, Martinez de la Torre Y, Vaporidi K, Lagoudaki ED, Ieronymaki E, Androulidaki A, Venihaki M, Margioris AN, Stathopoulos EN, Tsichlis PN, Tsatsanis C (2012) Akt1 and Akt2 protein kinases differentially contribute to macrophage polarization. Proc Natl Acad Sci U S A 109(24):9517–9522. https://doi.org/10.1073/pnas.1119038109

Atayde VD, Shi H, Franklin JB, Carriero N, Notton T, Lye LF, Owens K, Beverley SM, Tschudi C, Ullu E (2013) The structure and repertoire of small interfering RNAs in Leishmania (Viannia) braziliensis reveal diversification in the trypanosomatid RNAi pathway. Mol Microbiol 87(3): 580–593. https://doi.org/10.1111/mmi.12117

Bacellar O, Lessa H, Schriefer A, Machado P, Ribeiro de Jesus A, Dutra WO, Gollob KJ, Carvalho EM (2002) Up-regulation of Th1-type responses in mucosal leishmaniasis patients. Infect Immun 70(12):6734–6740. https://doi.org/10.1128/iai.70.12.6734-6740.2002

Bartoszewski R, Brewer JW, Rab A, Crossman DK, Bartoszewska S, Kapoor N, Fuller C, Collawn JF, Bebok Z (2011) The unfolded protein response (UPR)-activated transcription factor X-box-binding protein 1 (XBP1) induces microRNA-346 expression that targets the human antigen peptide transporter 1 (TAP1) mRNA and governs immune regulatory genes. J Biol Chem 286(48):41862–41870. https://doi.org/10.1074/jbc.M111.304956

Baumjohann D, Ansel KM (2013) MicroRNA-mediated regulation of T helper cell differentiation and plasticity. Nat Rev Immunol 13(9):666–678. https://doi.org/10.1038/nri3494

Bhagat TD, Zhou L, Sokol L, Kessel R, Caceres G, Gundabolu K, Tamari R, Gordon S, Mantzaris I, Jodlowski T, Yu Y, Jing X, Polineni R, Bhatia K, Pellagatti A, Boultwood J, Kambhampati S, Steidl U, Stein C, Ju W, Liu G, Kenny P, List A, Bitzer M, Verma A (2013) miR-21 mediates hematopoietic suppression in MDS by activating TGF-β signaling. Blood 121(15):2875–2881. https://doi.org/10.1182/blood-2011-12-397067

Bragato JP, Melo LM, Venturin GL, Rebech GT, Garcia LE, Lopes FL, de Lima VMF (2018) Relationship of peripheral blood mononuclear cells miRNA expression and parasitic load in canine visceral leishmaniasis. PLoS One 13(12):e0206876. https://doi.org/10.1371/journal.pone.0206876

Bruscia EM, Zhang PX, Satoh A, Caputo C, Medzhitov R, Shenoy A, Egan ME, Krause DS (2011) Abnormal trafficking and degradation of TLR4 underlie the elevated inflammatory response in cystic fibrosis. J Immunol 186(12):6990–6998. https://doi.org/10.4049/jimmunol.1100396

Bunn PT, Stanley AC, de Labastida Rivera F, Mulherin A, Sheel M, Alexander CE, Faleiro RJ, Amante FH, Montes De Oca M, Best SE, James KR, Kaye PM, Haque A, Engwerda CR (2014) Tissue requirements for establishing long-term CD4+ T cell-mediated immunity following Leishmania donovani infection. J Immunol 192(8):3709–3718. https://doi.org/10.4049/jimmunol.1300768

Cai Y, Yu X, Hu S, Yu J (2009) A brief review on the mechanisms of miRNA regulation. Genomics Proteomics Bioinformatics 7(4):147–154. https://doi.org/10.1016/S1672-0229(08)60044-3

Carthew RW, Sontheimer EJ (2009) Origins and mechanisms of miRNAs and siRNAs. Cell 136(4): 642–655. https://doi.org/10.1016/j.cell.2009.01.035

Cazalla D, Yario T, Steitz JA, Steitz J (2010) Down-regulation of a host microRNA by a Herpesvirus saimiri noncoding RNA. Science 328(5985):1563–1566. https://doi.org/10.1126/science.1187197

Chaussabel D, Semnani RT, McDowell MA, Sacks D, Sher A, Nutman TB (2003) Unique gene expression profiles of human macrophages and dendritic cells to phylogenetically distinct parasites. Blood 102(2):672–681. https://doi.org/10.1182/blood-2002-10-3232

Chen CZ, Li L, Lodish HF, Bartel DP (2004) MicroRNAs modulate hematopoietic lineage differentiation. Science 303(5654):83–86. https://doi.org/10.1126/science.1091903

Cheng AM, Byrom MW, Shelton J, Ford LP (2005) Antisense inhibition of human miRNAs and indications for an involvement of miRNA in cell growth and apoptosis. Nucleic Acids Res 33(4):1290–1297. https://doi.org/10.1093/nar/gki200

Cho J (2018) Transposon-derived non-coding RNAs and their function in plants. Front Plant Sci 9: 600. https://doi.org/10.3389/fpls.2018.00600

Corcoran L, Vremec D, Febbraio M, Baldwin T, Handman E (2002) Differential regulation of CD36 expression in antigen-presenting cells: Oct-2 dependence in B lymphocytes but not dendritic cells or macrophages. Int Immunol 14(10):1099–1104. https://doi.org/10.1093/intimm/dxf075

Costinean S, Zanesi N, Pekarsky Y, Tili E, Volinia S, Heerema N, Croce CM (2006) Pre-B cell proliferation and lymphoblastic leukemia/high-grade lymphoma in E(mu)-miR155 transgenic mice. Proc Natl Acad Sci U S A 103(18):7024–7029. https://doi.org/10.1073/pnas.0602266103

Cui J, You C, Chen X (2017) The evolution of microRNAs in plants. Curr Opin Plant Biol 35:61–67. https://doi.org/10.1016/j.pbi.2016.11.006

Cummings HE, Tuladhar R, Satoskar AR (2010) Cytokines and their STATs in cutaneous and visceral leishmaniasis. J Biomed Biotechnol 2010:294389. https://doi.org/10.1155/2010/294389

Curtale G, Mirolo M, Renzi TA, Rossato M, Bazzoni F, Locati M (2013) Negative regulation of Toll-like receptor 4 signaling by IL-10-dependent microRNA-146b. Proc Natl Acad Sci U S A 110(28):11499–11504. https://doi.org/10.1073/pnas.1219852110

Curtale G, Rubino M, Locati M (2019) MicroRNAs as molecular switches in macrophage activation. Front Immunol 10:799. https://doi.org/10.3389/fimmu.2019.00799

Da-Cruz AM, Bittar R, Mattos M, Oliveira-Neto MP, Nogueira R, Pinho-Ribeiro V, Azeredo-Coutinho RB, Coutinho SG (2002) T-cell-mediated immune responses in patients with cutaneous or mucosal leishmaniasis: long-term evaluation after therapy. Clin Diagn Lab Immunol 9(2):251–256. https://doi.org/10.1128/cdli.9.2.251-256.2002

Dang K, Myers KA (2015) The role of hypoxia-induced miR-210 in cancer progression. Int J Mol Sci 16(3):6353–6372. https://doi.org/10.3390/ijms16036353

Das S, Mukherjee S, Ali N (2021) Super enhancer-mediated transcription of miR146a-5p drives M2 polarization during Leishmania donovani infection. PLoS Pathog 17(2):e1009343. https://doi.org/10.1371/journal.ppat.1009343

de Mesquita TGR, Junior JDES, de Lacerda TC, Queiroz KLGD, Júnior CMDS, Neto JPM, Gomes LAM, de Souza MLG, Guerra MVF, Ramasawmy R (2021) Variants of MIRNA146A rs2910164 and MIRNA499 rs3746444 are associated with the development of cutaneous leishmaniasis caused by Leishmania guyanensis and with plasma chemokine IL-8. PLoS Negl Trop Dis 15(9):e0009795. https://doi.org/10.1371/journal.pntd.0009795

de Yébenes VG, Belver L, Pisano DG, González S, Villasante A, Croce C, He L, Ramiro AR (2008) miR-181b negatively regulates activation-induced cytidine deaminase in B cells. J Exp Med 205(10):2199–2206. https://doi.org/10.1084/jem.20080579

Diotallevi A, De Santi M, Buffi G, Ceccarelli M, Vitale F, Galluzzi L, Magnani M (2018) Infection induces microRNA hsa-miR-346 in human cell line-derived macrophages. Front Microbiol 9: 1019. https://doi.org/10.3389/fmicb.2018.01019

Dos Santos PL, de Oliveira FA, Santos ML, Cunha LC, Lino MT, de Oliveira MF, Bomfim MO, Silva AM, de Moura TR, de Jesus AR, Duthie MS, Reed SG, de Almeida RP (2016) The severity of visceral leishmaniasis correlates with elevated levels of Serum IL-6, IL-27 and sCD14. PLoS Negl Trop Dis 10(1):e0004375. https://doi.org/10.1371/journal.pntd.0004375

Du F, Yu F, Wang Y, Hui Y, Carnevale K, Fu M, Lu H, Fan D (2014) MicroRNA-155 deficiency results in decreased macrophage inflammation and attenuated atherogenesis in apolipoprotein E-deficient mice. Arterioscler Thromb Vasc Biol 34(4):759–767. https://doi.org/10.1161/ATVBAHA.113.302701

Eis PS, Tam W, Sun L, Chadburn A, Li Z, Gomez MF, Lund E, Dahlberg JE (2005) Accumulation of miR-155 and BIC RNA in human B cell lymphomas. Proc Natl Acad Sci U S A 102(10): 3627–3632. https://doi.org/10.1073/pnas.0500613102

Eren RO, Reverte M, Rossi M, Hartley MA, Castiglioni P, Prevel F, Martin R, Desponds C, Lye LF, Drexler SK, Reith W, Beverley SM, Ronet C, Fasel N (2016) Mammalian innate immune response to a Leishmania-resident RNA virus increases macrophage survival to promote parasite persistence. Cell Host Microbe 20(3):318–328. https://doi.org/10.1016/j.chom.2016.08.001

Fabian MR, Sonenberg N (2012) The mechanics of miRNA-mediated gene silencing: a look under the hood of miRISC. Nat Struct Mol Biol 19(6):586–593. https://doi.org/10.1038/nsmb.2296

Fernandes JCR, Aoki JI, Maia Acuña S, Zampieri RA, Markus RP, Floeter-Winter LM, Muxel SM (2019) Melatonin and Leishmania amazonensis Infection altered miR-294, miR-30e, and miR-302d impacting on Tnf, Mcp-1, and Nos 2 expression. Front Cell Infect Microbiol 9:60. https://doi.org/10.3389/fcimb.2019.00060

Frank B, Marcu A, de Oliveira Almeida Petersen AL, Weber H, Stigloher C, Mottram JC, Scholz CJ, Schurigt U (2015) Autophagic digestion of Leishmania major by host macrophages is associated with differential expression of BNIP3, CTSE, and the miRNAs miR-101c, miR-129, and miR-210. Parasit Vectors 8:404. https://doi.org/10.1186/s13071-015-0974-3

Gannavaram S, Bhattacharya P, Siddiqui A, Ismail N, Madhavan S, Nakhasi HL (2019) miR-21 expression determines the early vaccine immunity induced by. Front Immunol 10:2273. https://doi.org/10.3389/fimmu.2019.02273

Geiger A, Bossard G, Sereno D, Pissarra J, Lemesre JL, Vincendeau P, Holzmuller P (2016) Escaping deleterious immune response in their hosts: lessons from trypanosomatids. Front Immunol 7:212. https://doi.org/10.3389/fimmu.2016.00212

Georgakilas G, Perdikopanis N, Hatzigeorgiou AG (2018) Identifying pri-miRNA transcription start sites. Methods Mol Biol 1823:11–31. https://doi.org/10.1007/978-1-4939-8624-8_2

Geraci NS, Tan JC, McDowell MA (2015) Characterization of microRNA expression profiles in Leishmania-infected human phagocytes. Parasite Immunol 37(1):43–51. https://doi.org/10.1111/pim.12156

Gholamrezaei M, Rouhani S, Mohebali M, Mohammadi-Yeganeh S, Haji Molla Hoseini M, Haghighi A, Lasjerdi Z, Hamidi F, Sharifi-Yazdi MK (2020) MicroRNAs expression induces apoptosis of macrophages in response to. Iran J Parasitol 15(4):475–487. https://doi.org/10.18502/ijpa.v15i4.4851

Ghosh J, Bose M, Roy S, Bhattacharyya SN (2013) Leishmania donovani targets Dicer1 to downregulate miR-122, lower serum cholesterol, and facilitate murine liver infection. Cell Host Microbe 13(3):277–288. https://doi.org/10.1016/j.chom.2013.02.005

Gramzow L, Lobbes D, Innard N, Theißen G (2020) Independent origin of MIRNA genes controlling homologous target genes by partial inverted duplication of antisense-transcribed sequences. Plant J 101(2):401–419. https://doi.org/10.1111/tpj.14550

Hamidi F, Mohammadi-Yeganeh S, Haji Molla Hoseini M, Tabaei SJS, Taghipour N, Lasjerdi Z, Gholamrezaei M, Haghighi A (2021) Inhibition of anti-inflammatory cytokines, IL-10 and TGF-β, in Leishmania major infected macrophage by miRNAs: a new therapeutic modality against leishmaniasis. Microb Pathog 153:104777. https://doi.org/10.1016/j.micpath.2021.104777

Han J, LaVigne CA, Jones BT, Zhang H, Gillett F, Mendell JT (2020) A ubiquitin ligase mediates target-directed microRNA decay independently of tailing and trimming. Science 370(6523): eabc9546. https://doi.org/10.1126/science.abc9546

Hess C, Means TK, Autissier P, Woodberry T, Altfeld M, Addo MM, Frahm N, Brander C, Walker BD, Luster AD (2004) IL-8 responsiveness defines a subset of CD8 T cells poised to kill. Blood 104(12):3463–3471. https://doi.org/10.1182/blood-2004-03-1067

Hombach S, Kretz M (2016) Non-coding RNAs: classification, biology and functioning. Adv Exp Med Biol 937:3–17. https://doi.org/10.1007/978-3-319-42059-2_1

Huang V, Place RF, Portnoy V, Wang J, Qi Z, Jia Z, Yu A, Shuman M, Yu J, Li LC (2012) Upregulation of cyclin B1 by miRNA and its implications in cancer. Nucleic Acids Res 40(4): 1695–1707. https://doi.org/10.1093/nar/gkr934

Ishikawa H, Kaneko H, Watanabe H, Takagi A, Ming ZW (1987) Giant cell myocarditis in association with drug-induced skin eruption. Acta Pathol Jpn 37(4):639–644. https://doi.org/10.1111/j.1440-1827.1987.tb00398.x

Jin W, Wang J, Liu CP, Wang HW, Xu RM (2020) Structural basis for pri-miRNA recognition by drosha. Mol Cell 78(3):423–433.e5. https://doi.org/10.1016/j.molcel.2020.02.024

Johanson TM, Lew AM, Chong MM (2013) MicroRNA-independent roles of the RNase III enzymes drosha and dicer. Open Biol 3(10):130144. https://doi.org/10.1098/rsob.130144

Kautz-Neu K, Schwonberg K, Fischer MR, Schermann AI, von Stebut E (2012) Dendritic cells in Leishmania major infections: mechanisms of parasite uptake, cell activation and evidence for physiological relevance. Med Microbiol Immunol 201(4):581–592. https://doi.org/10.1007/s00430-012-0261-2

Kedzierski L, Evans KJ (2014) Immune responses during cutaneous and visceral leishmaniasis. Parasitology 141:1–19. https://doi.org/10.1017/S003118201400095X

Kelada S, Sethupathy P, Okoye IS, Kistasis E, Czieso S, White SD, Chou D, Martens C, Ricklefs SM, Virtaneva K, Sturdevant DE, Porcella SF, Belkaid Y, Wynn TA, Wilson MS (2013) miR-182 and miR-10a are key regulators of Treg specialisation and stability during schistosome and Leishmania-associated inflammation. PLoS Pathog 9(6):e1003451. https://doi.org/10.1371/journal.ppat.1003451

Kim VN (2005) Small RNAs: classification, biogenesis, and function. Mol Cells 19(1):1–15

Kim DH, Saetrom P, Snøve O, Rossi JJ (2008) MicroRNA-directed transcriptional gene silencing in mammalian cells. Proc Natl Acad Sci U S A 105(42):16230–16235. https://doi.org/10.1073/pnas.0808830105

Kingston ER, Bartel DP (2019) Global analyses of the dynamics of mammalian microRNA metabolism. Genome Res 29(11):1777–1790. https://doi.org/10.1101/gr.251421.119

Knoll M, Simmons S, Bouquet C, Grün JR, Melchers F (2013) miR-221 redirects precursor B cells to the BM and regulates their residence. Eur J Immunol 43(9):2497–2506. https://doi.org/10.1002/eji.201343367

Kondo M, Weissman IL, Akashi K (1997) Identification of clonogenic common lymphoid progenitors in mouse bone marrow. Cell 91(5):661–672. https://doi.org/10.1016/s0092-8674(00)80453-5

Koralov SB, Muljo SA, Galler GR, Krek A, Chakraborty T, Kanellopoulou C, Jensen K, Cobb BS, Merkenschlager M, Rajewsky N, Rajewsky K (2008) Dicer ablation affects antibody diversity and cell survival in the B lymphocyte lineage. Cell 132(5):860–874. https://doi.org/10.1016/j.cell.2008.02.020

Krol J, Busskamp V, Markiewicz I, Stadler MB, Ribi S, Richter J, Duebel J, Bicker S, Fehling HJ, Schübeler D, Oertner TG, Schratt G, Bibel M, Roska B, Filipowicz W (2010a) Characterizing light-regulated retinal microRNAs reveals rapid turnover as a common property of neuronal microRNAs. Cell 141(4):618–631. https://doi.org/10.1016/j.cell.2010.03.039

Krol J, Loedige I, Filipowicz W (2010b) The widespread regulation of microRNA biogenesis, function and decay. Nat Rev Genet 11(9):597–610. https://doi.org/10.1038/nrg2843

Kumar R, Nylén S (2012) Immunobiology of visceral leishmaniasis. Front Immunol 3:251. https://doi.org/10.3389/fimmu.2012.00251

Kumar V, Kumar A, Das S, Abhishek K, Verma S, Mandal A, Singh RK, Das P (2018) Activates hypoxia inducible factor-1α and miR-210 for survival in macrophages by downregulation of NF-κB mediated pro-inflammatory Immune response. Front Microbiol 9:385. https://doi.org/10.3389/fmicb.2018.00385

Kumar A, Vijaykumar S, Dikhit MR, Abhishek K, Mukherjee R, Sen A, Das P, Das S (2020a) Differential regulation of miRNA profiles of human cells experimentally infected by. Front Microbiol 11:1716. https://doi.org/10.3389/fmicb.2020.01716

Kumar V, Das S, Kumar A, Tiwari N, Abhishek K, Mandal A, Kumar M, Shafi T, Bamra T, Singh RK, Vijayakumar S, Sen A, Das P (2020b) Leishmania donovani infection induce differential miRNA expression in CD4+ T cells. Sci Rep 10(1):3523. https://doi.org/10.1038/s41598-020-60435-2

Lago TS, Silva JA, Lago EL, Carvalho EM, Zanette DL, Castellucci LC (2018) The miRNA 361-3p, a regulator of GZMB and TNF Is associated with therapeutic failure and longer time healing of cutaneous Leishmaniasis caused by. Front Immunol 9:2621. https://doi.org/10.3389/fimmu.2018.02621

Lasjerdi Z, Ghanbarian H, Mohammadi Yeganeh S, Seyyed Tabaei SJ, Mohebali M, Taghipour N, Koochaki A, Hamidi F, Gholamrezaei M, Haghighi A (2020) Comparative expression profile analysis of apoptosis-related miRNA and its target gene in. Iran J Parasitol 15(3):332–340. https://doi.org/10.18502/ijpa.v15i3.4197

Latchman DS (1996) The Oct-2 transcription factor. Int J Biochem Cell Biol 28(10):1081–1083. https://doi.org/10.1016/1357-2725(96)00050-7

Lemaire J, Mkannez G, Guerfali FZ, Gustin C, Attia H, Sghaier RM, Dellagi K, Laouini D, Renard P, Sysco-Consortium (2013) MicroRNA expression profile in human macrophages in response to Leishmania major infection. PLoS Negl Trop Dis 7(10):e2478. https://doi.org/10.1371/journal.pntd.0002478

Liao Y, Lönnerdal B (2010) Global microRNA characterization reveals that miR-103 is involved in IGF-1 stimulated mouse intestinal cell proliferation. PLoS One 5(9):e12976. https://doi.org/10.1371/journal.pone.0012976

Liao Y, Zhang M, Lönnerdal B (2013) Growth factor TGF-β induces intestinal epithelial cell (IEC-6) differentiation: miR-146b as a regulatory component in the negative feedback loop. Genes Nutr 8(1):69–78. https://doi.org/10.1007/s12263-012-0297-3

Liu Z, Wang J, Cheng H, Ke X, Sun L, Zhang QC, Wang HW (2018) Cryo-EM structure of human dicer and its complexes with a pre-miRNA substrate. Cell 173(5):1191–1203.e12. https://doi.org/10.1016/j.cell.2018.03.080

Lwin T, Lin J, Choi YS, Zhang X, Moscinski LC, Wright KL, Sotomayor EM, Dalton WS, Tao J (2010) Follicular dendritic cell-dependent drug resistance of non-Hodgkin lymphoma involves cell adhesion-mediated Bim down-regulation through induction of microRNA-181a. Blood 116(24):5228–5236. https://doi.org/10.1182/blood-2010-03-275925

Martinez-Nunez RT, Louafi F, Friedmann PS, Sanchez-Elsner T (2009) MicroRNA-155 modulates the pathogen binding ability of dendritic cells (DCs) by down-regulation of DC-specific intercellular adhesion molecule-3 grabbing non-integrin (DC-SIGN). J Biol Chem 284(24):16334–16342. https://doi.org/10.1074/jbc.M109.011601

Martinez-Nunez RT, Louafi F, Sanchez-Elsner T (2011) The interleukin 13 (IL-13) pathway in human macrophages is modulated by microRNA-155 via direct targeting of interleukin 13 receptor alpha1 (IL13Ralpha1). J Biol Chem 286(3):1786–1794. https://doi.org/10.1074/jbc.M110.169367

Matsui M, Chu Y, Zhang H, Gagnon KT, Shaikh S, Kuchimanchi S, Manoharan M, Corey DR, Janowski BA (2013) Promoter RNA links transcriptional regulation of inflammatory pathway genes. Nucleic Acids Res 41(22):10086–10109. https://doi.org/10.1093/nar/gkt777

Mauël J (1990) Macrophage-parasite interactions in Leishmania infections. J Leukoc Biol 47(2):187–193. https://doi.org/10.1002/jlb.47.2.187

McFarlane E, Carter KC, McKenzie AN, Kaye PM, Brombacher F, Alexander J (2011) Endogenous IL-13 plays a crucial role in liver granuloma maturation during Leishmania donovani infection, independent of IL-4Rα-responsive macrophages and neutrophils. J Infect Dis 204(1):36–43. https://doi.org/10.1093/infdis/jir080

Melo LM, Bragato JP, Venturin GL, Rebech GT, Costa SF, Garcia LE, Lopes FL, Eugênio FR, Patto Dos Santos PS, de Lima VMF (2019) Induction of miR 21 impairs the anti-Leishmania response through inhibition of IL-12 in canine splenic leukocytes. PLoS One 14(12):e0226192. https://doi.org/10.1371/journal.pone.0226192

Michlewski G, Cáceres JF (2019) Post-transcriptional control of miRNA biogenesis. RNA 25(1): 1–16. https://doi.org/10.1261/rna.068692.118

Mok Y, Schwierzeck V, Thomas DC, Vigorito E, Rayner TF, Jarvis LB, Prosser HM, Bradley A, Withers DR, Mårtensson IL, Corcoran LM, Blenkiron C, Miska EA, Lyons PA, Smith KGC (2013) MiR-210 is induced by Oct-2, regulates B cells, and inhibits autoantibody production. J Immunol 191(6):3037–3048. https://doi.org/10.4049/jimmunol.1301289

Mukherjee B, Paul J, Mukherjee S, Mukhopadhyay R, Das S, Naskar K, Sundar S, Dujardin JC, Saha B, Roy S (2015) Antimony-resistant Leishmania donovani exploits miR-466i to deactivate host Myd88 for regulating IL-10/IL-12 levels during early hours of infection. J Immunol 195(6): 2731–2742. https://doi.org/10.4049/jimmunol.1402585

Muxel SM, Laranjeira-Silva MF, Zampieri RA, Floeter-Winter LM (2017) Leishmania (Leishmania) amazonensis induces macrophage miR-294 and miR-721 expression and modulates infection by targeting NOS2 and L-arginine metabolism. Sci Rep 7:44141. https://doi.org/10.1038/srep44141

Nakamura M, Kanda T, Sasaki R, Haga Y, Jiang X, Wu S, Nakamoto S, Yokosuka O (2015) MicroRNA-122 inhibits the production of inflammatory cytokines by targeting the PKR activator PACT in human hepatic stellate cells. PLoS One 10(12):e0144295. https://doi.org/10.1371/journal.pone.0144295

Nimsarkar P, Ingale P, Singh S (2020) Systems studies uncover miR-146a as a target in. ACS Omega 5(21):12516–12526. https://doi.org/10.1021/acsomega.0c01502

Nunes S, Silva IB, Ampuero MR, de Noronha ALL, de Souza LCL, Correia TC, Khouri R, Boaventura VS, Barral A, Ramos PIP, Brodskyn C, Oliveira PRS, Tavares NM (2018) Integrated analysis reveals that miR-193b, miR-671, and TREM-1 correlate with a good response to treatment of human localized cutaneous leishmaniasis caused by. Front Immunol 9:640. https://doi.org/10.3389/fimmu.2018.00640

O'Connell RM, Rao DS, Chaudhuri AA, Boldin MP, Taganov KD, Nicoll J, Paquette RL, Baltimore D (2008) Sustained expression of microRNA-155 in hematopoietic stem cells causes a myeloproliferative disorder. J Exp Med 205(3):585–594. https://doi.org/10.1084/jem.20072108

O'Connell RM, Kahn D, Gibson WS, Round JL, Scholz RL, Chaudhuri AA, Kahn ME, Rao DS, Baltimore D (2010) MicroRNA-155 promotes autoimmune inflammation by enhancing inflammatory T cell development. Immunity 33(4):607–619. https://doi.org/10.1016/j.immuni.2010.09.009

Omoto Y, Yamanaka K, Tokime K, Kitano S, Kakeda M, Akeda T, Kurokawa I, Gabazza EC, Tsutsui H, Katayama N, Yamanishi K, Nakanishi K, Mizutani H (2010) Granzyme B is a novel interleukin-18 converting enzyme. J Dermatol Sci 59(2):129–135. https://doi.org/10.1016/j.jdermsci.2010.05.004

Ozbılge H, Aksoy N, Gurel MS, Yazar S (2006) IgG and IgG subclass antibodies in patients with active cutaneous leishmaniasis. J Med Microbiol 55(Pt 10):1329–1331. https://doi.org/10.1099/jmm.0.46667-0

Pacheco-Fernandez T, Volpedo G, Verma C, Satoskar AR (2021) Understanding the immune responses involved in mediating protection or immunopathology during leishmaniasis. Biochem Soc Trans 49(1):297–311. https://doi.org/10.1042/BST20200606

Pandey RK, Sundar S, Prajapati VK (2016) Differential expression of miRNA regulates T cell differentiation and plasticity during visceral leishmaniasis infection. Front Microbiol 7:206. https://doi.org/10.3389/fmicb.2016.00206

Pareyn M, Hendrickx R, Girma N, Hendrickx S, Van Bockstal L, Van Houtte N, Shibru S, Maes L, Leirs H, Caljon G (2020) Evaluation of a pan-Leishmania SL RNA qPCR assay for parasite detection in laboratory-reared and field-collected sand flies and reservoir hosts. Parasit Vectors 13(1):276. https://doi.org/10.1186/s13071-020-04141-y

Pekarsky Y, Santanam U, Cimmino A, Palamarchuk A, Efanov A, Maximov V, Volinia S, Alder H, Liu CG, Rassenti L, Calin GA, Hagan JP, Kipps T, Croce CM (2006) Tcl1 expression in chronic lymphocytic leukemia is regulated by miR-29 and miR-181. Cancer Res 66(24):11590–11593. https://doi.org/10.1158/0008-5472.CAN-06-3613

Piriyapongsa J, Jordan IK, Conley AB, Ronan T, Smalheiser NR (2011) Transcription factor binding sites are highly enriched within microRNA precursor sequences. Biol Direct 6:61. https://doi.org/10.1186/1745-6150-6-61

Place RF, Li LC, Pookot D, Noonan EJ, Dahiya R (2008) MicroRNA-373 induces expression of genes with complementary promoter sequences. Proc Natl Acad Sci U S A 105(5):1608–1613. https://doi.org/10.1073/pnas.0707594105

Pu M, Chen J, Tao Z, Miao L, Qi X, Wang Y, Ren J (2019) Regulatory network of miRNA on its target: coordination between transcriptional and post-transcriptional regulation of gene expression. Cell Mol Life Sci 76(3):441–451. https://doi.org/10.1007/s00018-018-2940-7

Ravindran R, Anam K, Bairagi BC, Saha B, Pramanik N, Guha SK, Goswami RP, Banerjee D, Ali N (2004) Characterization of immunoglobulin G and its subclass response to Indian kala-azar infection before and after chemotherapy. Infect Immun 72(2):863–870. https://doi.org/10.1128/IAI.72.2.863-870.2004

Reis LC, Ramos-Sanchez EM, Goto H (2013) The interactions and essential effects of intrinsic insulin-like growth factor-I on Leishmania (Leishmania) major growth within macrophages. Parasite Immunol 35(7–8):239–244. https://doi.org/10.1111/pim.12041

Ribeiro-de-Jesus A, Almeida RP, Lessa H, Bacellar O, Carvalho EM (1998) Cytokine profile and pathology in human leishmaniasis. Braz J Med Biol Res 31(1):143–148. https://doi.org/10.1590/s0100-879x1998000100020

Rodrigues V, Cordeiro-da-Silva A, Laforge M, Silvestre R, Estaquier J (2016) Regulation of immunity during visceral Leishmania infection. Parasit Vectors 9:118. https://doi.org/10.1186/s13071-016-1412-x

Rodriguez A, Vigorito E, Clare S, Warren MV, Couttet P, Soond DR, van Dongen S, Grocock RJ, Das PP, Miska EA, Vetrie D, Okkenhaug K, Enright AJ, Dougan G, Turner M, Bradley A (2007) Requirement of bic/microRNA-155 for normal immune function. Science 316(5824):608–611. https://doi.org/10.1126/science.1139253

Rothchild AC, Sissons JR, Shafiani S, Plaisier C, Min D, Mai D, Gilchrist M, Peschon J, Larson RP, Bergthaler A, Baliga NS, Urdahl KB, Aderem A (2016) MiR-155-regulated molecular network orchestrates cell fate in the innate and adaptive immune response to Mycobacterium tuberculosis. Proc Natl Acad Sci U S A 113(41):E6172–E6181. https://doi.org/10.1073/pnas.1608255113

Saliminejad K, Khorram Khorshid HR, Soleymani Fard S, Ghaffari SH (2019) An overview of microRNAs: biology, functions, therapeutics, and analysis methods. J Cell Physiol 234(5):5451–5465. https://doi.org/10.1002/jcp.27486

Santos Ca S, Boaventura V, Cardoso CR, Tavares N, Lordelo MJ, Noronha A, Costa J, Borges VM, de Oliveira CI, Van Weyenbergh J, Barral A, Barral-Netto M, Brodskyn CI (2013) CD8(+) granzyme B(+)-mediated tissue injury vs. CD4(+)IFNγ(+)-mediated parasite killing in human cutaneous leishmaniasis. J Invest Dermatol 133(6):1533–1540. https://doi.org/10.1038/jid.2013.4

Schmid-Hempel P (2009) Immune defence, parasite evasion strategies and their relevance for 'macroscopic phenomena' such as virulence. Philos Trans R Soc Lond Ser B Biol Sci 364(1513):85–98. https://doi.org/10.1098/rstb.2008.0157

Sheu-Gruttadauria J, Pawlica P, Klum SM, Wang S, Yario TA, Schirle Oakdale NT, Steitz JA, MacRae IJ (2019) Structural basis for target-directed microRNA degradation. Mol Cell 75(6):1243–1255.e7. https://doi.org/10.1016/j.molcel.2019.06.019

Singh AK, Pandey RK, Shaha C, Madhubala R (2016) MicroRNA expression profiling of Leishmania donovani-infected host cells uncovers the regulatory role of MIR30A-3p in host autophagy. Autophagy 12(10):1817–1831. https://doi.org/10.1080/15548627.2016.1203500

Siomi H, Siomi MC (2010) Posttranscriptional regulation of microRNA biogenesis in animals. Mol Cell 38(3):323–332. https://doi.org/10.1016/j.molcel.2010.03.013

Soares MF, Melo LM, Bragato JP, Furlan AO, Scaramele NF, Lopes FL, Lima VMF (2021) Differential expression of miRNAs in canine peripheral blood mononuclear cells (PBMC) exposed to Leishmania infantum in vitro. Res Vet Sci 134:58–63. https://doi.org/10.1016/j.rvsc.2020.11.021

Solano-Gallego L, Montserrat-Sangrà S, Ordeix L, Martínez-Orellana P (2016) Leishmania infantum-specific production of IFN-γ and IL-10 in stimulated blood from dogs with clinical leishmaniosis. Parasit Vectors 9(1):317. https://doi.org/10.1186/s13071-016-1598-y

Solov'eva NI, Volkova ZI (1990) Proteolytic activity of blood cells in rheumatoid arthritis. Vopr Med Khim 36(3):48–52

Song X, Li Y, Cao X, Qi Y (2019) MicroRNAs and their regulatory roles in plant-environment interactions. Annu Rev Plant Biol 70:489–525. https://doi.org/10.1146/annurev-arplant-050718-100334

Souza MA, Ramos-Sanchez EM, Muxel SM, Lagos D, Reis LC, Pereira VRA, Brito MEF, Zampieri RA, Kaye PM, Floeter-Winter LM, Goto H (2021) miR-548d-3p alters parasite growth and inflammation in. Front Cell Infect Microbiol 11:687647. https://doi.org/10.3389/fcimb.2021.687647

Staedel C, Darfeuille F (2013) MicroRNAs and bacterial infection. Cell Microbiol 15(9):1496–1507. https://doi.org/10.1111/cmi.12159

Stäger S, Alexander J, Carter KC, Brombacher F, Kaye PM (2003) Both interleukin-4 (IL-4) and IL-4 receptor alpha signaling contribute to the development of hepatic granulomas with optimal antileishmanial activity. Infect Immun 71(8):4804–4807. https://doi.org/10.1128/iai.71.8.4804-4807.2003

Stavast CJ, Erkeland SJ (2019) The non-canonical aspects of microRNAs: many roads to gene regulation. Cell 8(11):1465. https://doi.org/10.3390/cells8111465

Su Z, Yang Z, Xu Y, Chen Y, Yu Q (2015) MicroRNAs in apoptosis, autophagy and necroptosis. Oncotarget 6(11):8474–8490. https://doi.org/10.18632/oncotarget.3523

Tam W (2001) Identification and characterization of human BIC, a gene on chromosome 21 that encodes a noncoding RNA. Gene 274(1–2):157–167. https://doi.org/10.1016/s0378-1119(01)00612-6

Tarver JE, Donoghue PC, Peterson KJ (2012) Do miRNAs have a deep evolutionary history? BioEssays 34(10):857–866. https://doi.org/10.1002/bies.201200055

Thai TH, Calado DP, Casola S, Ansel KM, Xiao C, Xue Y, Murphy A, Frendewey D, Valenzuela D, Kutok JL, Schmidt-Supprian M, Rajewsky N, Yancopoulos G, Rao A, Rajewsky K (2007) Regulation of the germinal center response by microRNA-155. Science 316(5824):604–608. https://doi.org/10.1126/science.1141229

Tili E, Michaille JJ, Cimino A, Costinean S, Dumitru CD, Adair B, Fabbri M, Alder H, Liu CG, Calin GA, Croce CM (2007) Modulation of miR-155 and miR-125b levels following lipopolysaccharide/TNF-alpha stimulation and their possible roles in regulating the response to endotoxin shock. J Immunol 179(8):5082–5089. https://doi.org/10.4049/jimmunol.179.8.5082

Tiwari N, Kumar V, Gedda MR, Singh AK, Singh VK, Gannavaram S, Singh SP, Singh RK (2017) Identification and characterization of miRNAs in response to. Front Microbiol 8:314. https://doi.org/10.3389/fmicb.2017.00314

Treiber T, Treiber N, Meister G (2019) Regulation of microRNA biogenesis and its crosstalk with other cellular pathways. Nat Rev Mol Cell Biol 20(1):5–20. https://doi.org/10.1038/s41580-018-0059-1

Tserel L, Runnel T, Kisand K, Pihlap M, Bakhoff L, Kolde R, Peterson H, Vilo J, Peterson P, Rebane A (2011) MicroRNA expression profiles of human blood monocyte-derived dendritic cells and macrophages reveal miR-511 as putative positive regulator of Toll-like receptor 4. J Biol Chem 286(30):26487–26495. https://doi.org/10.1074/jbc.M110.213561

Varikuti S, Natarajan G, Volpedo G, Singh B, Hamza O, Messick GV, Guerau-de-Arellano M, Papenfuss TL, Oghumu S, Satoskar AR (2019) MicroRNA 155 contributes to host immunity against Leishmania donovani but is not essential for resolution of infection. Infect Immun 87(8): e00307–e00319. https://doi.org/10.1128/IAI.00307-19

Varikuti S, Verma C, Holcomb E, Jha BK, Viana A, Maryala R, Lamenza F, McElwain BK, Doni NY, Papenfuss T, Oghumu S, Gannavaram S, Nakhasi HL, Satoskar AR (2021a) MicroRNA-21 deficiency promotes the early Th1 immune response and resistance toward visceral leishmaniasis. J Immunol 207(5):1322–1332. https://doi.org/10.4049/jimmunol.2001099

Varikuti S, Verma C, Natarajan G, Oghumu S, Satoskar AR (2021b) MicroRNA155 plays a critical role in the pathogenesis of cutaneous Leishmania major infection by promoting a Th2 response and attenuating dendritic Cell activity. Am J Pathol 191(5):809–816. https://doi.org/10.1016/j.ajpath.2021.01.012

Vigorito E, Kohlhaas S, Lu D, Leyland R (2013) miR-155: an ancient regulator of the immune system. Immunol Rev 253(1):146–157. https://doi.org/10.1111/imr.12057

Vishnoi A, Rani S (2017) MiRNA biogenesis and regulation of diseases: an overview. Methods Mol Biol 1509:1–10. https://doi.org/10.1007/978-1-4939-6524-3_1

Volpedo G, Pacheco-Fernandez T, Bhattacharya P, Oljuskin T, Dey R, Gannavaram S, Satoskar AR, Nakhasi HL (2021) Determinants of innate immunity in visceral Leishmaniasis and their implication in vaccine development. Front Immunol 12:748325. https://doi.org/10.3389/fimmu.2021.748325

Von Stebut E (2007) Immunology of cutaneous leishmaniasis: the role of mast cells, phagocytes and dendritic cells for protective immunity. Eur J Dermatol 17(2):115–122. https://doi.org/10.1684/ejd.2007.0122

Wang S, Talukder A, Cha M, Li X, Hu H (2021) Computational annotation of miRNA transcription start sites. Brief Bioinform 22(1):380–392. https://doi.org/10.1093/bib/bbz178

Wechsler AS, Gordon MC, Dendorfer U, LeClair KP (1994) Induction of IL-8 expression in T cells uses the CD28 costimulatory pathway. J Immunol 153(6):2515–2523

Xiao C, Calado DP, Galler G, Thai TH, Patterson HC, Wang J, Rajewsky N, Bender TP, Rajewsky K (2007) MiR-150 controls B cell differentiation by targeting the transcription factor c-Myb. Cell 131(1):146–159. https://doi.org/10.1016/j.cell.2007.07.021

Xu H, Gu LN, Yang QY, Zhao DY, Liu F (2016) MiR-221 promotes IgE-mediated activation of mast cells degranulation by PI3K/Akt/PLCγ/Ca(2+) pathway. J Bioenerg Biomembr 48(3): 293–299. https://doi.org/10.1007/s10863-016-9659-7

Yeo JH, Chong MM (2011) Many routes to a micro RNA. IUBMB Life 63(11):972–978. https://doi.org/10.1002/iub.524

Ying SY, Chang DC, Lin SL (2008) The microRNA (miRNA): overview of the RNA genes that modulate gene function. Mol Biotechnol 38(3):257–268. https://doi.org/10.1007/s12033-007-9013-8

Yutiao M, Kinzie JD (1975) Consultation with the filipino boarding home: an after-care facility in Hawaii. Int J Soc Psychiatry 21(2):130–136. https://doi.org/10.1177/002076407502100208

Zhang XF, Yang Y, Yang XY, Tong Q (2018) MiR-188-3p upregulation results in the inhibition of macrophage proinflammatory activities and atherosclerosis in ApoE-deficient mice. Thromb Res 171:55–61. https://doi.org/10.1016/j.thromres.2018.09.043

Zhou B, Wang S, Mayr C, Bartel DP, Lodish HF (2007) miR-150, a microRNA expressed in mature B and T cells, blocks early B cell development when expressed prematurely. Proc Natl Acad Sci U S A 104(17):7080–7085. https://doi.org/10.1073/pnas.0702409104

Heat Shock Proteins as Emerging Therapeutic and Vaccine Targets Against Leishmaniasis

10

Shailendra Yadav, Apeksha Anand, and Neena Goyal

Abstract

Leishmania, an intracellular protozoa parasite, is the causative agent of a group of diseases called leishmaniasis. Clinical manifestations of leishmaniasis range from localized, self-healing cutaneous lesions to systemic infections that are generalized, progressive, and ultimately fatal. The major challenges in the control of leishmaniasis include a lack of approved vaccines, issues with current chemotherapy, and the emergence of drug resistance. The *Leishmania* parasite completes its life cycle in the gut of sand flies and the macrophages of mammalian hosts. During the digenetic life cycle, the parasite encounters various stress conditions, like high temperature, low pH, oxidative stress, change in nutritional availability, and host inflammatory response. Heat-shock proteins (HSPs), a subset of molecular chaperones, counteract these stresses by folding, assembly, secretion, and the regulation of other proteins, and hence, they play a critical role in the differentiation, virulence, and survival of *Leishmania* parasites within mammalian hosts. Heat shock proteins also play an essential role in the immune regulation of hosts and the parasite tolerance to chemotherapy. HSPs also appear to serve as an important antigen that can act as potent immunogens capable of elucidating cell-mediated as well as humoral responses and are thought to play a protective role and, hence, may reduce disease incidences. Therefore, HSPs of *Leishmania* may be employed as a potential vaccine and therapeutic target(s). The current review focuses on the diverse role and function of *Leishmania* HSPs and their application in the development of novel prophylactic vaccines and drug therapies for combating leishmaniasis.

S. Yadav · A. Anand · N. Goyal (✉)
Division of Biochemistry and Structural Biology, CSIR-Central Drug Research Institute, Lucknow, India

Academy of Scientific and Innovative Research (AcSIR), Gaziabaad, India

© The Author(s), under exclusive license to Springer Nature Singapore Pte Ltd. 2023

A. Selvapandiyan et al. (eds.), *Challenges and Solutions Against Visceral Leishmaniasis*, https://doi.org/10.1007/978-981-99-6999-9_10

Keywords

Leishmania · Therapeutics · Vaccine targets · Heat shock proteins · Immunogens

10.1 Introduction

Leishmaniasis is a group of diseases caused by protozoan parasites of the genus *Leishmania* and transmitted to the vertebrate hosts by female sandflies of the genus *Phlebotomus* in the Old World and *Lutzomyia* in the New World (Torres-Guerrero et al. 2017). The symptoms range from the most prevalent self-healing ulcers (cutaneous leishmaniasis (CL)) to the destruction of the mucous tissue of the nose, mouth, and throat (mucocutaneous leishmaniasis (MCL)) to life-threatening visceral leishmaniasis (VL/kala-azar) (Burza et al. 2018). The World Health Organization (WHO) in 2019 reported leishmaniasis in 98 countries (https://www.who.int/data/gho/data/themes/topics/gho-ntd-leishmaniasis). An estimated 12–15 million people are infected globally, with two million new cases reported each year, and 350 million people are thought to be at risk of developing leishmaniasis (Torres-Guerrero et al. 2017). The lack of approved vaccines, the limited treatment options, severe side effects of the current antileishmanial chemotherapy, drug resistance, cyclical epidemiological patterns, inadequate diagnostics, and a lack of a complete understanding of disease pathophysiology are major obstacles in global disease control (Ponte-Sucre et al. 2017; Kumari et al. 2021; Mann et al. 2021; Selvapandiyan et al. 2019). Thus, better health policies, epidemiologic monitoring, diagnostic tools, understanding the mechanism of drug resistance, and drug and vaccine development are the top priorities for the successful control and elimination of leishmaniasis.

Leishmania parasites complete their complex life cycle by transforming from a promastigote stage in the sand fly (vector) to an amastigote stage in the macrophage of a mammalian host (Teixeira et al. 2013). During this stage-specific differentiation process, the parasite encounters several stresses, like elevated temperature, hypoxia, nutrient deprivation, oxygen radicals, metabolic disruption, etc., that may cause the denaturation of certain essential cellular proteins (Hombach et al. 2014; Anas et al. 2019). As a result, the parasite deploys several mechanisms to adapt and survive within this hostile environment (Garlapati et al. 1999). One of these mechanisms is characterized by a complex cellular response that results in the induction of a family of proteins called heat shock proteins (HSPs) (Maresca and Carratu 1992). Heat shock proteins are the ubiquitous proteins expressed in all life forms, from bacteria to higher eukaryotes. They maintain the natural milieu of the cell by acting as molecular chaperones (Saibil 2013). HSPs bind to the newly synthesized proteins and protect them from denaturation and aggregation, thus helping the proteins to function properly (Saibil 2013). HSPs are also present near the demise of protein, i.e., proteasome-mediated degradation of the protein (Karademir et al. 2014). HSP family proteins have many other essential roles in the *Leishmania* parasite, like exosome packaging, cell cycle progression, the regulation of signal transduction, immune evasion, virulence, drug resistance, etc. (Requena et al. 2015). The main classes of HSP proteins, divided according to their molecular masses, are HSP100, HSP90/83, HSP70, HSP60/TCP1, HSP40, and small HSP (sHSP) (Requena et al. 2015). The functionality of many of these proteins in the *Leishmania* parasite has not yet been defined and thus has opened a new area for research to explore novel drug and vaccine targets for the treatment of Leishmaniasis.

The present report is focused on the functional role of the heat shock protein of *Leishmania* in parasite survival and virulence. Furthermore, it has also explored the prophylactic and therapeutic applications of the heat shock machinery of *Leishmania* toward the vaccine and drug development and associated challenges.

10.2 The Heat Shock Proteins of *Leishmania* and Their Functional Roles

The increased temperature with the acidification of growth media is sufficient to induce the conversion of promastigotes to amastigotes in several *Leishmania* spp.; this sparked the interest in studying the heat shock proteins of *Leishmania* parasite (Zilberstein and Shapira 1994; Barak et al. 2005). The heat shock proteins of the *Leishmania* parasite are involved in thermotolerance and the protection of the amastigote stage of the parasite from the harsh environment of phagolysosomes in the macrophage of the host cell (Hombach et al. 2014); this further fueled the curiosity to explore and investigate heat shock genes, their products, and their regulation in *Leishmania* spp. Crucial aspects of a few well-studied HSP families of the *Leishmania* parasites have been described to provide an in-depth understanding of the role of HSPs throughout different stages of parasite life. Table 10.1 shows a list of *Leishmania* HSPs and their functional roles.

10.2.1 HSP100/Clp Family

HSP100 family proteins are also called disaggregates as they have the ability to recognize protein aggregates that are fatal to the cells (Barends et al. 2010). They are known to work in tandem with HSP70 chaperone machinery to disaggregate protein aggregates in a nondestructive manner (Fig. 10.1) (Mogk et al. 2015). The HSP100/Clp chaperone family is a member of the AAA+ domain-containing ATPase superfamily. The AAA+ domain is a distinguishing feature of this superfamily, which is comprised of an α-β subdomain and a smaller, helical subdomain (Hanson and Whiteheart 2005).

There are two classes of HSP100. Class 1 consists of a hexamer subunit having two different nucleotide-binding sites, and class 2 consists of only one nucleotide-binding site (Richter et al. 2010). A class-1 HSP100/ClpB (LmjF29.1270) (the yeast homolog of HSP104) was first characterized in *L. major*, whose expression was increased at elevated temperature (Hubel et al. 1995). HSP100 was found to be nonessential for the growth of promastigotes and axenic amastigotes (Krobitsch and

Table 10.1 Heat shock proteins of *Leishmania* and their functional role

Family	Gene	Mass (KDa)	Identifier	Function/physiological role	References
HSP100 (AAA ATPase)	HSP100/ ClpB	96.9	LdBPK_291360.1	Essential for amastigote stage, exosome biogenesis, immune modulation	Krobitsch and Clos (1999), Silverman et al. (2010b), Hubel et al. (1997)
	HSP78	90.8	LdBPK_020680.1	Infection persistence	Das et al. (2020)
HSP90	HSP83	80.5	LdBPK_330360.1	Proliferation, cell cycle, interaction with MAP kinase I, casein kinase 1.2	Hombach and Clos (2014), Kaur et al. (2017), Hombach-Barrigah et al. (2019)
	LPG3/ GRP94	87	LdBPK_290790.1	Lipophosphoglycan synthesis	Descoteaux et al. (2002)
	HSP75/ TRAP1	72	LdBPK_332520.1	Exosome component	Silverman et al. (2010a)
HSP70	HSP70	71.6	LdBPK_282930 LdBPK_283000 LdBPK_282940 LdBPK_282970 LdBPK_282960 LdBPK_282950	Virulence, oxidative stress tolerance, antimony/miltefosine resistance, interaction with HSP90, MAP kinase I, casein kinase 1.2	Brandau et al. (1995), Miller et al. (2000), Folgueira et al. (2008), Kaur et al. (2017), Vacchina et al. (2016), Hombach-Barrigah et al. (2019)
	HSP70.4	70.4	LdBPK_261220.1	Metacyclic stage	Coulson and Smith (1990)
	HSP70-related (BiP)	71.8	LdBPK_281310.1	Uncharacterized	–
	HSP70-related protein1	71.8	LdBPK_302470.1 LdBPK_302480.1	Uncharacterized (mitochondrial precursor putative)	Searle et al. (1989)
HSP60 (CPN60)	CPN60.1	64.3	LdBPK_321940.1	Uncharacterized	Schluter et al. (2000)
	CPN60.2	60.5	LdBPK_362130.1	Mitochondrial localized, differentiation	Schluter et al. (2000)
	CPN60.3	59.3	LdBPK_362134.1	Uncharacterized	–
	CPN60.4	58.2	LdBPK_302830.1	Uncharacterized	Fraga et al. (2013)
	TCP1α	59.2	LdBPK_323470.1	Uncharacterized	–

HSP60 (TCP1)	TCP1β	57.7	LdBPK_271150.1	Uncharacterized	–
	TCP1γ	60.2	LdBPK_231460.1	Growth and infectivity, drug resistance	Bhaskar et al. (2015), Yadav et al. (2020a), Yadav et al. (2020b)
	TCP1δ	59.6	LdBPK_211330.1	Uncharacterized	–
	TCP1ε	59.2	LdBPK_321060.1	Uncharacterized	–
	TCP1ζ (TCP20)	58.8	LdBPK_131400.1	Uncharacterized	–
	TCP1η	61.6	LdBPK_353900.1	Uncharacterized	–
	TCP1θ	58.2	LdBPK_367240.1	Uncharacterized	–
HSP40	DnaJ		Numerous putative genes	Uncharacterized	–
HSP10 (CPN10)	CPN10	10.7	LdBPK_260590.1 LdBPK_260610.1	Parasite internalization, intracellular survival, an essential gene, interacts with CPN60	Zamora-Veyl et al. (2005), Colineau et al. (2017)
Small HSPs	HSP20	17.6	LdBPK_292560.1	Uncharacterized	Fraga et al. (2013), Montalvo-Alvarez et al. (2008)
	HSP23	22.9	LdBPK_340230.1	Infectivity, stress, and drug tolerance	Hombach et al. (2014), Batista et al. (2015)
Cochaperones	STI-1	29.0	LdBPK_360080.1	Interacts with HSP90, essential, phosphoprotein	Hombach et al. (2013), Morales et al. (2010)
	HOP	62.2	LdBPK_081020.1	Nonessential	Ommen et al. (2009)
	HIP	36.9	LmjF29.0320	Nonessential	Ommen et al. (2009)
	SGT	45.8	LdBPK_302740.1	Essential, interacts with HSP90	Ommen et al. (2010)
	P23	21.6	LdBPK_354540.1	Affects HSP90 ATPase domain	Hombach et al. (2015)
	AHA1	38.4	LdBPK_180210.1	Affects HSP90 ATPase domain	Bartsch et al. (2017)
	Cyclophili-n 40 (Cyp40)	38.5	LdBPK_354830.1	Essential for intracellular parasite stages, exosome biogenesis	Yau et al. (2014, 2016)

Family	Gene	Mass (KDa)	Identifier	Function/physiological role	References

CPN60: chaperonin 60 (mitochondrial HSP60); *CPN10*: mitochondrial chaperonin 10; *TCP1*: tailless complex polypeptide 1 (cytosolic HSP60); *STI-1*: stress-inducible protein; *SGT*: small glutamine-rich tetratricopeptide repeat protein; *AHA1*: activator of HSP90 ATPase

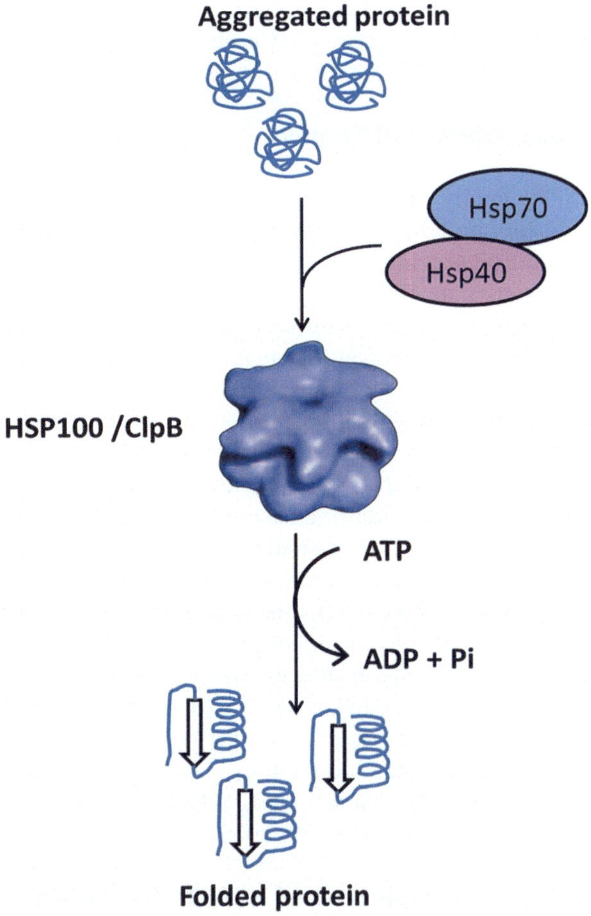

Fig. 10.1 HSP100/ClpB in cooperation with the HSP70 chaperone system to disaggregate aggregated protein. The HSP100/ClpB image is based on a cryo-electron-microscopy reconstruction study (Hanson and Whiteheart 2005)

Clos 1999). However, *L. donovani* HSP100-null mutants failed to infect and survive within mouse macrophages *in vitro* (Krobitsch and Clos 1999). Similarly, the *L. major* HSP100-null mutant was unable to infect BALB/c mice (Hubel et al. 1997). These observations indicated that HSP100 was indispensable for amastigote growth and infectivity. The mechanism was further explored by Silverman et al., and it showed that *Leishmania* HSP100 was involved in the sorting of various proteins into extracellular vesicles (EVs) (Silverman et al. 2010b). Thus, HSP100 plays a crucial role in the host's immune-modulatory function, hence in the establishment of an infection within the host cell (Silverman et al. 2010b). *Leishmania* class-2 HSP100 gene is encoded by HslU1 (LinJ.15.0090), HslU2 (LinJ.09.0380), and HslV (LinJ.36.4180). Among them, HslU1 and HslV are essential for the viability of promastigote, but HslU2 is nonessential (Chrobak et al. 2012). In *Escherichia coli*, HslU (ClpY) and HslV (ClpQ) form a complex called HslUV, which is involved in the proteolysis of misfolded protein (Missiakas et al. 1996), while in *L. donovani*, no such association is observed (Chrobak et al. 2012).

10.2.2 HSP83/90 Family

HSP90 family proteins are responsible for facilitating the folding of misfolded regulatory proteins like transcription factors and protein kinases (Pearl et al. 2008). They are present in very high abundance in the *Leishmania* promastigotes even in an unstressed state (Pearl et al. 2008). The members of this family, present in the cytoplasm, are mostly referred to as HSP83, while those found in the endoplasmic reticulum and mitochondrial matrix are referred to as Grp94 (gp96) and TRAP1 (HSP75), respectively (Johnson 2012). HSP83 (or HSP83/90) is encoded by 18 identical and tandem-repeated copies of the gene in *Leishmania* spp. (Kröber-Boncardo et al. 2020). The inhibition of HSP83/90 by geldanamycin or radicicol in *L. donovani* renders promastigotes conversion to amastigote-like axenic forms, induces the expression of amastigote-specific proteins, and arrests the cell cycle in the G2/M phase (Hombach and Clos 2014). Further, the inhibition of HSP83/90 also induces morphological changes, which clearly mimics the heat-shock-mediated differentiation process (Hombach and Clos 2014). This suggests that HSP83/90 plays a pivotal role in the stage-specific differentiation of parasites. *Leishmania* HSP83/90 is also involved in the signal transduction pathway as they are found to act as substrates for protein kinases like MAP kinase 1 and casein kinase 1.2 (Kaur et al. 2017; Hombach-Barrigah et al. 2019). The phosphorylation of HSP83/90 affects *in vitro* infectivity, morphology, and the proliferation of the promastigotes, suggesting that *Leishmania* uses HSP83/90 as an effector protein to control its life cycle (Kröber-Boncardo et al. 2020).

10.2.2.1 Cochaperones of HSP83/90 in Leishmania

HSP83/90 forms a foldosome complex with HSP70/HSP40 and cochaperones like STI1, SGT, and P23 to fold its client protein. STI1 (stress-inducible protein) or a mammalian orthologue HOP (HSP90/HSP70-organizing protein) has been implicated in the HSP83/90 chaperone cycle in two different ways. First, it binds to the C-terminus of HSP83/90 in the open conformation, where it acts as an ATPase inhibitor (Requena et al. 2015). Second, STI1 binds to HSP70 through its tetratricopeptide repeat (TPR) sites at the same time, when it binds to HSP83/90 (Requena et al. 2015). Another HSP90/HSP70 cochaperone, called HIP (HSC70 interacting protein), has been shown to be nonessential for *L. donovani* survival and infectivity (Ommen et al. 2009), while SGT (small glutamine-rich tetratricopeptide repeat), another cochaperone, was found to be essential for the growth of *L. donovani* promastigotes (Ommen et al. 2010). It directly interacts with HSP83/90, HSP70, and STI1 in the co-immunoprecipitation assay (Ommen et al. 2010). Figure 10.2 depicts the folding cycle of HSP83/90. In the first step, the HSP70-HSP40 chaperone binds to inactive/unfolded protein and interacts with HSP83/90, making a complex that is facilitated by STI1. In the next step, substrate protein is

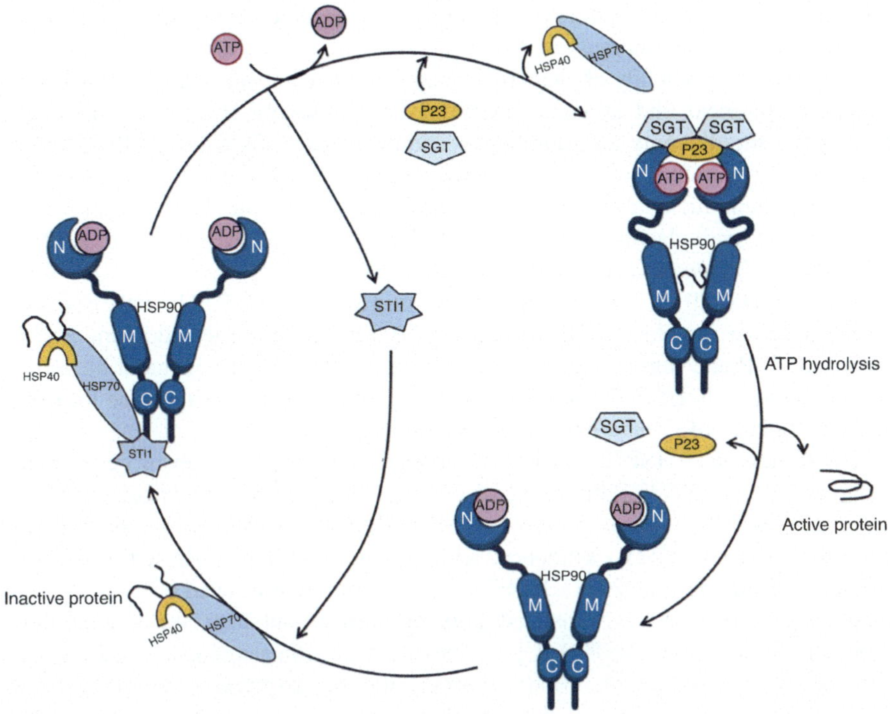

Fig. 10.2 The folding cycle of HSP83/90 chaperone in *L. donovani* parasite shows the foldosome complex formed by its cochaperone. N, M, and C denote the N-terminal, middle, and C-terminal domains of HSP83/90 (Requena et al. 2015)

transferred from HSP70-HSP40 to HSP83/90, releasing HSP70-HSP40 chaperones. This step is induced by ATP binding to the N-terminal of HSP83/90. The substrate-HSP90 complex is stabilized by the binding of SGT and p23 cochaperone, facilitating the folding of substrate protein (Ommen et al. 2010; Hombach et al. 2015). Finally, the active/folded substrate is released due to conformation changes in HSP83/90 induced by ATP hydrolysis.

10.2.3 HSP70 Family

HSP70 family proteins are involved in housekeeping functions like the folding and refolding of protein to maintain protein quality control, transmembrane transport of protein into organelles, and specific functions like protection against stress and apoptosis. HSP70 is a central player in protein quality control machinery and folds several substrates in collaboration with other ATP-dependent chaperones, such as HSP40, HSP83/90, and chaperonins (Kohler and Andreasson 2020). They also cooperate with proteins of the AAA+ family to dissociate aggregates of misfolded proteins (Kohler and Andreasson 2020).

HSP70 is a superfamily of chaperones comprising two main families classified as typical/prototypical and atypical HSP110/GRP170 family (Easton et al. 2000). Because of the diversity and number of coding genes, only a few HSP70 family proteins have been functionally characterized in *Leishmania* (Kröber-Boncardo et al. 2020). The genomic analysis showed the presence of cytosolic HSP70s, mitochondrial HSP70, endoplasmic reticulum (ER) HSP70 (BiP, GRP170), and HSP110 subfamily in *Leishmania* (Requena et al. 2015).

Atypical HSP70 in *Leishmania* consists of HSP70.4, HSP70.b, HSP70.c, and GRP170 group proteins (GRP170 and HSP70.a); endoplasmic reticulum HSP70 precursor; BiP and mitochondrial precursor; and mtHSP70 (Requena et al. 2015). Among them, only HSP70.4 has been characterized, which is found to be localized in the cytoplasm of *L. major* parasites (Searle and Smith 1993).

The prototypical HSP70 consists of tandemly organized six-seven gene copies (Folgueira et al. 2007; Ramirez et al. 2011). The first five identical genes are designated as HSP70-I and gene 6 as HSP70-II. The two vary primarily in the 3′ untranslated region (UTR) sequence, which governs HSP70 mRNA translational efficiency (Ramirez et al. 2011). HSP70-II transcripts are translated only during heat shock, while HSP70-I is translated both at normal and heat shock conditions (Folgueira et al. 2005). HSP70-II gene knockout in promastigotes caused reduced virulence (Folgueira et al. 2008). Further, the overexpression of HSP70-I in *L. chagasi* promastigotes showed resistance to macrophage-induced oxidative stress (Miller et al. 2000).

10.2.4 HSP60 Family Chaperones/Chaperonins

Chaperonins or HSP60 family chaperones are ubiquitous multimeric protein complexes present in all life forms and aid in protein folding by preventing the aggregation of misfolded or unfolded proteins (Lopez et al. 2015; Balchin et al. 2020). Chaperonins, like other heat shock proteins, are upregulated in response to cellular stresses and are thought to be the main component of molecular chaperone machinery (Saibil 2013). Chaperonin complexes are composed of homo- or hetero-oligomeric protein subunits forming a cage-like structure of high molecular weight (~1 MDa) to facilitate protein folding (Horwich et al. 2007). These proteins get energy for protein folding from ATP, which binds to the allosteric sites and gets hydrolyzed during the folding process (Saibil 2013). Chaperonins are classified into two groups: group I and group II. Chaperonins categorized in group I are present in prokaryotes and mitochondria/chloroplast of eukaryotes, while group II chaperonins are found in eukaryotic cytosol (Yebenes et al. 2011).

10.2.4.1 Group I Chaperonins

Group I chaperonins are known as GroEL in bacteria and Cpn60 in eukaryotic mitochondria, which require a cofactor called GroES and Cpn10, respectively

(Yebenes et al. 2011). GroEL, initially discovered in *E. coli*, has been demonstrated to assemble bacteriophage heads (Takano and Kakefuda 1972). *E. coli* GroEL is the most studied chaperonin, made up of two rings that are stacked back to back, each consisting of a ~57 kDa homo-oligomeric subunit. Each ring of GroEL is made up of seven identical subunits, and each subunit is composed of three domains, namely, the equatorial, intermediate, and apical domains (Xu et al. 1997). An apical domain that forms the entrance of the molecular cage is responsible for the binding of GroES and substrate protein, whereas the equatorial domain binds ATP to transmit conformational change activated by ATP binding and hydrolysis, and finally, the intermediate domain serves as a hinge connecting the other two domains (Saibil 2013). GroES is a ~10 kDa subunit, made up of a dome-shaped heptameric ring that binds to the end of the GroEL cylinder producing a hydrophobic cage-like structure in which the folding of nonnative polypeptide occurs (Xu et al. 1997). The cage volume of GroEL-ES is enough to hold a polypeptide of up to ~60 kDa in size (Russmann et al. 2012).

The kinetoplastid HSP60 was first reported by Giambiagi-deMarval et al. in *Trypanosoma cruzi* (Giambiagi-deMarval et al. 1993). It was localized throughout the mitochondria and near the kinetoplast in the parasite (Giambiagi-deMarval et al. 1996). Later, HSP60 was characterized in *L. major*, and the recombinant protein was shown to react very strongly with the sera of leishmaniasis patients, suggesting a humoral response (Rey-Ladino et al. 1997). In *L. donovani*, two different actively transcribing HSP60 genes, CPN60.1 and CPN60.2, were identified, but only CPN60.2 demonstrated constitutive expression in promastigotes, which increases by 2.5-fold under heat stress conditions or in axenic amastigotes (Schluter et al. 2000). A similar increase in the expression of cochaperonin, CPN10 of *L. donovani*, is also observed under heat stress and in axenic amastigotes (Zamora-Veyl et al. 2005). Both these proteins (CPN60 and CPN10) are colocalized in the mitochondria of the parasite, and CPN10 coprecipitates with CPN60 (Schluter et al. 2000; Zamora-Veyl et al. 2005).

10.2.4.2 Group II Chaperonins

Group II chaperonins have been reported in the cytosol of archaea (thermosome, Mm-cpn) and eukaryotes (T-complex-polypeptide-1 (TCP1) or chaperonin-containing TCP-1 (CCT) or TCP-1 ring complex (TRiC)). The major difference between group I and group II chaperonin is in the architecture of the apical domain, which comprises a built-in lid in group II chaperonin, which performs the same function as the GroES of group I chaperonin (Ditzel et al. 1998). The built-in lid is responsible for the closure of the inner chamber to accommodate the substrate (Booth et al. 2008). Yeast chaperonin TRiC/CCT/TCP1 contains eight paralogous subunits per ring, with 23–35% sequence identity between individual paralogous subunits (Frydman et al. 1992). The specificity of each subunit of TCP1 complex may differ toward their substrate and shows asymmetric ATP hydrolysis (Llorca et al. 2001; Rivenzon-Segal et al. 2005; Spiess et al. 2006; Amit et al. 2010). Each subunit of TCP1 has three domains: an apical domain that contains interacting sites with substrate protein, an equatorial domain that comprises the sites that specifically bind with ATP and are also involved in the intra- and interring interactions, and an intermediate domain that forms a connecting link between the two domains.

The function of TCP1 subunits as monomer or homo-oligomer: Amit et al. (2010) showed that a mutation in any of the individual TCP1 subunits out of the eight subunits causes a differential effect on cellular function (Amit et al. 2010). This observation potentiates the role of TCP1 subunits as a monomer and/or their hierarchical roles within the oligomer. In addition, the stoichiometric levels of TCP1 subunits are not present in yeast, which is evident in the increased cellular level of only two subunits, i.e., TCP1δ and TCP1θ, as compared to other subunits (Matalon et al. 2014). This further indicates that individual subunits may function as monomers or homo-oligomers. The overexpression of TCP1α and TCP1δ affect polyglutamine aggregation in yeast and mammalian cells (Tam et al. 2006), and the overexpression of TCP1ζ suppresses abnormal phenotypes in yeast (Kabir et al. 2005). Thus, individual subunits can also bind to misfolded proteins and act as a stabilizing interface. In the *in vitro* microtubule assembly assay, TCP1α, TCP1γ, TCP1ζ, and TCP1θ subunits were shown to be associated with microtubules (Roobol et al. 1999). The TCP1δ monomer subunit interacts with a component of the dynactin complex, $p150^{Glued}$, and plays a role in dynein-mediated transport along microtubules (Echbarthi et al. 2018). TCP1ε monomer interacts with myocardin-related transcription factor A (MRTF-A) and helps in actin polymerization (Elliott et al. 2015). Thus, the ratio of free or assembled subunits will be critical for the cell to balance between the individual or oligomeric function. Further, human TCP1δ and TCP1ε chaperonin subunits expressed in *E. coli* have been shown to form biologically active homo-oligomeric complexes (Sergeeva et al. 2013).

Leishmania group II chaperonin TCP1 γ, δ, η, and θ subunits were identified in the proteomic analysis of EVs released by the parasite. They were reported to be differentially upregulated in EVs isolated at 37 °C compared to 26 °C, and being highest among them was the TCP1γ subunit (Silverman et al. 2010a). Further, other transcriptomic (Das et al. 2013) and proteomic analyses (Vergnes et al. 2007; Biyani et al. 2011; Carnielli et al. 2014) of *Leishmania* drug resistance field isolates revealed a differential expression of various subunits of the TCP1 complex, indicating their possible role in drug resistance. The TCP1γ subunit was reported to be differentially expressed in various growth stages of promastigotes in the proteome analysis of *L. infantum* (Alcolea et al. 2011) and in *L. donovani* by western blot analysis (Bhaskar et al. 2012). In a recent study, the downregulation of the gamma subunit of *L. donovani* TCP1 has been shown to influence the extracellular vesicle proteome enrichment, resulting in the modulation of the macrophage microbicidal function (Yadav et al. 2022a).

L. donovani TCP1γ (LdTCP1γ) shows ~53% sequence identity with *Homo sapiens* at the protein level (Bhaskar et al. 2012). The homology model is used to predict the 3D structure of LdTCP1γ using the crystal structure of chain G of the *Bos taurus* CCT complex (PDB ID: 4B2T) as template (Fig. 10.3a–c). The biochemical characterization of the TCP1γ of *L. donovani* revealed that it is able to form a homo-oligomeric complex, made up of double-toroidal rings of seven subunits that exhibit ATP-dependent refolding activity (Fig. 10.3d, e) (Bhaskar et al. 2015). LdTCP1γ

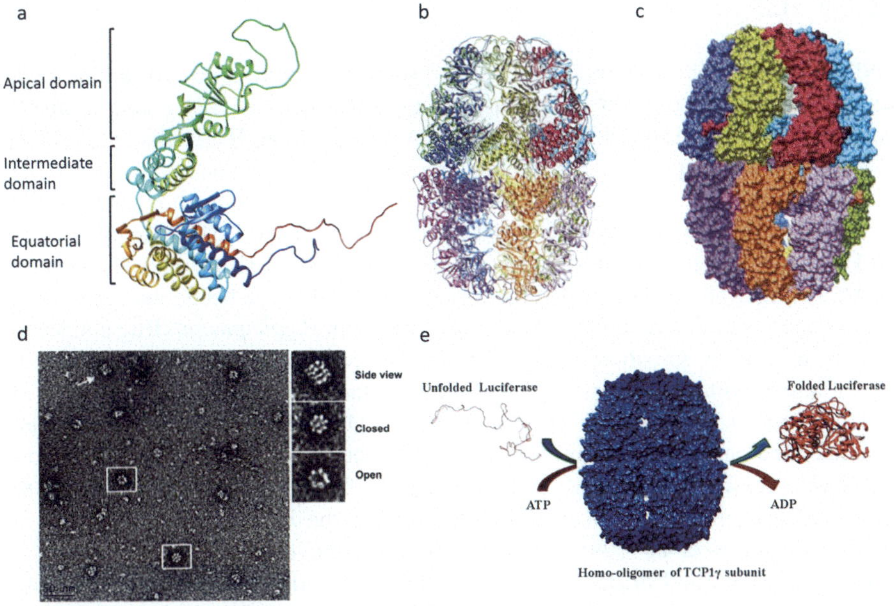

Fig. 10.3 TCP1γ of *L. donovani*: (**a**) Homology model of the monomeric structure of *L. donovani* TCP1γ. (**b** and **c**) The modeled 14-mer homooligomeric structure showing a double-ring LdTCP1γ complex (ribbon and surface representation, respectively). (**d**) The transmission electron microscopy (TEM) image shows the LdTCP1γ oligomer complex having sevenfold symmetry (box—top view). The double-ring structure of the complex is shown by an arrow (side view). The inset indicates a side view and top view (open and closed configuration). The bar indicates 50 nm. (**e**) The homo-tetradecamer complex of LdTCP1γ refolds denatured luciferase in an ATP-dependent manner (Bhaskar et al. 2015)

double-allele replacement mutants failed to survive, showing their essentiality for parasite survival (Yadav et al. 2020b). LdTCP1γ is also indispensable for the growth and infectivity of parasites and can be employed as a drug target (Yadav et al. 2020b). Furthermore, LdTCP1γ has been shown to be involved in miltefosine resistance by protecting the parasite from miltefosine-induced oxidative stress (Yadav et al. 2020a). *Leishmania* genome encodes for all the eight subunits of TCP1, and recently in a co-immunoprecipitation pullout study of TCP1γ, all the other seven subunits have been shown to interact with it (Yadav et al. 2022b). This shows that the parasite TCP1 complex may function as monomeric, homo-oligomeric, or complexes with two or more TCP1 subunits, as in the case of the eukaryotic TCP1 complex. However, future research involving a hetero-oligomeric complex formed by all eight subunits will shed light on the functional detail of this complex in the *Leishmania* parasite.

10.2.5 HSP40

HSP40 chaperonin (J-domain/DnaJ-like protein) comprises a 70 amino acid sequence called J domain. It includes four alpha-helices (I-IV) and a highly conserved His-Pro-Asp (HPD) motif between helices II and III (Botha et al. 2007). The J domain is responsible for the regulation of HSP70 activity and its functional diversity (Fan et al. 2003). The J domain of HSP40 upon interaction with HSP70 increases its ATPase activity and initiates the ATP-mediated HSP70 folding cycles. HSP40 also delivers the unfolded polypeptide to HSP70 (Hartl and Hayer-Hartl 2002; Young et al. 2004). The diverse variants of HSP40 have been found in the cell, six in *E. coli*, 41 in humans (Qiu et al. 2006), and 43 in *P. falciparum* (Rug and Maier 2011). In trypanosomatids, HSP40 exists in even more diverse forms, with 67 variants reported in *T. cruzi*, 65 in *T. brucei*, and 69 in *L. major* (Requena et al. 2015). Four of the *T. cruzi* HSP40 genes, TcJ1, TcJ2, TcJ3, and TcJ4, were characterized, and among them, only the TcJ2 gene responds to heat shock, whose mRNA expression upregulated on exposure to 37 °C for 3 h (Tibbetts et al. 1998). The TcJ2 protein has also been shown to increase the *in vitro* ATPase activity of HSP70 (Tibbetts et al. 1998). In trypanosomes, the mitochondrial homolog of cytosolic HSP40, and mtHSP40 in association with mtHSP70, plays an important role in the maintenance and replication of kDNA (kinetoplast DNA) (Tyc et al. 2015). However, to date, not a single HSP40 protein has been cloned and characterized in *Leishmania*, and our knowledge about the leishmanial HSP70-HSP40 complex is very limited. In a phosphoproteomic study on *L. infantum,* HSP40- J2 protein was identified and phosphorylated on its serine-89 residue. Further, the phosphorylation was 22-fold increased during differentiation from promastigotes to amastigotes (Tsigankov et al. 2014). Such drastic changes in phosphorylation suggest its prominent role in the stage-specific differentiation process of *Leishmania* parasites.

10.2.6 Small Heat Shock Proteins (sHSPs)

Small heat shock proteins (sHSPs) are the most divergent family of molecular chaperones (Basha et al. 2012). The basic structural feature of this protein family is the presence of a conserved α-crystallin domain (ACD) (Haslbeck et al. 2005). Typically, sHSPs act as holdases, keeping their cellular substrates in a near-native form to avoid stress-induced aggregation and to enable their refolding by ATP-dependent chaperones (Ungelenk et al. 2016). Small HSPs interact with the diverse set of substrate proteins and hence play a role in the regulation of a wide range of cellular functions (Nakamoto and Vigh 2007). These include the prevention of aggregation, membrane homeostasis, and stress tolerance (Nakamoto and Vigh 2007). Only three sHSPs, i.e., HSP20, HSP23, and P23, have been studied in *Leishmania* (Hombach et al. 2014, 2015; Requena et al. 2015). Leishmanial HSP20 has been shown to be suitable as a DNA vaccine (Montalvo-Alvarez et al. 2008). *L. braziliensis* recombinant P23 and HSP23 proteins have been found to negatively regulate the ATPase activity of HSP90 (Batista et al. 2015). Further, *L. donovani* P23 null mutants are found to be highly sensitive against HSP90 inhibitors like geldanamycin and radicicol, suggesting that P23 acts as a cochaperone (Hombach et al. 2015). By employing a reverse genetic approach, *L. donovani* HSP23 gene is shown to be involved in the stress response of parasites (Hombach et al. 2014). Further, its expression is also upregulated with the increase in temperature in the axenic amastigotes, as compared to promastigotes (Hombach et al. 2014; Tsigankov et al. 2014). HSP23 null mutants are neither able to survive nor differentiate into axenic amastigotes in vitro. Interestingly, HSP23 is also shown to have a role in antimony drug resistance as the null mutant promastigotes are more sensitive toward trivalent antimony Sb(III) (Tejera Nevado et al. 2016).

10.3 Targeting Chaperones as a Therapeutic Strategy Against Leishmaniasis

HSPs have been a very fascinating target in various types of cancer (Chatterjee and Burns 2017); however, in *Leishmania*, its relevance as a drug target is still under investigation. Recent studies revealed that the inhibition of *Leishmania* HSPs disrupts various cellular activities necessary for the survival, virulence, and pathogenesis of the parasites (Petersen et al. 2012; Varela et al. 2014; Das et al. 2020; Yadav et al. 2020b). Thus, these proteins can be employed as a promising drug target to develop novel antileishmanial drugs. A list of inhibitors effective against *Leishmania* HSP is depicted in Table 10.2.

Geldanamycin (GA), an ansamycin-benzoquinone antibiotic, competes with ATP for binding at the ATP pocket present in HSP83/90 and thus inhibits the activity of HSP83/90 (Ochel et al. 2001). Treatment with a high concentration of geldanamycin inhibits stage differentiation and leads to the growth arrest of the parasite in the G_2 phase of the cell cycle (Hombach and Clos 2014). Another HSP83/90 inhibitor, 17-(allylamino)-17-demethoxygeldanamycin (17-AAG), also shows antileishmanial activity against *L. major*, *L. amazonensis*, *L. panamensis*, and *L. infantum* promastigotes (Varela et al. 2014); (Petersen et al. 2012). 17-AAG is highly effective against the intracellular stage of the parasite as its IC_{50} is in the range of nanomolar concentration (Petersen et al. 2012). 17-AAG has shown a promising inhibitory effect on *L. braziliensis* promastigote growth, infectivity, and parasite multiplication, which potentiates its exploitation for the treatment of cutaneous leishmaniasis (Santos et al. 2014). However, due to poor pharmacokinetic parameters, it could not reach the stage of clinical trials (Jhaveri et al. 2012). Therefore, a liposomal formulation of 17-AAG has been prepared that shows an improved therapeutic index with better solubility (Petersen et al. 2018). Recently, Ap5A [P1, P5-di-(adenosine-5′)-penta-phosphate ammonium salt], an inhibitor of HSP78 (HSP100/ClpB family) protein, has very promisingly shown *in vivo* efficacy in the BALB/c mouse model for the treatment of VL (Das et al. 2020). EGCG [(−)-epigallocatechin 3-O-gallate], originally identified as an inhibitor of the HSP70 of cancer cells (Tran et al. 2010), also inhibits the growth of *L. braziliensis* under both *in vitro* and *in vivo* conditions

Table 10.2 Inhibitors targeting HSPs of *Leishmania*

Target	Inhibitor	*Leishmania* spp.	IC50	Reference(s)
HSP83	Geldanamycin	*L. amazonensis*	0.2 μM	Palma et al. (2019)
	Liposomal 17-AAG	*L. amazonensis*	0.006 nM	Petersen et al. (2018)
	17-AAG	*L. braziliensis*	65 nM	Santos et al. (2014)
		L. amazonensis	64.56 nM	Petersen et al. (2012)
		L. major	79.6 nM	Petersen et al. (2012)
		L. infantum	169.1 nM	Petersen et al. (2012)
	17-DMAG	*L. amazonensis*	0.0861 μM	Palma et al. (2019)
	Glb08	*L. braziliensis*	43 μM	Batista et al. (2020)
HSP70	EGCG	*L. braziliensis*	278.8 μM	Inacio et al. (2014)
	Polymyxin B	*L. amazonensis*	59.2 μM	Souza Ribeiro Costa et al. (2019)
	PBCANP-polymyxin B	*L. amazonensis*	4.6 μM	Souza Ribeiro Costa et al. (2019)
TCP1γ	Suramin	*L. donovani*	10.57 μM	Yadav et al. (2020b)
HSP78	Ap5A	*L. donovani*	15 μM	Das et al. (2020)
Cyp40	Cyclosporin A	*L. donovani*	15–20 μM	Yau et al. (2010)

EGCG: epigallocatechin-3-gallate; *17-AAG*: 17-allylamino-17-demethoxygeldanamycin; *Ap5A*: P1, P5-di(adenosine-5′)-penta-phosphate ammonium; *PBCANP*: poly(n-butylcyanoacrylate) nanoparticles; *17-DMAG*: 17-dimethylaminoethylamino-17-demethoxygeldanamycin

(Inacio et al. 2014). Similarly, polymyxin B, an inhibitor of *Plasmodium* HSP70 (Zininga et al. 2017), has been shown *in vitro* antileishmanial efficacy against *L. amazonensis* (Souza Ribeiro Costa et al. 2019). Through combined molecular modeling and *in vitro* assay strategy, several compounds, including Gbl08, have been identified that are capable of inhibiting the ATPase activity of HSP83/90 from *L. braziliensis* (Batista et al. 2020). Cyclosporin A (CsA) is found to inhibit a cochaperone, cyclophilin 40 (CyP40) of the *L. donovani* parasite, without showing substantial inhibition on host CyP40 (Yau et al. 2010). A cytosolic chaperonin, TCP1γ subunit, is inhibited by suramin, an antitrypanosomal drug (Yadav et al. 2020b). These studies together could pave the way for finding selective inhibitors of parasitic heat shock proteins as future drug development initiatives. However, few issues may complicate targeting this protein family, such as compensation of the inhibition of a HSP by altering the expression level of another chaperone and/or cochaperones (Bartsch et al. 2017; Prasanna and Upadhyay 2021).

10.4 *Leishmania* Heat Shock Proteins as Vaccine Candidates

In *Leishmania*, HSP70 is considered one of the potent immunogenic antigens. The potential of *L major* HSP70 in stimulating humoral immunity has been well demonstrated, along with its ability to stimulate both cytokines and chemokines, as well as dendritic cell maturation (Rafati et al. 2007). A proteomic analysis has revealed that HSP70 from *L. donovani* can act as an immune stimulator of the Th1-type response in cured and/or endemic VL patients (Gupta et al. 2007; Kumari et al. 2008). Further studies have shown that HSP70 can be employed as a DNA-based vaccine against *L. amazonensis* and *L. major* infections as it provides significant protection against both Old World and New World cutaneous leishmaniasis (Campbell et al. 2003). However, significant protection from *L. major* and partial protection from *L. amazonensis* are achieved with P4/HSP70 (Campbell et al. 2003). Further, three *L. infantum* antigens, namely, HSP-70, KMP-11, and PFR-2, are known to exhibit protection in experimentally infected dogs. Altogether, an enhanced lymphocyte proliferation as well as cytokine expressions particularly of interferon γ (IFN-γ) and tumor necrosis factor α (TNF-α) exemplify their potential role as a vaccine candidate against canine VL (Carrillo et al. 2008). Further, a Th1-type immune response is known to be generated by *L. infantum* HSP-70 in mice (Rasouli et al. 2014). Other reports suggest that *L. donovani* HSP70 may also act as an intrinsic adjuvant molecule when used with a polytope DNA vaccine. The evaluation of three vaccine formulations, namely, gp63 DNA vaccine, polytope DNA vaccine, and polytope DNA vaccine, fused to LdHSP70 has shown to trigger T-cell response, along with a significant reduction in parasite load (Sachdeva et al. 2009). This study was further extended by using two different adjuvants, MPL-A and ALD, in combination with the cocktail vaccine. gp63 + HSP70 + MPL-A demonstrates greater protection than those immunized with gp63 + HSP70 + ALD (Kaur et al. 2013). Moreover, LdHSP70 is also known to enhance the immunogenic potential of gp63 against *L. donovani* infection in BALB/c mice (Kaur et al. 2011b). Similarly, the administration of a fusion protein of the maltose-binding protein (MBP) and *L. infantum* HSP70 has shown a detectable anti-MBP humoral immune response in mice (Ommen and Clos 2010). Interestingly, *L. infantum* deletion mutant, lacking both HSP70-II alleles (ΔHSP70-II), is known to exhibit a protective immune response in BALB/c against *L. major* infection (Carrion et al. 2011). A similar kind of vaccination with this attenuated cell line promotes long-term immunity against both CL (Solana et al. 2017) and VL (Solana et al. 2020) infections. *In vitro* stimulation of lymphocytes, isolated from the PBMC of a mucosal leishmaniasis patient, in response to the recombinant HSP70 protein of *L. chagasi*, has demonstrated an increased T-cell proliferation (Kröber-Boncardo et al. 2020). A similar response is also observed with *L. braziliensis* HSP83/90 and HSP70. However, rLbhsp70 stimulates patients' PBMC proliferation but, unlike rLbhsp83, is unable to promote the proliferation of PBMC isolated from self-healing individuals. Thus, it appears that variations observed in PBMC stimulation and immune response between patients of ML versus self-healing CL by these two HSPs may be of particular importance in generating protective immune responses. It also demonstrates the potential role of parasite HSPs in T-cell activation. (Kröber-Boncardo et al. 2020).

An earlier study suggests that HSP70 and HSP83/90 function as both B-cell mitogens as well as immunoprophylactic agents (Abaza 2015). Interestingly, *L. infantum* HSP83 linked with mannose-binding protein (MBP) has shown a much stronger CMI and humoral response in mice as compared to MBP or HSP83 alone. The study suggests that LiHSP83/90 may be considered a promising candidate for the development of adjuvant-free fused antigen vaccine(s) for *Leishmania* (Echeverria et al. 2001). Further, the HSP70 and HSP83 of *L. donovani* in combination with different adjuvants, namely MPLA and ALD, show distinctive cytokine elicitation patterns and antibody switching to IgG2a, suggesting the potential role of these HSPs as protective vaccine molecules against VL infection (Kaur et al. 2011a).

Interestingly, the recombinant *L. donovani* HSP70 stimulates strong CMI immune responses not only in the lymphocytes of cured *Leishmania*-infected hamsters but also in the PBMCs of cured VL patients when administered either individually or in combination with other recombinant *L. donovani* proteins, such as elongation factor-2 (elf-2), protein disulfide isomerase (PDI), and triose phosphate isomerase (TPI) (Jaiswal et al. 2014). Consequently, investigators speculate that protective molecules that combine LdHSP70 and potent Th1-stimulating proteins may prove to be a better strategy for the development of a prophylactic vaccine against VL.

Leishmania HSP100 is also considered another vaccine candidate though requires more extensive studies. It has been observed that vaccination with the mutated HSP100 gene of *L. major* wild strain in C57/BL/6 and BALB/c mice models induces Th1-type polarization. This further promotes the development of a safe attenuated vaccine by employing single-gene replacements of genes that are essential for parasite survival (Reiling et al. 2006).

Another chaperonin, TCP20, a subunit of the cytosolic TCP-1 chaperonin (chaperonin containing t-complex polypeptide-1) of *L. donovani*, has been recognized as a Th1-stimulatory protein exhibiting immunogenic activity with a moderate CMI response in a hamster model against VL infection (Jaiswal et al. 2015).

The immunogenic potential and prophylactic efficacy of recombinant *L. donovani* HSP60 (rLdTCP1γ) have been established against VL infection in cured and immunized hamsters at different time intervals of the postinfection period (Anand et al. 2022). rLdTCP1γ shows the development of an effective CMI response in hamsters upon vaccination and induces upregulated expression levels of Th1 cytokines, conferring immunity to infection. This makes it another suitable target molecule for vaccine development (Anand et al. 2022). Table 10.3 shows a list of *Leishmania* HSPs as a vaccine target.

10.5 Chaperones and Drug Resistance in *Leishmania*

The use of pentavalent antimony, Sb(V), has been stopped as the first-line treatment in many parts of the world, including the Indian subcontinent, due to the emergence of resistance against it, resulting in a major hurdle in the cost-effective treatment of leishmaniasis (Ponte-Sucre et al. 2017). Unfortunately, resistant clinical cases have also been reported against second-line treatment, e.g., miltefosine (Srivastava et al. 2017; Thakur and Sinha 2020) and amphotericin B (Srivastava et al. 2011; Thakur

Table 10.3 *Leishmania* HSPs as a vaccine target

Antigen	Disease	Vaccine type	Animal model	Result/outcome	Reference
HSP70 (pcDNA-HSP70 and rHSP70)	CL and VL	DNA + recombinant protein (prime-boost strategy)	Mice and human patient samples	Humoral response	Rafati et al. (2007)
HSP70	VL (L.d)	Fractioned vaccine	Hamsters	Protection	Kumari et al. (2008)
HSP70	VL (*L.i*)	Recombinant vaccine	Mice	Th1 response	Rico et al. (1999)
HSP70	ML, VL, and CL	Recombinant vaccine	Human patient samples	γδ T-cell activation	Russo et al. (1993)
ΔHSP70-II	VL (*L.i*) CL (L.m)	Live attenuated vaccine	Mice	Protection	Solana et al. (2017, 2020)
P4/HSP70	CL (L.m)	DNA vaccine	Mice	Protection	Campbell et al. (2003)
HSP70, KMP11, and PFR2	VL (*L.i*)	Recombinant vaccine	Dogs	Th1 response	Carrillo et al. (2008)
Poly/HSP70/pcDNA	VL (*L.d*)	DNA vaccine	Mice	Protection	Sachdeva et al. (2009)
gp63 + HSP70 + (MPL-A/ALD)	VL (*L.d*)	Cocktail vaccine	Mice	Protection	Kaur et al. (2013)
gp63 + HSP70	VL (*L.d*)	Recombinant vaccine	Mice	Protection	Kaur et al. (2011b)
MBP + HSP70	VL (*L.i*)	Recombinant vaccine	Mice	Cellular and humoral response	Rico et al. (1998)
HSP70 with elf-2, PDI, and TPI	VL (*L.d*)	Recombinant vaccine	Hamsters	Protection and adjuvant property	Jaiswal et al. (2014)
rHSP70 and rHSP83	ML (*L. braz*)	Recombinant vaccine	Human patient samples	Immune response	Skeiky et al. (1995)
MBP + HSP70 HSP70 and HSP83	VL (*L.i*)	Recombinant vaccine	Mice	Immunoprophylactic targets and B-cell mitogens	Rico et al. (1999, 2002)

(continued)

Table 10.3 (continued)

Antigen	Disease	Vaccine type	Animal model	Result/outcome	Reference
MBP + HSP83	VL (*L.i*)	Fused antigen vaccine	Mice	Immune response	Echeverria et al. (2001)
HSP70 + HSP83	VL (*L.d*)	Recombinant vaccine	Mice	Protection	Kaur et al. (2011a)
HSP100	CL (*L.m*)	Attenuated vaccine	Mice	Th1 response	Reiling et al. (2006)
TCP20	VL (*L.d*)	Recombinant vaccine	Hamster	Protection	Jaiswal et al. (2015)
HSP60 (TCP1γ)	VL (*L.d*)	Recombinant vaccine	Hamster	Protection	Anand et al. (2022)

KMP11: kinetoplastida membrane protein 11; *PFR2*: paraflagellar rod protein 2; *MPL-A*: monophosphoryl lipid A; *MBP*: maltose-binding protein; *ALD*: autoclaved *Leishmania donovani*; *elf2*: elongation factor-2; *PDI*: protein disulfide isomerase; *TPI*: triose phosphate isomerase; *L.d*: *L. donovani*; *L.m*: *L. major*; *L.i*: *L. infantum*; *L. braz*: *L. braziliensis*

and Sinha 2020). The overexpression of HSPs has been shown to be associated with drug resistance in many infectious agents, including *Leishmania*, because drugs exert a type of chemical stress, which may be detrimental to the cells. *L. tarentolae* HSP70 is found to be fourfold upregulated in antimony-resistant parasites compared to wild-type parasites, and its overexpression in parasites causes resistance to pentavalent antimony (Brochu et al. 2004). Using a comparative proteomic approach, the upregulation of HSP70, HSP90, HSP100, TCP1β, and TCP1θ has been demonstrated in *L. donovani* clinical antimony-resistant isolates, suggesting their role in antimony resistance (Vergnes et al. 2007; Kumar et al. 2010; Biyani et al. 2011). The overexpression transfectants of HSP70 and HSP90 cause cross-resistance to miltefosine (Vergnes et al. 2007). Further, it has also been observed that HSP90 (Vergnes et al. 2007) and HSP70 (Brochu et al. 2004) negatively regulate the antimony- (Sereno et al. 2001) and miltefosine-induced (Paris et al. 2004) apoptosis in parasites. HSP60, HSP90, and HSP70 (cytosolic and mitochondrial) proteins have also been shown to be differentially expressed in the comparative proteomic study of the experimental AMB-resistant *L. infantum* parasite (Brotherton et al. 2014). HSPA9B, a cytosolic variant of HSP70, has been found to be downregulated in the experimental miltefosine-resistant *L. donovani* parasite (Vacchina et al. 2016). In addition, two separate comparative transcriptomic analyses of miltefosine-resistant *L. donovani* promastigotes and *L. infantum chagasi* clinical isolates have revealed the differential expression of the TCP1δ subunit (Das et al. 2013) and mitochondrial f60 (Carnielli et al. 2014), respectively. In a recent study, a cytosolic chaperonin, T-complex polypeptide-1 gamma subunit (TCP1γ), has been shown to protect the *L. donovani* parasites from miltefosine-induced oxidative stress by modulating the activity of trypanothione reductase and affecting the thiol levels inside the parasite (Yadav et al. 2020a, b). However, the detailed mechanistic role of many HSPs of the *Leishmania* parasite in drug resistance is still lacking and needs to be elucidated urgently for the employment of successful antileishmanial chemotherapeutic strategies.

10.6 Conclusion

HSPs are essential for a number of functions in the intracellular protozoan parasites, including protein folding, the prevention of protein aggregate formation, the modulation of host immune response, protein trafficking, cell adhesion, drug resistance, the export of virulence factor, the maintenance of protein homeostasis under both normal and stress conditions, etc. (Anas et al. 2019). HSPs are promising targets for antiparasite interventions because of their involvement in host-parasite interactions and host-immune modulation (Prasanna and Upadhyay 2021). Thus, targeting the HSPs of *Leishmania* parasites could provide novel strategies for combating leishmaniasis.

As a drug target, leishmanial HSP83 is the most studied HSP. The most promising geldanamycin derivative, 17-AAG, has been found effective against *Leishmania* spp. under both and conditions (Petersen et al. 2012). As 17-AAG exhibits low solubility and significant toxicity in human clinical trials (Jhaveri et al. 2012), liposomal 17-AAG has been developed, which has enhanced its therapeutic index and also pharmacokinetic properties (Petersen et al. 2018). However, there is still a long way to go before we can employ HSP inhibitors safely in clinical settings. Another concern is the compensation of the inhibited HSP by overexpression of other HSPs as in the case of HSP83 inhibition by Geldanamycin and radicicol which causes opposing effects by a cochaperone Aha1, an activator of HSP83 (Bartsch et al. 2017). Thus, the compensatory mechanism of other chaperones or cochaperones should be studied in detail before employing HSP inhibition.

The development of *Leishmania* vaccine has been proven to be difficult and challenging (Srivastava et al. 2016). Our current understanding regarding HSPs as vaccine targets is mostly based on experimental animal models conferring protection and has not been yet extrapolated to humans or dogs as it requires an understanding of pathophysiological and immunological complexities in the disease that differ according to the causative species and host immune response. Another major hurdle hampering the implementation of these HSPs in clinical trials is due to huge species diversity causing dynamics in virulence and their extent to induce an immune response. Thus, the application of more rationale-based techniques employing modifications in the drug delivery system, the use of fusion proteins (polyproteins), or defined epitopes with suitable adjuvants may captivate their progress toward vaccine development. These important issues will need to be addressed in future basic research. Furthermore, the unique features of leishmanial HSPs, as well as their functional roles and interactions with partner chaperones, cochaperones, and/or substrates, provide prospects for the development of novel vaccines and drugs to combat leishmaniasis.

References

Abaza S (2015) Heat shock proteins and parasitic diseases: Part II. Protozoa. Parasitol United J 8(1): 14

Alcolea PJ, Alonso A, Larraga V (2011) Proteome profiling of *Leishmania infantum* promastigotes. J Eukaryot Microbiol 58(4):352–358. https://doi.org/10.1111/j.1550-7408.2011.00549.x

Amit M, Weisberg SJ, Nadler-Holly M, McCormack EA, Feldmesser E, Kaganovich D et al (2010) Equivalent mutations in the eight subunits of the chaperonin CCT produce dramatically different cellular and gene expression phenotypes. J Mol Biol 401(3):532–543. https://doi.org/10.1016/j.jmb.2010.06.037

Anand A, Balodi DC, Ramalingam K, Yadav S, Goyal N (2022) Immunological characterization of rLdTCP1gamma for its prophylactic potential against visceral leishmaniasis in hamster model. Mol Immunol 141:33–42. https://doi.org/10.1016/j.molimm.2021.11.009

Anas M, Kumari V, Gupta N, Dube A, Kumar N (2019) Protein quality control machinery in intracellular protozoan parasites: hopes and challenges for therapeutic targeting. Cell Stress Chaperones 24(5):891–904. https://doi.org/10.1007/s12192-019-01016-9

Balchin D, Hayer-Hartl M, Hartl FU (2020) Recent advances in understanding catalysis of protein folding by molecular chaperones. FEBS Lett 594(17):2770–2781. https://doi.org/10.1002/1873-3468.13844

Barak E, Amin-Spector S, Gerliak E, Goyard S, Holland N, Zilberstein D (2005) Differentiation of *Leishmania donovani* in host-free system: analysis of signal perception and response. Mol Biochem Parasitol 141(1):99–108. https://doi.org/10.1016/j.molbiopara.2005.02.004

Barends TR, Werbeck ND, Reinstein J (2010) Disaggregases in 4 dimensions. Curr Opin Struct Biol 20(1):46–53. https://doi.org/10.1016/j.sbi.2009.12.014

Bartsch K, Hombach-Barrigah A, Clos J (2017) Hsp90 inhibitors radicicol and geldanamycin have opposing effects on *Leishmania* Aha1-dependent proliferation. Cell Stress Chaperones 22(5): 729–742. https://doi.org/10.1007/s12192-017-0800-2

Basha E, O'Neill H, Vierling E (2012) Small heat shock proteins and alpha-crystallins: dynamic proteins with flexible functions. Trends Biochem Sci 37(3):106–117. https://doi.org/10.1016/j.tibs.2011.11.005

Batista FA, Almeida GS, Seraphim TV, Silva KP, Murta SM, Barbosa LR et al (2015) Identification of two p23 co-chaperone isoforms in *Leishmania braziliensis* exhibiting similar structures and Hsp90 interaction properties despite divergent stabilities. FEBS J 282(2):388–406. https://doi.org/10.1111/febs.13141

Batista FAH, Ramos SL Jr, Tassone G, Leitao A, Montanari CA, Botta M et al (2020) Discovery of small molecule inhibitors of *Leishmania braziliensis* Hsp90 chaperone. J Enzyme Inhib Med Chem 35(1):639–649. https://doi.org/10.1080/14756366.2020.1726342

Bhaskar, Kumari N, Goyal N (2012) Cloning, characterization and sub-cellular localization of gamma subunit of T-complex protein-1 (chaperonin) from *Leishmania donovani*. Biochem Biophys Res Commun 429(1–2):70–74. https://doi.org/10.1016/j.bbrc.2012.10.090

Bhaskar, Mitra K, Kuldeep J, Siddiqi MI, Goyal N (2015) The TCP1gamma subunit of *Leishmania donovani* forms a biologically active homo-oligomeric complex. FEBS J 282(23):4607–4619. https://doi.org/10.1111/febs.13521

Biyani N, Singh AK, Mandal S, Chawla B, Madhubala R (2011) Differential expression of proteins in antimony-susceptible and -resistant isolates of *Leishmania donovani*. Mol Biochem Parasitol 179(2):91–99. https://doi.org/10.1016/j.molbiopara.2011.06.004

Booth CR, Meyer AS, Cong Y, Topf M, Sali A, Ludtke SJ et al (2008) Mechanism of lid closure in the eukaryotic chaperonin TRiC/CCT. Nat Struct Mol Biol 15(7):746–753. https://doi.org/10.1038/nsmb.1436

Botha M, Pesce ER, Blatch GL (2007) The Hsp40 proteins of Plasmodium falciparum and other apicomplexa: regulating chaperone power in the parasite and the host. Int J Biochem Cell Biol 39(10):1781–1803. https://doi.org/10.1016/j.biocel.2007.02.011

Brandau S, Dresel A, Clos J (1995) High constitutive levels of heat-shock proteins in human-pathogenic parasites of the genus *Leishmania*. Biochem J 310(Pt 1):225–232. https://doi.org/10.1042/bj3100225

Brochu C, Haimeur A, Ouellette M (2004) The heat shock protein HSP70 and heat shock cognate protein HSC70 contribute to antimony tolerance in the protozoan parasite leishmania. Cell Stress Chaperones 9(3):294–303

Brotherton MC, Bourassa S, Legare D, Poirier GG, Droit A, Ouellette M (2014) Quantitative proteomic analysis of amphotericin B resistance in *Leishmania infantum*. Int J Parasitol Drugs Drug Resist 4(2):126–132. https://doi.org/10.1016/j.ijpddr.2014.05.002

Burza S, Croft SL, Boelaert M (2018) Leishmaniasis. Lancet 392(10151):951–970. https://doi.org/10.1016/S0140-6736(18)31204-2

Campbell K, Diao H, Ji J, Soong L (2003) DNA immunization with the gene encoding P4 nuclease of *Leishmania amazonensis* protects mice against cutaneous Leishmaniasis. Infect Immun 71(11):6270–6278. https://doi.org/10.1128/IAI.71.11.6270-6278.2003

Carnielli JB, de Andrade HM, Pires SF, Chapeaurouge AD, Perales J, Monti-Rocha R et al (2014) Proteomic analysis of the soluble proteomes of miltefosine-sensitive and -resistant *Leishmania infantum* chagasi isolates obtained from Brazilian patients with different treatment outcomes. J Proteome 108:198–208. https://doi.org/10.1016/j.jprot.2014.05.010

Carrillo E, Crusat M, Nieto J, Chicharro C, Thomas Mdel C, Martinez E et al (2008) Immunogenicity of HSP-70, KMP-11 and PFR-2 leishmanial antigens in the experimental model of canine visceral leishmaniasis. Vaccine 26(15):1902–1911. https://doi.org/10.1016/j.vaccine.2008.01.042

Carrion J, Folgueira C, Soto M, Fresno M, Requena JM (2011) *Leishmania infantum* HSP70-II null mutant as candidate vaccine against leishmaniasis: a preliminary evaluation. Parasit Vectors 4: 150. https://doi.org/10.1186/1756-3305-4-150

Chatterjee S, Burns TF (2017) Targeting Heat Shock Proteins in Cancer: A Promising Therapeutic Approach. Int J Mol Sci 18(9). https://doi.org/10.3390/ijms18091978

Chrobak M, Forster S, Meisel S, Pfefferkorn R, Forster F, Clos J (2012) *Leishmania donovani* HslV does not interact stably with HslU proteins. Int J Parasitol 42(4):329–339. https://doi.org/10.1016/j.ijpara.2012.01.008

Colineau L, Clos J, Moon KM, Foster LJ, Reiner NE (2017) *Leishmania donovani* chaperonin 10 regulates parasite internalization and intracellular survival in human macrophages. Med Microbiol Immunol 206(3):235–257. https://doi.org/10.1007/s00430-017-0500-7

Coulson RM, Smith DF (1990) Isolation of genes showing increased or unique expression in the infective promastigotes of *Leishmania major*. Mol Biochem Parasitol 40(1):63–75. https://doi.org/10.1016/0166-6851(90)90080-6

Das M, Saudagar P, Sundar S, Dubey VK (2013) Miltefosine-unresponsive *Leishmania donovani* has a greater ability than miltefosine-responsive L. donovani to resist reactive oxygen species. FEBS J 280(19):4807–4815. https://doi.org/10.1111/febs.12449

Das S, Banerjee A, Kamran M, Ejazi SA, Asad M, Ali N et al (2020) A chemical inhibitor of heat shock protein 78 (HSP78) from *Leishmania donovani* represents a potential antileishmanial drug candidate. J Biol Chem 295(29):9934–9947. https://doi.org/10.1074/jbc.RA120.014587

Descoteaux A, Avila HA, Zhang K, Turco SJ, Beverley SM (2002) *Leishmania* LPG3 encodes a GRP94 homolog required for phosphoglycan synthesis implicated in parasite virulence but not viability. EMBO J 21(17):4458–4469. https://doi.org/10.1093/emboj/cdf447

Ditzel L, Lowe J, Stock D, Stetter KO, Huber H, Huber R et al (1998) Crystal structure of the thermosome, the archaeal chaperonin and homolog of CCT. Cell 93(1):125–138. https://doi.org/10.1016/s0092-8674(00)81152-6

Easton DP, Kaneko Y, Subjeck JR (2000) The hsp110 and Grp1 70 stress proteins: newly recognized relatives of the Hsp70s. Cell Stress Chaperones 5(4):276–290. https://doi.org/10.1379/1466-1268(2000)005<0276:thagsp>2.0.co;2

Echbarthi M, Vallin J, Grantham J (2018) Interactions between monomeric CCTdelta and p150 (Glued): A novel function for CCTdelta at the cell periphery distinct from the protein folding activity of the molecular chaperone CCT. Exp Cell Res 370(1):137–149. https://doi.org/10.1016/j.yexcr.2018.06.018

Echeverria P, Dran G, Pereda G, Rico AI, Requena JM, Alonso C et al (2001) Analysis of the adjuvant effect of recombinant *Leishmania infantum* Hsp83 protein as a tool for vaccination. Immunol Lett 76(2):107–110. https://doi.org/10.1016/s0165-2478(01)00179-1

Elliott KL, Svanstrom A, Spiess M, Karlsson R, Grantham J (2015) A novel function of the monomeric CCTepsilon subunit connects the serum response factor pathway to chaperone-mediated actin folding. Mol Biol Cell 26(15):2801–2809. https://doi.org/10.1091/mbc.E15-01-0048

Fan CY, Lee S, Cyr DM (2003) Mechanisms for regulation of Hsp70 function by Hsp40. Cell Stress Chaperones 8(4):309–316. https://doi.org/10.1379/1466-1268(2003)008<0309:mfrohf>2.0.co;2

Folgueira C, Quijada L, Soto M, Abanades DR, Alonso C, Requena JM (2005) The translational efficiencies of the two *Leishmania infantum* HSP70 mRNAs, differing in their 3′-untranslated regions, are affected by shifts in the temperature of growth through different mechanisms. J Biol Chem 280(42):35172–35183. https://doi.org/10.1074/jbc.M505559200

Folgueira C, Canavate C, Chicharro C, Requena JM (2007) Genomic organization and expression of the HSP70 locus in New and Old World *Leishmania* species. Parasitology 134(Pt 3): 369–377. https://doi.org/10.1017/S0031182006001570

Folgueira C, Carrion J, Moreno J, Saugar JM, Canavate C, Requena JM (2008) Effects of the disruption of the HSP70-II gene on the growth, morphology, and virulence of *Leishmania* infantum promastigotes. Int Microbiol 11(2):81–89

Fraga J, Montalvo AM, Van der Auwera G, Maes I, Dujardin JC, Requena JM (2013) Evolution and species discrimination according to the *Leishmania* heat-shock protein 20 gene. Infect Genet Evol 18:229–237. https://doi.org/10.1016/j.meegid.2013.05.020

Frydman J, Nimmesgern E, Erdjument-Bromage H, Wall JS, Tempst P, Hartl FU (1992) Function in protein folding of TRiC, a cytosolic ring complex containing TCP-1 and structurally related subunits. EMBO J 11(13):4767–4778

Garlapati S, Dahan E, Shapira M (1999) Effect of acidic pH on heat shock gene expression in *Leishmania*. Mol Biochem Parasitol 100(1):95–101. https://doi.org/10.1016/s0166-6851(99)00037-7

Giambiagi-deMarval M, Gottesdiener K, Rondinelli E, Van der Ploeg LH (1993) Predicted amino acid sequence and genomic organization of Trypanosoma cruzi hsp 60 genes. Mol Biochem Parasitol 58(1):25–31. https://doi.org/10.1016/0166-6851(93)90087-e

Giambiagi-deMarval M, Souto-Padron T, Rondinelli E (1996) Characterization and cellular distribution of heat-shock proteins HSP70 and HSP60 in Trypanosoma cruzi. Exp Parasitol 83(3): 335–345. https://doi.org/10.1006/expr.1996.0081

Gupta SK, Sisodia BS, Sinha S, Hajela K, Naik S, Shasany AK et al (2007) Proteomic approach for identification and characterization of novel immunostimulatory proteins from soluble antigens of *Leishmania donovani* promastigotes. Proteomics 7(5):816–823. https://doi.org/10.1002/pmic.200600725

Hanson PI, Whiteheart SW (2005) AAA+ proteins: have engine, will work. Nat Rev Mol Cell Biol 6(7):519–529. https://doi.org/10.1038/nrm1684

Hartl FU, Hayer-Hartl M (2002) Molecular chaperones in the cytosol: from nascent chain to folded protein. Science 295(5561):1852–1858. https://doi.org/10.1126/science.1068408

Haslbeck M, Franzmann T, Weinfurtner D, Buchner J (2005) Some like it hot: the structure and function of small heat-shock proteins. Nat Struct Mol Biol 12(10):842–846. https://doi.org/10.1038/nsmb993

Hombach A, Clos J (2014) No stress - Hsp90 and signal transduction in *Leishmania*. Parasitology 141(9):1156–1166. https://doi.org/10.1017/S0031182013002151

Hombach A, Ommen G, Chrobak M, Clos J (2013) The Hsp90-Sti1 interaction is critical for *Leishmania donovani* proliferation in both life cycle stages. Cell Microbiol 15(4):585–600. https://doi.org/10.1111/cmi.12057

Hombach A, Ommen G, MacDonald A, Clos J (2014) A small heat shock protein is essential for thermotolerance and intracellular survival of *Leishmania donovani*. J Cell Sci 127(Pt 21): 4762–4773. https://doi.org/10.1242/jcs.157297

Hombach A, Ommen G, Sattler V, Clos J (2015) *Leishmania donovani* P23 protects parasites against HSP90 inhibitor-mediated growth arrest. Cell Stress Chaperones 20(4):673–685. https://doi.org/10.1007/s12192-015-0595-y

Hombach-Barrigah A, Bartsch K, Smirlis D, Rosenqvist H, MacDonald A, Dingli F et al (2019) *Leishmania donovani* 90 kD heat shock protein - impact of phosphosites on parasite fitness, infectivity and casein kinase affinity. Sci Rep 9(1):5074. https://doi.org/10.1038/s41598-019-41640-0

Horwich AL, Fenton WA, Chapman E, Farr GW (2007) Two families of chaperonin: physiology and mechanism. Annu Rev Cell Dev Biol 23:115–145. https://doi.org/10.1146/annurev.cellbio.23.090506.123555

Hubel A, Brandau S, Dresel A, Clos J (1995) A member of the ClpB family of stress proteins is expressed during heat shock in *Leishmania* spp. Mol Biochem Parasitol 70(1–2):107–118. https://doi.org/10.1016/0166-6851(95)00012-p

Hubel A, Krobitsch S, Horauf A, Clos J (1997) *Leishmania major* Hsp100 is required chiefly in the mammalian stage of the parasite. Mol Cell Biol 17(10):5987–5995. https://doi.org/10.1128/MCB.17.10.5987

Inacio JD, Gervazoni L, Canto-Cavalheiro MM, Almeida-Amaral EE (2014) The effect of (-)-epigallocatechin 3-O--gallate in vitro and in vivo in *Leishmania braziliensis*: involvement of reactive oxygen species as a mechanism of action. PLoS Negl Trop Dis 8(8):e3093. https://doi.org/10.1371/journal.pntd.0003093

Jaiswal AK, Khare P, Joshi S, Kushawaha PK, Sundar S, Dube A (2014) Th1 stimulatory proteins of *Leishmania donovani*: comparative cellular and protective responses of rTriose phosphate isomerase, rProtein disulfide isomerase and rElongation factor-2 in combination with rHSP70 against visceral leishmaniasis. PLoS One 9(9):e108556. https://doi.org/10.1371/journal.pone.0108556

Jaiswal AK, Khare P, Joshi S, Rawat K, Yadav N, Sundar S et al (2015) Immunological consequences of stress-related proteins - cytosolic tryparedoxin peroxidase and chaperonin TCP20--identified in splenic amastigotes of *Leishmania donovani* as Th1 stimulatory, in experimental visceral leishmaniasis. Parasitology 142(5):728–744. https://doi.org/10.1017/S003118201400184X

Jhaveri K, Taldone T, Modi S, Chiosis G (2012) Advances in the clinical development of heat shock protein 90 (Hsp90) inhibitors in cancers. Biochim Biophys Acta 1823(3):742–755. https://doi.org/10.1016/j.bbamcr.2011.10.008

Johnson JL (2012) Evolution and function of diverse Hsp90 homologs and cochaperone proteins. Biochim Biophys Acta 1823(3):607–613. https://doi.org/10.1016/j.bbamcr.2011.09.020

Kabir MA, Kaminska J, Segel GB, Bethlendy G, Lin P, Della Seta F et al (2005) Physiological effects of unassembled chaperonin Cct subunits in the yeast Saccharomyces cerevisiae. Yeast 22(3):219–239. https://doi.org/10.1002/yea.1210

Karademir B, Bozaykut P, Kartal Ozer N (2014) Heat shock proteins and proteasomal degradation in normal and tumor cells. Free Radic Biol Med 75 Suppl 1:S35. https://doi.org/10.1016/j.freeradbiomed.2014.10.774

Kaur J, Kaur T, Kaur S (2011a) Studies on the protective efficacy and immunogenicity of Hsp70 and Hsp83 based vaccine formulations in *Leishmania donovani* infected BALB/c mice. Acta Trop 119(1):50–56. https://doi.org/10.1016/j.actatropica.2011.04.007

Kaur T, Sobti RC, Kaur S (2011b) Cocktail of gp63 and Hsp70 induces protection against *Leishmania donovani* in BALB/c mice. Parasite Immunol 33(2):95–103. https://doi.org/10.1111/j.1365-3024.2010.01253.x

Kaur T, Thakur A, Kaur S (2013) Protective immunity using MPL-A and autoclaved *Leishmania* donovani as adjuvants along with a cocktail vaccine in murine model of visceral leishmaniasis. J Parasit Dis 37(2):231–239. https://doi.org/10.1007/s12639-012-0171-7

Kaur P, Garg M, Hombach-Barrigah A, Clos J, Goyal N (2017) MAPK1 of *Leishmania donovani* interacts and phosphorylates HSP70 and HSP90 subunits of foldosome complex. Sci Rep 7(1):10202. https://doi.org/10.1038/s41598-017-09725-w

Kohler V, Andreasson C (2020) Hsp70-mediated quality control: should I stay or should I go? Biol Chem 401(11):1233–1248. https://doi.org/10.1515/hsz-2020-0187

Kröber-Boncardo C, Grünebast J, Clos J (2020) Heat shock proteins in *Leishmania* parasites. Springer, Dordrecht, pp 1–20

Krobitsch S, Clos J (1999) A novel role for 100 kD heat shock proteins in the parasite *Leishmania donovani*. Cell Stress Chaperones 4(3):191–198. https://doi.org/10.1379/1466-1268(1999)004<0191:anrfkh>2.3.co;2

Kumar A, Sisodia B, Misra P, Sundar S, Shasany AK, Dube A (2010) Proteome mapping of overexpressed membrane-enriched and cytosolic proteins in sodium antimony gluconate (SAG) resistant clinical isolate of *Leishmania donovani*. Br J Clin Pharmacol 70(4):609–617. https://doi.org/10.1111/j.1365-2125.2010.03716.x

Kumari S, Samant M, Khare P, Sundar S, Sinha S, Dube A (2008) Induction of Th1-type cellular responses in cured/exposed *Leishmania*-infected patients and hamsters against polyproteins of soluble *Leishmania donovani* promastigotes ranging from 89.9 to 97.1 kDa. Vaccine 26(37):4813–4818. https://doi.org/10.1016/j.vaccine.2008.06.102

Kumari D, Perveen S, Sharma R, Singh K (2021) Advancement in leishmaniasis diagnosis and therapeutics: an update. Eur J Pharmacol 910:174436. https://doi.org/10.1016/j.ejphar.2021.174436

Llorca O, Martin-Benito J, Gomez-Puertas P, Ritco-Vonsovici M, Willison KR, Carrascosa JL et al (2001) Analysis of the interaction between the eukaryotic chaperonin CCT and its substrates actin and tubulin. J Struct Biol 135(2):205–218. https://doi.org/10.1006/jsbi.2001.4359

Lopez T, Dalton K, Frydman J (2015) The mechanism and function of Group II chaperonins. J Mol Biol 427(18):2919–2930. https://doi.org/10.1016/j.jmb.2015.04.013

Mann S, Frasca K, Scherrer S, Henao-Martinez AF, Newman S, Ramanan P et al (2021) A review of leishmaniasis: current knowledge and future directions. Curr Trop Med Rep 8(2):121–132. https://doi.org/10.1007/s40475-021-00232-7

Maresca B, Carratu L (1992) The biology of the heat shock response in parasites. Parasitol Today 8(8):260–266. https://doi.org/10.1016/0169-4758(92)90137-q

Matalon O, Horovitz A, Levy ED (2014) Different subunits belonging to the same protein complex often exhibit discordant expression levels and evolutionary properties. Curr Opin Struct Biol 26: 113–120. https://doi.org/10.1016/j.sbi.2014.06.001

Miller MA, McGowan SE, Gantt KR, Champion M, Novick SL, Andersen KA et al (2000) Inducible resistance to oxidant stress in the protozoan *Leishmania chagasi*. J Biol Chem 275(43):33883–33889. https://doi.org/10.1074/jbc.M003671200

Missiakas D, Schwager F, Betton JM, Georgopoulos C, Raina S (1996) Identification and characterization of HsIV HsIU (ClpQ ClpY) proteins involved in overall proteolysis of misfolded proteins in Escherichia coli. EMBO J 15(24):6899–6909

Mogk A, Kummer E, Bukau B (2015) Cooperation of Hsp70 and Hsp100 chaperone machines in protein disaggregation. Front Mol Biosci 2:22. https://doi.org/10.3389/fmolb.2015.00022

Montalvo-Alvarez AM, Folgueira C, Carrion J, Monzote-Fidalgo L, Canavate C, Requena JM (2008) The *Leishmania* HSP20 is antigenic during natural infections, but, as DNA vaccine, it does not protect BALB/c mice against experimental L. amazonensis infection. J Biomed Biotechnol 2008:695432. https://doi.org/10.1155/2008/695432

Morales MA, Watanabe R, Dacher M, Chafey P, Osorio y Fortea, J., Scott, D. A., et al. (2010) Phosphoproteome dynamics reveal heat-shock protein complexes specific to the *Leishmania donovani* infectious stage. Proc Natl Acad Sci U S A 107(18):8381–8386. https://doi.org/10.1073/pnas.0914768107

Nakamoto H, Vigh L (2007) The small heat shock proteins and their clients. Cell Mol Life Sci 64(3):294–306. https://doi.org/10.1007/s00018-006-6321-2

Ochel HJ, Eichhorn K, Gademann G (2001) Geldanamycin: the prototype of a class of antitumor drugs targeting the heat shock protein 90 family of molecular chaperones. Cell Stress Chaperones 6(2):105–112. https://doi.org/10.1379/1466-1268(2001)006<0105:gtpoac>2.0.co;2

Ommen G, Clos J (2010) Heat shock proteins in protozoan parasites – *Leishmania* spp. In: Pockley AG, Calderwood SK, Santoro MG (eds) Prokaryotic and eukaryotic heat shock proteins in infectious disease. Springer, Dordrecht, pp 135–151

Ommen G, Lorenz S, Clos J (2009) One-step generation of double-allele gene replacement mutants in *Leishmania donovani*. Int J Parasitol 39(5):541–546. https://doi.org/10.1016/j.ijpara.2008.10.009

Ommen G, Chrobak M, Clos J (2010) The co-chaperone SGT of *Leishmania donovani* is essential for the parasite's viability. Cell Stress Chaperones 15(4):443–455. https://doi.org/10.1007/s12192-009-0160-7

Palma LC, Ferreira L, Petersen A, Dias BRS, Menezes JPB, Moreira DRM et al (2019) A docking-based structural analysis of geldanamycin-derived inhibitor binding to human or *Leishmania* Hsp90. Sci Rep 9(1):14756. https://doi.org/10.1038/s41598-019-51239-0

Paris C, Loiseau PM, Bories C, Breard J (2004) Miltefosine induces apoptosis-like death in *Leishmania donovani* promastigotes. Antimicrob Agents Chemother 48(3):852–859

Pearl LH, Prodromou C, Workman P (2008) The Hsp90 molecular chaperone: an open and shut case for treatment. Biochem J 410(3):439–453. https://doi.org/10.1042/BJ20071640

Petersen AL, Guedes CE, Versoza CL, Lima JG, de Freitas LA, Borges VM et al (2012) 17-AAG kills intracellular *Leishmania amazonensis* while reducing inflammatory responses in infected macrophages. PLoS One 7(11):e49496. https://doi.org/10.1371/journal.pone.0049496

Petersen A, Campos TA, Dantas D, Reboucas JS, da Silva JC, de Menezes JPB et al (2018) Encapsulation of the HSP-90 chaperone inhibitor 17-AAG in stable liposome allow increasing the therapeutic index as assessed, in vitro, on *Leishmania (L) amazonensis* amastigotes-hosted in mouse CBA macrophages. Front Cell Infect Microbiol 8:303. https://doi.org/10.3389/fcimb.2018.00303

Ponte-Sucre A, Gamarro F, Dujardin JC, Barrett MP, Lopez-Velez R, Garcia-Hernandez R et al (2017) Drug resistance and treatment failure in leishmaniasis: a 21st century challenge. PLoS Negl Trop Dis 11(12):e0006052. https://doi.org/10.1371/journal.pntd.0006052

Prasanna P, Upadhyay A (2021) Heat shock proteins as the druggable targets in leishmaniasis: promises and perils. Infect Immun 89(2). https://doi.org/10.1128/IAI.00559-20

Qiu XB, Shao YM, Miao S, Wang L (2006) The diversity of the DnaJ/Hsp40 family, the crucial partners for Hsp70 chaperones. Cell Mol Life Sci 63(22):2560–2570. https://doi.org/10.1007/s00018-006-6192-6

Rafati S, Gholami E, Hassani N, Ghaemimanesh F, Taslimi Y, Taheri T et al (2007) *Leishmania major* heat shock protein 70 (HSP70) is not protective in murine models of cutaneous leishmaniasis and stimulates strong humoral responses in cutaneous and visceral leishmaniasis patients. Vaccine 25(21):4159–4169. https://doi.org/10.1016/j.vaccine.2007.03.006

Ramirez CA, Requena JM, Puerta CJ (2011) Identification of the HSP70-II gene in *Leishmania braziliensis* HSP70 locus: genomic organization and UTRs characterization. Parasit Vectors 4:166. https://doi.org/10.1186/1756-3305-4-166

Rasouli M, Karimi MH, Kalani M, Ebrahimnezhad S, Namayandeh M, Moravej A (2014) Immunostimulatory effects of *Leishmania infantum* HSP70 recombinant protein on dendritic cells in vitro and in vivo. Immunotherapy 6(5):577–585. https://doi.org/10.2217/imt.14.28

Reiling L, Jacobs T, Kroemer M, Gaworski I, Graefe S, Clos J (2006) Spontaneous recovery of pathogenicity by *Leishmania major* hsp100−/− alters the immune response in mice. Infect Immun 74(11):6027–6036. https://doi.org/10.1128/IAI.00773-05

Requena JM, Montalvo AM, Fraga J (2015) Molecular chaperones of *Leishmania*: central players in many stress-related and -unrelated physiological processes. Biomed Res Int 2015:301326. https://doi.org/10.1155/2015/301326

Rey-Ladino JA, Joshi PB, Singh B, Gupta R, Reiner NE (1997) *Leishmania major*: molecular cloning, sequencing, and expression of the heat shock protein 60 gene reveals unique carboxy terminal peptide sequences. Exp Parasitol 85(3):249–263. https://doi.org/10.1006/expr.1996.4137

Richter K, Haslbeck M, Buchner J (2010) The heat shock response: life on the verge of death. Mol Cell 40(2):253–266. https://doi.org/10.1016/j.molcel.2010.10.006

Rico AI, Del Real G, Soto M, Quijada L, Martinez AC, Alonso C et al (1998) Characterization of the immunostimulatory properties of *Leishmania infantum* HSP70 by fusion to the Escherichia coli maltose-binding protein in normal and nu/nu BALB/c mice. Infect Immun 66(1):347–352. https://doi.org/10.1128/IAI.66.1.347-352.1998

Rico AI, Angel SO, Alonso C, Requena JM (1999) Immunostimulatory properties of the *Leishmania infantum* heat shock proteins HSP70 and HSP83. Mol Immunol 36(17):1131–1139. https://doi.org/10.1016/s0161-5890(99)00136-4

Rico AI, Girones N, Fresno M, Alonso C, Requena JM (2002) The heat shock proteins, Hsp70 and Hsp83, of *Leishmania infantum* are mitogens for mouse B cells. Cell Stress Chaperones 7(4):339–346. https://doi.org/10.1379/1466-1268(2002)007<0339:thspha>2.0.co;2

Rivenzon-Segal D, Wolf SG, Shimon L, Willison KR, Horovitz A (2005) Sequential ATP-induced allosteric transitions of the cytoplasmic chaperonin containing TCP-1 revealed by EM analysis. Nat Struct Mol Biol 12(3):233–237. https://doi.org/10.1038/nsmb901

Roobol A, Sahyoun ZP, Carden MJ (1999) Selected subunits of the cytosolic chaperonin associate with microtubules assembled in vitro. J Biol Chem 274(4):2408–2415. https://doi.org/10.1074/jbc.274.4.2408

Rug M, Maier AG (2011) The heat shock protein 40 family of the malaria parasite Plasmodium falciparum. IUBMB Life 63(12):1081–1086. https://doi.org/10.1002/iub.525

Russmann F, Stemp MJ, Monkemeyer L, Etchells SA, Bracher A, Hartl FU (2012) Folding of large multidomain proteins by partial encapsulation in the chaperonin TRiC/CCT. Proc Natl Acad Sci U S A 109(52):21208–21215. https://doi.org/10.1073/pnas.1218836109

Russo DM, Armitage RJ, Barral-Netto M, Barral A, Grabstein KH, Reed SG (1993) Antigen-reactive gamma delta T cells in human leishmaniasis. J Immunol 151(7):3712–3718

Sachdeva R, Banerjea AC, Malla N, Dubey ML (2009) Immunogenicity and efficacy of single antigen Gp63, polytope and polytopeHSP70 DNA vaccines against visceral Leishmaniasis in experimental mouse model. PLoS One 4(12):e7880. https://doi.org/10.1371/journal.pone.0007880

Saibil H (2013) Chaperone machines for protein folding, unfolding and disaggregation. Nat Rev Mol Cell Biol 14(10):630–642. https://doi.org/10.1038/nrm3658

Santos DM, Petersen AL, Celes FS, Borges VM, Veras PS, de Oliveira CI (2014) Chemotherapeutic potential of 17-AAG against cutaneous leishmaniasis caused by *Leishmania (Viannia) braziliensis*. PLoS Negl Trop Dis 8(10):e3275. https://doi.org/10.1371/journal.pntd.0003275

Schluter A, Wiesgigl M, Hoyer C, Fleischer S, Klaholz L, Schmetz C et al (2000) Expression and subcellular localization of cpn60 protein family members in *Leishmania donovani*. Biochim Biophys Acta 1491(1–3):65–74. https://doi.org/10.1016/s0167-4781(00)00028-2

Searle S, Smith DF (1993) *Leishmania major*: characterisation and expression of a cytoplasmic stress-related protein. Exp Parasitol 77(1):43–52. https://doi.org/10.1006/expr.1993.1059

Searle S, Campos AJ, Coulson RM, Spithill TW, Smith DF (1989) A family of heat shock protein 70-related genes are expressed in the promastigotes of *Leishmania major*. Nucleic Acids Res 17(13):5081–5095. https://doi.org/10.1093/nar/17.13.5081

Selvapandiyan A, Croft SL, Rijal S, Nakhasi HL, Ganguly NK (2019) Innovations for the elimination and control of visceral leishmaniasis. PLoS Negl Trop Dis 13(2019):e0007616

Sereno D, Holzmuller P, Mangot I, Cuny G, Ouaissi A, Lemesre JL (2001) Antimonial-mediated DNA fragmentation in *Leishmania infantum* amastigotes. Antimicrob Agents Chemother 45(7):2064–9. https://doi.org/10.1128/AAC.45.7.2064-2069.2001

Sergeeva OA, Chen B, Haase-Pettingell C, Ludtke SJ, Chiu W, King JA (2013) Human CCT4 and CCT5 chaperonin subunits expressed in Escherichia coli form biologically active homo-oligomers. J Biol Chem 288(24):17734–17744. https://doi.org/10.1074/jbc.M112.443929

Silverman JM, Clos J, de'Oliveira CC, Shirvani O, Fang Y, Wang C et al (2010a) An exosome-based secretion pathway is responsible for protein export from *Leishmania* and communication with macrophages. J Cell Sci 123(Pt 6):842–852. https://doi.org/10.1242/jcs.056465

Silverman JM, Clos J, Horakova E, Wang AY, Wiesgigl M, Kelly I et al (2010b) *Leishmania* exosomes modulate innate and adaptive immune responses through effects on monocytes and dendritic cells. J Immunol 185(9):5011–5022. https://doi.org/10.4049/jimmunol.1000541

Skeiky YA, Benson DR, Guderian JA, Whittle JA, Bacelar O, Carvalho EM et al (1995) Immune responses of leishmaniasis patients to heat shock proteins of *Leishmania* species and humans. Infect Immun 63(10):4105–4114. https://doi.org/10.1128/iai.63.10.4105-4114.1995

Solana JC, Ramirez L, Corvo L, de Oliveira CI, Barral-Netto M, Requena JM et al (2017) Vaccination with a *Leishmania infantum* HSP70-II null mutant confers long-term protective immunity against *Leishmania major* infection in two mice models. PLoS Negl Trop Dis 11(5):e0005644. https://doi.org/10.1371/journal.pntd.0005644

Solana JC, Ramirez L, Cook EC, Hernandez-Garcia E, Sacristan S, Martin ME et al (2020) Subcutaneous immunization of *Leishmania* HSP70-II null mutant line reduces the severity of the experimental visceral leishmaniasis in BALB/c mice. Vaccines (Basel) 8(1). https://doi.org/10.3390/vaccines8010141

Souza Ribeiro Costa J, Medeiros M, Yamashiro-Kanashiro EH, Rocha MC, Cotrim PC, Stephano MA et al (2019) Biodegradable nanocarriers coated with polymyxin B: evaluation of leishmanicidal and antibacterial potential. PLoS Negl Trop Dis 13(5):e0007388. https://doi.org/10.1371/journal.pntd.0007388

Spiess C, Miller EJ, McClellan AJ, Frydman J (2006) Identification of the TRiC/CCT substrate binding sites uncovers the function of subunit diversity in eukaryotic chaperonins. Mol Cell 24(1):25–37. https://doi.org/10.1016/j.molcel.2006.09.003

Srivastava P, Prajapati VK, Rai M, Sundar S (2011) Unusual case of resistance to amphotericin B in visceral leishmaniasis in a region in India where leishmaniasis is not endemic. J Clin Microbiol 49(8):3088–3091. https://doi.org/10.1128/JCM.00173-11

Srivastava S, Shankar P, Mishra J, Singh S (2016) Possibilities and challenges for developing a successful vaccine for leishmaniasis. Parasit Vectors 9(1):277. https://doi.org/10.1186/s13071-016-1553-y

Srivastava S, Mishra J, Gupta AK, Singh A, Shankar P, Singh S (2017) Laboratory confirmed miltefosine resistant cases of visceral leishmaniasis from India. Parasit Vectors 10(1):49. https://doi.org/10.1186/s13071-017-1969-z

Takano T, Kakefuda T (1972) Involvement of a bacterial factor in morphogenesis of bacteriophage capsid. Nat New Biol 239(89):34–37. https://doi.org/10.1038/newbio239034a0

Tam S, Geller R, Spiess C, Frydman J (2006) The chaperonin TRiC controls polyglutamine aggregation and toxicity through subunit-specific interactions. Nat Cell Biol 8(10):1155–1162. https://doi.org/10.1038/ncb1477

Teixeira DE, Benchimol M, Rodrigues JC, Crepaldi PH, Pimenta PF, de Souza W (2013) The cell biology of *Leishmania*: how to teach using animations. PLoS Pathog 9(10):e1003594. https://doi.org/10.1371/journal.ppat.1003594

Tejera Nevado P, Bifeld E, Hohn K, Clos J (2016) A telomeric cluster of antimony resistance genes on chromosome 34 of *Leishmania infantum*. Antimicrob Agents Chemother 60(9):5262–5275. https://doi.org/10.1128/AAC.00544-16

Thakur CP, Sinha K (2020) Observation on treatment of drug resistant Kala-azar patients with fungizone in Patna, Bihar. J Assoc Physicians India 68(6):30–31

Tibbetts RS, Jensen JL, Olson CL, Wang FD, Engman DM (1998) The DnaJ family of protein chaperones in Trypanosoma cruzi. Mol Biochem Parasitol 91(2):319–326. https://doi.org/10.1016/s0166-6851(97)00214-4

Torres-Guerrero E, Quintanilla-Cedillo MR, Ruiz-Esmenjaud J, Arenas R (2017) Leishmaniasis: a review. F1000Res 6:750. https://doi.org/10.12688/f1000research.11120.1

Tran PL, Kim SA, Choi HS, Yoon JH, Ahn SG (2010) Epigallocatechin-3-gallate suppresses the expression of HSP70 and HSP90 and exhibits anti-tumor activity in vitro and in vivo. BMC Cancer 10:276. https://doi.org/10.1186/1471-2407-10-276

Tsigankov P, Gherardini PF, Helmer-Citterich M, Spath GF, Myler PJ, Zilberstein D (2014) Regulation dynamics of *Leishmania* differentiation: deconvoluting signals and identifying phosphorylation trends. Mol Cell Proteomics 13(7):1787–1799. https://doi.org/10.1074/mcp.M114.037705

Tyc J, Klingbeil MM, Lukes J (2015) Mitochondrial heat shock protein machinery hsp70/hsp40 is indispensable for proper mitochondrial DNA maintenance and replication. mBio 6(1). https://doi.org/10.1128/mBio.02425-14

Ungelenk S, Moayed F, Ho CT, Grousl T, Scharf A, Mashaghi A et al (2016) Small heat shock proteins sequester misfolding proteins in near-native conformation for cellular protection and efficient refolding. Nat Commun 7:13673. https://doi.org/10.1038/ncomms13673

Vacchina P, Norris-Mullins B, Abengozar MA, Viamontes CG, Sarro J, Stephens MT et al (2016) Genomic appraisal of the multifactorial basis for in vitro acquisition of miltefosine resistance in *Leishmania donovani*. Antimicrob Agents Chemother 60(7):4089–4100. https://doi.org/10.1128/AAC.00478-16

Varela MR, Mollinedo-Gajate C, Muro A, Mollinedo F (2014) The HSP90 inhibitor 17-AAG potentiates the antileishmanial activity of the ether lipid edelfosine. Acta Trop 131:32–36. https://doi.org/10.1016/j.actatropica.2013.11.018

Vergnes B, Gourbal B, Girard I, Sundar S, Drummelsmith J, Ouellette M (2007) A proteomics screen implicates HSP83 and a small kinetoplastid calpain-related protein in drug resistance in

Leishmania donovani clinical field isolates by modulating drug-induced programmed cell death. Mol Cell Proteomics 6(1):88–101. https://doi.org/10.1074/mcp.M600319-MCP200

Xu Z, Horwich AL, Sigler PB (1997) The crystal structure of the asymmetric GroEL-GroES-(ADP) 7 chaperonin complex. Nature 388(6644):741–750. https://doi.org/10.1038/41944

Yadav S, Ali V, Singh Y, Kanojia S, Goyal N (2020a) *Leishmania donovani* chaperonin TCP1gamma subunit protects miltefosine induced oxidative damage. Int J Biol Macromol 165(Pt B):2607–2620. https://doi.org/10.1016/j.ijbiomac.2020.10.134

Yadav S, Kuldeep J, Siddiqi MI, Goyal N (2020b) TCP1gamma subunit is indispensable for growth and infectivity of *Leishmania donovani*. Antimicrob Agents Chemother 64(8). https://doi.org/10.1128/AAC.00669-20

Yadav S, Anand A, Balodi DC, Ramalingam K, Mitra K, Maras JS, Goyal N (2022a) Downregulation of gamma subunit of TCP1 chaperonin of *Leishmania donovani* modulates extracellular vesicles-mediated macrophage microbicidal function. Microb Pathog 169:105616. https://doi.org/10.1016/j.micpath.2022.105616

Yadav S, Anand A, Ramalingam K, Balodi DC, Maras JS, Goyal N (2022b) Unraveling of interacting protein network of chaperonin TCP1 gamma subunit of *Leishmania donovani*. Cell Stress Chaperones 27(3):205–222. https://doi.org/10.1007/s12192-022-01262-4

Yau WL, Blisnick T, Taly JF, Helmer-Citterich M, Schiene-Fischer C, Leclercq O et al (2010) Cyclosporin A treatment of *Leishmania donovani* reveals stage-specific functions of cyclophilins in parasite proliferation and viability. PLoS Negl Trop Dis 4(6):e729. https://doi.org/10.1371/journal.pntd.0000729

Yau WL, Pescher P, MacDonald A, Hem S, Zander D, Retzlaff S et al (2014) The *Leishmania donovani* chaperone cyclophilin 40 is essential for intracellular infection independent of its stage-specific phosphorylation status. Mol Microbiol 93(1):80–97. https://doi.org/10.1111/mmi.12639

Yau WL, Lambertz U, Colineau L, Pescher P, MacDonald A, Zander D et al (2016) Phenotypic characterization of a *Leishmania donovani* cyclophilin 40 null mutant. J Eukaryot Microbiol 63(6):823–833. https://doi.org/10.1111/jeu.12329

Yebenes H, Mesa P, Munoz IG, Montoya G, Valpuesta JM (2011) Chaperonins: two rings for folding. Trends Biochem Sci 36(8):424–432. https://doi.org/10.1016/j.tibs.2011.05.003

Young JC, Agashe VR, Siegers K, Hartl FU (2004) Pathways of chaperone-mediated protein folding in the cytosol. Nat Rev Mol Cell Biol 5(10):781–791. https://doi.org/10.1038/nrm1492

Zamora-Veyl FB, Kroemer M, Zander D, Clos J (2005) Stage-specific expression of the mitochondrial co-chaperonin of *Leishmania donovani*, CPN10. Kinetoplastid Biol Dis 4(1):3. https://doi.org/10.1186/1475-9292-4-3

Zilberstein D, Shapira M (1994) The role of pH and temperature in the development of *Leishmania* parasites. Annu Rev Microbiol 48:449–470. https://doi.org/10.1146/annurev.mi.48.100194.002313

Zininga T, Pooe OJ, Makhado PB, Ramatsui L, Prinsloo E, Achilonu I et al (2017) Polymyxin B inhibits the chaperone activity of Plasmodium falciparum Hsp70. Cell Stress Chaperones 22(5):707–715. https://doi.org/10.1007/s12192-017-0797-6

Advances in Antileishmanial Chemotherapy 11

Shaikh Shuhail, Saswata Das, Tirtharaj Datta, Priya Tyagi, Mohd Tasleem, Riya Singh, Parma Nand, and Garima Chouhan

Abstract

Leishmaniasis, an endemic tropical ailment, has invariably been categorized as one of the "neglected tropical diseases" and is constantly linked with poverty. However, under current circumstances, it is baffling that leishmaniasis is no longer endemic and poverty-restricted. Leishmaniasis has upheaved its impact by mushrooming outside of its endemic zones as well as occurring as a co-infection with many life-threatening infections. Thus, it is startling that even after afflicting a large chunk of the world's population, producing critical symptoms, and causing several thousands of deaths, efforts made to contain this disease fall much shorter than the stature of the disease. Though noteworthy breakthroughs have been made in the development of vaccines, none has been translated into commercial success, leaving chemotherapy as the sole alternative. Despite quantum leaps in the understanding of parasite biology, the advent of productive in silico drug discovery tools, and high throughput screening methods, there is still a dearth of promising antileishmanial drug candidates in the pipeline.

Shaikh Shuhail and Saswata Das contributed equally to this work.

S. Shuhail · S. Das · T. Datta · P. Tyagi · G. Chouhan (✉)
Department of Biotechnology, School of Engineering and Technology, Greater Noida, Uttar Pradesh, India
e-mail: garima.chauhan@sharda.ac.in

M. Tasleem
ICAR-NIPB, Pusa Campus, New Delhi, India

R. Singh
Amity Institute of Biotechnology, Amity University, Noida, Uttar Pradesh, India

P. Nand
School of Engineering and Technology, Greater Noida, Uttar Pradesh, India

© The Author(s), under exclusive license to Springer Nature Singapore Pte Ltd. 2023

A. Selvapandiyan et al. (eds.), *Challenges and Solutions Against Visceral Leishmaniasis*, https://doi.org/10.1007/978-981-99-6999-9_11

Thus, in this chapter, we cast light on existing antileishmanial therapy, discuss impressive prospects of target-based drug designing, and summarize recent advances made in the domain of antileishmanial drug discovery.

Keywords

Leishmaniasis · Drug targets · Antileishmanial · Antileishmanial chemotherapy · Antileishmanial drugs · Drug repurposing

11.1 Introduction

Leishmaniasis is a poverty-associated, vector-transmitted, protozoan-caused, neglected tropical disease that presents itself in distinct clinical patterns. At present, leishmaniasis afflicts 10–12 million people worldwide and is responsible for 2.4 million disability adjusted life years and the loss of 25,000 to 65,000 human lives annually (Dos Santos Meira and Gedamu 2019; WHO 2022a, b; Gabriel et al. 2019). Further, as estimated by World Health Organization (WHO), the population at risk for leishmaniasis exceeds 1 billion, and they are chiefly living in areas where the disease is endemic (WHO 2022a, b). Also, among the protozoan diseases, leishmaniasis accounts for the highest number of deaths after malaria.

The disease is caused by a protozoan parasite of the genus *Leishmania* that requires sandflies as vectors for the transmission of the disease. *Leishmania* is a digenetic unicellular parasite that exhibits two forms: flagellated motile promastigotes and aflagellated nonmotile amastigotes. There are around 30 species of *Leishmania* known to be pathogenic to humans; a few of the prominent species are *L. donovani*, *L. infantum*, *L. major*, *L. tropica*, *L. braziliensis*, *L. panamensis*, etc. Among the sandflies, the genera *Phlebotomus* and *Lutzomyia* transmit parasites in different geographical realms; *Phlebotomus* is the vector for Old World, whereas *Lutzomyia* transmits the disease in New World (Maroli et al. 2013). Leishmaniasis is a neglected tropical disease and displays a wide range of clinical patterns. These different clinical patterns arise due to differences in infective species, site of infection, and observable symptoms to manifest as either cutaneous, mucocutaneous or visceral infection.

11.1.1 Leishmaniasis: Clinical Patterns

Cutaneous leishmaniasis (CL) is the most usual of all forms. Though it is nonfatal, it causes lesions and ulcers on the skin, which can lead to severe disfiguration and lifelong scars, creating a social stigma in affected individuals. In the old world, CL is primly caused by *L. major*, *L. aethiopica*, and *L. tropica*, whereas in the new world, *L. braziliensis*, *L. panamensis*, and *L. mexicana* are chief species responsible for causing CL. The disease develops within a few weeks after the bite of an infected sandfly, showing early symptoms as a "sore" formation in the skin, which further develops into ulcers. Most of these lesions are painless, while some can be painful. Though CL heals by itself, it may take months, leaving scars all over the skin. CL lesions occur in the exposed body parts, such as the hands, legs, arms, face, etc. (Bilgic-Temel et al. 2019). It affects people of all age groups and gender while being more prominent in children. As per World Health Organization (WHO) data, 95% of CL cases occur in America, the Mediterranean region, the Middle East, and Asia (WHO 2022a, b). In addition, CL has two clinical forms, depending upon the infective species, known as local cutaneous leishmaniasis (LCL) and diffuse cutaneous leishmaniasis (DCL) (Alvar et al. 2012; Reithinger et al. 2007). In the former, ulcer skin lesions develop in the vicinity of the area bitten by the vector sandfly. In contrast, the latter refers to nonulcerative lesions, which develop in parts of the body other than those in close proximity to the insect bite site (Reithinger et al. 2007; Mcgwire and Satoskar 2014). DCL also appears more frequently among immunocompromised patients (Garrido-Jareño et al. 2020).

Mucocutaneous Leishmaniasis (MCL) is more consequential compared to CL as this disease produces grave symptoms. The severity of infection depends on the species responsible for causing the infection. Some of the species, such as *L. braziliensis*, can cause CL and MCL simultaneously, but this condition is rare. The transmission can occur through infected lesions on the skin to the mouth or nasal cavity, where the parasite may reside for months until the scabs are fallen. It is conclusive that MCL may occur in patients who have suffered from CL in the past. MCL develops in clinically resolved CL patients after a few months to years. Most commonly, MCL develops due to *L. braziliensis* infection, but other new world *Leishmania* species, such as *L. major*, *L. panamensis*, *L. amazonensis*, and *L. tropica*, are also capable of causing the disease. Host genetic factors play a role in the development of MCL, and the nasal cartilaginous septum is the first site to host *Leishmania* parasites. *Leishmania* parasites multiply and ulcerate the nasal septum, leading to its perforation and destruction. As the disease progresses, lymphatic fluid or blood carries the parasites to the oro-pharyngeal region and later to the upper respiratory tract (Mcgwire and Satoskar 2014; Daneshbod et al. 2011). Once the infection reaches the upper respiratory tract, the condition turns fatal. Under such circumstances, MCL patients become susceptible to other secondary infections and may fail to respond toward standard chemotherapy (Mcgwire and Satoskar 2014). The biggest share of MCL cases is in South America, while other prominent countries that report MCL are Brazil, Peru, and Bolivia.

Visceral leishmaniasis (VL), also known as kala-azar (black fever or Dumdum fever), is the most appalling of all clinical forms. Depending upon geographic diversities, the species causing VL differs. VL is caused chiefly by *L. donovani* and *L. infantum* in the Old World and New World, respectively. Where the causative species is *L. donovani*, the transmission is anthroponotic, whereas in the case of *L. infantum*, the transmission occurs in a zoonotic manner. In VL, the parasite infects the visceral organs of the body. The parasite incubates for around 2 to 6 months inside the host body before clinical symptoms, such as recurrent fever, pallor, weight loss, leucopenia, thrombocytopenia, hepato-splenomegaly, etc., start appearing (Lindoso et al. 2018; No 2016). Other symptoms include hypergammaglobulinemia and hyperpigmentation. The latter feature that leads to marked darkening of a patient's skin is characteristic of the Indian subcontinent, and the disease is nicknamed kala-azar after it (Varma and Naseem 2010). The dissemination of the parasite in the host further causes abdominal pain and facilitates sepsis condition. The dwindling of leucocytes makes the patient immunocompromised and vulnerable to secondary infections, ultimately leading to death.

VL displays posttreatment complications, such as PKDL (post kala-azar dermal leishmaniasis), known to occur as a relapse after recovery from VL. It is seen in VL caused by *L. donovani* (Zijlstra 2016) and is prominent in regions of Africa and India. In PKDL, macular, papular, and nodular rashes appear in patients who have clinically recovered from VL. The time of occurrence of PKDL after recovery from VL varies geographically. In India, it is observed to appear 2–7 years after VL, while in Sudan, it appears within 3–6 months of VL cure (Kumar et al. 2021). Macular and papular rashes were common in early PKDL, which appeared within 1 year of VL, while nodular rashes were prominent in Indian patients (Kumar et al. 2021; Zijlstra 2016). Nodular lesions bear a high parasitic load and serve as anthroponotic reservoirs of the parasite in between two epidemic cycles. Therefore, sandfly feeding in these nodular lesions are more likely to cause Leishmaniasis compared to macular lesions (Mondal et al. 2019; Selvapandiyan et al. 2019). There are 98 countries endemic to leishmaniasis among the 200 countries reporting to WHO, of which 71 countries are endemic to both VL and CL. VL is endemic to seven countries, and in 2020, 90% of the reported cases of VL were from just ten countries: Brazil, China, Ethiopia, Eritrea, India, Kenya, Somalia, South Sudan, Sudan, and Yemen (WHO 2022a, b). VL exists as a coinfection with other diseases caused by bacteria, protozoa, and viruses. Some of the known clinical outcomes have been observed as tuberculosis/VL coinfection, malaria/VL coinfection, and HIV/VL coinfection (Mohammed et al. 2016; Saini et al. 2022).

11.2 Therapy for Leishmaniasis

Effective therapy for leishmaniasis remains a major challenge in the absence of a vaccine as drug treatment often fails to eliminate the pathogen. *Leishmania* parasites can inhabit an already immunocompromised host or cause relapse, raising the need to revamp antiquated antileishmanial therapy. Among the available drugs, meglumine antimoniate and sodium stibogluconate (the antimonials) are the first line of drugs for treating all forms of leishmaniasis. The mechanism of action most likely involves the activation of the pentavalent form of the drug that acts as a prodrug and gets transformed into an active trivalent compound when administered in vivo (Frézard et al. 2009). However, antimonials are highly toxic in nature and pose numerous side effects, like nausea, vomiting, weakness, hepatotoxicity, and cardiotoxicity, with an additional limitation of parenteral (Frézard et al. 2009). Pentamidine, which once existed as a second line of treatment, is no longer used due to escalated side-effects and toxicity. It was available in two forms: pentamidine methanesulfonate and pentamidine isethionate salt (Ghorbani and Farhoudi 2017; Brindha et al. 2021). The use of pentamidine methanesulfonate has been discontinued, while pentamidine isethionate salts are used for the treatment of CL. The mechanism of action is not significant, but it has been reported that the drug enters *L. donovani* and accumulates in the mitochondria. Drug resistance is prevalent due to PRP1 (pentamidine resistance protein 1), an intracellular ABC (ATP-binding cassette) protein present in parasites that is responsible for drug resistance due to increased drug efflux and reduced uptake (Coelho et al. 2007; Brindha et al. 2021).

Amphotericin (AmB) is the second line of defense and is a broad antifungal agent repurposed for leishmaniasis. AmB is known to bind to the ergosterol and disrupts membrane integrity, resulting in cell lysis and the death of the cell. It is also administered intravenously and shows side effects, which range from moderate effects, such as fever and chills, to extreme ones, like hypokalemia, nephrotoxicity, and myocarditis. The liposomal formulation of AmB is now being administered as it minimizes serious side effects caused by AmB (Sundar and Chakravarty 2010). Miltefosine is the only approved drug that can be administered orally and is effective against VL. Miltefosine acts directly on the *Leishmania* parasite, inhibiting phosphatidylcholine and cytochrome c oxidase, causing apoptosis (Pinto-Martinez et al. 2018). It is advantageous due to its easy availability in primary care units, but its teratogenic nature restricts its use in pregnant and lactating females. Paromomycin has been repurposed from its actual actions toward treating amebiasis and varied fungal infections. It is administered intramuscularly, exhibits limited success, and is not free of cytotoxicity (Wiwanitkit 2012).

Table 11.1 summarises different commercially available antileishmanial drugs and their mechanism of action and toxicity. Apart from antimonials, none of the antileishmanial drugs has been approved originally for leishmaniasis. All of the drugs have been a result of piggyback chemotherapy or drug repurposing. In summary, the available therapies against leishmaniasis are expensive and toxic, exhibit drug resistance, and mostly require parenteral administration and hence hospital admission. Due to all these shortcomings, it is a dire need of the situation to study the parasite's biology, identify potential drug targets, and unravel novel molecules either coupled or uncoupled with drug carriers that can uplift already failing antileishmanial drug therapy.

11.3 Drug Targets

With the development of several drugs over a period, infectious agents have also evolved, with drug-resistant mechanisms being notorious against available drug molecules. Therefore, it becomes vital to investigate and discover novel drugs and

Table 11.1 Drugs available for the treatment of leishmaniasis

Drug	Original use/indication	*Leishmania species*	Mechanism of action	Toxic effects	Reference
Pentavalent antimonials	Antileishmanial	*L. donovani* *L. tropica* *L. mexicana* *L. braziliensis*	Binds to polypeptides. Defunctionalizes DNA topoisomerase & glycolytic enzymes. Hampers fatty acid beta-oxidation.	Musculoskeletal pain. Gastrointestinal disturbances. Mild to moderate headache. Elevation in liver and pancreatic enzymes.	Frézard et al. (2009)
Pentamidine	First used to treat trypanosomiasis	*L. donovani* *L. mexicana*	Interacts with trypanosomal DNA. Defunctionalizes RNA polymerase and ribosomes. Disrupts the activity of ornithine decarboxylase and thus ceases polyamine synthesis. Inhibits polypeptide and nucleic acid synthesis.	Hypotension after parenteral administration. Hypoglycemia. Nephrotoxicity causes pancreatitis in HIV-infected patients.	Bennett (2020); Aronson (2016)
AmB	Antifungal drug	*L. amazonensis* *L. tropica* *L. donovani* *L. infantum* *L. chagasi*	Binds to ergosterol in the membrane. Increases porosity of the membrane, leading to cell death.	Nausea. Fever. Hyper- or hypotension. Nephrotoxicity, which includes insufficient renal function, hypokalemia, and nephrogenic diabetes insipidus.	Sundar and Chakravarty (2010); Laniado-Laborín and Cabrales-Vargas (2009)
Miltefosine	Originally indicated for breast cancer and solid tumors	*L. donovani* *L. braziliensis* *L. guyanensis* *L. panamensis* *L. viannia*	Disrupts phosphatidylcholine synthesis. Inhibits decline in ATP synthesis and oxygen utilization by inhibiting cytochrome c oxidase. Hampers Ca^{2+} homeostasis.	Moderate gastrointestinal disturbances. Teratogenicity.	Dorlo et al. (2012); Pinto-Martinez et al. (2018)

Paromomycin	Antibiotic drug repurposed for leishmaniasis	*L. amazonensis* *L. donovani*	It is most likely suggested to alter membrane fluidity and interfere with ribosome association and mitochondrial membrane potential.	Nausea. Abdominal cramps. Diarrhea.	Chawla et al. (2011)

drug targets against different diseases. Drug target refers to the biomolecules (mostly proteins) present in the body that can bind to specific biochemical molecules (drugs) and evoke a therapeutic response. The target sites in these proteins are identified based on their unambiguity and high specificity. The potential drug target should be explicit to the parasite or nonidentical compared to the host homolog and a key requisite for the growth and survival of the parasite. For instance, a drug target could be an enzyme vital for the survival of disease-causing agents. The enzymes are specific molecules vital for various biosynthetic pathways and have molecule-binding pockets. Small therapeutic molecules can be designed to have antagonistic effects against these enzymes, or the active sites can be docked against some already known inhibitors or drugs that have shown therapeutic effects against other diseases. These enzymes (identified drug targets) should be easily assayed and available for high-throughput screening for wider fast-paced drug discovery and commercialization. The biochemical and metabolic pathways of *Leishmania* parasites are now being widely explored, and several novel drug targets have been identified and validated. Some of the critical biochemical pathways and key drug targets are discussed in the following sections.

11.3.1 Potential Drug Targets in *Leishmania* Parasite

11.3.1.1 Glycolytic Pathway

Parasites depend on the host for the carbon source. The Embden-Meyerhof-Parnas (EMP) pathway is the energy metabolism pathway crucial to both the host and the parasite. The EMP pathway is a ten-step process occurring in the cytosol in humans, dissimilar to *Leishmania*, where the first seven steps occur in the peroxisome-like organelles known as glycosomes. The glycolytic enzymes within the glycosomes in parasites are structurally divergent from that present in the mammalian host (Raj et al. 2020). The remaining three enzymes in the cytosol also vary significantly from that present in the host. As a result, among the ten enzymes of the glycolytic pathway, some enzymes have been extensively studied and categorized as putative drug targets. They are glucose-6-phosphate isomerase, glycerol-3-phosphate dehydrogenase, enolase, and aldolase.

Glucose-6-phosphate isomerase (G6P isomerase) (often referred to as phosphoglucose isomerase) is the second enzyme of the pathway responsible for converting glucose-6-phosphate into fructose-6-phosphate (Fig. 11.1). When comparing the structure of G6P isomerase in humans (PDB ID- 1JLH) and *L. mexicana* (PDB ID-1Q5O), the overall fold of *L. mexicana* and human G6P appeared to be similar, while a significant difference was observed when superimposing both structures. Superimposed images revealed the parasite's small domain to exhibit more open conformation, creating a larger active site cavity as compared to human G6P isomerase. This possibly hints at its functioning within the peroxisome-like compartment and could be a drug target site against the parasite (Cordeiro et al. 2004). However, an attempt to screen inhibitors against the G6P isomerase failed, and no significant inhibition was seen (Mota et al. 2018).

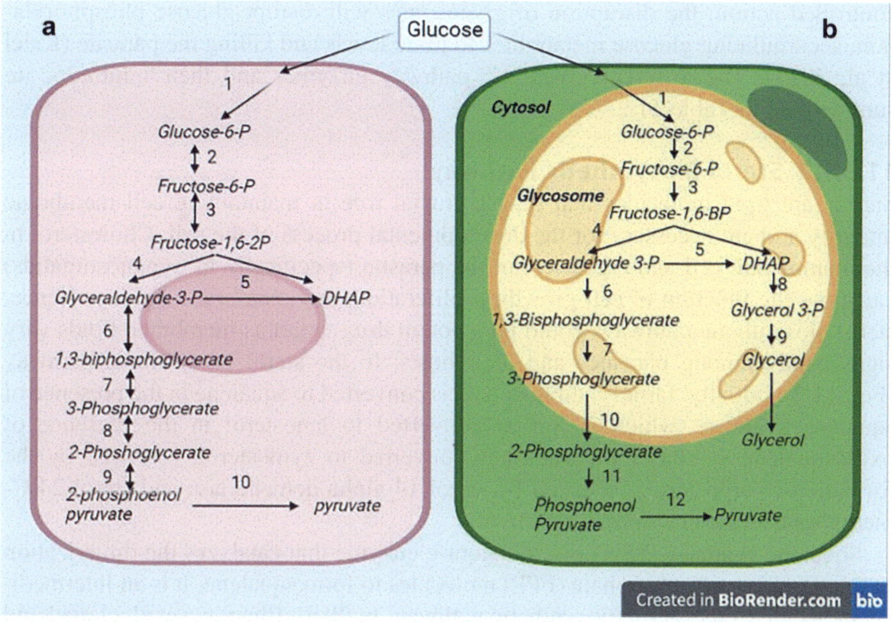

Fig. 11.1 Glycolytic pathway. (**a**) Glycolytic pathway occurring in the host: the glycolytic pathway inside the host animal cell; "P" represents phosphate; the characteristic enzymes are (1) hexokinase, (2) phosphoglucomutase, (3) phosphofructokinase, (4) aldolase, (5) triose phosphate isomerase, (6) glyceraldehyde 3-phosphate dehydrogenase, (7) phosphoglycerate kinase, (8) phosphoglyceromutase, (9) enolase, (10) pyruvate kinase. (**b**) Glycolytic pathway in *Leishmania*: the characteristic enzymes involved in process are (1) hexokinase, (2) phosphoglucomutase, (3) phosphofructokinase, (4) aldolase, (5) triose phosphate isomerase, (6) glyceraldehyde 3-phosphate dehydrogenase, (7) phosphoglycerate kinase, (8) glycerol 3-phosphate dehydrogenase, (9) glycerol-1-phosphatase, (10) phosphoglyceromutase, (11) enolase, (12) pyruvate kinase. "P" represents phosphate

Fructose-1,6-bisphosphatase is a gluconeogenic enzyme that converts fructose-1,6 bisphosphate to fructose-6-phosphate. It is present in the glycosomes or peroxisome-like organelles of *Leishmania major* and has an AMP allosteric binding site distinct to humans and hence can be a novel drug target against *L. major*, as demonstrated by Yuan et al. (2017).

Gupta et al. (2014) revealed experimentally that *L. donovani* enolase (*Ld*Eno) and *L. donovani* aldolase (*Ld*Ald) glycolytic enzymes are able to generate a Th1 immune response capable of regulating the disease. *Ld*Eno was further investigated in silico with a few compounds, among which triphosphate (PPPi)/[oxido(phosphonatooxy)-phosphoryl] phosphate showed high binding affinity, and the compound could be a potential lead for further in vitro evaluation (Vidhya et al. 2020). Other enzymes in the glycolytic pathway are required to be further explored to be established as putative drug targets. Another aspect of eliminating parasites may involve targeting proteins present on the glycosomal membrane. Glycosomes are essential for the parasite's survival as glycosomal enzymes are compartmentalized and under controlled action; the disruption of glycosomes will disrupt glucose phosphorylation, accumulating glucose metabolites to toxic levels and killing the parasite (Kalel et al. 2017). The different glycolytic pathway enzymes and their inhibitors are summarized in Table 11.2.

11.3.1.2 Sterol Biosynthetic Pathway

Sterols are lipid molecules that play a crucial role in maintaining cell membrane integrity and are necessary for the developmental process of the cell. Cholesterol in the mammalian cell and ergosterol in the parasite (specifically in trypanosomatids) carry out the function of cell growth, proliferation, and membrane rigidity. Hence, sterol biosynthetic pathways could be a potent drug target as membrane lipids vary significantly among parasites and their hosts. In the sterol biosynthetic pathway (Fig. 11.2), initially, farnesyl diphosphate is converted to squalene in the presence of squalene synthase, which is further converted to lanosterol in the presence of oxidosqualene synthase. Lanosterol is converted to zymosterol, followed by the formation of ergosterol, catalyzed by sterol 14-alpha demethylase and sterol 24-C-methyltransferase enzymes, respectively.

Squalene synthase (SSN) is a transferase enzyme that catalyzes the dimerization of two farnesyl pyrophosphate (FPP) molecules to form squalene. It is an intermediate molecule of the sterol biosynthetic pathway. In 2010, Bhargava et al. cloned and characterized *L. donovani* squalene synthase (*Ld*SSN), along with assaying zaragozic acid A, an inhibitor of mammalian SSN, to target *Ld*SSN. When measuring the activity, zaragozic acid A showed an inhibitory effect toward recombinant *Ld*SSN (Bhargava et al. 2010). Based upon these findings, zaragozic acid A warrants further investigation to understand its full antileishmanial potential.

C_{14}-demethylase (C14DM) is an intermediate enzyme of the sterol pathway that helps maintain membrane integrity. Azoles are primary inhibitors of C14DM. Antifungal agents such as itraconazole, ketoconazole, and posaconazole are among the inhibitors of C14DM (McCall et al. 2015). Other inhibitor molecules of C14DM are mentioned in the table below (Table 11.3). Sterol-24-C-methyltransferase (SMT) catalyzes the transfer of methyl group from S-adenosyl methionine to the C-24 position of the sterol side chain, thereby causing a distinguishable change in cholesterol and ergosterol synthesis. Mammalian cells lack SMT orthologs and hence are potent targets for antileishmanial drugs or vaccines (Mukherjee et al. 2019).

11.3.1.3 Trypanothione Reductase Enzyme

Trypanothione reductase belongs to the oxido-reductase class of enzymes and is a nicotinamide adenine dinucleotide phosphate (NADPH)-dependent flavoprotein. The enzyme is a homodimer and has a FAD-binding domain, an NADPH-binding domain, and an interface domain. The protein catalyzes the reduction of dithiol trypanothione (TS_2) to T $(SH)_2$ using NADPH. Trypanothione reductase structures are available for several species of *Leishmania* and *Trypanosoma*. All trypanosomatids share about 67% of the primary sequence, along with >82% sequence identity among *Leishmania* spp. and >80% identity among *Trypanosoma*

Table 11.2 Inhibitors of the glycolytic pathway in the *Leishmania* parasite

S. No.	Target enzyme	Compound/drug molecule	*Leishmania species*	Nature of action	References
1.	Enolase	Triphosphate (PPPi)/[oxido (phosphonatooxy) phosphoryl] phosphate, SF2312, FPAH	*L. donovani*	In silico prediction suggests that the compound is a potential lead for in vitro evaluation.	Vidhya et al. (2020)
2.	Glycerol-3-phosphate dehydrogenase	2′,4′-dihydroxychalcone	*L. amazonensis*	In silico study has proven that the presence of chalcone is important for enzyme inhibition.	Passalacqua et al. (2015)
3.	Pyruvate phosphate dikinase	Z220582104	*L. donovani*	In silico studies predicted compound binding to pyruvate phosphate dikinase. In silico studies were complimented with in vitro assays, which also reflected potent antileishmanial activity and low toxicity.	Kashif et al. (2019)
4.	Hexokinase, glyceraldehyde 3-phosphate dehydrogenase, phosphoglycerate kinase, and succinate dehydrogenase	3-Bromopyruvate	*L. amazonensis*	Inhibition of enzymes from the glycolytic pathway and oxidative phosphorylation.	Gomes et al. (2021)
5.	Pyruvate kinase	Zinc sulfate	*L. donovani*	In vitro study showed inhibitory activity on *Leishmania* promastigotes due to zinc metal binding sites.	Sirajuddin et al. (2023)
6.	Phosphoglycerate mutase B	Decavanadate	*L. tarentolae*	In vitro study showed decavanadate as a potent inhibitor of phosphoglycerate mutase B.	Turner et al. (2012)

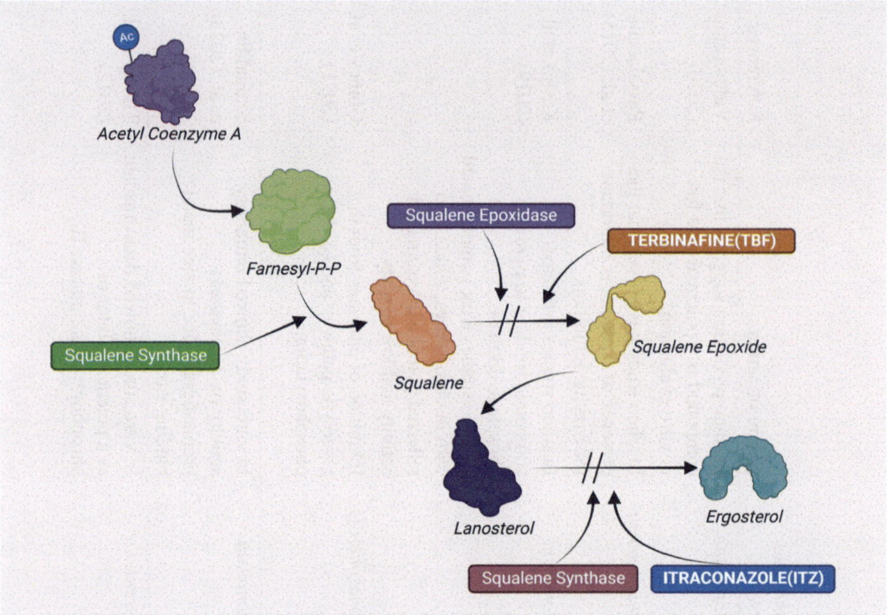

Fig. 11.2 Sterol pathway in leishmaniasis. The production of ergosterol is duly essential for the survival of the parasite, which is synthesized using the sterol biosynthetic pathway

spp., along with a similarity of 100% in the substrate-binding region (Battista et al. 2020).

The host elicits immune response against the parasite, creating an unfavorable environment by producing reactive oxygen species (ROS) to kill the protozoa. The parasite has developed defense against the host through the trypanothione reductase enzyme, which is essential and exclusive to trypanosomatids. It acts as a detoxifying agent and neutralizes the ROS produced by the host macrophages, thereby promoting the survival of the parasite (Turcano et al. 2018). Thus, the enzyme is rudimentary to *Leishmania*'s survival against host defense and a key target against the parasite.

Due to high specificity and distinguishing variability between the human homolog glutathione reductase (GR) and trypanothione reductase (TR), the enzyme poses as a pharmacological target. Though both enzymes share about 38% sequence identity, they differ in the substrate and its binding site (Battista et al. 2020). Trypanothione and glutathione are the substrate of TR and GR, respectively, and are analogous to each other. Trypanothione has a net positive charge due to spermidine moiety, whereas glutathione is negatively charged at physiological pH. Hence, the binding sites of trypanothione and glutathione are positively and negatively charged, respectively. The interaction sites in TR are Glu18, Trp21, Ser109, Tyr110, Met114, and spermidine moiety, which is conserved and specific

Table 11.3 Inhibitors of the sterol biosynthesis pathway in the *Leishmania* parasite

S. No.	Target enzyme	Drug molecule	*Leishmania species*	Nature of action	References
1.	Squalene synthase	Ancistrotanzanine B (phytochemical)	*L. donovani*	In silico analysis suggested strong-binding with lowest- binding energy –9.83 kcal/mol.	Wadanambi and Mannapperuma (2021)
2.	Squalene epoxidase	Tolnaftate (antifungal)	*L. amazonensis*, *L. braziliensis*, *L. infantum*	In vitro assay confirmed significant reduction in ergosterol.	Yamamoto et al. (2020)
3.	Sterol 14alpha-demethylase (CYP51)	Ketoconazole, itraconazole, posaconazole (antifungal)	*L. major*, *L. donovani*	In vitro suppression of parasite growth mediated via a CYP51-specific mechanism.	McCall et al. (2015)
4.	Sterol alpha-14 demethylase (LdSDM)	Zafirlukast and Avodart (asthma and prostate enlargement treatment, respectively)	*L. donovani*	In vitro assay confirmed intramacrophagic amastigote inhibition by Avodart.	Tabrez et al. (2021a, b)
5.	Sterol methyltransferase (SMT)	Simeprevir (antiviral drug)	*L. donovani*	In silico analysis confirmed significant binding to SMT and inhibited the growth of promastigotes, inducing ROS generation in vitro.	Tabrez et al. (2021a, b)
6.	*Ld*SMT	Gigantol, flavan-3-ol, and parthenolide	*L. donovani*	In silico study predicted gigantol as the most potent inhibitor among other compounds. Inhibition of promastigote and amastigote growth in vitro. Generation of ROS.	Rahman et al. (2021)
7.	Sterol content specifically (C-24 methyltransferase)	Aglycone tomatidine (tomato spp.)	*L. amazonensis*	In vitro inhibition of parasite growth. Alterations in sterol content. Disruption of mitochondrial membrane potential. Reduction in membrane permeability, leading to reduced cell death.	Medina et al. (2012)

(continued)

Table 11.3 (continued)

S. No.	Target enzyme	Drug molecule	*Leishmania species*	Nature of action	References
8.	C-24 methyltransferase	Imipramine (antidepressant) and miconazole (antifungal) combination	*L. amazonensis*	Antipromastigote activity in culture. In vitro suppression of amastigote growth.	Andrade-Neto et al. (2016)

to TR only and not GR (Battista et al. 2020). Therefore, the enzyme is a potential drug candidate against parasite infection and is extensively studied (Table 11.4).

Maamri et al. (2021) investigated the potential of two plant-derived compounds, 3-methoxycarpachromene and masticadienonic acid, using in silico tools, along with ADMET prediction, and proposed them to be therapeutically active against the *Leishmania* parasite, having high oral viability and low toxicity. In 2018, Turcano et al. screened about 120,000 compounds as TR inhibitors. Among these, one compound was identified to bind at the catalytic site of NADPH. The binding site was unique, and an in vitro validation of the compound revealed a dose-dependent antiproliferative effect on *L. infantum* promastigotes, validating the in silico prediction. In 2012, trypanothione reductase was targeted by an antiarthritic drug named gold(I)-coated auranofin. As TR has dithiol motifs present on its active site and gold is also highly thiophilic in nature, gold-coated auranofin is expected to inhibit the action of TR. Ilari et al. (2012) established the inhibitory effect of gold(I)-coated auranofin through enzymatic assay against the TR, along with killing actions against the promastigote form of parasites.

11.3.1.4 Polyamines Biosynthetic Pathway

Polyamines are essential for all cells, including protozoans, and assist the differentiation of *Leishmania* from promastigotes to the amastigote form. L-arginine is an essential amino acid for *Leishmania* and is scavenged from the host cytoplasm with the assistance of AAP3 (amino acid permease 3). L-arginine is converted into ornithine by L-arginase. Ornithine decarboxylase (ODC) enzyme catalyzes the synthesis of putrescine from ornithine. Putrescine transforms into spermidine, which further produces spermine aided by spermidine synthase and spermine synthase, respectively. S-adenosylmethionine decarboxylases present in the host convert S-adenosylmethionine to decarboxylated S-adenosylmethionine. This acts as a substrate for spermidine synthase as well as spermine synthase to produce spermidine and spermine. Spermidine is catalyzed by trypanothione synthase to yield trypanothione (Fig. 11.3a). It neutralizes the ROS produced by the host (Aoki et al. 2018). L-arginine uptake enzyme AAP3 and other enzymes, such as L-arginase, ODC, spermidine synthase, S-adenosylmethionine decarboxylase, spermine synthase, and trypanothione synthase in this pathway and are considered to be promising drug targets that can regulate *Leishmania* infection as their pathway, biosynthesis and utilization are distinct to human host (Fig. 11.3b).

Polyamines enable the growth, differentiation, and survival of parasitic protozoa. Arginase, a metallo-hydrolase, is the first enzyme of the polyamine biosynthetic pathway. Four inhibitors have been reported against LmARG (*Leishmania mexicana* arginase), 2(S)-amino-6-boronohexanoic acid (ABH), *S*-(2-boronoethyl)-l-cysteine (BEC), Nω-hydroxy-l-arginine (NOHA), and the hydroxylated nor-Nω-hydroxy-l-arginine (nor-NOHA). Among these two drugs, ABH and nor-NOHA were observed to be the most potent inhibitors of arginase (D'Antonio et al. 2013). In vitro analysis has shown that ABH and nor-NOHA reduce concentrations of putrescine, spermidine, and trypanothione levels and inhibit the growth of promastigotes.

Table 11.4 Inhibitors of *Leishmania* trypanothione reductase

S. No.	Compound/drug	Target enzyme	*Leishmania* spp.	Nature of action	References
1.	Chalcone-based compound (chalcone 6)	TR	*L. major*, *L. tropica*, *L. infantum*	Inhibition of promastigote growth and low toxicity to host cells.	Ortalli et al. (2018)
2.	Epigallocatechin 3-O-gallate	TR	*L. infantum*	In silico prediction unveiled binding to TR. Inhibition of promastigote proliferation in vitro.	Inacio et al. (2021)
3.	Water-soluble ferrocenylquinoline-based preclinical drug	TR	*L. donovani*	In vitro analysis exhibited activity against amastigotes, and in silico confirmed strong ferrocenylquinoline binding to TR for leishmanicidal activity.	Mukherjee et al. (2020)
4.	6-chloro-3-[3-(3-hydroxy-2-piperidyl)-2-oxo-propyl]-7-(4-pyridyl) quinazolin-4-one	TR	*L. donovani*	In silico analysis revealed high binding affinity to TR.	Pandey et al. (2017)
5.	Nitrothiophene-2-carboxamides 6e and 8	TR	*Leishmania* spp.	In vitro leishmanicidal activity.	Battista et al. (2022)
6.	ZINC12151998	TR	*L. mexicana*	Computational analysis showed inhibition activity, and in vitro assay proved leishmanicidal property.	Matadamas-Martínez et al. (2019)
7.	Mesoionic 1,3,4-thiadiazolium-2-aminide derivatives MI-4-NO2	TR	*L. infantum*, *L. braziliensis*, *L. amazonensis*	In silico studies showed noncompetitive inhibition. Mesoionic compound was most effective against *L. infantum*.	Rodrigues et al. (2012)
8.	4-((1-(4-ethylphenyl)-2-methyl-5-(4-(methylthio)phenyl)-1*H*-pyrrol-3-yl) methyl)thiomorpholine (compound 1)	TR	*L. infantum*	In silico study showed competitive inhibition.	Baiocco et al. (2013)
9.	Imidazole-phenyl-thiazole, compounds 3e and 3f	TR	*L. infantum*	In vitro analysis showed enzyme inhibition. Both extracellular and intracellular parasite killing.	Revuelto et al. (2019)

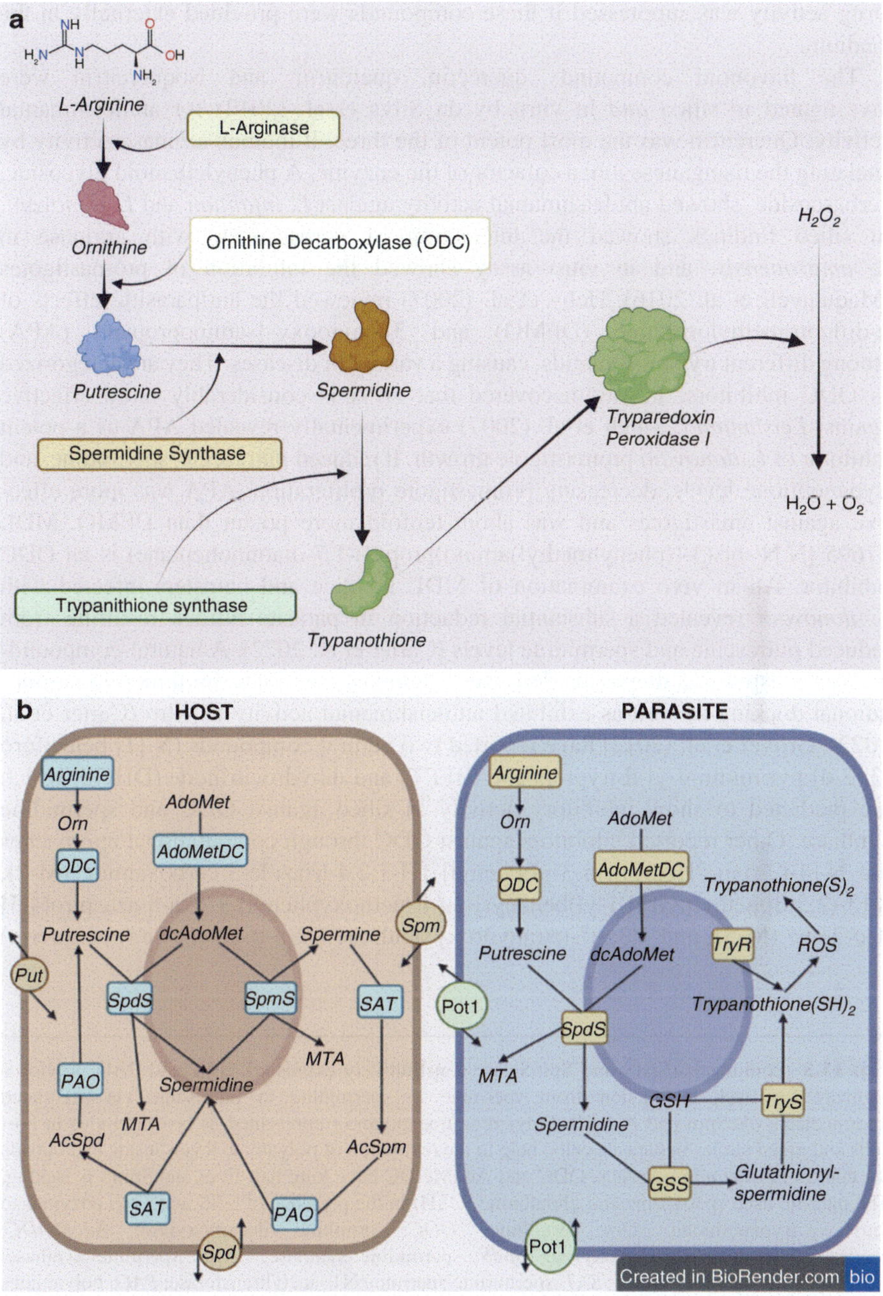

Fig. 11.3 Transport biosynthesis and utilization of polyamines. (**a**) Trypanothione synthesis pathway: trypanothione is exclusive to trypanosomatids. It facilitates the neutralization of hydrogen peroxide by macrophages. It is synthesized by trypanothione synthase and then reduced by trypanothione reductase. (**b**) Trypanothione utilization in host cells and parasites: arginase is essential for the synthesis of polyamine precursor ornithine (Orn). In human cells, ODC and *Ado*MetDC (decarboxylases) are highly regulated, having shortened half-lives, whereas synthases

Drug activity was suppressed if these compounds were provided externally in the medium.

The flavonoid compounds quercetin, quercitrin, and isoquercitrin were investigated in silico *and* in vitro by da Silva et al. (2012) for antileishmanial activity. Quercitrin was the most potent of the three. It inhibits arginase activity by chelating the manganese ion, a cofactor of the enzyme. A phenylethanoid glycoside, verbascoside, showed antileishmanial activity against *L. infantum* and *L. donovani*. In silico findings showed the interaction of verbascoside with arginase in *L. amazonensis*, and in vitro assay showed the inhibition of promastigotes (Maquiaveli et al. 2016). Heby et al. (2007) reviewed the antiparasitic effects of α-difluoromethylornithine (DFMO) and 3-aminooxy-1-aminopropane (APA) among different trypanosomatids, causing a variety of diseases. They are categorized as ODC inhibitors. It was discovered that APA is considerably more effective against *Leishmania*. Singh et al. (2007) experimentally revealed APA as a potent inhibitor of *L. donovani* promastigote growth. It reduced putrescine, spermidine, and trypanothione levels, decreasing promastigote proliferation. APA was more effective against amastigotes and was about tenfold more potent than DFMO. MDL 27695 (N,N′-bis(3-((phenylmethyl)amino)propyl)-1,7-diaminoheptane) is an ODC inhibitor. An in vivo examination of MDL in mice and hamsters infected with *L. donovani* revealed a substantial reduction in parasite burden resulting from reduced putrescine and spermidine levels (Carter et al. 2022). A natural-compound-derived diospyrin, a diepoxide derivative, demonstrated ODC inhibition in computational docking as well as exhibited antileishmanial activity in vitro (Carter et al. 2022). Grover et al. (2012) have reported two natural compounds (N-[**1**] benzofuro [3, 2-d] pyrimidin-4-yl-d-tryptophane (BFPT) and dihydrocitrinone (DHC)), which are predicted to show inhibitory activity in silico against ODC and spermidine synthase. Other reported inhibitors against ODC through computational approaches are N-[4-(2-oxo-2Hchromen-3-yl)phenyl]-1H-1,2,4-triazole-3-carboxamide (M-2); 8-[3-(2,5-dimethylpyrrol-1-yl)benzoyl]-3-(4-methoxyphenyl)-1-oxa-8-azaspiro[4.5] dec-2-ene (M-5); and 1,3,6,7-tetrahydroxyxanthone C2-β-D-glucoside (mangiferin)

Fig. 11.3 (continued) (SpdS and SpmS) are constitutively expressed. SAT and PAO provide a pathway for back conversion from spermine to spermidine to putrescine via acetylated intermediates (AcSpm and AcSpd). Polyamine transporters (represented as brown circles in host cells and green circles in parasite cells) help in the regulation of polyamine levels inside and outside the cell. In *Leishmania* parasites, ODC and AdoMetDC have long half-lives and SpmS is lacking. The parasite used spermidine and glutathione (GSH) in the presence of GSS and TryS enzymes to retrieve trypanothione. *Orn*: ornithine; *ODC*: ornithine decarboxylase; *AdoMetDC*: S-adenosylmethionine decarboxylase; *SpdS*: spermidine synthase; *SpmS*: spermine synthase; *POT1*: polyamine transporter 1; *SAT*: spermidine/spermine N1 -acetyltransferase; *PAO*: polyamines oxidase; *AcSpm*: acetylated spermine; *AcSpd*: acetylated spermidine; *GSS* glutathione spermidine synthase; *TryS*: trypanothione synthase; *TryR*: trypanothione reductase; *Put*: putrescine; *GSH*: glutathione; *dcAdoMet*: decarboxylated S-adenosylmethionine; *MTA*: 5′-deoxy-5′--methylthioadenosine; *ROS*: reactive oxygen species (adapted and modified from Heby et al. 2007) make sure this reference is included in reference section as well

(Das et al. 2016; Carter et al. 2022). Anthraquinone hypericin is a natural compound derived from *Hypericum perforatum*, which demonstrated a favorable in silico score and, on further investigation, showed reduced spermidine and trypanothione levels, inhibiting the growth of promastigotes.

11.3.1.5 Fe-Super Oxide Dismutase (Fe-SODA)

Reactive oxygen species (ROS), by-products of metabolic reactions, are produced by macrophages in response to parasite phagocytosis and assist in clearing the parasite. The superoxide dismutase enzyme converts superoxide (O_2^-) to hydrogen peroxide (H_2O_2), which helps in cell differentiation. A study by Mittra et al. (2013) showed the necessity of iron uptake for activation of SOD. SOD is further responsible for enhancement of parasite virulence and differentiation of promastigotes to amastigotes. Getachew and Gedamu (2012) investigated and discovered that in *L. donovani*, *Ld*FeSODA protects the parasite from oxidative stress and miltefosine-induced cytotoxicity. *Ld*FeSODA also played a crucial role in miltefosine-resistant strains of parasites (Veronica et al. 2019). Hence, *Ld*FeSODA can be an effective and promising novel drug target to cure VL and other clinical forms of leishmaniasis. Mannich base-type compounds reportedly show high selectivity in amastigotes affecting the parasite infectivity rate and are more active than glucantime. These compounds alter glucose catabolism, leading to greater levels of iron superoxide dismutase inhibition (Martin-Montes et al. 2017).

11.3.1.6 N-myristoyltransferase (NMT)

The human genome project opened the gates and paved the way for understanding each organism's genome. The advancement in genome sequencing provided an insightful view of the genetic makeup of the organism and helped interpret the specific genes or proteins determining novel drug targets. One such enzymatic target is NMT, which catalyzes the posttranslational addition of myristic acid to the N-terminal glycine of a specific protein (Fig. 11.4). NMT is a validated drug target in the *Leishmania* parasite. It is presumed essential for the correct cellular localization of modified proteins and their functioning.

Rackham et al. (2015) computationally screened several compounds and revealed the action of a specific compound as most potent against LdNMT. The IMP-105 compound was reported to be most effective and selective toward *Leishmania* NMT over human NMT in computational screening (Bell et al. 2020). The curzerene compound, derived from *Eugenia uniflora* present in essential oil, showed remarkable binding with NMT, as analyzed using in silico tools (Nunes et al. 2021). The group further tested the compound in vitro and demonstrated noteworthy antipromastigote and antiamastigote activity with low macrophagic toxicity (Nunes et al. 2021).

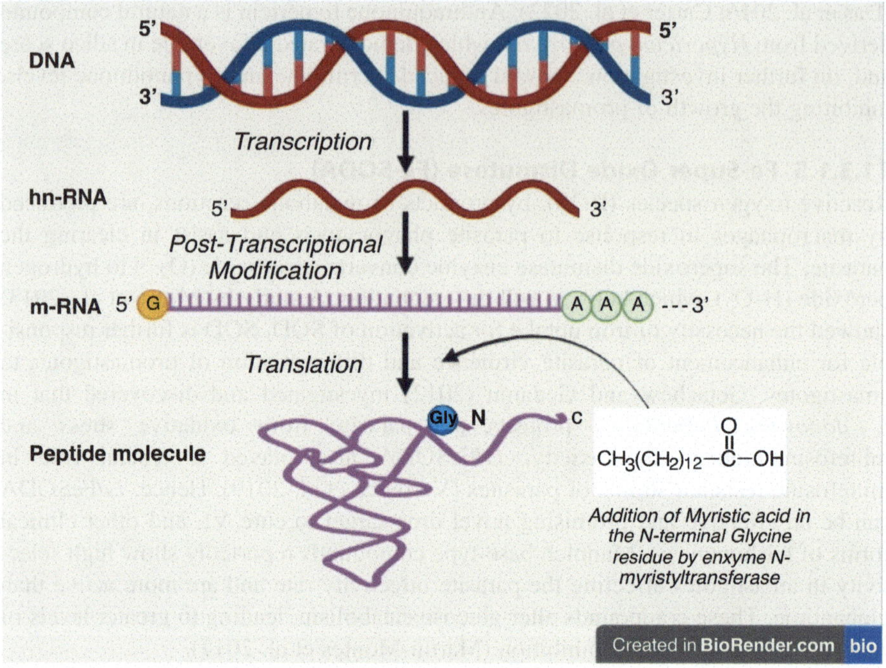

Fig. 11.4 Activity of N-myristoyltransferase enzyme. NMT catalyzes the addition of myristic acid in the N-terminal of newly synthesized polypeptide at glycine residue

11.3.1.7 Folate Biosynthetic Pathway

Folate, also known as vitamin B9 (folic acid), consists of three moieties: pterin, *p*ABA (p-amino benzoic acid), and glutamate. Eukaryotes are unable to synthesize folic acid de novo. Folate is provided as an essential dietary supplement. Its uptake is mediated by membrane transport protein. The DHFR (dihydrofolate reductase) enzyme catalyzes the conversion of folate to dihydrofolate (H2F), followed by a further reduction to H4F (tetrahydrofolate). Another enzyme, PTR1 (pteridine reductase 1), also reduces folate to H2F and H4F (Vickers and Beverley 2011). Tetrahydrofolate is essential for the biosynthesis of purines, thymidylate, pantothenate, RNA, etc. (Bertacine Dias et al. 2018). DHFR is an essential enzyme of the folate biosynthetic pathway and thymidine synthesis and is thus classified as a validated drug target (Bertacine Dias et al. 2018, Sharma and Bharatam 2021).

Several studies have revealed compounds as potential drug candidates against DHFR-TS and PTR1. Methotrexate, cycloguanil, and trimethoprim are established therapeutics that inhibit the DHFR (Hendrickx et al. 2019). Sharma and Bharatam (2021) screened compounds in silico against *Ld*DHFR (*L. donovani* DHFR) and discovered that ZINC57774418 (Z18), ZINC69844431 (Z31), ZINC71746025 (Z25), and D11596 (DB96) are four possible leads against *Ld*DHFR based on in silico interaction and stability in the binding of ligand-substrate. *Leishmania* has a bifunctional enzyme (DHFR-thymidylate synthase) DHFR-TS distinct to humans where DHFR and TS are separate (das Neves et al. 2019).

Using in vitro and in silico techniques, Hermann et al. (2017) discovered 15 natural compounds with strong antileishmanial activity. Sophoraflavanone G was discovered to be the most efficacious of all. The 2,4-diaminoquinazoline derivative is a particular inhibitor of *L. chagasi* PRT1, whereas the 2,4-diaminopyrimidine derivatives substituted at the sixth position were shown to be competitive inhibitors of both enzymes (*Lc*PTR1) (Teixeira et al. 2019). Other dual inhibitors of DHFR and PRT1 reported are 2-(4-((4-nitrobenzyl)oxy)phenyl)-1H-benzo[d]imidazole and 2-(4-((2,4-dichlorobenzyl)oxy)phenyl)-1H-benzo[d]oxazolemolecules (Kapil et al. 2019) (Teixeira et al. 2019).

11.4 Advances in Antileishmanial Chemotherapy

11.4.1 Drug Repurposing

Therapeutic switching, referred to as drug repositioning/repurposing, drug retasking, drug reprofiling, drug rescuing, drug recycling, or drug redirection, is an alternative method of discovery for novel drugs (Shakya et al. 2011). This methodology is cost-effective, has a decreased time frame, and is much less vulnerable to failure. In the context of leishmaniasis, all available antileishmanial drugs, except for antimonials, are a fruit of drug repurposing. As stated earlier, AmB was originally developed as an antifungal agent. Paromomycin was an antibiotic developed during the mid-twentieth century, while miltefosine is a known anticancer agent.

Studies have revealed that kinases such as cyclin-dependent kinases (CDKs), referred to as CRKs (CDC2-related kinases), in *the Leishmania* parasite and mitogen-activated kinases (MAPK) are essential for cell viability and proliferation (Efstathiou and Smirlis 2021). The US FDA-approved anticancer protein kinase inhibitor molecules, sunitinib, sorafenib, and lapatinib, have displayed remarkable antileishmanial activity (Sanderson et al. 2014). The in vivo studies exhibited activity against *L. donovani* amastigotes, showing potency similar to miltefosine and no toxic effects on human cells (Sanderson et al. 2014). Raloxifene, an estrogen receptor modulator developed as an anticancer drug for breast tumors, is reported for antileishmanial activity. In vitro and in vivo analyses against various species has demonstrated that raloxifene is effective against promastigotes as well as intracellular amastigotes (Reimão et al. 2014). It was found to cause functional disorder in the plasma membrane of *L. amazonensis* promastigotes and leads to the disruption of the mitochondrial membrane, leading to cell death (Reimão et al. 2014). Therefore, therapeutic switching could be a promising approach toward the treatment of leishmaniasis and may change the landscape of antileishmanial drug discovery. Various studies demonstrating the use of therapeutic drugs for leishmaniasis are listed in Table 11.5.

Table 11.5 Therapeutic switching compounds

S. No.	Drug	Original use/indication	Leishmania species	Mechanism of action	Reference
1.	Fluconazole, posaconazole, and itraconazole	Antifungal drugs	*L. donovani*	In vitro study showed the inhibition of lanosterol 14-a-demethylase.	Braga (2019)
2.	Simeprevir	Antiviral drug	*L. donovani*	In vitro study showed antipromastigote activity.	Tabrez et al. (2021a, b)
3.	Butanefine (B-SNEDDS)	Antifungal	*L. infantum*	Demonstrated lower cytotoxicity and higher efficacy against amastigotes & promastigotes.	Bezerra-Souza et al. (2019)
4.	Glyburide	Antifungal	*L. donovani*	Binds to LdTrySyn, inhibiting its activity and thereby restricting parasite growth.	Rub et al. (2019)
5.	Triphenylethylene	Multiple pharmacological purposes	*L. major*, *L. donovani*, and *L. braziliensis*	High consistent activity against a wide variety of *Leishmania* spp.	Montoya and Krysan (2018)
6.	Suramin	Antitrypanosomatid drug	*L. donovani*	In vivo administration of suramin decreased parasite load on BALB/c mice and elevated host immune responses.	Khanra et al. (2020)
7.	Spiro-indolone NITD609 E	Antimalarial drug	*L. infantum*	In vitro studies showed the killing of promastigotes and amastigotes without having much cytotoxicity. Drug also affects the sterol pathway.	Scala et al. (2014)
8.	Spiro[indole-3,3′-pyrrolizidine]-2-one	Antimalarial drug	*L. donovani*	Obstructs drug-DNA-enzyme covalent complex formation, killing both wild-type and drug-resistant parasite strains in vitro. In vivo studies displayed potent antileishmanial activity in BALB/c mice by the inhibition of bi-subunit topoisomerase-IB	Saha et al. (2016)

9.	JS87	Antimalarial drug	*L. braziliensis*	Acts through the alteration of parasite internal regulation. Also affects sterols.	Leañez et al. (2019)
10.	Aryl quinoline	Antifungal drug	*L. braziliensis*	Disrupts the mitochondrial electrochemical potential, thereby blocking the ergosterol pathway and thereby killing the promastigotes of the parasite	Chanquia et al. (2019)
11.	6-ethyl-2-phenylquinoline	Antifungal drug	*L. braziliensis*	Disrupts parasite bioenergetics by the inhibition of acidocalcisomes and the modification of the ergosterol biosynthetic pathway	Bompart et al. (2013)
12..	Fenarimol	Herbicide	*L. major*	In silico analysis done by the docking of the compound against the parasite's catalytic binding site. 4-lanosterol demethylase revealed antileishmanial action of the compound.	Zeiman et al. (2008)
	Glycyrrhizic acid	Plant compound	*L. donovani*	Mechanism involves PGE2 inhibition and NO production for restricting parasite growth.	Richard (2021)
	Methotrexate	Anticancerous drug	*L. donovani*	In silico analysis of docking with compound (5-(3-(octyloxy)-benzyl) pyrimidine-2,4-diamine) against LdDHFR enzyme showed the inhibition of LdDHFR functionality and conformational changes in the folate biosynthesis pathway.	Sharma and Bharatam (2021)
	Sophoraflavanone G	Anti-inflammatory and antimicrobial drug	*L. donovani*	Works by inhibiting the folate biosynthetic pathway (in silico analysis).	Herrmann et al. (2017)
	RDS 777	Functional against respiratory tract infection Antimalarial and antidengue drug	*L. donovani*	In silico analysis showed the binding of the compound to the catalytic site of TR, thereby restricting the expression of T[SH]2 required for energy production in the parasite.	Saccoliti et al. (2017)

(continued)

Table 11.5 (continued)

S. No.	Drug	Original use/indication	Leishmania species	Mechanism of action	Reference
	Ammonium trichloro [1,2-ethanediolato-O,O′]-tellurate	Immunomodulatory drug	*L. donovani*	Forms thiol bonds with the cysteine residues of TR in promastigotes, causing death via ROS-mediated programmed cell death. Also blocks the integrin-dependent PI3K/Akt pathway, activating host MAPK's instigating Th1 response.	Vishwakarma et al. (2018)
	Thiosemicarbazones	Antibacterial drug	*L. donovani*	In silico analysis revealed tying up of the compound with TR, thereby inhibiting the trypanothione reductase pathway.	de Aquino et al. (2022)
	Masticadienonic acid	Reported for antitumor activity	*L. infantum*	Binds in the active site of TR, thereby inhibiting the TR pathway and, in turn, parasite inactivation.	Maamri et al. (2022)
	β-sitosterol	Chloroform extract of *Corchorus capsulari*	*L. donovani*	Injects apoptotic activity in the parasite by ROS regulation and also stops the TR pathway, thereby causing parasite inactivation.	Pramanik et al. (2020)

11.4.2 Use of Drug Delivery Vehicles

Drug delivery vehicles enable the formulation, manufacturing, storage, and transport of a pharmaceutical compound to a desired site inside the host to achieve a therapeutic outcome. Drug delivery vehicles act as packaging agents that safeguard drugs and ensure they travel smoothly to their target site inside the body. They alter the pharmacokinetic and pharmacodynamic properties of drugs they carry and considerably improve their therapeutic efficacy while reducing toxicity. Commonly, micelles, liposomes, and nanoparticles have been used as drug delivery vehicles, and out of these, remarkable advances have been made to develop nano drugs against leishmaniasis. Typically, materials ranging from 1 to 100 nm in size are classified as nanoparticles.

Nanoparticles can vary in nature based on their starting material and can be categorized as metal-based nanoparticles, polymeric nanoparticles, nanoemulsions, solid lipid nanoparticles, etc. (Ribeiro et al. 2014). Metal nanoparticles are submicron-scale entities that are synthesized from pure metals or their compounds. Metallic nanoparticles exhibit properties such as high surface area, mechanical strength, magnetic properties, optical properties, and a low melting point (Baig et al. 2021). Particles that carry a polymer core are classified as polymeric nanoparticles. The biggest advantage of polymeric nanoparticles is their biodegradable, low-toxic nature, and they are evidently the drug delivery vehicles of choice (Zielińska et al. 2020). On the other hand, nanoemulsions are a stable clear mixture of two immiscible phases with a droplet size ranging between 20 and 500 nm (Gupta 2020). Nanoemulsions are sturdy and can easily deliver hydrophobic pharmaceutical agents (Usama Ashhar et al. 2022; Gupta 2020). Solid-lipid nanoparticles (SLN) are submicron-sized vesicles made up of solid lipids (Scioli Montoto et al. 2020). SLN possesses a large surface area, presents a high drug-loading capacity, and is receiving major attention for the intravenous delivery of drugs (Newton and Kaur 2019; Mukherjee et al. 2009). A wide range of metallic nanoparticles, polymeric nanoparticles, and SLN explored for the treatment of leishmaniasis are summarized in Table 11.6.

Liposomes are other drug-delivery vesicles that have been researched and are in current commercial use for leishmaniasis. A liposome is a spherical vesicle containing one or more lipid bilayers. It is used as a transport method for drugs to effectively facilitate absorption directly through the mouth or by preventing degradation by gastric acid. They have the ability to encapsulate both hydrophobic and hydrophilic active substances in the lipid layer or in the aqueous center. The size of these spheres, like lipid vesicles, can vary. Liposomes used for medical purposes are about 50–450 nm in size (Bozzuto and Molinari 2015). These are well-studied nanocarriers for the intended therapy of macrophages, making them a very successful delivery technique for advanced antileishmanial treatment. Liposomal AmB was developed to overcome the massive side effects caused by AmB and has been in use for more than 20 years. Liposomal AmB is a drug of choice in cases when the host is immunocompromised, especially in cases where there is coinfection of *Leishmania* with HIV (Sundar and Chakravarty 2010).

Table 11.6 Nanoformulations with antileishmanial activity

Nanoparticle system	Nanoparticle type	Mechanism of action	Reference
Polymeric	Rifampicin & pentamidine-coloaded polymeric nanoparticle (RFP-PTM-PNP)	The composition helped in the sustained release of rifampicin and pentamidine drugs. In vivo analysis showed higher skin penetration of the composition, showing reduced or no skin irritant, along with higher macrophage uptake and increased promastigote inhibition.	Khan et al. (2023)
	Chalcone (CH8)-loaded microparticles with modified polymeric matrix (PLGA-PVP)	New formulations increase the in situ release of drug and allows longer retention and antileishmanial effect for 60 days. In vitro analysis in BALB/c mice showed 98% efficacy for CH8/ PLGA-PVP, reducing the parasite load.	Sousa-Batista et al. (2022)
	Amphotericin B-pentamidine loaded-niosomes (AmB-PTM-NIO)	The topical drug composition showed prolonged release with increased skin penetration and higher inhibitory effect. The composition exhibited to be safe, displaying no skin irritation. The phagocytosis results in the release of AmB and PTM within the cell, triggering ROS generation and mitochondrial damage, leading to the cell death of the parasite.	Anjum et al. (2023)
	Ivermectin plus polymeric micelles (IVE/M)	In vivo analysis in BALB/c mice infected with *L. amazonensis*. The composition elicited immune response showing increased IFN-γ and IL-12.	Freitas et al. (2023)

(continued)

Table 11.6 (continued)

Nanoparticle system	Nanoparticle type	Mechanism of action	Reference
		Significant reduction in lesion size as well as parasitic load in the spleen and liver were observed.	
	PEGylated liposomal AmB (PEG-Lip-AmB)	Increased the efficacy of AmB against *L. major* in vitro. In vivo analysis shows reduction in parasitic burden, along with reduced lesion size. In vivo results demonstrated reduced renal toxicity of AmB. AmB effect increased against both promastigotes and amastigotes (by 2.7-folds).	Boshrouyeh et al. (2023)
	Imiquimod and terbinafine-coloaded mannosylated transethosomes (IMQ-TER-MTES)	In vivo analysis in BALB/c mice with the gel composition was performed. It showed a higher selectivity and inhibitory effect toward *L. major* amastigotes. It increased the healing potential of CL, reducing the lesion size significantly.	Jamshaid et al. (2023)
	Carvacrol-containing hydrogels obtained from polymer P407/P188	The hydrogel composition of carvacrol showed increased selectivity toward parasite-infected macrophages as well as reduced toxicity of carvacrol toward mammalian cells. Showed increased antileishmanial activity against both promastigote and amastigote forms of *L. amazonensis*.	Costa et al. (2023)

(continued)

Table 11.6 (continued)

Nanoparticle system	Nanoparticle type	Mechanism of action	Reference
	Miltefosine-loaded chitosan NPs (MCNPs)	The in vitro analysis manifested significant antileishmanial activity for both promastigotes and axenic amastigotes with less hemolytic activity.	Khokhar et al. (2022)
	Rutin-loaded PLGA NPs	The composition has been effective against the promastigote form of *L. infantum* during the in vitro assay, showing leishmanicidal activity.	Çetin and Ateş (2022)
	Carboxymethyl chitosan-loaded AmB	The conjugation enhanced hydrophilicity and increased the bioavailability of AmB. The nanoformulation showed a 97% inhibition of amastigotes in vitro, whereas in vivo studies demonstrated a reduction in liver parasitic burden with no toxic effects in oral administration.	Singh et al. (2022)
	Sorafenib-loaded polycaprolactone (PCL) and cellulose acetate (CA)	The formulation hybrid showed increased antileishmanial activity in vivo, demonstrating lesion healing and a reduction in parasitic burden, which are efficient against CL.	Alemomen et al. (2023)
	AmB-loaded chitosan NPs	High effectivity against both promastigotes and amastigotes was observed for the in vitro analysis of *L. tropica*.	Sohail et al. (2021)
	Risedronate-Eudragit E NPs	The in vivo assay of the composed hydrogel reduced lesion size in BALB/c mice infected with *L. amazonensis* with no necrosis effect. Bisphosphonates (risedronate) showed	Peralta et al. (2021)

(continued)

Table 11.6 (continued)

Nanoparticle system	Nanoparticle type	Mechanism of action	Reference
		increased antileishmanial efficacy with the Eudragit polymer with high parasitic inhibition.	
	Miltefosine-loaded transfersomes	Higher efficacy of the formulation was observed in vitro against the leishmanial parasite with significantly lower IC50 and in higher uptake of vesicles by macrophages. In vivo assay in BALB/c mice showed a significant drop in parasite load.	Batool et al. (2021)
	Rifampin-vancomycin-coloaded transfersomes	The in vitro study exhibited increased leishmanicidal activity against both promastigotes and amastigotes, while the in vivo assay showed a significant reduction in lesion size. Ex vivo analysis revealed that the permeation of drug improved with the formulation.	Salim et al. (2020)
	Paromomycin-loaded mannosylated chitosan NPs	The formulation increased the cellular uptake and reduced the parasite burden significantly in THP-1 cells, affecting both the stages of parasite, specifically amastigotes with minimal cytotoxicity at a high concentration of the formulation.	Esfandiari et al. (2019)
	Chitosan and its derivatives	High-molecular-weight chitosan shown to be most effective against *L. major* and *L. mexicana* promastigotes and	Loiseau et al. (2020)

(continued)

Table 11.6 (continued)

Nanoparticle system	Nanoparticle type	Mechanism of action	Reference
		amastigotes in vitro. Antileishmanial activity is pH dependent with maximum activity observed at pH 6.5. Chitosan-activated NO and ROS but antileishmanial activity not mediated by these molecules. Antileishmanial activity reported as a result of the uptake of chitosan by pinocytosis and further accumulation in parasitophorous vacuole of *Leishmania*-infected macrophages.	
	Low-molecular-weight chitosan	Inhibition of the growth of *L. major* promastigotes. *Reduction* in the dermal lesions of *L. major*-infected BALB/cmice.	Esboei et al. (2018)
	Pentamidine conjugated with hyaluronic acid	Antipromastigote effect. Antiamastigote activity against *L. major* amastigotes internalized in macrophages. No toxicity toward primary macrophages.	Micale et al. (2015)
	Nigella sativa oil-entrapped poly-ε-caprolactone nanoparticles	In vitro inhibition of *L. infantum* promastigotes and intramacrophagic amastigotes. Upregulation of NO production.	Abamor et al. (2018)
	AmB-loaded polycaprolactone (PCL) nanoparticles	IC_{50} of nanoformulation significantly lower in comparison to free drug against *L. tropica* KWH23 and *L. donovani* amastigotes.	Saqib et al. (2020)
	Mesosporous Silica encapsulated Pentamidine nanoparticles	MSN-COOH NP demonstrated controlled	Peretti et al. (2018)

(continued)

Table 11.6 (continued)

Nanoparticle system	Nanoparticle type	Mechanism of action	Reference
		release of PTM and reduced nephrotoxicity.	
	PLGA-nanotized curcumin and miltefosine	Synergistic effect of drugs was observed in nanoconjugated form. Potent antipromastigote and antiamastigote. Enhanced reduction of parasite load in vivo. Immunomodulatory effect was also observed in terms of lymphocyte proliferation and the production of RNS/ROS species.	Tiwari et al. (2017)
	G8-loaded PMMA nanoparticles	Decreased IC_{50} with no toxicity on human RBCs.	Feuser et al. (2019)
	Oleanolic acid (OA)-loaded PLGA-TPGS nanoparticles	Increased solubility and bioavailability of drugs. Effective against drug-resistant strains. In vivo study of *L. donovani* shows that the amastigote load in the spleen of BALB/c mice was decreased.	Ghosh et al. (2016)
	PLGA-PEG-encapsulated miltefosine	Decreased dose and enhanced activity. Selective targeting. Inhibited amastigotes in splenic tissue is more traditional than miltefosine and AmB.	Kumar et al. (2016)
	Mannose-anchored thiolated-chitosan-loaded AmB nanocarriers	Induced high macrophage viability and more drug absorption than native drug. Suppressed the growth of intramacrophagic amastigotes in vitro. Declined parasite burden in *L. donovani*-infected BALB/c mice.	Shahnaz et al. (2017)
	Curcumin-loaded mannose-functionalized chitosan nanoparticles (Cur-MCN)	In vivo studies in hamsters decreased parasitic load in the	Chaubey et al. (2018)

(continued)

Table 11.6 (continued)

Nanoparticle system	Nanoparticle type	Mechanism of action	Reference
		spleen. In vitro cytotoxicity assay demonstrated nontoxicity against macrophages cells.	
	Amphotericin-B-loaded on mannan-functionalized PLGA nanospheres	Rapid phagocytosis by macrophages and dendritic cells. Increased cytotoxic T-cell response. Increased expression of MHC class II and CD86, CD80, and CD40. Increase of MR expression on BMM-ϕ due to the presence of IL-4. Decrease in toxicity on human and murine macrophages as compared to the traditional Amb. Decreased parasite load on the liver and splenic tissue.	Barros et al. (2015)
	Doxorubicin (DOX)-loaded chondroitin nanocapsule (NCAP) DOX-NCAP	Cell cycle arrest at G1 and S phases, inducing apoptosis. Enhanced Th1 and decreased Th2 response. Improved in vitro antileishmanial activity as compared to DOX. Increased encapsulation efficiency, Reduced cytotoxicity as compared to traditional DOX. Enhanced assimilation of DOX-NCAP in macrophages as compared to DOX. Improved in vivo parasite inhibition capability compared to free DOX done on golden hamster. Upregulated expression	Chaurasia et al. (2015)

(continued)

Table 11.6 (continued)

Nanoparticle system	Nanoparticle type	Mechanism of action	Reference
		of TNF-α, INF-γ, and IL-12.	
	Artemisinin-loaded PLGA (ALPLGA)	Elevated IgG2a levels and pro-inflammatory cytokines (IFN-γ and IL-2) and suppressed Th2 cytokines (IL-10 and IL-4). Suppressed parasite load and hepatosplenomegaly in liver as compared to free artemisinin.	Want et al. (2015)
	AmB-loaded chitosan (Cs)-chondroitin sulfate (ChS) complex (NQC-AmB)	Reduced IC_{50} against *L. amazonensis.* Low toxicity in murine macrophages. No hemolytic activity in human blood. Declined parasite burden significantly.	Ribeiro et al. (2014)
	AmB-loaded PLGA nanospheres (AmB-PLGA-NS)	Preferential accumulation in the visceral organs. Enhanced CD8 T-cell response. Decline in splenic parasite load. Reduced toxicity was observed.	Costa Lima et al. (2024)
	Mannose-conjugated chitosan nanoparticles loaded with rifampicin	Greater macrophage uptake using mannose receptors. Increased accumulation of nanoparticles in macrophage-rich organs (liver and spleen). Increased accumulation of rifampicin was observed. Higher ex vivo uptake of rifampicin-conjugated nanoparticles than normal unconjugated CNPs.	Chaubey and Mishra (2014)
	1,2-diacyl-sn-glycero-3-phospho-L-serine (PS)-coated	Increased uptake of PS-AmB GNPs in	Khatik et al. (2014)

(continued)

Table 11.6 (continued)

Nanoparticle system	Nanoparticle type	Mechanism of action	Reference
	gelatin nanoparticles (GNPs) loaded with AmB PS-AmB-GNPs	macrophage cell lines. Enhanced antileishmanial activity. In vivo study on Wistar rats showed increased accumulation in macrophage-rich organs (liver & spleen).	
	AmB-loaded PLGA	In vivo study shows that it decreased *Leishmania donovani* liver parasite burden as compared to the traditional AmB. Much better decline in liver and splenic parasite burden as compared to AmB. No sign of nephrotoxicity observed.	Italia et al. (2012)
	Nonionic surfactant vesicles (NIV) loaded with AmB (NIV-AmB)	Suppressed *L. donovani* parasite burden, but lesion size in *L. major*-infected BALB/c mice remained unaffected	Alsaadi et al. (2012)
	Caffeic acid phenylethyl ester conjugated with PLGA NPs $(CAPE)_{PLGA}$	Decreased proliferation rate of *Leishmania infantum* promastigotes. Stimulated NO production in macrophage cells. $(CAPE)_{PLGA}$ up to 30μg/ml exhibited no toxicity.	Abamor (2017)
	Pegylated poly (lactic acid) (PLA) conjugated with bisnaphthalimidopropyl (BNIP)	Decreased metabolic activity of macrophage cells inside the host cells, thereby inhibiting growth. Efficiently suppressing parasite growth inside human macrophage than in murine cells.	Costa Lima et al. (2012)
Metallic	Curcumin-coated gold	In vitro assay resulted in the inhibition of promastigotes and intracellular amastigotes from the macrophages.	Amini et al. (2023)

(continued)

Table 11.6 (continued)

Nanoparticle system	Nanoparticle type	Mechanism of action	Reference
		In vivo assay in BALB/c mice showed a significant reduction in parasitic load and no harmful toxic effects. Induced Th1 responses and augmented host cellular immune responses, reducing leishmania parasite burden.	
	Mannosylated thiolated chitosan-Ag nanoparticles (MTC-Ag)	In vitro assay demonstrated inhibitory effect against intracellular amastigotes. IC_{50} reduced to fourfolds. 83% macrophage cells were viable in the cytotoxicity assay.	Shakeel et al. (2023)
	Calcium oxide and magnesium oxide nanoparticles (CaONPs/MgONPs)	In vitro and in vivo analysis demonstrated that MgONPs is highly efficient against both promastigote and amastigote forms of *L. major*. MgONPs reduced the lesions significantly in BALB/c *Leishmania*-infected mice.	Ghaffari et al. (2023)
	Silver (I) and 1,10-phenanthroline-5,6-dione Copper (II) and 1,10-phenanthroline-5,6-dione	It showed a high inhibition of the growth of promastigote forms of *L. amazonensis* and *L. chagasi*, along with alterations to the morphology of these parasites. It also showed a potential effect against the amastigote forms of both *Leishmania* species, with reduced toxicity to the host cell.	Oliveira et al. (2023)

(continued)

Table 11.6 (continued)

Nanoparticle system	Nanoparticle type	Mechanism of action	Reference
	CaO NPs	CaO NPs showed great antileishmanial effects on both *L. tropica* and *L. infantum* species promastigote form. Higher inhibitory effects were observed in *L. infantum* during in vitro assay. CaO NPs exhibit great susceptibility to amastigotes and illicit the generation of NO and other antileishmanial compounds.	Dalir Ghaffari et al. (2023)
	Gold-chrysin	Increased antileishmanial activity. Decreased parasite burden in mammalian macrophages.	Raj et al. (2022)
	Gold-4′,7-dihydroxyflavone Silver-4′,7-dihydroxyflavone	Potent inhibitory activity *against L. donovani* promastigotes. ROS-mediated leishmanicidal effect.	Sasidharan and Saudagar (2022)
	Silver-curcumin	Significantly lowered the growth of promastigotes and amastigotes. Decrease in lesion length within the footpads, and *Leishmania* burden in each lymph nodes had been reduced significantly.	Badirzadeh et al. (2022)
	CdO/chitosan core shell NPS	The nano composition reduced the proliferation of promastigotes, along with the upregulation of Th1. It also induced apoptosis in the parasite.	Bahraminegad et al. (2021)
	Fe3O4 piroctone olamine (PO) magnetic NPs	The in vitro analysis revealed antileishmanial effect against intracellular amastigotes	Albalawi et al. (2021)

(continued)

Table 11.6 (continued)

Nanoparticle system	Nanoparticle type	Mechanism of action	Reference
		with greater selectivity toward the parasite and low cytotoxicity for macrophages J774A.1. In vivo assay in BALB/c mice reduced the lesion caused by *L. major* and promoted the release of NO.	
	AmB-loaded glycine-coated iron oxide NPs	The composition was twofold effective against the *L. donovani* parasite during the in vitro study. In vivo studies in hamsters showed higher efficacy toward amastigotes.	Kumar et al. (2017)
	Gold-amphotericin B	Enhanced uptake by macrophages in comparison to free AmB. Significant improvement in antileishmanial activity in comparison to free AmB. Nanoformulation was immunomodulatory in nature and upregulated IL-12 and IFN-γ levels. Reduction in ergosterol content.	Kumar et al. (2019)
	Gold-andrographolide	Enhanced antileishmanial activity against *L. donovani axenic* and intramacrophagic amastigotes. Reduced toxicity in macrophages in vitro.	Das et al. (2018)
	Biogenic silver nanoparticles (Ag NPs) synthesized from *Fusarium oxysporum*	Induction of apoptosis in *L. amazonensis* promastigotes, as evidenced by ROS production, loss of mitochondrial membrane potential, phosphatidylserine	Fanti et al. (2018)

(continued)

Table 11.6 (continued)

Nanoparticle system	Nanoparticle type	Mechanism of action	Reference
		externalization, and disruption of membrane integrity. Antiamastigote activity observed.	
	TiAgNps loaded with melgumine antimonate	Led to the elimination of *L. tropica* promastigotes and amastigotes in the culture. Parasite growth suppressed in *L. tropica*-infected BALB/c mice.	Abamor et al. (2017)
	Titanium-oxide- conjugated chitosan-loaded glucantime nanoparticles	More effective in the reduction of *L. major* promastigote and amastigote growth in vitro in comparison to free drug.	Varshosaz et al. (2018)
	Ag-decanethiol	AgNps-SCH shows remarkable leishmanicidal effects. Higher killing of amastigotes was observed at minimal concentration. The NP showed low toxicity and 20,000-folds greater therapeutic index.	Isaac-Márquez et al. (2018)
	Au NPs synthesized from *Maytenus royleanus*	Substantial reduction in the growth of *Leishmania* parasites.	Ahmad et al. (2016)
	PEGylated Ag doped with zinc oxide (ZnO) (DSNs)	Photo-dynamic therapy elicited ROS generation. In daylight, PEGylated Ag doped with ZnO showed 0.28 quantum yield, while nondoped shows 0.13. Antileishmanial activity observed only in daylight and not in the dark.	Nadhman et al. (2014)
	Biogenic selenium nanoparticle (Se NPs) biosynthesized by *Bacillus* sp. MSh-1	Declined *L. major* promastigote and amastigote proliferation in vitro.	Beheshti et al. (2013)

(continued)

Table 11.6 (continued)

Nanoparticle system	Nanoparticle type	Mechanism of action	Reference
		Induced DNA fragmentation in *L. major* promastigotes. Cured cutaneous lesions in *L. major*-infected BALB/c mice.	
	Quercetin-mediated Au NPs	Remarkable decline in the growth of axenic and intracellular *L. donovani* amastigote viability. Paromomycin- and sodium-stibogluconate-resistant axenic and intramacrophagic amastigotes were also susceptible. Nanoformulation was found to be safe in nature.	Das et al. (2013)
	Antimony sulfide NPs (Sb2S5)	Reduction in *L. infantum* promastigote and amastigote proliferation in vitro. Dose-dependent cytotoxicity observed at higher concentrations. Phosphatidylserine externalization observed in treated promastigotes.	Soflaei et al. (2012)
	Biofabricated silver nanoparticles (AgNPs) using extracts of *Mentha arvensis*	Dose-dependent cytotoxicity against *L. tropica* significantly reduced parasitic burden. NP formulation is biocompatible and safer for human use.	Javed et al. (2020)
	Manganese oxide NPs (Mn2O3)	In vitro analysis exhibited a reduction in *L. major* promastigotes. In vivo studies revealed a significant reduction in ulcer size and increased survival of mice.	Tavakoli et al. (2019)
	Gold nanoparticle with lipoic acid GNP-LA	Increased leishmanicidal activity	Kumar et al. (2019)

(continued)

Table 11.6 (continued)

Nanoparticle system	Nanoparticle type	Mechanism of action	Reference
	(GL) amphotericin B (GL-AmB)	against *L. donovani* promastigotes and amastigotes in vitro and GL-AmB are significantly less cytotoxic compared to AmB.	
	Biogenic antimony sulfide nanoparticles obtained from *Serratia marcescens*	Amplified concentration of NPs increased leishmanicidal activity. Maximum activity was observed at 62µg/ml and was noncytotoxic in vitro. Lesion size reduced in BALB/c mice during treatment.	Mohtasebi et al. (2019)
	Titanium oxide (TiO_2) anatase & UV	Antileishmanial effects of TiO_2 increased in the presence of UV in *L. major*. Anatase induces a rapid decrease in promastigote number, whereas rutile is less toxic to parasites and the human skin, but leishmanicidal effects are increased significantly on synergism with UV radiation.	Dolat et al. (2020)
	Titanium oxide (TiO_2) rutile & UV		
Nonmetallic nanoparticles	Andrographolide-soya-L-α-phosphatidylcholine (ANSPC)	In vitro MTT assay on *L. donovani* strain demonstrated that the ANSPC complex can cross the parasite plasma membrane and was toxic to the parasite.	Pingle et al. (2023)
	SeNPs (selenium nanoparticles)	Antileishmanial activity demonstrated different concentrations of SeNPs, showing inhibitory effects in vitro on *L. major* promastigotes.	Norouzi et al. (2022)
	Amphotericin-B-loaded polymer-reduced graphene oxides (rGO-PEI-AmB)	In the presence of near-infrared light, it enhanced the release of	Vitorino et al. (2022)

(continued)

Table 11.6 (continued)

Nanoparticle system	Nanoparticle type	Mechanism of action	Reference
		AmB from the formulation. rGO-PEI-AmB showed high leishmanicidal activity against *L. amazonensis* promastigotes in vitro.	
	Buparvaquone-loaded silicon NPs	The drug formulation is more selective and efficient toward intracellular parasites with no cytotoxicity toward macrophages. In vivo results showed a reduction in parasitic burden near 100%.	Thapa et al. (2021)
	Graphene-CNT-composite with AmB	In vitro reduction of *L. donovani* amastigote burden in J774A.1 macrophages. In vivo decline of parasite burden in *L. donovani*- infected golden Syrian hamsters. No cytotoxicity against J774A.1 macrophages. No hepatic or renal toxicity observed in the hamster model of disease.	Gedda et al. (2020)
Nanoemulsions	Quercetin-loaded nanoemulsion (QTNE)	In vitro studies demonstrated a significant reduction in promastigotes and intracellular amastigotes in *L. donovani*. Antileishmanial activity was exhibited by the increased production of ROS and NO, inducing membrane disintegration and phosphatidylserine release, damaging the parasitic DNA.	Das et al. (2022)
	Nanoemulsion (NE) with *Pterodon emarginatus* Vogel oleoresin (SO-NE)	In vivo analysis of *L. amazonensis* resulted in reduced parasitic load and decreased lesion	Kawakami et al. (2021)

(continued)

Table 11.6 (continued)

Nanoparticle system	Nanoparticle type	Mechanism of action	Reference
		size with minimal or no inflammatory response.	
	Nanoemulsion (NE) with chloroaluminium phthalocyanine (ClAlPc) and paromomycin sulfate (PM) (NE-ClAlPc-PM)	In vitro study of *L. braziliensis* unveiled significant inhibition in both promastigotes and amastigotes in the presence of light. However, in vivo results of the followed protocol was ineffective on the BALB/c mice model.	Leal Pinto et al. (2021)
	Trans-chalcone-loaded NE	In vitro study showed higher inhibition of *L. amazonensis* amastigotes. Accumulation of chalcones was observed in epidermal and dermal layers. Demonstrated cytotoxic effects on THP-1 cells and low selectivity compared to free drug forms against the parasite.	Coelho et al. (2021)
	Nanoemulsion (NE) with aromatic/heterocyclic sulphonamides	In vitro study of *L. infantum* and *L. amazonensis* showed significant inhibition of promastigotes at varying concentrations, and improved NE formulation improved drug bioavailability.	da Silva Cardoso et al. (2018)
	Copaifera paupera oleoresin nanoemulsion (copaiba oleoresin and α-copane)	Increased inhibition of promastigotes observed in *L. infantum* and *L. amazonensis.*	Rodrigues et al. (2018)
	Lavandula angustifolia and *Rosmarinus officinalis* nanoemulsion	Significant inhibition of *L. major* promastigotes and amastigotes with no cytotoxicity to the host cells.	Shokri et al. (2017)
Solid lipid	Solid lipid NP oryzalin (LNP)	Reduction in parasitic burden in the spleen and liver of *L. infantum*-infected mice was observed.	Lopes et al. (2014)

(continued)

Table 11.6 (continued)

Nanoparticle system	Nanoparticle type	Mechanism of action	Reference
	Nanostructured lipid carriers (NLCs) carvacrol	Increased leishmanicidal activity was observed in amastigote forms in *L. amazonensis* and relatively lower cytotoxicity compared to free carvacrol.	Galvão et al. (2020)
	Chitosan-coated AmB-loaded SLNs (AmB-SLNs)	Highly potent against *L. donovani* amastigotes, showing elevation in TNF-α and IL-12, along with NO generation in vitro. In vivo study and histological assay displayed no toxicity or degeneration of kidney and liver, demonstrating formulations to be safer.	Jain et al. (2014)
	2-hydroxypropyl-β-cyclodextrin (HPCD)-modified (SLN) of (Amb) & (PM)–(HPCD)-modified dual-drug-loaded (SLN) (mDDSLN)	Maximum inhibition was observed against *L. donovani* intracellular amastigotes. In vitro and in vivo analyses revealed a significant reduction in parasitic burden in the murine model.	Parvez et al. (2020)
Liposomal	Meglumine antimoniate-phosphatidylserine liposomes (MA-LPs)	Effective inhibition of about 23-folds observed in intracellular amastigotes of *L. infantum*. Pharmacokinetics study in mice models revealed that LPs enhanced the uptake of drugs by liver and spleen.	Borborema et al. (2018); Tuon et al. (2022)
	N-methylglucamine antimoniate (NMG) associated with photodynamic therapy with liposomal chloroaluminium phthalocyanine (AlClPC)	In vivo study in C57BL6 mice infected with *L. amazonensis* decreased parasitic load and toxicity.	Ribeiro et al. (2016)
	Berberine	Berberine loaded in liposomes enhanced the selectivity of parasites by sevenfold and reduced cytotoxicity to	Calvo et al. (2020)

(continued)

Table 11.6 (continued)

Nanoparticle system	Nanoparticle type	Mechanism of action	Reference
		macrophages in *L. infantum*, and in vivo results exhibited a significant reduction in parasitic burden in the liver and spleen.	
	Miltefosine	Inhibition of *L. major* promastigotes and intracellular amastigotes observed in in vitro analysis.	Kavian et al. (2019)
	Liposomal phospholipase A2 (Asp49-PLA2)	Parasitic load significantly reduced in BALB/c mice infected with *L. amazonensis* with increased production of TNF-α and nitrites responsible for leishmanicidal activity.	de Barros et al. (2016)

11.5 Conclusion

Leishmaniasis containment and elimination have been a global effort wherein various governments have pooled resources and developed integrated methods for vector control and disease suppression. However, these efforts for the containment and elimination of leishmaniasis have yet to be fruitful as the disease continues to expand outside its endemic domains. In the absence of an efficacious vaccine, antileishmanial drug discovery needs the impetus it currently lacks. Efforts should be made in the direction of "target-based drug discovery" and "therapeutic switching" so that the likelihood of finding promising drug candidates is more. Efforts shall also be made to develop successful nano- or lipid-based drug carriers that can improve drug bioavailability and efficacy and, at the same time, reduce toxicity. In totality, drug discovery against leishmaniasis needs substantial enhancement that can be achieved through diverse rational drug-designing approaches and the development of novel drug-delivery vehicles.

Acknowledgement This work was supported by research grant from ICMR, Government of India, # 6/9-7(241)/2020-ECD-II.

References

Abamor ES (2017) Antileishmanial activities of caffeic acid phenethyl ester loaded PLGA nanoparticles against *Leishmania infantum* promastigotes and amastigotes *in vitro*. Asian Pac J Trop Med 10(1):25–34. https://doi.org/10.1016/j.apjtm.2016.12.006

Abamor ES, Allahverdiyev AM, Bagirova M, Rafailovich M (2017) Meglumine antimoniate-TiO2@Ag nanoparticle combinations reduce toxicity of the drug while enhancing its antileishmanial effect. Acta Trop 169:30–42. https://doi.org/10.1016/j.actatropica.2017.01.005

Abamor ES, Tosyali OA, Bagirova M, Allahverdiyev A (2018) *Nigella sativa* oil entrapped polycaprolactone nanoparticles for leishmaniasis treatment. IET Nanobiotechnol 12(8): 1018–1026. https://doi.org/10.1049/iet-nbt.2018.5115

Ahmad A, Syed F, Imran M, Khan AU, Tahir K, Khan ZUH, Yuan Q (2016) Phytosynthesis and antileishmanial activity of gold nanoparticles by *Maytenus Royleanus*. J Food Biochem 40(4): 420–427. https://doi.org/10.1111/jfbc.12232

Albalawi AE, Khalaf AK, Alyousif MS, Alanazi AD, Baharvand P, Shakibaie M, Mahmoudvand H (2021) Fe3O4@piroctone olamine magnetic nanoparticles: synthesize and therapeutic potential in cutaneous leishmaniasis. Biomed Pharmacother. 139:111566. https://doi.org/10.1016/j.biopha.2021.111566

Alemomen M, Taymouri S, Saberi S, Varshosaz J (2023) Preparation, optimization, and in vitro-in vivo evaluation of sorafenib-loaded polycaprolactone and cellulose acetate nanofibers for the treatment of cutaneous leishmaniasis. Drug Deliv Transl Res 13(3):862–882. https://doi.org/10.1007/s13346-022-01250-2

Alsaadi M, Italia JL, Mullen AB, Ravi Kumar MN, Candlish AA, Williams RA, Shaw CD, Al Gawhari F, Coombs GH, Wiese M, Thomson AH, Puig-Sellart M, Wallace J, Sharp A, Wheeler L, Warn P, Carter KC (2012) The efficacy of aerosol treatment with non-ionic surfactant vesicles containing amphotericin B in rodent models of leishmaniasis and pulmonary aspergillosis infection. J Control Release 160(3):685–691. https://doi.org/10.1016/j.jconrel.2012.04.004

Alvar J, Vélez ID, Bern C, Herrero M, Desjeux P, Cano J, Jannin J, den Boer M, Leishmaniasis Control Team WHO (2012) Leishmaniasis worldwide and global estimates of its incidence. PLoS One 7(5):e35671. https://doi.org/10.1371/journal.pone.0035671

Amini SM, Hadighi R, Najm M, Alipour M, Hasanpour H, Vosoogh M, Vosough A, Hajizadeh M, Badirzadeh A (2023) The therapeutic effects of curcumin-coated gold nanoparticle against leishmania major causative agent of zoonotic cutaneous leishmaniasis (ZCL): an in vitro and in vivo study. Curr Microbiol 80(4):104. https://doi.org/10.1007/s00284-022-03172-1

Andrade-Neto VV, Pereira TM, Canto-Cavalheiro MD, Torres-Santos EC (2016) Imipramine alters the sterol profile in *Leishmania amazonensis* and increases its sensitivity to miconazole. Parasit Vectors 9:183. https://doi.org/10.1186/s13071-016-1467-8

Anjum A, Shabbir K, Din FU, Shafique S, Zaidi SS, Almari AH, Alqahtani T, Maryiam A, Moneeb Khan M, Al Fatease A, Bashir S, Khan GM (2023) Co-delivery of amphotericin B and pentamidine loaded niosomal gel for the treatment of Cutaneous leishmaniasis. Drug Deliv 30(1):2173335. https://doi.org/10.1080/10717544.2023.2173335

Aoki JI, Muxel SM, RibeiroFernandes JC, Floeter-Winter LM (2018) The polyamine pathway as a potential target for leishmaniases chemotherapy. In: Afrin F, Hemeg H (eds) Leishmaniases as re-emerging diseases. IntechOpen. https://doi.org/10.5772/intechopen.75867

Aronson JK (ed) (2016) Meyer's side effects of drugs: *The International Encyclopedia of Adverse Reactions and Interactions*

Badirzadeh A, Alipour M, Najm M, Vosoogh A, Vosoogh M, Samadian H, Hashemi AS, Farsangi ZJ, Amini SM (2022) Potential therapeutic effects of curcumin coated silver nanoparticle in the treatment of cutaneous leishmaniasis due to *Leishmania major in-vitro* and in a murine model. J Drug Deliv Sci Technol 74:103576

Bahraminegad S, Pardakhty A, Sharifi I, Ranjbar M (2021) The assessment of apoptosis, toxicity effects and anti-leishmanial study of Chitosan/CdO core-shell nanoparticles, eco-friendly synthesis and evaluation. Arab J Chem 14(4):103085

Baig N, Kammakakam I, Falath W (2021) Nanomaterials: A review of synthesis methods, properties, recent progress, and challenges. Mater Adv 2(6):1821–1871

Baiocco P, Poce G, Alfonso S, Cocozza M, Porretta GC, Colotti G, Biava M, Moraca F, Botta M, Yardley V, Fiorillo A, Lantella A, Malatesta F, Ilari A (2013) Inhibition of *Leishmania infantum* trypanothione reductase by azole-based compounds: a comparative analysis with its physiological substrate by X-ray crystallography. ChemMedChem 8(7):1175–1183. https://doi.org/10.1002/cmdc.201300176

Batool S, Zahid F, Ud-Din F, Naz SS, Dar MJ, Khan MW, Zeb A, Khan GM (2021) Macrophage targeting with the novel carbopol-based miltefosine-loaded transfersomal gel for the treatment of cutaneous leishmaniasis: in vitro and in vivo analyses. Drug Dev Ind Pharm 47(3):440–453. https://doi.org/10.1080/03639045.2021.1890768

Battista T, Colotti G, Ilari A, Fiorillo A (2020) Targeting trypanothione reductase, a key enzyme in the redox trypanosomatid metabolism, to develop new drugs against leishmaniasis and trypanosomiasis. Molecules (Basel, Switzerland). 25(8):1924. https://doi.org/10.3390/molecules25081924

Battista T, Federico S, Brogi S, Pozzetti L, Khan T, Butini S, Ramunno A, Fiorentino E, Orsini S, Di Muccio T, Fiorillo A, Exertier C, Di Risola D, Colotti G, Gemma S, Ilari A, Campiani G (2022) Optimization of potent and specific trypanothione reductase inhibitors: a structure-based drug discovery approach. ACS Infect Dis 8(8):1687–1699. https://doi.org/10.1021/acsinfecdis.2c00325

Beheshti N, Soflaei S, Shakibaie M, Yazdi MH, Ghaffarifar F, Dalimi A, Shahverdi AR (2013) Efficacy of biogenic selenium nanoparticles against *Leishmania major: in vitro* and *in vivo* studies. J Trace Elements Med Biol 27(3):203–207. https://doi.org/10.1016/j.jtemb.2012.11.002

Bell AS, Yu Z, Hutton JA, Wright MH, Brannigan JA, Paape D, Roberts SM, Sutherell CL, Ritzefeld M, Wilkinson AJ, Smith DF, Leatherbarrow RJ, Tate EW (2020) Novel thienopyrimidine inhibitors of *Leishmania N*-Myristoyltransferase with on-target activity in intracellular amastigotes. J Med Chem 63(14):7740–7765. https://doi.org/10.1021/acs.jmedchem.0c00570

Bennett JE (2020) Drugs for protozoal infections other than malaria. In: McCarthy JS, Wortmann GW, Kirchhoff LV (eds) Mandell, Douglas and Bennett's principles and practices of infectious disease*s*, 9th edn. Elsevier, Amsterdam, pp 535–543

Bertacine Dias MV, Santos JC, Libreros-Zúñiga GA, Ribeiro JA, Chavez-Pacheco SM (2018) Folate biosynthesis pathway: mechanisms and insights into drug design for infectious diseases. Future Med Chem 10(8):935–959. https://doi.org/10.4155/fmc-2017-0168

Bezerra-Souza A, Fernandez-Garcia R, Rodrigues GF, Bolas-Fernandez F, Dalastra Laurenti M, Passero LF, Lalatsa A, Serrano DR (2019) Repurposing butenafine as an oral nanomedicine for visceral leishmaniasis. Pharmaceutics 11(7):353. https://doi.org/10.3390/pharmaceutics11070353

Bhargava P, Kumar K, Chaudhaery SS, Saxena AK, Roy U (2010) Cloning, overexpression and characterization of *Leishmania donovani* squalene synthase. FEMS Microbiol Lett 311(1):82–92. https://doi.org/10.1111/j.1574-6968.2010.02071.x

Bilgic-Temel A, Murrell DF, Uzun S (2019) Cutaneous leishmaniasis: a neglected disfiguring disease for women. Int J Women's Dermatol 5(3):158–165. https://doi.org/10.1016/j.ijwd.2019.01.002

Bompart D, Núñez-Durán J, Rodríguez D, Kouznetsov VV, Meléndez Gómez CM, Sojo F, Arvelo F, Visbal G, Alvarez A, Serrano-Martín X, García-Marchán Y (2013) Anti-leishmanial evaluation of C2-aryl quinolines: mechanistic insight on bioenergetics and sterol biosynthetic pathway of *Leishmania braziliensis*. Bioorg Med Chem 21(14):4426–4431. https://doi.org/10.1016/j.bmc.2013.04.063

Borborema SET, Osso Junior JA, Tempone AG, de Andrade Junior HF, do Nascimento, N. (2018) Pharmacokinetic of meglumine antimoniate encapsulated in phosphatidylserine-liposomes in mice model: A candidate formulation for visceral leishmaniasis. Biomed Pharmacother. 103: 1609–1616. https://doi.org/10.1016/j.biopha.2018.05.004

Boshrouyeh R, Amari S, Boshrouyeh Ghandashtani M et al (2023) A topical gel nanoformulation of amphotericin B (AmB) for the treatment of cutaneous leishmaniasis (CL). J Sol-Gel Sci Technol 105:768–780. https://doi.org/10.1007/s10971-023-06041-w

Bozzuto G, Molinari A (2015) Liposomes as nanomedical devices. Int J Nanomedicine 10:975–999. https://doi.org/10.2147/IJN.S68861

Braga SS (2019) Multi-target drugs active against *leishmaniasis*: a paradigm of drug repurposing. Eur J Med Chem 183:111660. https://doi.org/10.1016/j.ejmech.2019.111660

Brindha J, Balamurali MM, Chanda K (2021) An overview on the therapeutics of neglected infectious diseases-leishmaniasis and chagas diseases. Front Chem 9:622286. https://doi.org/10.3389/fchem.2021.622286

Calvo A, Moreno E, Larrea E, Sanmartín C, Irache JM, Espuelas S (2020) Berberine-loaded liposomes for the treatment of *Leishmania infantum*-infected BALB/c mice. Pharmaceutics 12(9):858. https://doi.org/10.3390/pharmaceutics12090858

Carter NS, Kawasaki Y, Nahata SS, Elikaee S, Rajab S, Salam L, Alabdulal MY, Broessel KK, Foroghi F, Abbas A, Poormohamadian R, Roberts SC (2022) Polyamine metabolism in leishmania parasites: a promising therapeutic target. Med Sci (Basel, Switzerland) 10(2):24. https://doi.org/10.3390/medsci10020024

Çetin FK, Ateş SC (2022) In vitro investigation of Rutin-loaded PLGA nanoparticles on leishmania infantum promastigotes. Süleyman Demirel Üniversitesi Fen Bilimleri Enstitüsü Dergisi 26(3): 427–434

Chanquia SN, Larregui F, Puente V, Labriola C, Lombardo E, García Liñares G (2019) Synthesis and biological evaluation of new quinoline derivatives as antileishmanial and antitrypanosomal agents. Bioorg Chem 83:526–534. https://doi.org/10.1016/j.bioorg.2018.10.053

Chaubey P, Mishra B (2014) Mannose-conjugated chitosan nanoparticles loaded with rifampicin for the treatment of visceral leishmaniasis. Carbohydr Polym 101:1101–1108. https://doi.org/10.1016/j.carbpol.2013.10.044

Chaubey P, Mishra B, Mudavath SL, Patel RR, Chaurasia S, Sundar S, Suvarna V, Monteiro M (2018) Mannose-conjugated curcumin-chitosan nanoparticles: Efficacy and toxicity assessments against *Leishmania donovani*. Int J Biol Macromol 111:109–120. https://doi.org/10.1016/j.ijbiomac.2017.12.143

Chaurasia M, Pawar VK, Jaiswal AK, Dube A, Paliwal SK, Chourasia MK (2015) Chondroitin nanocapsules enhanced doxorubicin induced apoptosis against leishmaniasis via Th1 immune response. Int J Biol Macromol 79:27–36. https://doi.org/10.1016/j.ijbiomac.2015.04.043

Chawla B, Jhingran A, Panigrahi A, Stuart KD, Madhubala R (2011) Paromomycin affects translation and vesicle-mediated trafficking as revealed by proteomics of paromomycin -susceptible -resistant Leishmania donovani. PLoS one 6(10):e26660. https://doi.org/10.1371/journal.pone.0026660

Coelho AC, Messier N, Ouellette M, Cotrim PC (2007) Role of the ABC transporter PRP1 (ABCC7) in pentamidine resistance in *Leishmania* amastigotes. Antimicrob Agents Chemother 51(8):3030–3032. https://doi.org/10.1128/AAC.00404-07

Coelho D, Veleirinho B, Mazzarino L, Alberti T, Buzanello E, Oliveira RE, Yunes RA, Moraes M, Steindel M, Maraschin M (2021) Polyvinyl alcohol-based electrospun matrix as a delivery system for nanoemulsion containing chalcone against *Leishmania (Leishmania) amazonensis*. Colloids Surf B: Biointerfaces 198:111390. https://doi.org/10.1016/j.colsurfb.2020.111390

Cordeiro AT, Michels PA, Delboni LF, Thiemann OH (2004) The crystal structure of glucose-6-phosphate isomerase from *Leishmania mexicana* reveals novel active site features. Eur J Biochem 271(13):2765–2772. https://doi.org/10.1111/j.1432-1033.2004.04205.x

Costa Lima S, Rodrigues V, Garrido J, Borges F, Kong Thoo Lin P, Cordeiro da Silva A (2012) *In vitro* evaluation of bisnaphthalimidopropyl derivatives loaded into pegylated nanoparticles against *Leishmania infantum* protozoa. Int J Antimicrob Agents 39(5):424–430. https://doi.org/10.1016/j.ijantimicag.2012.01.003

Costa Lima SA, Silvestre R, Barros D, Cunha J, Baltazar MT, Dinis-Oliveira RJ, Cordeiro-da-Silva A (2014) Crucial CD8(+) T-lymphocyte cytotoxic role in amphotericin B nanospheres efficacy against experimental visceral leishmaniasis. Nanomedicine 10(5):1021–1030. https://doi.org/10.1016/j.nano.2013.12.013

Costa AMB, Silva ARST, Santos AJ, Galvão JG, Andrade-Neto VV, Torres-Santos EC, Ueki MM, Almeida LE, Sarmento VHV, Dolabella SS, Scher R, Lira AAM, Nunes RS (2023) Thermosensitive system formed by poloxamers containing carvacrol: an effective carrier system against Leishmania amazonensis. Acta Trop 237:106744. https://doi.org/10.1016/j.actatropica.2022.106744

D'Antonio EL, Ullman B, Roberts SC, Dixit UG, Wilson ME, Hai Y, Christianson DW (2013) Crystal structure of arginase from *Leishmania mexicana* and implications for the inhibition of polyamine biosynthesis in parasitic infections. Arch Biochem Biophys 535(2):163–176. https://doi.org/10.1016/j.abb.2013.03.015

da Silva Cardoso V, Vermelho AB, Ricci Junior E, Almeida Rodrigues I, Mazotto AM, Supuran CT (2018) Antileishmanial activity of sulphonamide nanoemulsions targeting the β-carbonic anhydrase from *Leishmania* species. J Enzyme Inhib Med Chem 33(1):850–857. https://doi.org/10.1080/14756366.2018.1463221

da Silva ER, Maquiaveli CC, Magalhães PP (2012) The leishmanicidal flavonols quercetin and quercitrin target *Leishmania (Leishmania) amazonensis* arginase. Exp Parasitol 130(3):183–188. https://doi.org/10.1016/j.exppara.2012.01.015

da Silva SS, Mizokami SS, Fanti JR, Costa IN, Bordignon J, Felipe I, Pavanelli WR, Verri WA Jr, Conchon Costa I (2018) Glucantime reduces mechanical hyperalgesia in cutaneous leishmaniasis and complete Freund's adjuvant models of chronic inflammatory pain. J Pharm Pharmacol 70(6):768–777. https://doi.org/10.1111/jphp.12896

Dalir Ghaffari A, Barati M, Ghaffarifar F, Pirestani M, Ebrahimi M, KarimiPourSaryazdi A (2023) Investigation of antileishmanial activities of CaO nanoparticles on L. tropica and L. infantum parasites, in vitro. J Parasitic Dis 47(1):73–81. https://doi.org/10.1007/s12639-022-01539-6

Daneshbod Y, Oryan A, Davarmanesh M, Shirian S, Negahban S, Aledavood A, Davarpanah MA, Soleimanpoor H, Daneshbod K (2011) Clinical, histopathologic, and cytologic diagnosis of mucosal leishmaniasis and literature review. Arch Pathol Lab Med 135(4):478–482. https://doi.org/10.5858/2010-0069-OA.1

das Neves GM, Kagami LP, Gonçalves IL, Eifler-Lima VL (2019) Targeting pteridine reductase 1 and dihydrofolate reductase: the old is a new trend for leishmaniasis drug discovery. Future Med Chem 11(16):2107–2130. https://doi.org/10.4155/fmc-2018-0512

Das S, Roy P, Mondal S, Bera T, Mukherjee A (2013) One pot synthesis of gold nanoparticles and application in chemotherapy of wild and resistant type visceral leishmaniasis. Colloids Surf B: Biointerfaces 107:27–34. https://doi.org/10.1016/j.colsurfb.2013.01.061

Das M, Singh S, Dubey VK (2016) Novel inhibitors of ornithine decarboxylase of *Leishmania* parasite (LdODC): the parasite resists LdODC inhibition by overexpression of spermidine synthase. Chem Biol Drug Des 87(3):352–360. https://doi.org/10.1111/cbdd.12665

Das S, Halder A, Mandal S, Mazumder MAJ, Bera T, Mukherjee A, Roy P (2018) Andrographolide engineered gold nanoparticle to overcome drug resistant visceral leishmaniasis. Artificial Cells Nanomed Biotechnol. 46(1sup):751–762. https://doi.org/10.1080/21691401.2018.1435549

Das SS, Dubey AK, Verma PRP, Singh SK, Singh SK (2022) Therapeutic potential of quercetin-loaded nanoemulsion against experimental visceral leishmaniasis: *in vitro/ex vivo* studies and mechanistic insights. Mol Pharm 19(9):3367–3384. https://doi.org/10.1021/acs.molpharmaceut.2c00492

de Aquino TM, França PHB, Rodrigues ÉEES, Nascimento IJS, Santos-Júnior PFS, Aquino PGV, Santos MS, Queiroz AC, Araújo MV, Alexandre-Moreira MS, Rodrigues RRL, Rodrigues KAF, Freitas JD, Bricard J, Meneghetti MR, Bourguignon JJ, Schmitt M, da Silva-Júnior EF, de Araújo-Júnior JX (2022) Synthesis, antileishmanial activity and *in silico* studies of aminoguanidine hydrazones (AGH) and thiosemicarbazones (TSC) against *Leishmania chagasi* amastigotes. Med Chem (Shariqah (United Arab Emirates)) 18(2):151–169. https://doi.org/10.2174/1573406417666210216154428

de Barros NB, Macedo SRA, Ferreira AS, Tagliari MP, Zanchi FB, Kayano AM, Soares AM, Nicolete R (2016) Liposomes containing an ASP49-phospholipase A2 from *Bothrops jararacussu* snake venom as experimental therapy against cutaneous leishmaniasis. Int Immunopharmacol 36:225–231. https://doi.org/10.1016/j.intimp.2016.04.02

Dolat E, Salarabadi SS, Layegh P, Jaafari MR, Sazgarnia S, Sazgarnia A. The effect of UV radiation in the presence of TiO2-NPs on Leishmania major promastigotes. Biochim Biophys Acta Gen Subj. 2020;1864(6):129558. https://doi.org/10.1016/j.bbagen.2020.129558

Dorlo TP, Huitema AD, Beijnen JH, de Vries PJ (2012) Optimal dosing of miltefosine in children and adults with visceral leishmaniasis. Antimicrob Agents Chemother 56(7):3864–3872. https://doi.org/10.1128/AAC.00292-12

Dos Santos Meira C, Gedamu L (2019) Protective or detrimental? Understanding the role of host immunity in leishmaniasis. Microorganisms 7(12):695. https://doi.org/10.3390/microorganisms7120695

Efstathiou A, Smirlis D (2021) *Leishmania* protein kinases: important regulators of the parasite life cycle and molecular targets for treating leishmaniasis. Microorganisms 9(4):691. https://doi.org/10.3390/microorganisms9040691

Esboei BR, Mohebali M, Mousavi P, Fakhar M, Akhoundi B (2018) Potent antileishmanial activity of chitosan against Iranian strain of *Leishmania major* (MRHO/IR/75/ER): *In vitro* and *in vivo* assay. J Vector Borne Dis 55(2):111–115. https://doi.org/10.4103/0972-9062.242557

Esfandiari F, Motazedian MH, Asgari Q, Morowvat MH, Molaei M, Heli H (2019) Paromomycin-loaded mannosylated chitosan nanoparticles: synthesis, characterization and targeted drug delivery against leishmaniasis. Acta Trop 197:105072. https://doi.org/10.1016/j.actatropica.2019.105072

Fanti JR, Tomiotto-Pellissier F, Miranda-Sapla MM, Cataneo AHD, Andrade CGTJ, Panis C, Rodrigues JHDS, Wowk PF, Kuczera D, Costa IN, Nakamura CV, Nakazato G, Durán N, Pavanelli WR, Conchon-Costa I (2018) Biogenic silver nanoparticles inducing *Leishmania amazonensis* promastigote and amastigote death *in vitro*. Acta Trop 178:46–54. https://doi.org/10.1016/j.actatropica.2017.10.027

Feuser PE, Tonini ML, Jacques AV, Santos da Silva MC, Steindel M, Sayer C, Hermes de Araújo PH (2019) Increased in vitro *leishmanicidal* activity of octyl gallate loaded poly(methyl methacrylate) nanoparticles. Pharm Dev Technol 24(5):593–599. https://doi.org/10.1080/10837450.2018.1547747

Freitas CS, Lage DP, Machado AS, Vale DL, Martins VT, Cardoso JMO, Oliveira-da-Silva JA, Reis TAR, Tavares GSV, Ramos FF, Ludolf F, Pereira IAG, Bandeira RS, Fujiwara RT, Bueno LL, Roatt BM, Chávez-Fumagalli MA, Coelho EAF (2023) Exploring drug repositioning for leishmaniasis treatment: ivermectin plus polymeric micelles induce immunological response and protection against tegumentary leishmaniasis. Cytokine. 164:156143. https://doi.org/10.1016/j.cyto.2023.156143. Advance online publication

Frézard F, Demicheli C, Ribeiro RR (2009) Pentavalent antimonials: new perspectives for old drugs. Molecules (Basel, Switzerland) 14(7):2317–2336. https://doi.org/10.3390/molecules14072317

Gabriel Á, Valério-Bolas A, Palma-Marques J, Mourata-Gonçalves P, Ruas P, Dias-Guerreiro T, Santos-Gomes G (2019) Cutaneous leishmaniasis: the complexity of host's effective immune response against a polymorphic parasitic disease. J Immunol Res 2019:2603730. https://doi.org/10.1155/2019/2603730

Galvão JG, Santos RL, Silva ARST, Santos JS, Costa AMB, Chandasana H, Andrade-Neto VV, Torres-Santos EC, Lira AAM, Dolabella S, Scher R, Kima PE, Derendorf H, Nunes RS (2020) Carvacrol loaded nanostructured lipid carriers as a promising parenteral formulation for leishmaniasis treatment. Eur J Pharm Sci 150:105335. https://doi.org/10.1016/j.ejps.2020.105335

Garrido-Jareño M, Sahuquillo-Torralba A, Chouman-Arcas R, Castro-Hernández I, Molina-Moreno JM, Llavador-Ros M, Gómez-Ruiz MD, López-Hontangas JL, Botella-Estrada R, Salavert-Lleti M, Pemán-García J (2020) Cutaneous and mucocutaneous leishmaniasis: experience of a Mediterranean hospital. Parasit Vectors 13(1):24. https://doi.org/10.1186/s13071-020-3901-1

Gedda MR, Madhukar P, Vishwakarma AK, Verma V, Kushwaha AK, Yadagiri G, Mudavath SL, Singh OP, Srivastava ON, Sundar S (2020) Evaluation of safety and antileishmanial efficacy of amine functionalized carbon-based composite nanoparticle appended with amphotericin B: an *in vitro* and preclinical study. Front Chem 8:510. https://doi.org/10.3389/fchem.2020.00510

Getachew F, Gedamu L (2012) *Leishmania donovani* mitochondrial iron superoxide dismutase A is released into the cytosol during miltefosine induced programmed cell death. Mol Biochem Parasitol 183(1):42–51. https://doi.org/10.1016/j.molbiopara.2012.01.005

Ghaffari AD, Barati M, KarimiPourSaryazdi A, Ghaffarifar F, Pirestani M, Ebrahimi M (2023) In vitro and in vivo study on antiprotozoal activity of calcium oxide (CaO) and magnesium oxide (MgO) nanoparticles on promastigote and amastigote forms of Leishmania major. Acta Trop 238:106788. https://doi.org/10.1016/j.actatropica.2022.106788

Ghorbani M, Farhoudi R (2017) Leishmaniasis in humans: drug or vaccine therapy? Drug Des Devel Ther 12:25–40. https://doi.org/10.2147/DDDT.S146521

Ghosh S, Kar N, Bera T (2016) Oleanolic acid loaded poly lactic co- glycolic acid- vitamin E TPGS nanoparticles for the treatment of Leishmania donovani infected visceral leishmaniasis. Int J Biol Macromol 93(PtA):961–970. https://doi.org/10.1016/j.ijbiomac.2016.09.014

Gomes MT, Paes-Vieira L, Gomes-Vieira AL, Cosentino-Gomes D, da Silva APP, Giarola NLL, Da Silva D, Sola-Penna M, Galina A, Meyer-Fernandes JR (2021) 3-Bromopyruvate: a new strategy for inhibition of glycolytic enzymes in *Leishmania amazonensis*. Exp Parasitol 229: 108154. https://doi.org/10.1016/j.exppara.2021.108154

Grover A, Katiyar SP, Singh SK, Dubey VK, Sundar D (2012) A leishmaniasis study: structure-based screening and molecular dynamics mechanistic analysis for discovering potent inhibitors of spermidine synthase. Biochim Biophys Acta 1824(12):1476–1483. https://doi.org/10.1016/j.bbapap.2012.05.016

Gupta A (2020) Nanoemulsions. In: Nanoparticles for biomedical applications. Elsevier, Amsterdam, pp 371–384

Gupta R, Kumar V, Kushawaha PK, Tripathi CP, Joshi S, Sahasrabuddhe AA, Mitra K, Sundar S, Siddiqi MI, Dube A (2014) Characterization of glycolytic enzymes--rAldolase and rEnolase of *Leishmania donovani*, identified as Th1 stimulatory proteins, for their immunogenicity and immunoprophylactic efficacies against experimental visceral leishmaniasis. PLoS One 9(1): e86073. https://doi.org/10.1371/journal.pone.0086073

Heby O, Persson L, Rentala M (2007) Targeting the polyamine biosynthetic enzymes: a promising approach to therapy of African sleeping sickness, Chagas' disease, and leishmaniasis. Amino Acids 33(2):359–366. https://doi.org/10.1007/s00726-007-0537-9

Hendrickx S, Caljon G, Maes L (2019) Need for sustainable approaches in antileishmanial drug discovery. Parasitol Res 118(10):2743–2752. https://doi.org/10.1007/s00436-019-06443-2

Herrmann FC, Sivakumar N, Jose J, Costi MP, Pozzi C, Schmidt TJ (2017) In silico identification and in vitro evaluation of natural inhibitors of *Leishmania* major pteridine reductase I. Molecules (Basel, Switzerland) 22(12):2166. https://doi.org/10.3390/molecules22122166

Ilari A, Baiocco P, Messori L, Fiorillo A, Boffi A, Gramiccia M, Di Muccio T, Colotti G (2012) A gold-containing drug against parasitic polyamine metabolism: the X-ray structure of trypanothione reductase from *Leishmania infantum* in complex with auranofin reveals a dual mechanism of enzyme inhibition. Amino Acids 42(2–3):803–811. https://doi.org/10.1007/s00726-011-0997-9

Inacio JDF, Fonseca MS, Limaverde-Sousa G, Tomas AM, Castro H, Almeida-Amaral EE (2021) Epigallocathechin-*O*-3-gallate inhibits trypanothione reductase of *Leishmania infantum*, causing alterations in redox balance and leading to parasite death. Front Cell Infect Microbiol 11: 640561. https://doi.org/10.3389/fcimb.2021.640561

Isaac-Márquez AP, Talamás-Rohana P, Galindo-Sevilla N, Gaitan-Puch SE, Díaz-Díaz NA, Hernández-Ballina GA, Lezama-Dávila CM (2018) Decanethiol functionalized silver nanoparticles are new powerful leishmanicidals in vitro. World J Microbiol Biotechnol 34(3): 38. https://doi.org/10.1007/s11274-018-2420-0

Italia JL, Kumar MN, Carter KC (2012) Evaluating the potential of polyester nanoparticles for per oral delivery of amphotericin B in treating visceral leishmaniasis. J Biomed Nanotechnol 8(4): 695–702. https://doi.org/10.1166/jbn.2012.1414

Jain V, Gupta A, Pawar VK, Asthana S, Jaiswal AK, Dube A, Chourasia MK (2014) Chitosan-assisted immunotherapy for intervention of experimental leishmaniasis via amphotericin B-loaded solid lipid nanoparticles. Appl Biochem Biotechnol 174(4):1309–1330. https://doi.org/10.1007/s12010-014-1084-y

Jamshaid H, Din FU, Nousheen K, Khan SU, Fatima A, Khan S, Choi HG, Khan GM (2023) Mannosylated imiquimod-terbinafine co-loaded transethosomes for cutaneous leishmaniasis; assessment of its anti-leishmanial potential, in vivo safety and immune response modulation. Biomater Adv 145:213266. https://doi.org/10.1016/j.bioadv.2022.213266

Javed B, Raja NI, Nadhman A, Mashwani ZUR (2020) Understanding the potential of bio-fabricated non-oxidative silver nanoparticles to eradicate *Leishmania* and plant bacterial pathogens. Appl Nanosci 10(6):2057–2067. https://doi.org/10.1007/s13204-020-01355-5

Kalel VC, Emmanouilidis L, Dawidowski M, Schliebs W, Sattler M, Popowicz GM, Erdmann R (2017) Inhibitors of glycosomal protein import provide new leads against trypanosomiasis. Microbial Cell (Graz, Austria) 4(7):229–232. https://doi.org/10.15698/mic2017.07.581

Kashif M, Hira SK, Upadhyaya A, Gupta U, Singh R, Paladhi A, Khan FI, Rub A, Manna PP (2019) *In silico* studies and evaluation of antiparasitic role of a novel pyruvate phosphate dikinase inhibitor in *Leishmania donovani* infected macrophages. Int J Antimicrob Agents 53(4): 508–514. https://doi.org/10.1016/j.ijantimicag.2018.12.011

Kavian Z, Alavizadeh SH, Golmohamadzadeh S, Badiee A, Khamesipour A, Jaafari MR (2019) Development of topical liposomes containing miltefosine for the treatment of *Leishmania major* infection in susceptible BALB/c mice. Acta Trop 196:142–149. https://doi.org/10.1016/j.actatropica.2019.05.018

Kawakami MYM, Zamora LO, Araújo RS, Fernandes CP, Ricotta TQN, de Oliveira LG, Queiroz-Junior CM, Fernandes AP, da Conceição EC, Ferreira LAM, Barros ALB, Aguiar MG, Oliveira AEMFM (2021) Efficacy of nanoemulsion with Pterodon emarginatus Vogel oleoresin for topical treatment of cutaneous leishmaniasis. Biomed Pharmacother. 134:111109. https://doi.org/10.1016/j.biopha.2020.111109

Khan MM, Zaidi SS, Siyal FJ, Khan SU, Ishrat G, Batool S, Mustapha O, Khan S, ud Din, F. (2023) Statistical optimization of co-loaded rifampicin and pentamidine polymeric nanoparticles for the treatment of cutaneous leishmaniasis. J Drug Deliv Sci Technol 79:104005. https://doi.org/10.1016/j.jddst.2022.104005

Khanra S, Juin SK, Jawed JJ, Ghosh S, Dutta S, Nabi SA, Dash J, Dasgupta D, Majumdar S, Banerjee R (2020) In vivo experiments demonstrate the potent antileishmanial efficacy of repurposed suramin in visceral leishmaniasis. PLoS Negl Trop Dis 14(8):e0008575. https://doi.org/10.1371/journal.pntd.0008575

Khatik R, Dwivedi P, Khare P, Kansal S, Dube A, Mishra PR, Dwivedi AK (2014) Development of targeted 1,2-diacyl-sn-glycero-3-phospho-l-serine-coated gelatin nanoparticles loaded with amphotericin B for improved *in vitro* and *in vivo* effect in leishmaniasis. Expert Opin Drug Deliv 11(5):633–646. https://doi.org/10.1517/17425247.2014.889678

Khokhar M, Shereen MA, Khan M, Khan RU, Sohail A, Khan IU, Khan IU, Khattak S (2022) In vitro efficacy of polymer coated miltefosine drug against leishmania tropica. J Parasitic Dis 46(2):366–376. https://doi.org/10.1007/s12639-021-01452-4

Kumar R, Sahoo GC, Pandey K, Das VNR, Topno RK, Ansari MY, Rana S, Das P (2016) Development of PLGA-PEG encapsulated miltefosine based drug delivery system against visceral leishmaniasis. Mater Sci Eng C Mater Biol Appl 59:748–753. https://doi.org/10.1016/j.msec.2015.10.083

Kumar R, Pandey K, Sahoo GC, Das S, Das VNR, Topno RK, Das P (2017) Development of high efficacy peptide coated iron oxide nanoparticles encapsulated amphotericin B drug delivery system against visceral leishmaniasis. Mater Sci Eng C 75:1465–1471

Kumar P, Shivam P, Mandal S, Prasanna P, Kumar S, Prasad SR, Kumar A, Das P, Ali V, Singh SK, Mandal D (2019) Synthesis, characterization, and mechanistic studies of a gold nanoparticle-amphotericin B covalent conjugate with enhanced antileishmanial efficacy and reduced cytotoxicity. Int J Nanomedicine 14:6073–6101. https://doi.org/10.2147/IJN.S196421

Kumar P, Chatterjee M, Das NK (2021) Post Kala-Azar Dermal Leishmaniasis: clinical features and differential diagnosis. Indian J Dermatol 66(1):24–33. https://doi.org/10.4103/ijd.IJD_602_20

Laniado-Laborín R, Cabrales-Vargas MN (2009) Amphotericin B: side effects and toxicity. Rev Iberoamericana Micol 26(4):223–227. https://doi.org/10.1016/j.riam.2009.06.003

Leal Pinto SM, Muehlmann LA, Ojeda LLM, Vera Arias AM, Cordero MVR, Santos MFMA, Azevedo RB, Rivero PE (2021) Nanoemulsions with chloroaluminium phthalocyanine and paromomycin for combined photodynamic and antibiotic therapy for cutaneous leishmaniasis. Infect Chemother 53(2):342–354. https://doi.org/10.3947/ic.2021.0010

Leañez J, Nuñez J, García-Marchan Y, Sojo F, Arvelo F, Rodriguez D, Buscema I, Alvarez-Aular A, Bello Forero JS, Kouznetsov VV, Serrano-Martín X (2019) Anti-leishmanial effect of spiro dihydroquinoline-oxindoles on volume regulation decrease and sterol biosynthesis of *Leishmania braziliensis*. Exp Parasitol. 198:31–38. https://doi.org/10.1016/j.exppara.2019.01.011

Lindoso JAL, Moreira CHV, Cunha MA, Queiroz IT (2018) Visceral leishmaniasis and HIV coinfection: current perspectives. HIV/AIDS (Auckland, N.Z.). 10:193–201. https://doi.org/10.2147/HIV.S143929

Loiseau PM, Pomel S, Croft SL (2020) Chitosan contribution to therapeutic and vaccinal approaches for the control of *Leishmaniasis*. Molecules (Basel, Switzerland) 25(18):4123. https://doi.org/10.3390/molecules25184123

Lopes RM, Gaspar MM, Pereira J, Eleutério CV, Carvalheiro M, Almeida AJ, Cruz ME (2014) Liposomes versus lipid nanoparticles: comparative study of lipid-based systems as oryzalin carriers for the treatment of leishmaniasis. J Biomed Nanotechnol 10(12):3647–3657. https://doi.org/10.1166/jbn.2014.1874

Maamri S, Benarous K, Yousfi M (2021) Identification of 3-methoxycarpachromene and masticadienonic acid as new target inhibitors against trypanothione reductase from *Leishmania infantum* using molecular docking and ADMET prediction. Molecules (Basel, Switzerland) 26(11):3335. https://doi.org/10.3390/molecules26113335

Maquiaveli CC, Lucon-Júnior JF, Brogi S, Campiani G, Gemma S, Vieira PC, Silva ER (2016) Verbascoside inhibits promastigote growth and arginase activity of *Leishmania amazonensis*. J Nat Products 79(5):1459–1463. https://doi.org/10.1021/acs.jnatprod.5b00875

Maroli M, Feliciangeli MD, Bichaud L, Charrel RN, Gradoni L (2013) *Phlebotomine sandflies* and the spreading of leishmaniases and other diseases of public health concern. Med Vet Entomol 27(2):123–147. https://doi.org/10.1111/j.1365-2915.2012.01034.x

Martin-Montes A, Santivañez-Veliz M, Moreno-Viguri E, Martín-Escolano R, Jiménez-Montes C, Lopez-Gonzalez C, Marín C, Sanmartín C, Gutiérrez Sánchez R, Sánchez-Moreno M, Pérez-Silanes S (2017) *In vitro* antileishmanial activity and iron superoxide dismutase inhibition of arylamine Mannich base derivatives. Parasitology 144(13):1783–1790. https://doi.org/10.1017/S0031182017001123

Matadamas-Martínez F, Hernández-Campos A, Téllez-Valencia A, Vázquez-Raygoza A, Comparán-Alarcón S, Yépez-Mulia L, Castillo R (2019) *Leishmania mexicana* trypanothione reductase inhibitors: computational and biological studies. Molecules (Basel, Switzerland) 24(18):3216. https://doi.org/10.3390/molecules24

McCall LI, El Aroussi A, Choi JY, Vieira DF, De Muylder G, Johnston JB, Chen S, Kellar D, Siqueira-Neto JL, Roush WR, Podust LM, McKerrow JH (2015) Targeting Ergosterol biosynthesis in *Leishmania donovani*: essentiality of sterol 14 alpha-demethylase. PLoS Negl Trop Dis 9(3):e0003588. https://doi.org/10.1371/journal.pntd.0003588

McGwire BS, Satoskar AR (2014) Leishmaniasis: clinical syndromes and treatment. QJM Monthly J Assoc Phys 107(1):7–14. https://doi.org/10.1093/qjmed/hct116

Medina JM, Rodrigues JC, De Souza W, Atella GC, Barrabin H (2012) Tomatidine promotes the inhibition of 24-alkylated sterol biosynthesis and mitochondrial dysfunction in *Leishmania amazonensis* promastigotes. Parasitology 139(10):1253–1265. https://doi.org/10.1017/S0031182012000522

Micale N, Piperno A, Mahfoudh N, Schurigt U, Schultheis M, Mineo PG, Schirmeister T, Scala A, Grassi G (2015) A hyaluronic acid–pentamidine bioconjugate as a macrophage mediated drug targeting delivery system for the treatment of leishmaniasis. RSC Adv 5(116):95545–95550

Mittra B, Cortez M, Haydock A, Ramasamy G, Myler PJ, Andrews NW (2013) Iron uptake controls the generation of *Leishmania* infective forms through regulation of ROS levels. J Exp Med 210(2):401–416. https://doi.org/10.1084/jem.20121368

Mohammed HB, Ali AA, Idriss MI, Gasmelseid KM, Yousif MM, Ibrahem AA, Abdallah TM (2016) Prevalence of Hepatitis B, Hepatitis C, HIV and malaria co infection among patients infected with visceral Leishmaniasis in Gedarif, Eastern Sudan. Global J Infect Dis Clin Res 2(1):021–024

Mohtasebi S, Mohebali M, Elikaee S, Akhoundi B, Foroushani AR, Teimouri A, Yarizadeh H (2019) *In vitro* and *in vivo* anti-parasitic activity of biogenic antimony sulfide nanoparticles on *Leishmania major* (MRHO/IR/75/ER). Parasitol Res 118(9):2669–2678. https://doi.org/10.1007/s00436-019-06382-y

Mondal D, Bern C, Ghosh D, Rashid M, Molina R, Chowdhury R, Nath R, Ghosh P, Chapman LAC, Alim A, Bilbe G, Alvar J (2019) Quantifying the infectiousness of Post-Kala-Azar Dermal Leishmaniasis toward sand flies. Clin Infect Dis 69(2):251–258. https://doi.org/10.1093/cid/ciy891

Montoya MC, Krysan DJ (2018) Repurposing estrogen receptor antagonists for the treatment of infectious disease. mBio 9(6):e02272–e02218. https://doi.org/10.1128/mBio.02272-18

Mota SGR, Mercaldi GF, Pereira JGC, Oliveira PSL, Rodriguez A, Cordeiro AT. First Nonphosphorylated Inhibitors of Phosphoglucose Isomerase Identified by Chemical Library Screening. SLAS Discov. 2018;23(10):1051-1059. https://doi.org/10.1177/2472555218787468

Mukherjee S, Ray S, Thakur RS (2009) Solid lipid nanoparticles: a modern formulation approach in drug delivery system. Indian J Pharm Sci 71(4):349–358. https://doi.org/10.4103/0250-474X.57282

Mukherjee S, Xu W, Hsu FF, Patel J, Huang J, Zhang K (2019) Sterol methyltransferase is required for optimal mitochondrial function and virulence in *Leishmania major*. Mol Microbiol 111(1):65–81. https://doi.org/10.1111/mmi.14139

Mukherjee D, Yousuf M, Dey S, Chakraborty S, Chaudhuri A, Kumar V, Sarkar B, Nath S, Hussain A, Dutta A, Mishra T, Roy BG, Singh S, Chakraborty S, Adhikari S, Pal C (2020) Targeting the trypanothione reductase of tissue-residing *Leishmania* in hosts' reticuloendothelial system: a flexible water-soluble ferrocenylquinoline-based preclinical drug candidate. J Med Chem 63(24):15621–15638. https://doi.org/10.1021/acs.jmedchem.0c00690

Nadhman A, Nazir S, Khan MI, Arooj S, Bakhtiar M, Shahnaz G, Yasinzai M (2014) PEGylated silver doped zinc oxide nanoparticles as novel photosensitizers for photodynamic therapy against Leishmania. Free Radic Biol Med 77:230–238. https://doi.org/10.1016/j.freeradbiomed.2014.09.005

Newton AM, Kaur S (2019) Solid lipid nanoparticles for skin and drug delivery: methods of preparation and characterization techniques and applications. In: Nanoarchitectonics in biomedicine. William Andrew Publishing, Norwich, pp 295–334

No JH (2016) Visceral leishmaniasis: Revisiting current treatments and approaches for future discoveries. Acta Trop 155:113–123. https://doi.org/10.1016/j.actatropica.2015.12.016

Norouzi R, Mirzaei F, Siyadatpanah A, Sadati SJA (2022) Evaluation of anti-leishmanial effect of selenium nanoparticles on *Leishmania major* promastigotes in vitro. Avicenna J Clin Microbiol Infect 9(3):119–123

Nunes TAL, Santos MM, de Oliveira MS, de Sousa JMS, Rodrigues RRL, Sousa PSA, de Araújo AR, Pereira ACTDC, Ferreira GP, Rocha JA, Rodrigues Junior V, da Silva MV, Rodrigues KADF (2021) Curzerene antileishmania activity: effects on *Leishmania amazonensis* and possible action mechanisms. Int Immunopharmacol 100:108130. https://doi.org/10.1016/j.intimp.2021.108130

Oliveira SSC, Santos VS, Devereux M, McCann M, Santos ALS, Branquinha MH (2023) The anti-leishmania amazonensis and anti-leishmania chagasi action of Copper(II) and Silver(I) 1,10-Phenanthroline-5,6-dione coordination compounds. Pathogens (Basel, Switzerland) 12(1):70. https://doi.org/10.3390/pathogens12010070

Ortalli M, Ilari A, Colotti G, De Ionna I, Battista T, Bisi A, Gobbi S, Rampa A, Di Martino RMC, Gentilomi GA, Varani S, Belluti F (2018) Identification of chalcone-based antileishmanial agents targeting trypanothione reductase. Eur J Med Chem 152:527–541. https://doi.org/10.1016/j.ejmech.2018.04.057

Pandey RK, Kumbhar BV, Srivastava S, Malik R, Sundar S, Kunwar A, Prajapati VK (2017) Febrifugine analogues as *Leishmania donovani* trypanothione reductase inhibitors: binding energy analysis assisted by molecular docking, ADMET and molecular dynamics simulation. J Biomol Struct Dyn 35(1):141–158. https://doi.org/10.1080/07391102.2015.1135298

Parvez S, Yadagiri G, Gedda MR, Singh A, Singh OP, Verma A, Sundar S, Mudavath SL (2020) Modified solid lipid nanoparticles encapsulated with Amphotericin B and Paromomycin: an effective oral combination against experimental murine visceral leishmaniasis. Sci Rep 10(1): 12243. https://doi.org/10.1038/s41598-020-69276-5

Passalacqua TG, Torres FA, Nogueira CT, de Almeida L, Del Cistia ML, dos Santos MB, Dutra LA, Bolzani VS, Regasini LO, Graminha MA, Marchetto R, Zottis A (2015) The 2′,4′--dihydroxychalcone could be explored to develop new inhibitors against the glycerol-3-phosphate dehydrogenase from *Leishmania* species. Bioorg Med Chem Lett 25(17):3564–3568. https://doi.org/10.1016/j.bmcl.2015.06.085

Peralta MF, Guzman ML, Bracamonte ME, Marco JD, Olivera ME, Carrer DC, Barroso PA (2021) Efficacy of topical risedronate and risedronate - Eudragit E complex in a model of cutaneous leishmaniasis induced by Leishmania (Leishmania) amazonensis. Heliyon 7(5):e07136. https://doi.org/10.1016/j.heliyon.2021.e07136

Peretti E, Miletto I, Stella B, Rocco F, Berlier G, Arpicco S (2018) Strategies to obtain encapsulation and controlled release of pentamidine in mesoporous silica nanoparticles. Pharmaceutics 10(4):195. https://doi.org/10.3390/pharmaceutics10040195

Pingle P, Mourya A, Namdeo M, Babu KC, Veerabomma H, Maurya R, Singh PK, Mehra NK, Srivastava S, Madan J (2023) Andrographolide-Soya-L-α-phosphatidyl choline complex augmented solubility and drug delivery in leishmania donovani, a causative agent for cutaneous and visceral leishmaniasis. AAPS PharmSciTech 24(1):46. https://doi.org/10.1208/s12249-023-02507-w

Pinto-Martinez AK, Rodriguez-Durán J, Serrano-Martin X, Hernandez-Rodriguez V, Benaim G (2018) Mechanism of action of miltefosine on *Leishmania donovani* involves the impairment of acidocalcisome function and the activation of the sphingosine-dependent plasma membrane Ca2+ channel. Antimicrob Agents Chemother 62(1):e01614–e01617. https://doi.org/10.1128/AAC.01614-17

Pramanik PK, Chakraborti S, Bagchi A, Chakraborti T (2020) Bioassay-based Corchorus capsularis L. leaf-derived β-sitosterol exerts antileishmanial effects against *Leishmania donovani* by targeting trypanothione reductase. Sci Rep. 10(1):20440. https://doi.org/10.1038/s41598-020-77066-2

Rackham MD, Yu Z, Brannigan JA, Heal WP, Paape D, Barker KV, Wilkinson AJ, Smith DF, Leatherbarrow RJ, Tate EW (2015) Discovery of high affinity inhibitors of *Leishmania donovani N*-myristoyltransferase. MedChemComm 6(10):1761–1766. https://doi.org/10.1039/c5md00241a

Rahman F, Tabrez S, Ali R, Akand SK, Zahid M, Alaidarous MA, Alsaweed M, Alshehri BM, Banawas S, Bin Dukhyil AA, Rub A (2021) Virtual screening of natural compounds for potential inhibitors of Sterol C-24 methyltransferase of *Leishmania donovani* to overcomeleishmaniasis. J Cell Biochem. https://doi.org/10.1002/jcb.29944. Advance online publication

Raj S, Sasidharan S, Balaji SN, Saudagar P (2020) An overview of biochemically characterized drug targets in metabolic pathways of *Leishmania* parasite. Parasitol Res 119(7):2025–2037. https://doi.org/10.1007/s00436-020-06736-x

Raj S, Sasidharan S, Tripathi T, Saudagar P (2022) Biofunctionalized Chrysin-conjugated gold nanoparticles neutralize *Leishmania* parasites with high efficacy. Int J Biol Macromol 205:211–219. https://doi.org/10.1016/j.ijbiomac.2022.02.047

Reimão JQ, Miguel DC, Taniwaki NN, Trinconi CT, Yokoyama-Yasunaka JK, Uliana SR (2014) Antileishmanial activity of the estrogen receptor modulator raloxifene. PLoS Negl Trop Dis 8(5):e2842. https://doi.org/10.1371/journal.pntd.0002842

Reithinger R, Dujardin JC, Louzir H, Pirmez C, Alexander B, Brooker S (2007) Cutaneous leishmaniasis. Lancet Infect Dis 7(9):581–596. https://doi.org/10.1016/S1473-3099(07)70209-8

Revuelto A, Ruiz-Santaquiteria M, de Lucio H, Gamo A, Carriles AA, Gutiérrez KJ, Sánchez-Murcia PA, Hermoso JA, Gago F, Camarasa MJ, Jiménez-Ruiz A, Velázquez S (2019) Pyrrolopyrimidine vs imidazole-phenyl-thiazole scaffolds in nonpeptidic dimerization inhibitors of *Leishmania infantum* trypanothione reductase. ACS Infect Dis 5(6):873–891. https://doi.org/10.1021/acsinfecdis.8b00355

Ribeiro TG, Chávez-Fumagalli MA, Valadares DG, França JR, Rodrigues LB, Duarte MC, Lage PS, Andrade PH, Lage DP, Arruda LV, Abánades DR, Costa LE, Martins VT, Tavares CA, Castilho RO, Coelho EA, Faraco AA (2014) Novel targeting using nanoparticles: an approach to the development of an effective anti-leishmanial drug-delivery system. Int J Nanomedicine 9:877–890. https://doi.org/10.2147/IJN.S55678

Ribeiro JBP, Miranda-Vilela AL, Graziani D, de Aguiar Gomes MR, Amorim AAS, Garcia RD, de Souza Filho J, Tedesco AC, Primo FL, Moreira JR, Lima AV, Sampaio RNR (2016) Evaluation of the efficacy of systemic miltefosine associated with photodynamic therapy with liposomal chloroaluminium phthalocyanine in the treatment of cutaneous leishmaniasis caused by Leishmania (L.) amazonensis in C57BL/6 mice. Photodiagn Photodyn Ther 13:282–290

Richard SA (2021) Exploring the pivotal immunomodulatory and anti-inflammatory potentials of glycyrrhizic and glycyrrhetinic acids. Mediat Inflamm 2021:6699560. https://doi.org/10.1155/2021/6699560

Rodrigues RF, Castro-Pinto D, Echevarria A, dos Reis CM, Del Cistia CN, Sant'Anna CM, Teixeira F, Castro H, Canto-Cavalheiro M, Leon LL, Tomás A (2012) Investigation of trypanothione reductase inhibitory activity by 1,3,4-thiadiazolium-2-aminide derivatives and molecular docking studies. Bioorg Med Chem 20(5):1760–1766. https://doi.org/10.1016/j.bmc.2012.01.009

Rodrigues IA, Ramos AS, Falcão DQ, Ferreira JLP, Basso SL, Silva JRA, Amaral ACF (2018) Development of nanoemulsions to enhance the antileishmanial activity of *Copaifera paupera* oleoresins. Biomed Res Int 2018:9781724. https://doi.org/10.1155/2018/9781724

Rub A, Shaker K, Kashif M, Arish M, Dukhyil AAB, Alshehri BM, Alaidarous MA, Banawas S, Amir K (2019) Repurposing glyburide as antileishmanial agent to fight against *Leishmaniasis*. Protein Pept Lett 26(5):371–376. https://doi.org/10.2174/0929866526666190301114012

Saccoliti F, Angiulli G, Pupo G, Pescatori L, Madia VN, Messore A, Colotti G, Fiorillo A, Scipione L, Gramiccia M, Di Muccio T, Di Santo R, Costi R, Ilari A (2017) Inhibition of *Leishmania infantum* trypanothione reductase by diaryl sulfide derivatives. J Enzyme Inhib Med Chem 32(1):304–310. https://doi.org/10.1080/14756366.2016.1250755

Saha S, Acharya C, Pal U, Chowdhury SR, Sarkar K, Maiti NC, Jaisankar P, Majumder HK (2016) A novel spirooxindole derivative inhibits the growth of *Leishmania donovani* parasites both in vitro and in vivo by targeting type IB topoisomerase. Antimicrob Agents Chemother 60(10):6281–6293. https://doi.org/10.1128/AAC.00352-16

Saini I, Joshi J, Kaur S (2022) Unwelcome prevalence of leishmaniasis with several other infectious diseases. Int Immunopharmacol 110:109059. https://doi.org/10.1016/j.intimp.2022.109059

Salim MW, Shabbir K, Yousaf AM, Choi HG, Khan GM (2020) Preparation, in-vitro and in-vivo evaluation of Rifampicin and Vancomycin Co-loaded transfersomal gel for the treatment of cutaneous leishmaniasis. J Drug Deliv Sci Technol 60:101996

Sanderson L, Yardley V, Croft SL (2014) Activity of anti-cancer protein kinase inhibitors against *Leishmania* spp. J Antimicrob Chemother 69(7):1888–1891. https://doi.org/10.1093/jac/dku069

Saqib M, Ali Bhatti AS, Ahmad NM, Ahmed N, Shahnaz G, Lebaz N, Elaissari A (2020) Amphotericin B Loaded polymeric nanoparticles for treatment of *Leishmania* infections (Basel, Switzerland). Nanomaterials 10(6):1152. https://doi.org/10.3390/nano10061152

Sasidharan S, Saudagar P (2022) Gold and silver nanoparticles functionalized with 4′,7-dihydroxyflavone exhibit activity against *Leishmania donovani*. Acta Trop 231:106448. https://doi.org/10.1016/j.actatropica.2022.106448

Scala A, Cordaro M, Grassi G, Piperno A, Barberi G, Cascio A, Risitano F (2014) Direct synthesis of C3-mono-functionalized oxindoles from N-unprotected 2-oxindole and their antileishmanial activity. Bioorg Med Chem 22(3):1063–1069. https://doi.org/10.1016/j.bmc.2013.12.039

Scioli Montoto S, Muraca G, Ruiz ME (2020) Solid lipid nanoparticles for drug delivery: pharmacological and biopharmaceutical aspects. Front Mol Biosci 7:587997. https://doi.org/10.3389/fmolb.2020.587997

Selvapandiyan A, Croft SL, Rijal S, Nakhasi HL, Ganguly NK (2019) Innovations for the elimination and control of visceral leishmaniasis. PLoS Negl Trop Dis 13(9):e0007616. https://doi.org/10.1371/journal.pntd.0007616

Shahnaz G, Edagwa BJ, McMillan J, Akhtar S, Raza A, Qureshi NA, Yasinzai M, Gendelman HE (2017) Development of mannose-anchored thiolated amphotericin B nanocarriers for treatment of visceral leishmaniasis. Nanomedicine (Lond) 12(2):99–115. https://doi.org/10.2217/nnm-2016-0325

Shakeel M, Kiani MH, Sarwar HS, Akhtar S, Rauf A, Ibrahim IM, Ajalli N, Shahnaz G, Rahdar A, Díez-Pascual AM (2023) Emulgel-loaded mannosylated thiolated chitosan-coated silver nanoparticles for the treatment of cutaneous leishmaniasis. Int J Biol Macromol 227:1293–1304. https://doi.org/10.1016/j.ijbiomac.2022.11.326

Shakya N, Bajpai P, Gupta S (2011) Therapeutic switching in *Leishmania* chemotherapy: a distinct approach towards unsatisfied treatment needs. J Parasitic Dis 35(2):104–112. https://doi.org/10.1007/s12639-011-0040-9

Sharma VK, Bharatam PV (2021) Identification of selective inhibitors of *Ld*DHFR enzyme using pharmacoinformatic methods. J Comput Biol 28(1):43–59. https://doi.org/10.1089/cmb.2019.0332

Shokri A, Saeedi M, Fakhar M, Morteza-Semnani K, Keighobadi M, Hosseini Teshnizi S, Kelidari HR, Sadjadi S (2017) Antileishmanial activity of *Lavandula angustifolia* and *Rosmarinus Officinalis* essential oils and nano-emulsions on *Leishmania major* (MRHO/IR/75/ER). Iran J Parasitol 12(4):622–631

Singh S, Mukherjee A, Khomutov AR, Persson L, Heby O, Chatterjee M, Madhubala R (2007) Antileishmanial effect of 3-aminooxy-1-aminopropane is due to polyamine depletion. Antimicrob Agents Chemother 51(2):528–534. https://doi.org/10.1128/AAC.01055-06

Singh A, Yadagiri G, Negi M, Kushwaha AK, Singh OP, Sundar S, Mudavath SL (2022) Carboxymethyl chitosan modified lipid nanoformulations as a highly efficacious and biocompatible oral anti-leishmanial drug carrier system. Int J Biol Macromol 204:373–385. https://doi.org/10.1016/j.ijbiomac.2022.02.006

Sirajuddin M, Mahdi AA, Srivastava A (2023) Assessment of Leishmania donovani Inhibition by Zinc Sulfate: an In vitro and In silico Investigation. Proc Natl Acad Sci India Sect B Biol Sci 93:91–96. https://doi.org/10.1007/s40011-022-01400-z

Soflaei S, Dalimi A, Ghaffarifar F, Shakibaie M, Shahverdi AR, Shafiepour M (2012) *In vitro* antiparasitic and apoptotic effects of antimony sulfide nanoparticles on *Leishmania infantum*. J Parasitol Res 2012:756568. https://doi.org/10.1155/2012/756568

Sohail A, Khan RU, Khan M, Khokhar M, Ullah S, Ali A, Bilal H, Khattak S, Khan M, Ahmad B (2021) Comparative efficacy of amphotericin B-loaded chitosan nanoparticles and free amphotericin B drug against Leishmania tropica. Bull Natl Res Centre 45(1):1–9

Sousa-Batista AJ, Arruda-Costa N, Pacienza-Lima W, Carvalho-Gondim F, Santos RF, Da-Silva SAG, Ré MI, Rossi-Bergmann B (2022) In vivo safety and efficacy of chalcone-loaded microparticles with modified polymeric matrix against cutaneous leishmaniasis. Pharmaceutics 15(1):51. https://doi.org/10.3390/pharmaceutics15010051

Sundar S, Chakravarty J (2010) Liposomal amphotericin B and leishmaniasis: dose and response. J Global Infect Dis 2(2):159–166. https://doi.org/10.4103/0974-777X.62886

Tabrez S, Rahman F, Ali R, Akand SK, Alaidarous MA, Alshehri BM, Banawas S, Dukhyil AAB, Rub A (2021a) Targeting sterol alpha-14 demethylase of *Leishmania donovani* to fight against leishmaniasis. J Cell Biochem. https://doi.org/10.1002/jcb.29922. Advance online publication

Tabrez S, Rahman F, Ali R, Muhammad F, Alshehri BM, Alaidarous MA, Banawas S, Dukhyil AAB, Rub A (2021b) Repurposing of FDA-approved drugs as inhibitors of sterol C-24 methyltransferase of *Leishmania donovani* to fight against leishmaniasis. Drug Dev Res 82(8):1154–1161. https://doi.org/10.1002/ddr.21820

Tavakoli P, Ghaffarifar F, Delavari H, Shahpari N (2019) Efficacy of manganese oxide (Mn2O3) nanoparticles against *Leishmania major in vitro* and *in vivo*. J Trace Elements Med Biol Organ Soc Minerals Trace Elements (GMS) 56:162–168. https://doi.org/10.1016/j.jtemb.2019.08.003

Teixeira BVF, Teles ALB, Silva SGD, Brito CCB, Freitas HF, Pires ABL, Froes TQ, Castilho MS (2019) Dual and selective inhibitors of pteridine reductase 1 (PTR1) and dihydrofolate reductase-thymidylate synthase (DHFR-TS) from *Leishmania chagasi*. J Enzyme Inhib Med Chem 34(1):1439–1450. https://doi.org/10.1080/14756366.2019.1651311

Thapa R, Mondal S, Riikonen J, Rantanen J, Näkki S, Nissinen T, Närvänen A, Lehto VP (2021) Biogenic nanoporous silicon carrier improves the efficacy of buparvaquone against resistant visceral leishmaniasis. PLoS Negl Trop Dis 15(6):e0009533. https://doi.org/10.1371/journal.pntd.0009533

Tiwari B, Pahuja R, Kumar P, Rath SK, Gupta KC, Goyal N (2017) Nanotized curcumin and miltefosine, a potential combination for treatment of experimental visceral *Leishmaniasis*. Antimicrob Agents Chemother 61(3):e01169–e01116. https://doi.org/10.1128/AAC.01169-16

Tuon FF, Dantas LR, de Souza RM, Ribeiro VST, Amato VS (2022) Liposomal drug delivery systems for the treatment of leishmaniasis. Parasitol Res 121(11):3073–3082. https://doi.org/10.1007/s00436-022-07659-5

Turcano L, Torrente E, Missineo A, Andreini M, Gramiccia M, Di Muccio T, Genovese I, Fiorillo A, Harper S, Bresciani A, Colotti G, Ilari A (2018) Identification and binding mode of a novel *Leishmania* Trypanothione reductase inhibitor from high throughput screening. PLoS Negl Trop Dis 12(11):e0006969. https://doi.org/10.1371/journal.pntd.0006969

Turner TL, Nguyen VH, McLauchlan CC, Dymon Z, Dorsey BM, Hooker JD, Jones MA (2012) Inhibitory effects of decavanadate on several enzymes and *Leishmania tarentolae* in vitro. J Inorg Biochem 108:96–104. https://doi.org/10.1016/j.jinorgbio.2011.09.009

Usama Ashhar M, Vyas P, Vohora D, Kumar Sahoo P, Nigam K, Dang S, Ali J, Baboota S (2022) Amelioration of oxidative stress utilizing nanoemulsion loaded with bromocriptine and glutathione for the management of Parkinson's disease. Int J Pharm 618:121683. https://doi.org/10.1016/j.ijpharm.2022.121683

Varma N, Naseem S (2010) Hematologic changes in visceral leishmaniasis/kala azar. Indian J Hematol Blood Transfusion 26(3):78–82. https://doi.org/10.1007/s12288-010-0027-1

Varshosaz J, Arbabi B, Pestehchian N, Saberi S, Delavari M (2018) Chitosan-titanium dioxide-glucantime nanoassemblies effects on promastigote and amastigote of *Leishmania major*. Int J Biol Macromol. 107(PtA):212–221. https://doi.org/10.1016/j.ijbiomac.2017.08.177

Veronica J, Chandrasekaran S, Dayakar A, Devender M, Prajapati VK, Sundar S, Maurya R (2019) Iron superoxide dismutase contributes to miltefosine resistance in *Leishmania donovani*. FEBS J 286(17):3488–3503. https://doi.org/10.1111/febs.14923

Vickers TJ, Beverley SM (2011) Folate metabolic pathways in *Leishmania*. Essays Biochem 51: 63–80. https://doi.org/10.1042/bse0510063

Vidhya VM, Lakshmi BS, Ponnuraj K (2020) *In silico* prediction of a new lead compound targeting enolase of *trypanosomatids* through structure-based virtual screening and molecular dynamic studies. J Mol Model 26(2):23. https://doi.org/10.1007/s00894-019-4284-0

Vishwakarma P, Parmar N, Chandrakar P, Sharma T, Kathuria M, Agnihotri PK, Siddiqi MI, Mitra K, Kar S (2018) Ammonium trichloro [1, 2-ethanediolato-O, O′]-tellurate cures experimental visceral leishmaniasis by redox modulation of Leishmania donovani trypanothione reductase and inhibiting host integrin linked PI3K/Akt pathway. Cell Mol Life Sci 75(3): 563–588

Vitorino LS, Dos Santos TC, Bessa IAA, Santos ECS, Verçoza BRF, de Oliveira LAS, Rodrigues JCF, Ronconi CM (2022) Amphotericin-B-loaded polymer-functionalized reduced graphene oxides for Leishmania amazonensis chemo-photothermal therapy. Colloids Surf B Biointerfaces. 209(1 Pt):112169. https://doi.org/10.1016/j.colsurfb.2021.112169

Wadanambi PM, Mannapperuma U (2021) Computational study to discover potent phytochemical inhibitors against drug target, squalene synthase from *Leishmania donovani*. Heliyon 7(6): e07178. https://doi.org/10.1016/j.heliyon.2021.e07178

Want MY, Islamuddin M, Chouhan G, Ozbak HA, Hemeg HA, Dasgupta AK, Chattopadhyay AP, Afrin F (2015) Therapeutic efficacy of artemisinin-loaded nanoparticles in experimental visceral leishmaniasis. Colloids Surf B: Biointerfaces 130:215–221. https://doi.org/10.1016/j.colsurfb.2015.04.013

WHO (2022a). https://www.who.int/news-room/fact-sheets/detail/leishmaniasis. Last Accessed on 26th November, 2022

WHO (2022b). https://www.who.int/data/gho/data/themes/topics/topic-details/GHO/leishmaniasis – Data of Sep 21: Retrieved on Feb 19th 2022

Wiwanitkit V (2012) Interest in paromomycin for the treatment of visceral leishmaniasis (kala-azar). Ther Clin Risk Manag 8:323–328. https://doi.org/10.2147/TCRM.S30139

Yamamoto ES, de Jesus JA, Bezerra-Souza A, Brito JR, Lago JHG, Laurenti MD, Passero LFD (2020) Tolnaftate inhibits ergosterol production and impacts cell viability of *Leishmania sp.* Bioorg Chem. 102:104056. https://doi.org/10.1016/j.bioorg.2020.104056

Yuan M, Vásquez-Valdivieso MG, McNae IW, Michels PAM, Fothergill-Gilmore LA, Walkinshaw MD (2017) Structures of *Leishmania* Fructose-1,6-bisphosphatase reveal species-specific differences in the mechanism of allosteric inhibition. J Mol Biol 429(20): 3075–3089. https://doi.org/10.1016/j.jmb.2017.08.010

Zeiman E, Greenblatt CL, Elgavish S, Khozin-Goldberg I, Golenser J (2008) Mode of action of fenarimol against *Leishmania spp.* J Parasitol 94(1):280–286. https://doi.org/10.1645/GE-1259.1

Zielińska A, Carreiró F, Oliveira AM, Neves A, Pires B, Venkatesh DN, Durazzo A, Lucarini M, Eder P, Silva AM, Santini A, Souto EB (2020) Polymeric nanoparticles: production, characterization, toxicology and ecotoxicology. Molecules (Basel, Switzerland) 25(16):3731. https://doi.org/10.3390/molecules25163731

Zijlstra EE (2016) The immunology of post-kala-azar dermal leishmaniasis (PKDL). Parasit Vectors 9(1):464. https://doi.org/10.1186/s13071-016-1721-0

Miltefosine Unresponsiveness in Visceral Leishmaniasis 12

Ruchi Singh, Aditya Verma, Sushmita Ghosh, Vinay Kumar, Angamuthu Selvapandiyan, and Poonam Salotra

Abstract

Treatment failure in visceral leishmaniasis (VL; kala-azar) is an increasing problem worldwide, primarily due to drug resistance. Pentavalent antimonials were the mainstay of treatment for decades. However, during the last 20 years, their use has been superseded by other drugs in the Indian subcontinent (ISC) due to the emergence of resistance. Miltefosine, the only oral drug, was registered to treat visceral leishmaniasis (VL) in 2002 in ISC. Soon after its introduction, the drug was considered a treatment of choice in the kala-azar elimination program in the ISC. However, reports of relapses and a decline in its efficacy have raised significant concerns in recent years. Currently, a single-dose regimen of liposomal amphotericin B is the first-line treatment for VL in India.

Understanding the factors responsible for the development of drug resistance will help manage the disease, especially in endemic areas. Treatment outcome is governed by several factors related to host-parasite and drug kinetics. Host-related factors include host immunity or host nutritional status; for example, enhanced Th1-type response is crucial for effective treatment outcomes. Factors related to the drug, such as long half-life, inappropriate dosing, or self-administration of the drug, may lead to the exposure of parasites to the subtherapeutic dosage and thus be responsible for the selection of drug-resistant parasites. Similarly, several parasite-related factors impact the outcome of the treatment. *Leishmania* parasites can adapt to various environmental shifts, including

R. Singh (✉) · A. Verma · S. Ghosh · V. Kumar · P. Salotra
ICMR-National Institute of Pathology, Safdarjung Hospital Campus, New Delhi, India
e-mail: ruchisingh.nip@gov.in

A. Selvapandiyan
Department of Molecular Medicine, Jamia Hamdard, New Delhi, India
e-mail: selvapandiyan@jamiahamdard.ac.in

© The Author(s), under exclusive license to Springer Nature Singapore Pte Ltd. 2023

A. Selvapandiyan et al. (eds.), *Challenges and Solutions Against Visceral Leishmaniasis*, https://doi.org/10.1007/978-981-99-6999-9_12

exposure to the drug in the field. For example, parasite virulence, fitness to host immune responses, oxidative stress evasion, host defense mechanism modulation, metabolic adaptations, and other unknown factors contribute to drug susceptibility and treatment response. This chapter covers various factors contributing to declining efficacy and relapses following miltefosine treatment. Further, the genomic and transcriptomic alterations responsible for the phenotypic alterations in miltefosine-resistant parasites are discussed.

Keywords

Visceral leishmaniasis · Miltefosine · Miltefosine resistance · Drug resistance · Parasite factors

12.1 Introduction

Leishmania parasites are obligate intracellular pathogens responsible for a group of diseases ranging from self-healing cutaneous (cutaneous leishmaniasis (CL)) lesions to potentially fatal visceral (visceral leishmaniasis (VL)) disease. Thirty species of this protozoan parasite cause infection in mammals, out of which about 21 are pathogenic to humans (Centers for Disease Control and Prevention; DPDx 2017). Without any vaccine for human use and effective vector control, current strategies to control leishmaniasis primarily depend on early diagnosis and chemotherapy (Selvapandiyan et al. 2019); however, very few therapeutic options are available. Pentavalent antimonials (SbVs), the mainstay treatment for VL until the year 2000, have lost efficacy in regions with high endemicity in India (northern Bihar) and adjoining areas of Nepal, where about two-thirds of patients were refractory to SbV treatment (Stauch et al. 2012). The drug is no longer recommended in the Indian subcontinent (ISC) for VL treatment due to the emergence of resistance. However, the drug is used in combination with paromomycin as a first-line treatment for VL in East Africa (Sundar and Singh 2018; WHO 2010).

Amphotericin B, a macrolide polyene antifungal agent, was found highly effective for VL treatment (Alvar et al. 2006). The disadvantage of severe adverse reactions and prolonged hospitalization for the administration of amphotericin B pushed the development of a liposomal formulation of amphotericin B (L-AmB). L-AmB showed improved efficacy and safety profile. The World Health Organization (WHO) has recommended a single dose of 10 mg/kg of L-AmB as a first-line treatment in ISC, which showed a more than 95% cure rate (WHO 2010). Although the drug is at low risk of developing resistance in *Leishmania* parasites, there are several reports of in vitro resistance selection (Al-Mohammed et al. 2005; Mbongo et al. 1998; Pountain et al. 2019; Purkait et al. 2012). Additionally, treatment failures with amphotericin B (Eichenberger et al. 2017; Morizot et al. 2016) and the isolation of amphotericin-B-resistant *Leishmania* parasites (Srivastava et al. 2011) have also been reported.

Miltefosine (hexadecylphosphocholine) is the only effective oral drug available for leishmaniasis treatment. The drug was initially developed as an anticancer agent that has proven effective against VL (Jha et al. 1999; Sundar et al. 2002) and CL (Machado et al. 2010; Sindermann and Engel 2006; Soto et al. 2004). In 2005, miltefosine monotherapy was recommended as a first line of treatment in the ISC in the kala-azar elimination program (Dhillon et al. 2008); however, due to a decline in the final cure rate and an increase in relapses (Sundar et al. 2012), the drug has been abandoned as the primary treatment of choice. Treatment failure and a drop in the efficacy of miltefosine have also been reported in studies from Nepal (Ostyn et al. 2014; Rijal et al. 2013), although only a few completely miltefosine-resistant isolates have been identified to date. Further, a lower efficacy of 43% with a standard 28-day treatment was reported in *Leishmania infantum* infections in Brazil due to natural miltefosine resistance in parasites (Carnielli et al. 2019).

Aminoglycoside antibiotic paromomycin was registered for VL treatment in India in 2006. The drug showed a final cure rate of 94.6%, with no evidence of relapse of up to 6 months of follow-up. The drug has increasingly been tested in combination and has shown promising efficacy with antimonials (Kimutai et al. 2017), L-AmB, and miltefosine (Rahman et al. 2017). Resistance to paromomycin has not been reported yet; however, the in vitro selection of paromomycin-resistant isolates has been reported in several studies (Hendrickx et al. 2012; Shaw et al. 2019; Verma et al. 2017).

Drug resistance in *Leishmania* depends on several factors, which are inadequately understood. It is imperative to explore the causes responsible for the emergence of drug-resistant phenotypes as this will help manage the disease. Here, we discuss the various factors, such as drug kinetics and parasite- or host-related factors, which govern the development of miltefosine resistance in *Leishmania* parasites.

12.2 Worldwide VL Epidemiology

The global burden of VL remains challenging to determine accurately due to various factors, such as poor surveillance, underreporting, and limited data credibility. Nevertheless, studies by WHO's Leishmaniasis Control Program and others have attempted to estimate the incidence and distribution of VL globally, highlighting the need for improved surveillance systems and data collection methods to better understand the true burden of this neglected tropical disease. VL is endemic to some geographic regions, despite being encountered on every continent except Oceania (Torres-Guerrero et al. 2017). The three major eco-epidemiological hotspots for VL were East Africa (57% of the total cases worldwide), the Indian subcontinent (18% of the total cases worldwide), and Brazil (16% of the total cases worldwide). In 2020, six countries (Brazil, Ethiopia, Eritrea, Kenya, India, and Sudan) reported more than 1000 VL cases, accounting for 79% of all cases globally. The number of cases decreased to less than 2500, a 94% decrease (2295/36 920) compared to previous years. The trend of decreasing VL cases in Africa continued in 2020, despite outbreaks in Chad and Kenya. Since 2018, the African region has reported more VL cases than the Southeast Asian region for 3 consecutive years (WHO 2021). Although China reported 166 and 215 VL cases in 2019 and 2020, respectively, representing less than 1% of the global VL burden, it is still considered a high-burden country based on the operational definition (WHO 2021).

12.3 Miltefosine Treatment and Relapses

Miltefosine was the first-line treatment of choice in the ISC for 10 years (2002–2012). Later, after a decade of its use, several reports of decreased efficacy and relapses followed miltefosine treatment. Various trials on miltefosine efficacy revealed diverse outcomes (ranging from 0% to 20%) based on clinical manifestations, *Leishmania* species, dosage, and clinical protocols, among other factors in the Indian subcontinent (Table 12.1).

These studies are potentially intriguing from a clinical standpoint, and only a handful furnishes valuable epidemiological data. To begin with, investigations in the ISC by Sundar et al. (2012) and Rijal et al. (2013) showed a rise in the rate of miltefosine unresponsiveness in VL cases attributable to *L. (L.) donovani* due to a decade of usage of miltefosine (for a similar dosage). In Nepal, extending the follow-up period from 6 to 12 months revealed the doubling of the relapse rate from 10.8% to 20%, implying potential underreporting in studies with a 6-month follow-up. Ritmeijer et al. (2006) in Ethiopia reported a rise in the drug unresponsiveness rate in VL cases, which depends on the patient's immune status (5% treatment failure in HIV-negative vs. 18% treatment failure in HIV-positive patients), showcasing the intermingling of the epidemiology of *Leishmania*-HIV coinfection and treatment failure in leishmaniasis, a characteristic that has previously been explained with some other drugs (Cota et al. 2014; Ritmeijer et al. 2006).

In Brazil, in a phase II, open-label, dose-escalation study on the efficacy and safety of oral miltefosine against VL due to *L. infantum*, the cure rate was only 42% with a recommended treatment of 28 days and 68% with an extended treatment of 42 days (Carnielli et al. 2019) (Table 12.1).

In the context of sodium stibogluconate (SSG), it was evident that drug-resistant parasites had played a significant (albeit not exclusive) role in treatment failure, specifically in ISC. The escalating rate of miltefosine treatment failure has yet to be linked to parasite resistance in ISC. In the case of miltefosine resistance, the scenario is slightly different. *L. donovani* strains harvested from miltefosine relapse VL patients showed lower susceptibility to miltefosine at the intracellular amastigote stage (Bhandari et al. 2012). Further, isolates obtained from post kala-azar dermal leishmaniasis (PKDL) patients that relapsed after treatment with miltefosine also revealed lower susceptibility to the drug (IC_{50} nearly 16.13 μM in PKDL relapse vs. 2.43 μM in cured patients) (Bhandari et al. 2012). In *L infantum* isolates from Brazilian VL patients, in vitro miltefosine susceptibility of the pretreatment isolates from cured (mean $IC_{50} \sim 5$ μM) and relapsed patients (mean $IC_{50} \sim 13$ μM) showed a positive correlation with the clinical outcome (Carnielli et al. 2019). Another recent study demonstrated miltefosine IC_{50} ranging from 5.89 and 23.7 μM in 73 *L. infantum* strains from different parts of Brazil (Espada et al. 2021).

Table 12.1 Various studies reporting clinical efficacy of miltefosine treatment in visceral leishmaniasis (VL)

Country	Patient treated	Therapeutic regimen	Year of recruitment	Percentage of treatment failure	References
Indian subcontinent					
India	291 (adult)	For 25 kg 50 mg daily; >25 kg 100 mg daily, 28 days	1999–2000	3% (6 months)	Sundar et al. (2002)
India	39 (children)	Group A = 21 patients; 1.5 mg/kg daily, 28 days	1999–2000	9.5% (6 months)	Sundar et al. (2003)
		Group B = 18 patients; 2.5 mg/kg daily, 28 days		9.5% (6 months)	
India	79 (children)	2.5 mg/kg daily for 28 days	2001–2002	5% (6 months)	Bhattacharya et al. (2004)
Bangladesh	977(adult and children)	≤12 years, 2.5 mg/kg; >12 years of age (<25 kg), 50 mg/day (evening); >12 years (≥25 kg), 100 mg/day for 28 days	2006–2007	2.45% (initial month)	Rahman et al. (2011)
				9.72% (6 months)	
India	71	2.5 mg/kg daily for 28 days	2007–2009	7%	Patra et al. (2012)
India	551 (adult and children)	2.5 mg/kg daily (<12 years) for 28 days 50 mg daily (25 kg) or 100 mg daily (>25 kg)	2009–2010	6.9% (6 months)	Sundar et al. (2012)
Nepal	115	For 28 days, ≥12 years, >25 kg: 100 mg daily; <25 kg: 50 mg daily; children 2–11 years: 2.5 mg/kg daily,	2008–2011	10.8% (6 months) 20% (12 months)	Rijal et al. (2013)
East Africa					
Ethiopia (HIV-neg)	106	100 mg daily for 28 days,	1997–	0% (6 months)	Ritmeijer et al. (2006)
Ethiopia (HIV +ve)	45	100 mg daily for 28 days	1997–	35.5% (6 months)	Ritmeijer et al. (2006)
Sudan & Kenya (HIV-neg)	49	2.5 mg/kg daily for 28 days	2010–2012	28% (6 months)	Wasunna et al. (2016)
South America					

(continued)

Table 12.1 (continued)

Country	Patient treated	Therapeutic regimen	Year of recruitment	Percentage of treatment failure	References
Brazil	42	2.5 mg/kg daily; children, 100 mg daily; adults for 28 days (14 patients) or 42 days (28 patients)	2005–2007	4.8% initial failure 35.5% (6 months)	Carnielli et al. (2019)

However, there is no evidence of complete miltefosine resistance to date. Only a few completely miltefosine-resistant isolates have been identified so far, which include two miltefosine-resistant *L. infantum* strains obtained from HIV-positive patients in France and two miltefosine-resistant *L. donovani* strains obtained from Indian patients, as shown in Table 12.2 (Cojean et al. 2012; Mondelaers et al. 2016; Srivastava et al. 2017).

The paucity of miltefosine relapse/resistant field isolates and the easy selection of resistance to miltefosine experimentally in the laboratory point to an increased likelihood of miltefosine drug resistance development in immune-compromised individuals, most likely as a result of successively longer exposures to inadequate drug concentrations.

12.4 Possible Causes for Relapse/Unresponsiveness

Several drug-/parasite-related factors and host immune responses discussed here may aid treatment failure (Fig. 12.1).

12.5 Drug-Related Factors

Drug pharmacokinetics (including absorption, distribution, metabolism, and the excretion of the drug from the host body) define the fate of the drug in the host body over time. The drug should be absorbed into the infected host cells after administration to the host to deploy its activity. There are shreds of evidence that *Leishmania* parasites also infect and survive within fibroblasts (Bogdan et al. 2000), keratinocytes (Scorza et al. 2017), hepatocytes (Gangneux et al. 2005), and stem cells (Carvalho-Gontijo et al. 2018), in addition to the different macrophages. Recently, quiescent *Leishmania* parasites have been observed in mesothelium tissues (collagen-rich) that do not accumulate miltefosine and thus may provide a safe environment for quiescent parasites during miltefosine treatment (Kloehn et al. 2021). These parasites may lead to the generation of drug-tolerant phenotypes. Such

Table 12.2 In vitro miltefosine susceptibility of *Leishmania* clinical isolates originated from patients of VL

Leishmania Species	Country	Patient employed, clinical status, period of recruitment	*Leishmania* Stage	Miltefosine susceptibility IC_{50}(μM)	Genomic changes	References
L. donovani	Nepal	24, VL, 2002–2004	Amastigote	6.3	Not available	Yardley et al. (2005)
L. donovani	India	−13, VL, cured, 2009–2010	Amastigote	2.4	No SNPs	Bhandari et al. (2012)
		−3, relapses, VL, 2009–2010		4.7		
		−22, VL, 2009–2010		2.6		
L. donovani	India	−24, VL, pre-tx, 2009–2010	Promastigote	3.7	Not available	Prajapati et al. (2013)
		−26, VL, post-tx, 2009–2010		6.1		
L. donovani	Nepal	17, VL, cure, 2008–2011	Promastigote	4.3	Not available	Rijal et al. (2013)
		26, VL, relapse, 2008–2011		2.6		
L. donovani	India	2, VL, failure, 2011	Promastigote	>100	SNPs in LdMT (2 nonsynonymous)	Srivastava et al. (2017)
L. infantum	France	1, VL, HIV,1995	Promastigote	>40	2 nt INDEL in LdMT	Mondelaers et al. (2016)
			Amastigote	>20		
		1, VL, HIV, 1995, 2005	Promastigote	>40	Frameshift mutation in LiMT and 1 nt INDEL in LiRos3	
			Amastigote	>20		
L. infantum	Brazil	14 cured, pretreatment, 2005	Amastigote	5	No correlation with miltefosine susceptibility locus	Carnielli et al. (2019)
		12 relapse, pretreatment 2005	Amastigote	13		
L. infantum	Brazil	60 isolates from VL patients, 2002–2015	Promastigote	5.89–23.7	No Correlation with miltefosine susceptibility locus	Espada et al. (2021)
		10 VL, 2005–2011	Amastigote	1.41–4.57		

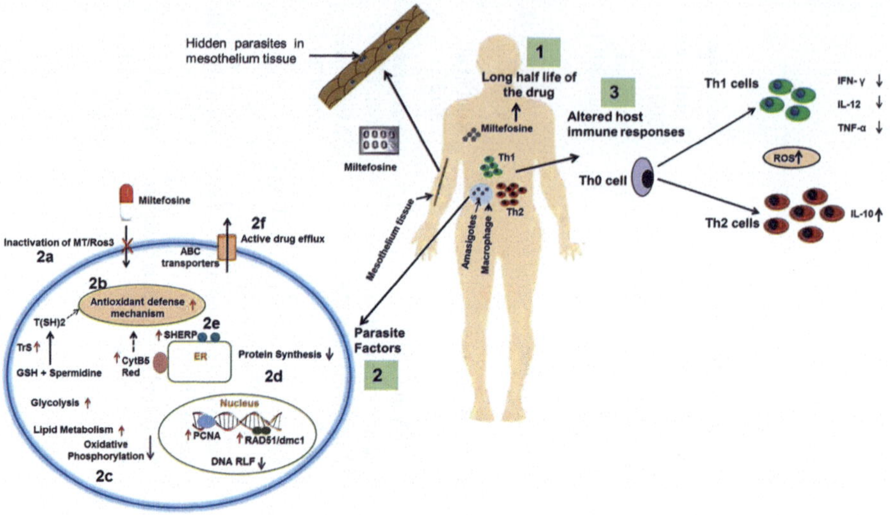

Fig. 12.1 Representation of factors responsible for miltefosine resistance in *Leishmania* parasites—(1) drug-related factors: hidden *Leishmania* parasites in the mesothelium tissue that do not accumulate miltefosine get more prolonged exposure to the drug having a long half-life and may become more tolerant to the drug. (2) Parasite-related factors: adaptations in *Leishmania* parasites include (2a). The inactivation of miltefosine transporter MT/Ros3 reduces drug internalization (2b). Increased tolerance toward oxidative stress caused by miltefosine drug (2c). An altered way of energy generation may help the parasites survive under drug pressure (2d). Altered DNA synthesis and reduced protein synthesis (2e). Increased metacyclogenesis and infectivity (2f). Active drug efflux results in reduced drug accumulation within the parasite (3) Modulation of host immune responses: miltefosine-resistant parasites are better equipped than sensitive ones to favor their survival within the host

situations explain the importance of drug pharmacokinetics and bioavailability and their role in developing drug resistance.

Clinical pharmacokinetics remains to be largely unknown in the case of antileishmanial drugs. Before miltefosine was licensed for the treatment of VL, research on its clinical pharmacokinetics was scarce (Dorlo et al. 2012; Verrest and Dorlo 2017). Upon clinical development, some investigations reported the drug's plasma concentrations in VL, shedding light on its clinical pharmacokinetics (de Valliere et al. 2009; de Vries et al. 2006). Further, in healthy people, no pharmacokinetic assessment was conducted. Dose-finding investigations were ventured without detailed pharmacokinetic analyses, and only a small amount of pharmacokinetic data from these studies, which were never published in the first place, could be found in registration paperwork (Dorlo et al. 2012). The pharmacokinetics of miltefosine were different in pediatric and adult patients in ISC based on their treatment regime. In adults who were given 100 mg of miltefosine, it was found that on day 23 of treatment, they had a maximum median drug concentration (70 μg/ml), indicating a clearance half-life between 150 and 200 h (7 days). However, in children, the reported mean predose concentration of miltefosine when administered with 2.5 mg/kg was 24 mg/ml, with a clearance half-life of 180 h between days 23 and 28 of treatment (Dorlo et al. 2012). Considerable efforts were made in the marketing and clinical development of miltefosine, after which many reports were published describing miltefosine plasma concentrations in VL patients (de Vries et al. 2006). Elimination of miltefosine from the body follows a two-compartmental disposition model as the drug has exceptionally long elimination half-lives having a primary elimination half-life of roughly 7 days (Berman 2005) and a terminal half-life of 31 days (Dorlo et al. 2008). The drug's long half-life is responsible for retaining subtherapeutic doses of miltefosine for a longer time, which may lead to the exposure of surviving *Leishmania* parasites to the drug for a longer duration and may result in the emergence of drug-resistant parasites. Compared to the median of 30 days, a 1-day reduction in miltefosine plasma concentration over the 10X EC_{50} (mean half-maximal effective concentration) was related to a 1.08-fold rise in the risks of therapeutic failure in VL (Dorlo et al. 2014). It has been reported that miltefosine accumulates inside PBMCs, which could alter miltefosine exposure at its site of action, although a noncompartmental analysis revealed no significant link with the therapeutic outcome (Kip et al. 2015).

Drug-related factors for drug unresponsiveness also involve the poor pharmaceutical formulation of the drug available in the market, which leads to exposure to subtherapeutic doses, resulting in treatment failure and drug resistance. The counterfeit formulation of miltefosine in the tablet named "Miltefos" by LC-MS analysis was reported in 10 and 50 mg packaging (Dorlo et al. 2012).

12.5.1 Parasite-Related Factors

Various parasite-related features that contribute to the emergence of miltefosine-resistant parasites are discussed below.

12.5.1.1 Virulence

The successful survival of *Leishmania* parasites within host macrophages is brought about by several phenotypic adaptations that help parasites to persist, replicate, and transmit to establish the infection (Brandonisio et al. 2000; Contreras et al. 2010). Virulence is defined as the total injury caused by the parasites to the host. There are plenty of virulence factors that help parasites to interfere with signaling pathways in the host, modulating cytokine production and inhibiting the generation of reactive oxygen species (ROS) and nitric oxide (NO) (Forget et al. 2006; Soulat and Bogdan 2017). Various studies showed that the acquisition of miltefosine resistance in vitro led to increased parasite fitness, infectivity, and metacyclogenesis, which are drivers of parasite virulence (Kulshrestha et al. 2014; Turner et al. 2015). In addition, studies including isolates from VL and PKDL patients that relapsed after miltefosine treatment revealed that miltefosine unresponsiveness was associated with metacyclogenesis and infectivity (Deep et al. 2017; Rai et al. 2013).

12.5.1.2 Drug Accumulation

Intracellular drug accumulation is a prerequisite for the antileishmanial mechanism of miltefosine. Reduced drug accumulation has always been defined as one of the major causes of miltefosine resistance in *Leishmania*. Two independent mechanisms bring about the lower miltefosine accumulation within *Leishmania* parasites, (i) reduced uptake of the drug, and (ii) active drug efflux.

Miltefosine bind to the outer leaflet of the plasma membrane (Rakotomanga et al. 2005), and their translocation to the inner leaflet of the plasma membrane is energy dependent, the protein-mediated mechanism that involves two proteins: miltefosine transporter (MT) and its beta subunit Ros3, which together form the miltefosine translocation machinery. Poor drug uptake (more than 95%) in miltefosine-resistant promastigotes has been repeatedly linked to mutations in the miltefosine transporter genes MT and/or Ros3, which lead to drug unresponsiveness (Mondelaers et al. 2016; Perez-Victoria et al. 2003; Shaw et al. 2016). However, most of the isolates from Indian VL/PKDL patients that relapsed following miltefosine treatment did not show mutation(s) or altered expression in MT and/or Ros3 (Bhandari et al. 2012), although in a few isolates, such as *L. infantum* strains obtained from France and *L. donovani* strains obtained from India, single nucleotide polymorphism (SNP) was observed in the MT gene (Cojean et al. 2012; Srivastava et al. 2017).

The active efflux of miltefosine involves a higher expression of ABC (ATP-binding cassette) transporters, which mediate the movement of molecules across the membrane in an energy-dependent manner. The potential role of P-glycoprotein MDR1 and MRPA has been defined in antileishmanial drug resistance. The overexpression of MDR1 in *L. tropica* resulted in miltefosine-resistant lines (Perez-Victoria et al. 2001). Another ABC transporter, ABCG4, was also found to be involved in miltefosine resistance (Castanys-Munoz et al. 2008).

12.5.1.3 Fitness to Host Defense Mechanisms

Parasite fitness is a critical phenotypic trait defined as the complex interplay between numerous factors that ensure successful survival, reproduction, and transmission within the host. In the context of drug resistance in *Leishmania*, parasite fitness is debatable and depends on the species of parasite and particular drug. For example, SAG-resistant *Leishmania* parasites have shown more tolerance toward macrophage-killing mechanisms (Carter et al. 2005). As far as miltefosine resistance is concerned, an in-vitro-selected miltefosine-resistant parasite was shown to tolerate oxidative stress (Mishra and Singh 2013). However, in another study, when the in vitro selection was carried out at the intracellular amastigote stage, the drug-resistant parasites showed a decrease in parasite fitness in terms of reduced metacyclogenesis and stress tolerance (Hendrickx et al. 2016). Further, *L. donovani* parasites isolated from patients that relapsed following miltefosine treatment showed a survival advantage over isolates from patients prior to the treatment, as evidenced by lower miltefosine-induced reactive oxygen species (ROS) generation and elevated intracellular thiol content (Deep et al. 2017). However, miltefosine-resistant *Leishmania* parasites did not tolerate nitrosative stress (Hendrickx et al. 2016). *L. infantum* parasite isolates from patients who relapsed after miltefosine treatment and lacked the miltefosine susceptibility locus (MSL)-modulated nitric oxide accumulation in host macrophages (Carnielli et al. 2022).

12.5.1.4 Modulations of Host Immune Responses

Leishmania parasites, an obligate intracellular pathogen, successfully survive, replicate, and subsequently transmit within the host by evading the host's immune responses (Rossi and Fasel 2018; Vanaerschot et al. 2014). *Leishmania* infection is settled within the host macrophages, which also play an important role in the clearance of the infection (Liu and Uzonna 2012). Protective immune response by the host during *Leishmania* infection includes the induction of interferon- γ (IFN-γ) secreting Th1 cells by interleukin-12 (IL-12), which are produced by antigen-presenting cells (APCs) (Kumar et al. 2014; Muller et al. 1989). This results in the production of ROS and nitric oxide (NO), which take part in killing intracellular amastigotes. However, these parasites are very successful in manipulating host immune responses in order to favor their survival within the host. Several studies suggest that drug-resistant parasites are better equipped to manipulate host immune responses than drug-sensitive parasites (Ghosh et al. 2013; Guha et al. 2014; Mukherjee et al. 2013). As far as miltefosine-resistant parasites are concerned, miltefosine-resistant clinical isolates showed modulated levels of IL-10, IL-12, and TNF-α produced in infected macrophages upon miltefosine treatment as compared to miltefosine-sensitive parasites. Further, miltefosine drug-induced decreased levels of ROS, and NO in macrophages infected with miltefosine-resistant clinical isolates were also observed (Deep et al. 2017; Khanra et al. 2017). Poor T-cell proliferation in response to leishmanial antigen was also observed in the hamster model infected with miltefosine-resistant parasites (Khanra et al. 2017). These observations led to the inference that, to an extent, miltefosine unresponsiveness in leishmaniasis is also dependent on host immune modulation.

12.5.2 Changes in the Genetic Makeup

Leishmania's ~32 MB genome comprises over 8300 protein-coding genes (Ivens et al. 2005; Peacock et al. 2007). In order to survive stressful environmental conditions, *Leishmania* parasites regulate gene expression by DNA copy number variation (CNV), which includes gene amplification/deletion or aneuploidy. In addition, single nucleotide polymorphism (SNP) also helps the parasites adapt to environmental stress, such as drug pressure. Changes in the genetic makeup of *Leishmania* parasites, particularly in gene-encoding drug targets, drug transporters, metabolic proteins, or stress-associated proteins, help the parasites resist being killed by the drug and thus lead to drug resistance. Several studies report the altered genetic structure of *Leishmania* parasites in the context of miltefosine unresponsiveness.

A single point mutation in the MT gene was associated with the resistant phenotype by inactivating the transporter protein (Seifert et al. 2003; Seifert et al. 2007). Further, mutations in MT (particularly $_{420}$Thr → Asn, $_{856}$Leu → Pro, and $_{832}$Leu → Phe) have been linked to an enhanced rate of resistance in vitro and in vivo

(Cojean et al. 2012; Kulshrestha et al. 2014; Perez-Victoria et al. 2003). These alterations cause lower absorption, greater efflux, a quicker metabolism, and alteration in the lipid composition of the parasite membrane (Perez-Victoria et al. 2006; Seifert et al. 2007). In addition, Kulshrestha et al. identified SNP $_{527}$T → A, which leads to the substitution of $_{176}$Val → Asp in the MT gene in experimentally selected miltefosine-resistant isolates. Two unique SNPs in LdMT, $_{354}$Tyr → Phe and $_{1078}$Phe → Tyr, were discovered in publically accessible LdMT (3294 bp) and LdRos (1142 bp) sequences (Kulshrestha et al. 2014). The phosphorylation of tyrosine and phenylalanine is vital for cell growth and division (Nascimento et al. 2003). These SNPs (nonsynonymous) might induce a change in conformation in this protein, causing changes in miltefosine intake by the parasite and, eventually, miltefosine resistance.

In *Leishmania*, genetic changes identified in ABC transporter genes cause enhanced expression of ABCG6 and ABCG4, the two ABC transporter proteins found in the parasite's plasma membrane and the flagellar pocket, and were linked to greater resistance to various alkyl-lysophospholipid analogs (BoseDasgupta et al. 2008; Castanys-Munoz et al. 2007; Castanys-Munoz et al. 2008), resulting in the efflux of the drug across the plasma membrane.

A research study by Shaw et al. reported that nonsynonymous SNPs at various sites of MT, the omission of the complete gene, and variations in the chromosome number, i.e., ploidy changes in chromosome 13 carrying MT, all lead to miltefosine resistance (Shaw et al. 2016). Another study revealed an alteration in the ploidy of chromosome nos. 5 (pentasomic), 26 (trisomic), and 31 (pentasomic) in the miltefosine-resistant parasite (Rastrojo et al. 2018). Additionally, Bhattacharya et al. reported an alteration in chromosomal ploidy in chromosomes 9, 12, 13, 22, 23, 26, and 31 in the miltefosine-resistant *Leishmania* parasite (Bhattacharya et al. 2019). Aneuploidy is a characteristic feature of *L. donovani* and is used as a lifestyle to make the variation in its genetic content (Downing et al. 2011; Mannaert et al. 2012), and this capability to change the chromosome copy number could be a fast way out in response to stringent drug pressures, which may act as a transitional phase to achieve the fitness state continued till more useful mutations happen (Covert 3rd et al. 2013).

A study led by Carnielli et al. (2018) showed the association of genetically stable miltefosine sensitivity locus (MSL) with miltefosine treatment failure in *L infantum* isolates from Brazil. The MSL comprises four genes encoding helicase-like protein 3′-nucleotidase (NUC1)/nucleases (NUC2) (tandem duplicates) and 3,2-trans enoyl-CoA isomerase. The complete absence of MSL was linked to a ninefold increase in the treatment failure rate (Carnielli et al. 2018). The deletion of both NUC1 and NUC2 from the MSL resulted in a significant decrease in miltefosine susceptibility in the clinical isolates of *L. infantum*. The deletion of MSL also impacted the parasite lipid content, and these parasites were more resistant to lipid metabolism perturbation caused by miltefosine (Carnielli et al. 2022).

Leishmania parasites cunningly alter drug interaction with the external lipid membrane of the parasite. Miltefosine is an amphiphilic and zwitterionic molecule;

hence, the membrane fluidity of the *Leishmania* plasma membrane is of great concern. Membrane fluidity depends on the level of unsaturation in the alkyl chain of the fatty in the membrane and the alkylation of sterols in the side chain, as well as the length of unsaturated fatty acid (Mbongo et al. 1998). Partial inactivation of desaturases causes a decrease in unsaturated alkyl chain content, leading to reduced ergosterol levels and decreased membrane fluidity (Rosenthal 1987). This might also impact drug-membrane interactions (Rakotomanga et al. 2005); nevertheless, this concept needs to be validated in the case of miltefosine unresponsiveness.

12.5.3 Alterations in Transcriptome

Leishmania parasites constitute a polycistronic transcription unit, where protein-coding genes are organized in a large cluster. These polycistronic units are transcribed into polycistronic pre-RNA that are processed into mature RNA by trans-splicing and polyadenylation, and in that way, gene expression is regulated (Clayton 2019). Multiple studies have suggested that *Leishmania* parasites selected for drug resistance under the pressure of a particular drug exhibited global changes in gene expression profiling (Andrade et al. 2020; Ghosh et al. 2020; Kulshrestha et al. 2014; Patino et al. 2019; Rastrojo et al. 2018; Verma et al. 2017).

Kulshrestha et al. (2014) reported altered expression of 311 genes in miltefosine-resistant parasites using microarray technology. The study suggested that genes involved in antioxidant defense mechanisms, energy generation, DNA replication, translation machinery, and membrane transport systems were differentially expressed in miltefosine-resistant parasites. Another study by Rastrojo et al. (2018) revealed a twofold expression difference in 18 transcripts in miltefosine-resistant lines by RNA sequence analysis.

Miltefosine resistance in *Leishmania* parasites has been found to be associated with the downregulated expression of LdMT and LdRos 3 (Kulshrestha et al. 2014) and the upregulated expression of ABC transporters, such as MDR1 (ABCB4) (Perez-Victoria et al. 2001), ABCG4 (Castanys-Munoz et al. 2007), and ABCA7, ABCG5, and ABCG2 (Kulshrestha et al. 2014), which lead to reduced drug accumulation. Other transporters, such as folate/biopterin transporter, ABCF2 (an ABC transporter), and amino acid transporter, were found to be downregulated in miltefosine-resistant parasites (Deep et al. 2017; Rastrojo et al. 2018; Vacchina et al. 2016).

Metacyclogenesis-associated transcripts, such as META1 and SHERP (small hydrophilic endoplasmic reticulum-associated protein) were observed to exhibit downregulated expression in miltefosine-resistant parasites (Rastrojo et al. 2018). On the contrary, SHERP showed upregulated expression in miltefosine-resistant parasites in an earlier study (Kulshrestha et al. 2014). However, the adaptation levels of *L. donovani* parasites differed in the two studies: 73 μM vs 8 μM. It is plausible that at an earlier level of adaptation, the drug-resistant parasites showed reduced parasite metacyclogenesis and thus infectivity, and at a higher drug concentration adaptation, the parasites showed increased infectivity.

Miltefosine is known to kill *Leishmania* parasites in an apoptosis-like manner mediated by reactive oxygen species (ROS) (Mishra and Singh 2013; Paris et al. 2004; Verma and Dey 2004). The drug-resistant parasites utilize various ways to overcome oxidative damage. For example, higher mRNA expression of trypanothione synthetase, cytosolic tryparedoxin peroxidase, and cytochrome b5 reductase was observed to play an important role in antioxidant defense mechanism in miltefosine-resistant parasites (Deep et al. 2017). HSP70 and HSP83, which have a role in reduced production of superoxide radicals, showed higher expression in miltefosine-resistant parasites; thus, these may help protect against cell damage (Kulshrestha et al. 2014). The overexpression of a chaperonin protein T-complex protein-1 that is involved in ATP-dependent protein folding was also observed to be associated with miltefosine resistance, and in the mutant cell, overexpressing the T-complex protein-1 increased thiol levels, and an upregulated expression of tryparedoxin peroxidase was observed that was responsible for ROS neutralization (Yadav et al. 2020).

Miltefosine-resistant parasites exhibited altered DNA synthesis, as depicted by the downregulated expression of DNA replication licensing factor and the upregulated expression of PCNA and RAD51/dmc1. The DNA replication licensing factor is a DNA-dependent ATPase required for initiating DNA replication in eukaryotes, whereas PCNA and RAD51/dmc1 are enzymes required for DNA repair. The drug-resistant parasites also showed reduced protein synthesis as there was a downregulated expression of aminoacyl-tRNA synthetase, chaperon proteins, and ubiquitin-activating enzymes (Kulshrestha et al. 2014).

An altered way of energy generation was also evident in miltefosine-resistant parasites. The energy generated by oxidative phosphorylation was reduced in miltefosine-resistant parasites, as indicated by the downregulated expression of ATP synthase. However, these parasites depend on lipid catabolism for energy generation as there was an upregulated expression of lipase and lipase precursor-like protein in lab-generated parasites as well as parasites isolated from patients who relapsed after miltefosine treatment (Deep et al. 2017; Kulshrestha et al. 2014). Another gene encoding for phosphoglucomutase involved in the conversion of glucose-1-phosphate to glucose-6-phosphate (in glycolysis and pentose phosphate pathway) showed up-regulated expression in lab-generated resistant parasite and field isolate of relapse patients (Deep et al. 2017).

Overall, miltefosine-resistant parasites have a vastly different transcriptome, which may be exploited in various ways, such as the identification of diagnostic markers or chemotherapeutic targets to combat miltefosine unresponsiveness in the field.

12.6 Conclusion

Drug unresponsiveness or treatment failure is a continuous hindrance in the control of visceral leishmaniasis, especially in endemic areas. Oral antileishmanial drug miltefosine is an effective treatment option, but reports of relapses following treatment have been a significant concern during the past decade. The exact reason for relapses or drug resistance is not fully understood; some potential reasons for drug unresponsiveness or treatment failure in VL include (1) parasite drug resistance: the parasite may have evolved mechanisms to resist the drugs used for treatment, leading to treatment failure; (2) drug toxicity: in some cases, the drugs used for treatment may cause adverse side effects, leading to treatment discontinuation or reduced efficacy; (3) host immune response: the host immune response plays a critical role in controlling parasite load and may influence the outcome of treatment; in some cases, a weak or dysregulated immune response may lead to treatment failure; (4) drug pharmacokinetics: the way the drug is absorbed, distributed, metabolized, and excreted may also play a role in determining treatment outcomes.

To overcome these challenges, it is crucial to understand the underlying factors contributing to drug unresponsiveness and to develop alternative treatments and combination therapies to improve treatment outcomes in VL. Research in this area is ongoing and will likely lead to improved treatment options in the future.

References

Al-Mohammed HI, Chance ML, Bates PA (2005) Production and characterization of stable amphotericin-resistant amastigotes and promastigotes of Leishmania mexicana. Antimicrob Agents Chemother 49(8):3274–3280. https://doi.org/10.1128/AAC.49.8.3274-3280.2005

Alvar J, Croft S, Olliaro P (2006) Chemotherapy in the treatment and control of leishmaniasis. Adv Parasitol 61:223–274. https://doi.org/10.1016/S0065-308X(05)61006-8

Andrade JM, Goncalves LO, Liarte DB, Lima DA, Guimaraes FG, de Melo Resende D, Santi AMM, de Oliveira LM, Velloso JPL, Delfino RG, Pescher P, Spath GF, Ruiz JC, Murta SMF (2020) Comparative transcriptomic analysis of antimony resistant and susceptible Leishmania infantum lines. Parasit Vectors 13(1):600. https://doi.org/10.1186/s13071-020-04486-4

Berman J (2005) Miltefosine to treat leishmaniasis. Expert Opin Pharmacother 6(8):1381–1388. https://doi.org/10.1517/14656566.6.8.1381

Bhandari V, Kulshrestha A, Deep DK, Stark O, Prajapati VK, Ramesh V, Sundar S, Schonian G, Dujardin JC, Salotra P (2012) Drug susceptibility in Leishmania isolates following miltefosine treatment in cases of visceral leishmaniasis and post kala-azar dermal leishmaniasis. PLoS Negl Trop Dis 6(5):e1657. https://doi.org/10.1371/journal.pntd.0001657

Bhattacharya A, Leprohon P, Bigot S, Padmanabhan PK, Mukherjee A, Roy G, Gingras H, Mestdagh A, Papadopoulou B, Ouellette M (2019) Coupling chemical mutagenesis to next generation sequencing for the identification of drug resistance mutations in Leishmania. Nat Commun 10(1):5627. https://doi.org/10.1038/s41467-019-13344-6

Bogdan C, Donhauser N, Doring R, Rollinghoff M, Diefenbach A, Rittig MG (2000) Fibroblasts as host cells in latent leishmaniosis. J Exp Med 191(12):2121–2130. https://doi.org/10.1084/jem.191.12.2121

BoseDasgupta S, Ganguly A, Roy A, Mukherjee T, Majumder HK (2008) A novel ATP-binding cassette transporter, ABCG6 is involved in chemoresistance of Leishmania. Mol Biochem Parasitol 158(2):176–188. https://doi.org/10.1016/j.molbiopara.2007.12.007

Brandonisio O, Panaro MA, Sisto M, Acquafredda A, Fumarola L, Leogrande D (2000) Interactions between Leishmania parasites and host cells. Parassitologia 42(3–4):183–190. https://www.ncbi.nlm.nih.gov/pubmed/11686076

Carnielli JBT, Monti-Rocha R, Costa DL, Molina Sesana A, Pansini LNN, Segatto M, Mottram JC, Costa CHN, Carvalho SFG, Dietze R (2019) Natural resistance of leishmania infantum to miltefosine contributes to the low efficacy in the treatment of visceral leishmaniasis in Brazil. Am J Trop Med Hyg 101(4):789–794. https://doi.org/10.4269/ajtmh.18-0949

Carnielli JBT, Dave A, Romano A, Forrester S, de Faria PR, Monti-Rocha R, Costa CHN, Dietze R, Graham IA, Mottram JC (2022) 3'Nucleotidase/nuclease is required for Leishmania infantum clinical isolate susceptibility to miltefosine. EBioMedicine 86:104378. https://doi.org/10.1016/j.ebiom.2022.104378

Carnielli JBT, Crouch K, Forrester S, Silva VC, Carvalho SFG, Damasceno JD, Brown E, Dickens NJ, Costa DL, Costa CHN, Dietze R, Jeffares DC, Mottram JC (2018) A Leishmania infantum genetic marker associated with miltefosine treatment failure for visceral leishmaniasis. EBioMedicine 36:83–91. https://doi.org/10.1016/j.ebiom.2018.09.029

Carter KC, Hutchison S, Boitelle A, Murray HW, Sundar S, Mullen AB (2005) Sodium stibogluconate resistance in Leishmania donovani correlates with greater tolerance to macrophage antileishmanial responses and trivalent antimony therapy. Parasitology 131(Pt 6): 747–757. https://doi.org/10.1017/S0031182005008486

Carvalho-Gontijo R, Moreira DR, Resende M, Costa-Silva MF, Peruhype-Magalhaes V, Ribeiro CMF, Ribeiro DD, Silvestre R, Cordeiro-da-Silva A, Martins-Filho OA, Teixeira-Carvalho A (2018) Infection of hematopoietic stem cells by Leishmania infantum increases erythropoiesis and alters the phenotypic and functional profiles of progeny. Cell Immunol 326:77–85. https://doi.org/10.1016/j.cellimm.2017.10.016

Castanys-Munoz E, Alder-Baerens N, Pomorski T, Gamarro F, Castanys S (2007) A novel ATP-binding cassette transporter from Leishmania is involved in transport of phosphatidylcholine analogues and resistance to alkyl-phospholipids. Mol Microbiol 64(5):1141–1153. https://doi.org/10.1111/j.1365-2958.2007.05653.x

Castanys-Munoz E, Perez-Victoria JM, Gamarro F, Castanys S (2008) Characterization of an ABCG-like transporter from the protozoan parasite Leishmania with a role in drug resistance and transbilayer lipid movement. Antimicrob Agents Chemother 52(10):3573–3579. https://doi.org/10.1128/AAC.00587-08

Centers for Disease Control and Prevention; DPDx (2017, December 2017). Retrieved 31.01.23 from

Clayton C (2019) Regulation of gene expression in trypanosomatids: living with polycistronic transcription. Open Biol 9(6):190072. https://doi.org/10.1098/rsob.190072

Cojean S, Houze S, Haouchine D, Huteau F, Lariven S, Hubert V, Michard F, Bories C, Pratlong F, Le Bras J, Loiseau PM, Matheron S (2012) Leishmania resistance to miltefosine associated with genetic marker. Emerg Infect Dis 18(4):704–706. https://doi.org/10.3201/eid1804.110841

Contreras I, Gomez MA, Nguyen O, Shio MT, McMaster RW, Olivier M (2010) Leishmania-induced inactivation of the macrophage transcription factor AP-1 is mediated by the parasite metalloprotease GP63. PLoS Pathog 6(10):e1001148. https://doi.org/10.1371/journal.ppat.1001148

Cota GF, de Sousa MR, de Mendonca AL, Patrocinio A, Assuncao LS, de Faria SR, Rabello A (2014) Leishmania-HIV co-infection: clinical presentation and outcomes in an urban area in Brazil. PLoS Negl Trop Dis 8(4):e2816. https://doi.org/10.1371/journal.pntd.0002816

Covert AW 3rd, Lenski RE, Wilke CO, Ofria C (2013) Experiments on the role of deleterious mutations as stepping stones in adaptive evolution. Proc Natl Acad Sci U S A 110(34):E3171–E3178. https://doi.org/10.1073/pnas.1313424110

de Valliere S, Mary C, Joneberg JE, Rotman S, Bullani R, Greub G, Gillmore JD, Buffet PA, Tarr PE (2009) AA-amyloidosis caused by visceral leishmaniasis in a human immunodeficiency virus-infected patient. Am J Trop Med Hyg 81(2):209–212. https://www.ncbi.nlm.nih.gov/pubmed/19635871

de Vries PJ, van der Meide WF, Godfried MH, Schallig HD, Dinant HJ, Faber WR (2006) Quantification of the response to miltefosine treatment for visceral leishmaniasis by QT-NASBA. Trans R Soc Trop Med Hyg 100(12):1183–1186. https://doi.org/10.1016/j.trstmh.2006.01.002

Deep DK, Singh R, Bhandari V, Verma A, Sharma V, Wajid S, Sundar S, Ramesh V, Dujardin JC, Salotra P (2017) Increased miltefosine tolerance in clinical isolates of Leishmania donovani is associated with reduced drug accumulation, increased infectivity and resistance to oxidative stress. PLoS Negl Trop Dis 11(6):e0005641. https://doi.org/10.1371/journal.pntd.0005641

Dhillon GP, Sharma SN, Nair B (2008) Kala-azar elimination programme in India. J Indian Med Assoc 106(10):664, 666–668. https://www.ncbi.nlm.nih.gov/pubmed/19552101

Dorlo TP, van Thiel PP, Huitema AD, Keizer RJ, de Vries HJ, Beijnen JH, de Vries PJ (2008) Pharmacokinetics of miltefosine in Old World cutaneous leishmaniasis patients. Antimicrob Agents Chemother 52(8):2855–2860. https://doi.org/10.1128/AAC.00014-08

Dorlo TP, Eggelte TA, de Vries PJ, Beijnen JH (2012) Characterization and identification of suspected counterfeit miltefosine capsules. Analyst 137(5):1265–1274. https://doi.org/10.1039/c2an15641e

Dorlo TP, Rijal S, Ostyn B, de Vries PJ, Singh R, Bhattarai N, Uranw S, Dujardin JC, Boelaert M, Beijnen JH, Huitema AD (2014) Failure of miltefosine in visceral leishmaniasis is associated with low drug exposure. J Infect Dis 210(1):146–153. https://doi.org/10.1093/infdis/jiu039

Downing T, Imamura H, Decuypere S, Clark TG, Coombs GH, Cotton JA, Hilley JD, de Doncker S, Maes I, Mottram JC, Quail MA, Rijal S, Sanders M, Schonian G, Stark O, Sundar S, Vanaerschot M, Hertz-Fowler C, Dujardin JC, Berriman M (2011) Whole genome sequencing of multiple Leishmania donovani clinical isolates provides insights into population structure and mechanisms of drug resistance. Genome Res 21(12):2143–2156. https://doi.org/10.1101/gr.123430.111

Eichenberger A, Buechi AE, Neumayr A, Hatz C, Rauch A, Huguenot M, Diamantis-Karamitopoulou E, Staehelin C (2017) A severe case of visceral leishmaniasis and liposomal amphotericin B treatment failure in an immunosuppressed patient 15 years after exposure. BMC Infect Dis 17(1):81. https://doi.org/10.1186/s12879-017-2192-4

Espada CR, Levatti EVC, Boite MC, Lamounier D, Alvar J, Cupolillo E, Costa CHN, Rode J, Uliana SRB (2021) In vitro susceptibility to miltefosine of leishmania infantum (syn. L. chagasi) Isolates from different geographical areas in Brazil. Microorganisms 9(6). https://doi.org/10.3390/microorganisms9061228

Forget G, Gregory DJ, Whitcombe LA, Olivier M (2006) Role of host protein tyrosine phosphatase SHP-1 in Leishmania donovani-induced inhibition of nitric oxide production. Infect Immun 74(11):6272–6279. https://doi.org/10.1128/IAI.00853-05

Gangneux JP, Lemenand O, Reinhard Y, Guiguen C, Guguen-Guillouzo C, Gripon P (2005) In vitro and ex vivo permissivity of hepatocytes for Leishmania donovani. J Eukaryot Microbiol 52(6):489–491. https://doi.org/10.1111/j.1550-7408.2005.00055.x

Ghosh M, Roy K, Roy S (2013) Immunomodulatory effects of antileishmanial drugs. J Antimicrob Chemother 68(12):2834–2838. https://doi.org/10.1093/jac/dkt262

Ghosh S, Verma A, Kumar V, Pradhan D, Selvapandiyan A, Salotra P, Singh R (2020) Genomic and transcriptomic analysis for identification of genes and interlinked pathways mediating artemisinin resistance in Leishmania donovani. Genes (Basel) 11(11). https://doi.org/10.3390/genes11111362

Guha R, Das S, Ghosh J, Sundar S, Dujardin JC, Roy S (2014) Antimony resistant Leishmania donovani but not sensitive ones drives greater frequency of potent T-regulatory cells upon interaction with human PBMCs: role of IL-10 and TGF-beta in early immune response. PLoS Negl Trop Dis 8(7):e2995. https://doi.org/10.1371/journal.pntd.0002995

Hendrickx S, Inocencio da Luz RA, Bhandari V, Kuypers K, Shaw CD, Lonchamp J, Salotra P, Carter K, Sundar S, Rijal S, Dujardin JC, Cos P, Maes L (2012) Experimental induction of paromomycin resistance in antimony-resistant strains of L. donovani: outcome dependent on in vitro selection protocol. PLoS Negl Trop Dis 6(5):e1664. https://doi.org/10.1371/journal.pntd.0001664

Hendrickx S, Beyers J, Mondelaers A, Eberhardt E, Lachaud L, Delputte P, Cos P, Maes L (2016) Evidence of a drug-specific impact of experimentally selected paromomycin and miltefosine resistance on parasite fitness in Leishmania infantum. J Antimicrob Chemother 71(7): 1914–1921. https://doi.org/10.1093/jac/dkw096

Ivens AC, Peacock CS, Worthey EA, Murphy L, Aggarwal G, Berriman M, Sisk E, Rajandream MA, Adlem E, Aert R, Anupama A, Apostolou Z, Attipoe P, Bason N, Bauser C, Beck A, Beverley SM, Bianchettin G, Borzym K, Bothe G, Bruschi CV, Collins M, Cadag E, Ciarloni L, Clayton C, Coulson RM, Cronin A, Cruz AK, Davies RM, De Gaudenzi J, Dobson DE, Duesterhoeft A, Fazelina G, Fosker N, Frasch AC, Fraser A, Fuchs M, Gabel C, Goble A, Goffeau A, Harris D, Hertz-Fowler C, Hilbert H, Horn D, Huang Y, Klages S, Knights A, Kube M, Larke N, Litvin L, Lord A, Louie T, Marra M, Masuy D, Matthews K, Michaeli S, Mottram JC, Müller-Auer S, Munden H, Nelson S, Norbertczak H, Oliver K, O'neil S, Pentony M, Pohl TM, Price C, Purnelle B, Quail MA, Rabbinowitsch E, Reinhardt R, Rieger M, Rinta J, Robben J, Robertson L, Ruiz JC, Rutter S, Saunders D, Schäfer M, Schein J, Schwartz DC, Seeger K, Seyler A, Sharp S, Shin H, Sivam D, Squares R, Squares S, Tosato V, Vogt C, Volckaert G, Wambutt R, Warren T, Wedler H, Woodward J, Zhou S, Zimmermann W, Smith DF, Blackwell JM, Stuart KD, Barrell B, Myler PJ (2005) The genome of the kinetoplastid parasite, Leishmania major. Science 309(5733):436–442. https://doi.org/10.1126/science.1112680

Jha TK, Sundar S, Thakur CP, Bachmann P, Karbwang J, Fischer C, Voss A, Berman J (1999) Miltefosine, an oral agent, for the treatment of Indian visceral leishmaniasis. N Engl J Med 341(24):1795–1800. https://doi.org/10.1056/NEJM199912093412403

Khanra S, Sarraf NR, Das AK, Roy S, Manna M (2017) Miltefosine resistant field isolate from Indian Kala-Azar patient shows similar phenotype in experimental infection. Sci Rep 7(1): 10330. https://doi.org/10.1038/s41598-017-09720-1

Kimutai R, Musa AM, Njoroge S, Omollo R, Alves F, Hailu A, Khalil EA, Diro E, Soipei P, Musa B, Salman K, Ritmeijer K, Chappuis F, Rashid J, Mohammed R, Jameneh A, Makonnen E, Olobo J, Okello L, Sagaki P, Strub N, Ellis S, Alvar J, Balasegaram M, Alirol E, Wasunna M (2017) Safety and effectiveness of sodium stibogluconate and paromomycin combination for the treatment of visceral leishmaniasis in Eastern Africa: results from a pharmacovigilance programme. Clin Drug Investig 37(3):259–272. https://doi.org/10.1007/s40261-016-0481-0

Kip AE, Rosing H, Hillebrand MJ, Castro MM, Gomez MA, Schellens JH, Beijnen JH, Dorlo TP (2015) Quantification of miltefosine in peripheral blood mononuclear cells by high-performance liquid chromatography-tandem mass spectrometry. J Chromatogr B 62:998–999. https://doi.org/10.1016/j.jchromb.2015.06.017

Kloehn J, Boughton BA, Saunders EC, O'Callaghan S, Binger KJ, McConville MJ (2021) Identification of metabolically quiescent leishmania mexicana parasites in peripheral and cured dermal granulomas using stable isotope tracing imaging mass spectrometry. mBio:*12*(2). https://doi.org/10.1128/mBio.00129-21

Kulshrestha A, Sharma V, Singh R, Salotra P (2014) Comparative transcript expression analysis of miltefosine-sensitive and miltefosine-resistant Leishmania donovani. Parasitol Res 113(3): 1171–1184. https://doi.org/10.1007/s00436-014-3755-6

Kumar R, Singh N, Gautam S, Singh OP, Gidwani K, Rai M, Sacks D, Sundar S, Nylen S (2014) Leishmania specific CD4 T cells release IFNgamma that limits parasite replication in patients with visceral leishmaniasis. PLoS Negl Trop Dis 8(10):e3198. https://doi.org/10.1371/journal.pntd.0003198

Liu D, Uzonna JE (2012) The early interaction of Leishmania with macrophages and dendritic cells and its influence on the host immune response. Front Cell Infect Microbiol 2:83. https://doi.org/10.3389/fcimb.2012.00083

Machado PR, Ampuero J, Guimaraes LH, Villasboas L, Rocha AT, Schriefer A, Sousa RS, Talhari A, Penna G, Carvalho EM (2010) Miltefosine in the treatment of cutaneous leishmaniasis caused by Leishmania braziliensis in Brazil: a randomized and controlled trial. PLoS Negl Trop Dis 4(12):e912. https://doi.org/10.1371/journal.pntd.0000912

Mannaert A, Downing T, Imamura H, Dujardin JC (2012) Adaptive mechanisms in pathogens: universal aneuploidy in Leishmania. Trends Parasitol 28(9):370–376. https://doi.org/10.1016/j.pt.2012.06.003

Mbongo N, Loiseau PM, Billion MA, Robert-Gero M (1998) Mechanism of amphotericin B resistance in Leishmania donovani promastigotes. Antimicrob Agents Chemother 42(2): 352–357. https://doi.org/10.1128/AAC.42.2.352

Mishra J, Singh S (2013) Miltefosine resistance in Leishmania donovani involves suppression of oxidative stress-induced programmed cell death. Exp Parasitol 135(2):397–406. https://doi.org/10.1016/j.exppara.2013.08.004

Mondelaers A, Sanchez-Canete MP, Hendrickx S, Eberhardt E, Garcia-Hernandez R, Lachaud L, Cotton J, Sanders M, Cuypers B, Imamura H, Dujardin JC, Delputte P, Cos P, Caljon G, Gamarro F, Castanys S, Maes L (2016) Genomic and molecular characterization of miltefosine resistance in leishmania infantum strains with either natural or acquired resistance through experimental selection of intracellular amastigotes. PLoS One 11(4):e0154101. https://doi.org/10.1371/journal.pone.0154101

Morizot G, Jouffroy R, Faye A, Chabert P, Belhouari K, Calin R, Charlier C, Miailhes P, Siriez JY, Mouri O, Yera H, Gilquin J, Tubiana R, Lanternier F, Mamzer MF, Legendre C, Peyramond D, Caumes E, Lortholary O, Buffet P (2016) Antimony to cure visceral leishmaniasis unresponsive to liposomal amphotericin B. PLoS Negl Trop Dis 10(1):e0004304. https://doi.org/10.1371/journal.pntd.0004304

Mukherjee B, Mukhopadhyay R, Bannerjee B, Chowdhury S, Mukherjee S, Naskar K, Allam US, Chakravortty D, Sundar S, Dujardin JC, Roy S (2013) Antimony-resistant but not antimony-sensitive Leishmania donovani up-regulates host IL-10 to overexpress multidrug-resistant protein 1. Proc Natl Acad Sci U S A 110(7):E575–E582. https://doi.org/10.1073/pnas.1213839110

Muller I, Pedrazzini T, Farrell JP, Louis J (1989) T-cell responses and immunity to experimental infection with leishmania major. Annu Rev Immunol 7:561–578. https://doi.org/10.1146/annurev.iy.07.040189.003021

Nascimento M, Abourjeily N, Ghosh A, Zhang WW, Matlashewski G (2003) Heterologous expression of a mammalian protein tyrosine phosphatase gene in Leishmania: effect on differentiation. Mol Microbiol 50(5):1517–1526. https://doi.org/10.1046/j.1365-2958.2003.03811.x

Ostyn B, Hasker E, Dorlo TP, Rijal S, Sundar S, Dujardin JC, Boelaert M (2014) Failure of miltefosine treatment for visceral leishmaniasis in children and men in South-East Asia. PLoS One 9(6):e100220. https://doi.org/10.1371/journal.pone.0100220

Paris C, Loiseau PM, Bories C, Breard J (2004) Miltefosine induces apoptosis-like death in Leishmania donovani promastigotes. Antimicrob Agents Chemother 48(3):852–859. https://doi.org/10.1128/AAC.48.3.852-859.2004

Patino LH, Imamura H, Cruz-Saavedra L, Pavia P, Muskus C, Mendez C, Dujardin JC, Ramirez JD (2019) Major changes in chromosomal somy, gene expression and gene dosage driven by Sb (III) in Leishmania braziliensis and Leishmania panamensis. Sci Rep 9(1):9485. https://doi.org/10.1038/s41598-019-45538-9

Patra P, Guha SK, Maji AK, Saha P, Ganguly S, Chakraborty A, Kundu PK, Sarker S, Ray K (2012) Efficacy of oral miltefosine in visceral leishmaniasis in rural West Bengal India. Indian J Pharm 44(4):500. https://doi.org/10.4103/0253-7613.99326

Peacock CS, Seeger K, Harris D, Murphy L, Ruiz JC, Quail MA, Peters N, Adlem E, Tivey A, Aslett M, Kerhornou A, Ivens A, Fraser A, Rajandream MA, Carver T, Norbertczak H, Chillingworth T, Hance Z, Jagels K, Moule S, Ormond D, Rutter S, Squares R, Whitehead S, Rabbinowitsch E, Arrowsmith C, White B, Thurston S, Bringaud F, Baldauf SL, Faulconbridge A, Jeffares D, Depledge DP, Oyola SO, Hilley JD, Brito LO, Tosi LR, Barrell B, Cruz AK, Mottram JC, Smith DF, Berriman M (2007) Comparative genomic analysis of three Leishmania species that cause diverse human disease. Nat Genet 39(7):839–847. https://doi.org/10.1038/ng2053

Perez-Victoria JM, Parodi-Talice A, Torres C, Gamarro F, Castanys S (2001) ABC transporters in the protozoan parasite Leishmania. Int Microbiol 4(3):159–166. https://doi.org/10.1007/s10123-001-0031-2

Perez-Victoria FJ, Castanys S, Gamarro F (2003) Leishmania donovani resistance to miltefosine involves a defective inward translocation of the drug. Antimicrob Agents Chemother 47(8): 2397–2403. https://doi.org/10.1128/AAC.47.8.2397-2403.2003

Perez-Victoria FJ, Sanchez-Canete MP, Castanys S, Gamarro F (2006) Phospholipid translocation and miltefosine potency require both L. donovani miltefosine transporter and the new protein LdRos3 in Leishmania parasites. J Biol Chem 281(33):23766–23775. https://doi.org/10.1074/jbc.M605214200

Pountain AW, Weidt SK, Regnault C, Bates PA, Donachie AM, Dickens NJ, Barrett MP (2019) Genomic instability at the locus of sterol C24-methyltransferase promotes amphotericin B resistance in Leishmania parasites. PLoS Negl Trop Dis 13(2):e0007052. https://doi.org/10.1371/journal.pntd.0007052

Prajapati VK, Sharma S, Rai M, Ostyn B, Salotra P, Vanaerschot M, Dujardin JC, Sundar S (2013) In vitro Susceptibility of Leishmania donovani to Miltefosine in Indian Visceral Leishmaniasis. The American Journal of Tropical Medicine and Hygiene 89(4):750–754. https://doi.org/10.4269/ajtmh.13-0096

Purkait B, Kumar A, Nandi N, Sardar AH, Das S, Kumar S, Pandey K, Ravidas V, Kumar M, De T, Singh D, Das P (2012) Mechanism of amphotericin B resistance in clinical isolates of Leishmania donovani. Antimicrob Agents Chemother 56(2):1031–1041. https://doi.org/10.1128/AAC.00030-11

Rahman M, Ahmed BN, Faiz MA, Chowdhury MZ, Islam QT, Sayeedur R, Rahman MR, Hossain M, Bangali AM, Ahmad Z, Islam MN, Mascie-Taylor CG, Berman J, Arana B (2011) Phase IV Trial of Miltefosine in Adults and Children for Treatment of Visceral Leishmaniasis (Kala-Azar) in Bangladesh. The American Journal of Tropical Medicine and Hygiene 85(1):66–69. https://doi.org/10.4269/ajtmh.2011.10-0661

Rahman R, Goyal V, Haque R, Jamil K, Faiz A, Samad R, Ellis S, Balasegaram M, Boer MD, Rijal S, Strub-Wourgaft N, Alves F, Alvar J, Sharma B (2017) Safety and efficacy of short course combination regimens with AmBisome, miltefosine and paromomycin for the treatment of visceral leishmaniasis (VL) in Bangladesh. PLoS Negl Trop Dis 11(5):e0005635. https://doi.org/10.1371/journal.pntd.0005635

Rai K, Cuypers B, Bhattarai NR, Uranw S, Berg M, Ostyn B, Dujardin JC, Rijal S, Vanaerschot M (2013) Relapse after treatment with miltefosine for visceral leishmaniasis is associated with increased infectivity of the infecting Leishmania donovani strain. mBio 4(5):e00611–e00613. https://doi.org/10.1128/mBio.00611-13

Rakotomanga M, Saint-Pierre-Chazalet M, Loiseau PM (2005) Alteration of fatty acid and sterol metabolism in miltefosine-resistant Leishmania donovani promastigotes and consequences for drug-membrane interactions. Antimicrob Agents Chemother 49(7):2677–2686. https://doi.org/10.1128/AAC.49.7.2677-2686.2005

Rastrojo A, Garcia-Hernandez R, Vargas P, Camacho E, Corvo L, Imamura H, Dujardin JC, Castanys S, Aguado B, Gamarro F, Requena JM (2018) Genomic and transcriptomic alterations in Leishmania donovani lines experimentally resistant to antileishmanial drugs. Int J Parasitol Drugs Drug Resist 8(2):246–264. https://doi.org/10.1016/j.ijpddr.2018.04.002

Rijal S, Ostyn B, Uranw S, Rai K, Bhattarai NR, Dorlo TP, Beijnen JH, Vanaerschot M, Decuypere S, Dhakal SS, Das ML, Karki P, Singh R, Boelaert M, Dujardin JC (2013) Increasing failure of miltefosine in the treatment of Kala-azar in Nepal and the potential role of parasite drug resistance, reinfection, or noncompliance. Clin Infect Dis 56(11):1530–1538. https://doi.org/10.1093/cid/cit102

Ritmeijer K, Dejenie A, Assefa Y, Hundie TB, Mesure J, Boots G, den Boer M, Davidson RN (2006) A comparison of miltefosine and sodium stibogluconate for treatment of visceral leishmaniasis in an Ethiopian population with high prevalence of HIV infection. Clin Infect Dis 43(3):357–364. https://doi.org/10.1086/505217

Rosenthal MD (1987) Fatty acid metabolism of isolated mammalian cells. Prog Lipid Res 26(2): 87–124. https://doi.org/10.1016/0163-7827(87)90009-9

Rossi M, Fasel N (2018) How to master the host immune system? Leishmania parasites have the solutions! Int Immunol 30(3):103–111. https://doi.org/10.1093/intimm/dxx075

Scorza BM, Wacker MA, Messingham K, Kim P, Klingelhutz A, Fairley J, Wilson ME (2017) Differential activation of human keratinocytes by leishmania species causing localized or disseminated disease. J Invest Dermatol 137(10):2149–2156. https://doi.org/10.1016/j.jid.2017.05.028

Seifert K, Matu S, Javier Perez-Victoria F, Castanys S, Gamarro F, Croft SL (2003) Characterisation of Leishmania donovani promastigotes resistant to hexadecylphosphocholine (miltefosine). Int J Antimicrob Agents 22(4):380–387. https://doi.org/10.1016/s0924-8579(03)00125-0

Seifert K, Perez-Victoria FJ, Stettler M, Sanchez-Canete MP, Castanys S, Gamarro F, Croft SL (2007) Inactivation of the miltefosine transporter, LdMT, causes miltefosine resistance that is conferred to the amastigote stage of Leishmania donovani and persists in vivo. Int J Antimicrob Agents 30(3):229–235. https://doi.org/10.1016/j.ijantimicag.2007.05.007

Selvapandiyan A, Croft SL, Rijal S, Nakhasi HL, Ganguly NK (2019) Innovations for the elimination and control of visceral leishmaniasis. PLoS Negl Trop Dis 13(9):e0007616. https://doi.org/10.1371/journal.pntd.0007616

Shaw CD, Lonchamp J, Downing T, Imamura H, Freeman TM, Cotton JA, Sanders M, Blackburn G, Dujardin JC, Rijal S, Khanal B, Illingworth CJ, Coombs GH, Carter KC (2016) In vitro selection of miltefosine resistance in promastigotes of Leishmania donovani from Nepal: genomic and metabolomic characterization. Mol Microbiol 99(6):1134–1148. https://doi.org/10.1111/mmi.13291

Shaw CD, Imamura H, Downing T, Blackburn G, Westrop GD, Cotton JA, Berriman M, Sanders M, Rijal S, Coombs GH, Dujardin JC, Carter KC (2019) Genomic and metabolomic polymorphism among experimentally selected paromomycin-resistant leishmania donovani strains. Antimicrob Agents Chemother 64(1). https://doi.org/10.1128/AAC.00904-19

Sindermann H, Engel J (2006) Development of miltefosine as an oral treatment for leishmaniasis. Trans R Soc Trop Med Hyg 100(Suppl 1):S17–S20. https://doi.org/10.1016/j.trstmh.2006.02.010

Soto J, Arana BA, Toledo J, Rizzo N, Vega JC, Diaz A, Luz M, Gutierrez P, Arboleda M, Berman JD, Junge K, Engel J, Sindermann H (2004) Miltefosine for new world cutaneous leishmaniasis. Clin Infect Dis 38(9):1266–1272. https://doi.org/10.1086/383321

Soulat D, Bogdan C (2017) Function of macrophage and parasite phosphatases in leishmaniasis. Front Immunol 8:1838. https://doi.org/10.3389/fimmu.2017.01838

Srivastava P, Prajapati VK, Rai M, Sundar S (2011) Unusual case of resistance to amphotericin B in visceral leishmaniasis in a region in India where leishmaniasis is not endemic. J Clin Microbiol 49(8):3088–3091. https://doi.org/10.1128/JCM.00173-11

Srivastava S, Mishra J, Gupta AK, Singh A, Shankar P, Singh S (2017) Laboratory confirmed miltefosine resistant cases of visceral leishmaniasis from India. Parasit Vectors 10(1):49. https://doi.org/10.1186/s13071-017-1969-z

Stauch A, Duerr HP, Dujardin JC, Vanaerschot M, Sundar S, Eichner M (2012) Treatment of visceral leishmaniasis: model-based analyses on the spread of antimony-resistant L. donovani in Bihar, India. PLoS Negl Trop Dis 6(12):e1973. https://doi.org/10.1371/journal.pntd.0001973

Sundar S, Jha TK, Sindermann H, Junge K, Bachmann P, Berman J (2003) Oral miltefosine treatment in children with mild to moderate Indian visceral leishmaniasis. Pediatr Infect Dis J 22(5):434–438. https://doi.org/10.1097/01.inf.0000066877.72624.cb

Sundar S, Singh A (2018) Chemotherapeutics of visceral leishmaniasis: present and future developments. Parasitology 145(4):481–489. https://doi.org/10.1017/S0031182017002116

Sundar S, Jha TK, Thakur CP, Engel J, Sindermann H, Fischer C, Junge K, Bryceson A, Berman J (2002) Oral miltefosine for Indian visceral leishmaniasis. N Engl J Med 347(22):1739–1746. https://doi.org/10.1056/NEJMoa021556

Sundar S, Singh A, Rai M, Prajapati VK, Singh AK, Ostyn B, Boelaert M, Dujardin JC, Chakravarty J (2012) Efficacy of miltefosine in the treatment of visceral leishmaniasis in India after a decade of use. Clin Infect Dis 55(4):543–550. https://doi.org/10.1093/cid/cis474

Torres-Guerrero E, Quintanilla-Cedillo MR, Ruiz-Esmenjaud J, Arenas R (2017) Leishmaniasis: a review. F1000Res. 6:750. https://doi.org/10.12688/f1000research.11120.1

Turner KG, Vacchina P, Robles-Murguia M, Wadsworth M, McDowell MA, Morales MA (2015) Fitness and phenotypic characterization of miltefosine-resistant leishmania major. PLoS Negl Trop Dis 9(7):e0003948. https://doi.org/10.1371/journal.pntd.0003948

Vacchina P, Norris-Mullins B, Abengozar MA, Viamontes CG, Sarro J, Stephens MT, Pfrender ME, Rivas L, Morales MA (2016) Genomic appraisal of the multifactorial basis for in vitro acquisition of miltefosine resistance in Leishmania donovani. Antimicrob Agents Chemother 60(7):4089–4100. https://doi.org/10.1128/AAC.00478-16

Vanaerschot M, Huijben S, Van den Broeck F, Dujardin JC (2014) Drug resistance in vectorborne parasites: multiple actors and scenarios for an evolutionary arms race. FEMS Microbiol Rev 38(1):41–55. https://doi.org/10.1111/1574-6976.12032

Verma NK, Dey CS (2004) Possible mechanism of miltefosine-mediated death of Leishmania donovani. Antimicrob Agents Chemother 48(8):3010–3015. https://doi.org/10.1128/AAC.48.8.3010-3015.2004

Verma A, Bhandari V, Deep DK, Sundar S, Dujardin JC, Singh R, Salotra P (2017) Transcriptome profiling identifies genes/pathways associated with experimental resistance to paromomycin in Leishmania donovani. Int J Parasitol Drugs Drug Resist 7(3):370–377. https://doi.org/10.1016/j.ijpddr.2017.10.004

Verrest L, Dorlo TPC (2017) Lack of clinical pharmacokinetic studies to optimize the treatment of neglected tropical diseases: a systematic review. Clin Pharmacokinet 56(6):583–606. https://doi.org/10.1007/s40262-016-0467-3

Wasunna M, Njenga S, Balasegaram M, Alexander N, Omollo R, Edwards T, Dorlo TP, Musa B, Ali MH, Elamin MY, Kirigi G, Juma R, Kip AE, Schoone GJ, Hailu A, Olobo J, Ellis S, Kimutai R, Wells S, Khalil EA, Strub Wourgaft N, Alves F, Musa A (2016) Efficacy and Safety of AmBisome in Combination with Sodium Stibogluconate or Miltefosine and Miltefosine Monotherapy for African Visceral Leishmaniasis: Phase II Randomized Trial PLOS Neglected Tropical Diseases. 10(9):e0004880. https://doi.org/10.1371/journal.pntd.0004880

WHO (2010) *Control of the leishmaniasis: report of a meeting of the WHO Expert Committee on the Control of Leishmaniases, Geneva, 22–26 March 2010.* (949). (WHO technical report series, Issue

WHO (2021) *Global leishmaniasis surveillance: 2019–2020, a baseline for the 2030 roadmap* (Weekly Epidemiological Record., Issue

Yadav S, Ali V, Singh Y, Kanojia S, Goyal N (2020) Leishmania donovani chaperonin TCP1gamma subunit protects miltefosine induced oxidative damage. Int J Biol Macromol 165(Pt B):2607–2620. https://doi.org/10.1016/j.ijbiomac.2020.10.134

Yardley V, Croft SL, De Doncker S, Dujardin JC, Koirala S, Rijal S, Miranda C, Llanos-Cuentas A, Chappuis F (2005) The sensitivity of clinical isolates of Leishmania from Peru and Nepal to miltefosine. Am J Trop Med Hyg 73(2):272–275

Toward a Safe and Efficacious Pan-*Leishmania* Vaccine

13

Parna Bhattacharya, Greta Volpedo, Thalia Pacheco-Fernandez, Ranadhir Dey, Greg Matlashewski, Abhay R. Satoskar, Sanjay Singh, Sreenivas Gannavaram, and Hira L. Nakhasi

Abstract

Leishmaniasis has been designated by the World Health Organization (WHO) as one of the top neglected tropical diseases with significant morbidity and mortality in low- and middle-income countries (LMICs). With increasing migration from and ecotourism to endemic countries, cases of leishmaniasis have been spreading into nonendemic countries. Currently, there is a lack of effective strategies to control this disease and to achieve elimination by the 2030 target set forth by the WHO. Vaccination can be an effective measure to control leishmaniasis and has the potential to achieve disease elimination. Previously, efforts focused on the development of vaccines from killed parasites with or without adjuvants, subunit vaccines, and DNA-based vaccines that were immunogenic and efficacious in animal models against virulent infection through needle injection. However, when tested against natural sandfly infection or in clinical settings, they were

Parna Bhattacharya, Greta Volpedo, and Thalia Pacheco-Fernandez contributed equally to this work.

P. Bhattacharya · R. Dey · S. Gannavaram · H. L. Nakhasi (✉)
Division of Emerging and Transfusion Transmitted Diseases, Office of Blood Research and Review, Center for Biologics Evaluation and Research, Food and Drug Administration, Silver Spring, MD, USA
e-mail: Hira.Nakhasi@fda.hhs.gov

G. Volpedo · T. Pacheco-Fernandez · A. R. Satoskar
Departments of Pathology and Microbiology, Wexner Medical Center, The Ohio State University, Columbus, OH, USA

G. Matlashewski
Department of Microbiology and Immunology, McGill University, Montreal, QC, Canada

S. Singh
Gennova Biopharmaceuticals, Pune, India

© The Author(s), under exclusive license to Springer Nature Singapore Pte Ltd. 2023

A. Selvapandiyan et al. (eds.), *Challenges and Solutions Against Visceral Leishmaniasis*, https://doi.org/10.1007/978-981-99-6999-9_13

determined not to be efficacious. People who have recovered from leishmaniasis are protected for life against future infections, suggesting that an effective vaccine is possible against this disease. Exposure to, or deliberate infection with, wild-type *Leishmania (L.) major* (leishmanization), was shown to be effective against both visceral leishmaniasis (VL) and cutaneous leishmaniasis (CL). However, the practice of deliberate infection raises safety concerns and has not been pursued. Therefore, a well-characterized genetically modified parasite that is safe, but as efficacious as leishmanization, could be an alternative. The presence of both CL and VL in many LMICs makes it desirable to have a vaccine that can protect against both forms of the disease. In this chapter, we will describe the development of a vaccine candidate that has been genetically modified by deletion of a *centrin* gene using CRISPR-Cas9 technology in a dermotropic *L. major* parasite strain (*LmCen*$^{-/-}$). Centrin plays an essential role in the basal body duplication of the amastigote form of the *Leishmania* parasite and is necessary for its cell division. The *LmCen*$^{-/-}$ vaccine manufactured under Good Laboratory Practice has been shown to be safe and immunogenic and provide long-term protection against both CL and VL through infected sandfly vector exposure in preclinical studies. Taken together, these observations suggest that a genetically modified parasite could be a viable vaccine candidate for evaluation in clinical trials.

Keywords

Leishmaniasis · Leishmanization · Vaccine development · Live-attenuated vaccine · Gene deletion · Safety and protective immunity

13.1 Introduction

Over 1.2 million cases of leishmaniasis are estimated annually to affect tropical regions, where this disease is endemic (CDC 2020), and the eradication of leishmaniasis requires the implementation of appropriate diagnostic methods, treatments, and prevention of infection (Sundar et al. 2018). Despite the quality of life and economic costs of leishmaniasis (Pires et al. 2019; Okwor and Uzonna 2016), currently there is no vaccine available for human use. Experimental vaccines (reviewed elsewhere (Volpedo et al. 2021a) such as DNA vaccines (Domínguez-Bernal et al. 2015), peptide vaccines (Petitdidier et al. 2019), subunit vaccines (Duthie et al. 2017; Duthie and Reed 2014; Lari et al. 2022), and live recombinant vaccine (Salari et al. 2020) have been explored, albeit only a few candidates have made it into clinical trials (Moafi et al. 2019). Nevertheless, vaccination is known to be possible as natural *Leishmania* infection provides long-term immune protection against reinfection (reviewed in Pacheco-Fernandez et al. 2021). Based on this observation, the deliberate intradermal inoculation of live *Leishmania (L.) major* has been successfully used as a human vaccination method in the Middle East and the former Soviet Union states for decades, a process named leishmanization (Khamesipour et al. 2005). However, just as in a naturally occurring infection, complications of leishmanization, including nonhealing lesions, may occur in some people. Therefore, the practice of leishmanization has been discontinued due to safety concerns (Pacheco-Fernandez et al. 2021; Seyed et al. 2018). Viewed in that respect, live-attenuated *Leishmania* parasites represent a safer, yet efficient, alternative to leishmanization, with their genetic modifications to genes critical for parasite virulence and growth allowing the establishment of a long-term protective immune response while avoiding the development of the disease.

Several live-attenuated vaccines have been developed in the past, which include *L. donovani* (Ld) mutants, such as the *Biopterin transporter 1* (*BT1*)-deletion mutant parasites. Infection with *LdBT1* deletion mutants resulted in protective immunity, characterized by interferon (IFN)-γ production, while being less infective than LdWT parasites. Nevertheless, LdBT1 parasites were still identified in the liver and spleen after 4 weeks of immunization (Papadopoulou et al. 2002). Similarly, inoculation with the heat shock protein-null (Δ*Hsp70-II*) *L. infantum* parasites generated protective inflammatory responses and the reduction of parasitic loads after a challenge infection with *L. major*, and it was successful in protecting against the development of cutaneous leishmaniasis (CL), although it failed to control the infection in the spleen and bone marrow in a murine model of visceral leishmaniasis (VL) induced by infective *L. infantum* challenge (Carrión et al. 2011; Solana et al. 2017, 2020; Soto et al. 2021;). A different approach that targets stage-specific genes, which would allow the proliferation of the parasite in the promastigote stage but impede survival in the amastigote stage, has been successful, such as *L. donovani* mutants lacking the *centrin* ($LdCen^{-/-}$) or p27 ($Ldp27^{-/-}$) gene, the latter encoding an amastigote-specific protein part of the cytochrome *c* oxidase complex (Selvapandiyan et al. 2014). $Ldp27^{-/-}$ amastigotes survived in the macrophages while generating a long-term protective immune response against CL and VL (Dey et al. 2010; Dey et al. 2013). Similarly, the *KHARON1* null mutant in *L. infantum* (Δ*Likh1*) causes cell-cycle arrest in the parasite. Nevertheless, Δ*Likh1* was incapable of sustaining macrophage infection while inducing protection against challenge with virulent *L. infantum* (Santi et al. 2018). Further, the efficacy of attenuated *L. donovani* clonal parasites as an immunoprophylactic and immunotherapeutic agent against *L. donovani* infection in the murine model has been demonstrated (Bhaumik et al. 2009). Despite the success of live-attenuated vaccines in experimental models of leishmaniasis, none of these mutants were tested in clinical trials.

For the purpose of brevity, this chapter focuses on the *centrin*-deleted (*cen−/−*) *Leishmania* mutants, a vaccine candidate with the knockout of a stage-specific gene in *Leishmania*. The experimental steps taken to demonstrate the safety, immunogenicity, and efficacy of *centrin* knockouts against WT *Leishmania* infection are discussed, as well as its potential as the first human pan-*Leishmania* vaccine.

13.2 Amastigote-Specific Gene Mutants of *Leishmania* as Candidate Vaccines for Leishmaniasis

It is known that individuals who recover from *Leishmania* infection, either after drug treatment or self-healing in the case of CL, develop long-lasting protection against future infections (Pacheco-Fernandez et al. 2021). Based on this observation, we and others have developed live-attenuated vaccines, where genetic modifications target a stage-specific gene that allows for the survival of promastigote parasites but causes no disease once they transform into amastigotes (Selvapandiyan et al. 2014).

In the case of *LdCen*$^{-/-}$, *Leishmania-centrin-1* is deleted, which is one of the five different *centrin* genes encoded by *Leishmania*, and it is necessary to sustain amastigote replication. Accordingly, its deficiency results in cell growth arrest at the G2/M phase and the induction of apoptosis (Selvapandiyan et al. 2001). Thus, the loss of *centrin-1* led to an impairment in the growth of *LdCen*$^{-/-}$ axenic amastigotes and in the infected macrophages in vivo, making it a nonvirulent parasite (Selvapandiyan et al. 2004). It is important to highlight that *LdCen*$^{-/-}$ promastigote growth in vitro was not affected; therefore, this allows for a large-scale culture of the parasite, which is desired for vaccine production (Selvapandiyan et al. 2014).

13.3 Immunogenicity and Efficacy Characteristics of *LdCen*$^{-/-}$ Parasites

The immune mechanisms that allow *LdCen*$^{-/-}$ to be an efficacious vaccine have been characterized by our group during the last few years. First, *LdCen*$^{-/-}$ activates the innate immune response by inducing proinflammatory responses in macrophages and neutrophils (Bhattacharya et al. 2015, 2020). Neutrophils are the first immune cells to be recruited to the *Leishmania* infection site and therefore determine the quality of immune response (Ribeiro-Gomes and Sacks 2012). Immunization with *LdCen*$^{-/-}$ increased the early influx of neutrophils to the infection site, more than that observed in infection with LdWT. Neutrophil recruitment was observed to be mediated by CCL2 and CCL3 expression, induced by *LdCen*$^{-/-}$, and these neutrophils showed an elevated expression of costimulatory molecules (CD80 and MHC-II), compared to those recruited by WT infection (Fig. 13.1.1). Two distinct neutrophil subtypes (Nα and Nβ) were identified in *Leishmania* infection in ear and lymph node tissues with the Nα population, inducing a strong proliferation of T-cells ex vivo compared to Nβ. Higher numbers of Nα neutrophils were observed during infection with *LdCen*$^{-/-}$, compared to its WT counterpart. Importantly, these neutrophil responses are necessary for the protection elicited by *LdCen*$^{-/-}$, as shown by the neutrophil depletion experiments, where protection elicited by *LdCen*$^{-/-}$ was compromised as a consequence of Ag-specific Th1 response abrogation due to the lack of neutrophils (Bhattacharya et al. 2020). *LdCen*$^{-/-}$ infection generates a different expression profile than *LdWT* and induces macrophage polarization toward a pro-inflammatory M1 phenotype, in contrast to the downregulation of such

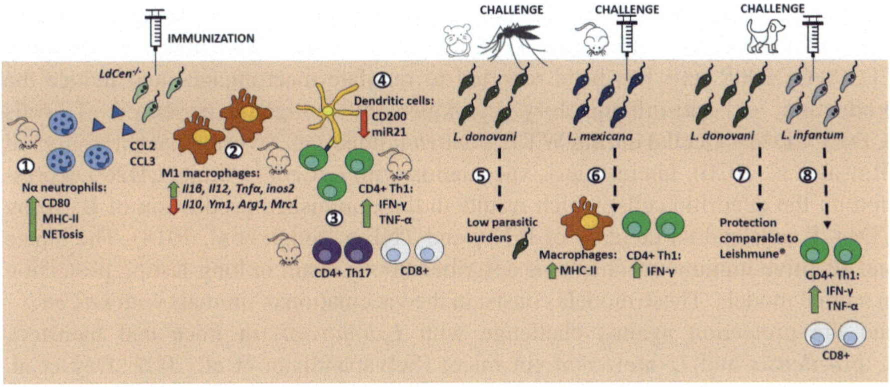

Fig. 13.1 *LdCen*$^{-/-}$-mediated mechanisms of protection. (1) In mice, *LdCen*$^{-/-}$ induces CCL2, and CCL3 mediates the early influx of neutrophils, with upregulated CD80, MHC-II, and NETosis. (2) In mice, *LdCen*$^{-/-}$ mediates M1 polarization in macrophages, characterized by the upregulation of inflammatory-related genes such as *Il1β*, *Il12*, *Tnfα*, and *inos2* and the downregulation of anti-inflammatory-related genes such as *Il10*, *Ym1*, *Arg1*, and *Mrc1*. (3) In mice, *LdCen*$^{-/-}$ promotes the accumulation of protective CD8+, as well as CD4+ Th17 and Th1 T cells. (4) In mice, *LdCen*$^{-/-}$ limits CD200 expression in dendritic cells, disrupting the CD200-CD200R axis. This results in increased IFN-y and TNF-a expression by CD4+ T cells. miR21 also goes down in immunized mice. (5) In hamsters, immunization with *LdCen*$^{-/-}$ leads to significantly lower parasitic burdens after sandfly *L. donovani* challenge, compared to nonimmunized hamsters. (6) In mice, immunization with *LdCen*$^{-/-}$ leads to the recruitment of MHC-II-expressing macrophages and IFN-y-expressing CD4+ Th1 cells after needle *L. mexicana* challenge, compared to nonimmunized mice. (7) In dogs, immunization with *LdCen*$^{-/-}$ leads to protection against sandfly *L. donovani* challenge, comparable to Leishmune®. (8) In dogs, immunization with *LdCen*$^{-/-}$ leads to protection against sandfly *L. infantum* challenge, characterized by increased numbers of CD8+, as well as IFN-y- and TNF-a-producing CD4+ T cells

immune response induced by *LdWT* parasites (Bhattacharya et al. 2015; Gannavaram et al. 2019). Particularly in human macrophages and murine dendritic cells, *LdCen*$^{-/-}$ suppresses the expression of MicroRNA (MiR)-21, which negatively regulates IL-12. Therefore, infection with *LdCen*$^{-/-}$ promotes the production of IL-12 by the presenter cells in higher concentrations than infection with *LdWT*, hence favoring CD4+ T-cell proliferation (Gannavaram et al. 2019). Consequently, the M1 phenotype (characterized by the upregulation of inflammatory-related genes such as *Il1β*, *Il12*, *Tnfα*, and *inos2* and the downregulation of anti-inflammatory-related genes such as *Il10*, *Ym1*, *Arg1*, and *Mrc1*) activates the antiparasitic Th1 response (Fig. 13.1.2) (Bhattacharya et al. 2015; Tomiotto-Pellissier et al. 2018). Studies analyzing adaptive immunity showed that vaccination with *LdCen*$^{-/-}$ elicits a combination of Th17 and Th1 immune responses, both shown to be necessary to promote antiparasitic activities (Fig. 13.1.3) (Selvapandiyan et al. 2009; Banerjee et al. 2018). The production of IL-17 was sustained through IL-1β, IL-6, TGF-β, and IL-23 (Banerjee et al. 2018), while the Th1 response was characterized by a higher IFN-γ/IL-10-producing T-cell ratio, as well as an increase in cytotoxic CD8+ T cells, and IgG levels, especially of Th1-promoting IgG2a.[33] This response is possible because of the downregulatory effect that *LdCen*$^{-/-}$ has on the immune inhibitory CD200/CD200R axis (Fig. 13.1.4) (Singh et al. 2018). The immune inhibitory CD200/CD200R axis has been reported to regulate macrophages and induce the production of anti-inflammatory cytokines by antigen-experienced T cells (CD4 + CD44+ T cells) during WT *L. donovani* infection, favoring parasitic survival (Rawat et al. 2020). Interestingly, vaccination with *LdCen*$^{-/-}$ limits CD200 expression on the dendritic cells, which results in the diminished production of IL-10 by CD4+ T cells and an increase of IFN-γ and TNF-α (Singh et al. 2018). The innate and adaptive immune mechanisms described above result in long-lasting protection in animal models. These models consist in the vaccination of animals with *LdCen*$^{-/-}$ and the protection against challenge with *L.donovani* (in mice and hamster), *L. brazilensis* and *L. mexicana* (in mice) (Selvapandiyan et al. 2009; Dey et al. 2014; Medkour et al. 2019; Kaszak et al. 2015). While *L. donovani* causes VL and is endemic in Old World countries, *L. mexicana*, a causative agent of CL, and *L. braziliensis*, a causative agent of MCL, are endemic in the New World (Selvapandiyan et al. 2009; de Oliveira and Brodskyn 2012; Ready 2014).

Immunization with *LdCen*$^{-/-}$ also provided protection against the American CL caused by *L. mexicana*. In the case of *L. mexicana* challenge, there was an absence of dermal lesions, with a significant recruitment of MHC-II-expressing macrophages to the challenge site and a strong inflammatory cytokine response characterized by IFN-γ-producing T cells and diminished IL-4 and IL-13 production (Fig. 13.1.6) (Dey et al. 2014). The development of a pan-*Leishmania* vaccine would require a demonstration of protection against different strains of *Leishmania* parasites that cause distinct clinical manifestations. Thus, the protection observed against *Leishmania* species prevalent in different geographic regions that cause distinct pathologies highlights the potential of *LdCen*$^{-/-}$ as a pan-leishmania vaccine.

Toward understating the tripartite interactions between the vector, the parasite, and the host, studies have been undertaken to explore the role of sandfly vector in vaccine immunity. The infectious inoculum of sandfly is very complex and is composed of parasites and vector-gut-derived components (Lestinova et al. 2017). This infectious inoculum induces a distinct host immune response critical for parasite dissemination, which is not fully recapitulated in a needle-initiated infection (Rogers 2012). Needle inoculation commonly used in vaccine studies does not fully recapitulate these critical factors. For instance, the enhanced virulence observed in sandfly-mediated transmission has been attributed to a sustained recruitment of neutrophils to bite sites, where they protect captured parasites early after transmission (Peters et al. 2008). To address the efficacy more rigorously, *LdCen*$^{-/-}$ immunized hamsters were challenged with *L. donovani*-infected sandflies. Nine months post challenge, parasitic burdens in the liver and spleen were significantly low in the immunized groups compared to the nonimmunized group, indicating that *LdCen*$^{-/-}$ immunization-induced protection is robust (Fig. 13.1.5). In addition, vaccination also protected against mortality due to *L. donovani* virulent infection (unpublished data, personal communication, Ranadhir Dey).

To warrant the efficacy of a vaccine, it should also provide protection to vulnerable groups, including aged individuals who have reduced immunity, and target potential reservoirs in endemic areas. The $LdCen^{-/-}$ vaccine has been demonstrated to be safe and efficacious against *L. donovani* infection in aged mice, where there was a protective response similar to that observed in younger mice (Bhattacharya et al. 2016). On the other hand, the vaccination of reservoirs is a realistic strategy to eradicate the disease. $LdCen^{-/-}$ has been shown to provide protection in different reservoirs: asymptomatic infections in murine models and dogs. However, in the Indian subcontinent, the spread of *Leishmania* is considered to be due to anthroponotic transmission (Le Rutte et al. 2016). Asymptomatic infections are usually not clinically diagnosed due to the lack of evident symptoms; therefore, asymptomatic patients do not receive drug treatment, allowing for further transmission of the disease (Singh et al. 2014). Moreover, asymptomatic-infected individuals will be the predominant population to receive the vaccine as asymptomatic cases are more frequent than symptomatic ones (Andrade-Narvaez et al. 2016; Bern et al. 2007; Topno et al. 2010; Chakravarty et al. 2019). In a murine model of asymptomatic infection, the $LdCen^{-/-}$ vaccine showed to be immunogenic and protective against a WT *L. donovani* challenge, and the memory cell response was very similar to the one elicited in naïve mice after vaccination. Hence, $LdCen^{-/-}$ mutants could be used in asymptomatically infected individuals in the endemic regions (Ismail et al. 2017).

A second group considered a reservoir of concern is patients with post-kala-azar dermal leishmaniasis (PKDL) (Gedda et al. 2020). PDKL is a dermal manifestation of leishmaniasis that occurs in patients apparently cured of VL. Five to 15% and up to 50% of VL patients develop PKDL in India and Sudan, respectively (WHO 2013). Just as discussed before, human reservoirs sustain anthroponotic transmission of the disease, and it is imperative to target them for a successful VL elimination (Gedda et al. 2020). To test the potential of $LdCen^{-/-}$ parasites in inducing protection in ex vivo human studies from endemic regions, peripheral blood mononuclear cells (PBMCs) of patients from healed VL and PKDL were used. Another genetic mutant of *L. donovani* that lacked the p27 gene ($Ldp27^{-/-}$), necessary for amastigote replication, was also tested in these studies (Dey et al. 2010; Avishek et al. 2016). PBMC-derived macrophages from patients with healed VL or PKDL showed elevated production of IFN-γ, TNF-α, IL-2, IL-6, IL-12 and IL-17 after infection with both $LdCen^{-/-}$ and $Ldp27^{-/-}$. Moreover, there was an increase in IFN-γ-producing allogenic T cells (both CD4+ and CD8+) induced by both mutants (Avishek et al. 2016), therefore demonstrating that both live-attenuated vaccines are immunogenic in healed VL and PKDL individuals, highlighting the potential of $LdCen^{-/-}$ as a vaccine, which can be used against various phenotypes of visceral disease.

The protective response elicited by $LdCen^{-/-}$ was also demonstrated in dogs, the main reservoir for *L. infantum* in endemic regions of South America and the Mediterranean (Medkour et al. 2019; Kaszak et al. 2015). In canine VL, it is crucial to prevent the disease in dogs as the drug treatments available are the same as those used for humans, and therefore, treating infected dogs can lead to the generation and propagation of drug-resistant parasites. Additionally, canine VL is characterized by high rates of relapse and low frequency of clinical cure (Gonçalves et al. 2019). The recombinant vaccine Leishmune® was the first licensed vaccine for use in canine VL prevention (Palatnik-de-Sousa and Nico 2020). *LdCen*$^{-/-}$ has shown comparable protection to that elicited by Leishmune® in dogs (Fiuza et al. 2015). Interestingly, immunization with *LdCen*$^{-/-}$ mutants was also capable of generating cross-protection against *L. infantum*: in particular, a strong T-cell response characterized by both CD4+ and CD8+ T-cell proliferation as well as IFN-γ, TNF-α, and IL-12/IL-23p40 production, together with a decrease in IL-4 secretion (Fiuza et al. 2013). In the case of both Leishmune® and *LdCen*$^{-/-}$ vaccines, long-term protection (24 months) was demonstrated against *L. infantum,* characterized by the reduction of parasitic burden in the bone marrow (Fiuza et al. 2015). Nevertheless, there was an increase in the Th1 response, an elevated reduction of IL-4 secretion, and CD8+ T-cell proliferation in dogs vaccinated with *LdCen*$^{-/-}$, compared to those vaccinated with Leishmune® (Fig. 13.1.7) (Fiuza et al. 2015; Viana et al. 2016). Hence, *LdCen*$^{-/-}$ could be effective in treating reservoirs such as dogs for parasites causing VL.

Overall, studies in preclinical animal models and ex vivo human studies reviewed in this section demonstrated the immunogenicity and efficacy of *LdCen*$^{-/-}$ to protect against multiple *Leishmania* strains, which highlights its potential as a potential pan-*Leishmania* vaccine (Fig. 13.1.1).

13.4 Evaluation of Safety Characteristics of *LdCen*$^{-/-}$ Parasites

To demonstrate that the safety of *LdCen*$^{-/-}$ was solely dependent on the mutant's inability to proliferate inside the host, immunization experiments were performed in severe combined immunodeficient (SCID) mice. *LdCen−/−* inoculation in such mice resulted in no parasites detected in liver and spleen after 12 weeks of infection, showing how parasite clearance is not dependent on the host immune response (Selvapandiyan et al. 2009).

Despite the efficacy demonstrated by *LdCen*$^{-/-}$ parasites, advancing these mutants to clinical trials has not been straightforward due to the presence of antibiotic resistance genes used to develop the vaccine strain through homologous recombination. These genes are inserted into the pathogen's genome to be used as markers for the selection of the genetically engineered mutants of interest. The presence of antibiotic resistance genes in vaccine candidates raises safety concerns in human use. However, this is easy to remedy due to the recent development of advanced genetic engineering techniques that allow for the generation of specific targeted mutations without the insertion of an antibiotic resistance gene. For instance, CRISPR-Cas9 gene editing technology has been optimized by our group and others to develop live-attenuated *Leishmania* parasites (Ishemgulova et al. 2018; Beneke et al. 2017; Baker et al. 2021; Shrivastava et al. 2019; Zhang et al. 2020a; Salehi Sangani et al. 2019a, 2019b; Adaui et al. 2020; ; Martel et al. 2017; Espada et al. 2021). However, to our knowledge, *Cen*$^{-/-}$ parasites generated by our group are the only mutants created without the integration of any antibiotic resistance genes (Zhang et al. 2020b; Volpedo et al. 2022).

Another factor delaying the advancement of *LdCen*$^{-/-}$ parasites to clinical trials is the theoretical concern of visceralizing the potential of these mutants. Although difficult to demonstrate, the hypothetical risk of reversion to virulent phenotype still remains a rare possibility. The use of a dermotropic strain, such as *L. major*, is more appropriate for advancing to clinical trials. Moreover, *L. major* causes self-healing lesions (Volpedo et al. 2021b), which could easily be treated via FDA-cleared nonpharmacological approaches, such as a device that provides radiofrequency-induced heat therapy (Bumb et al. 2013). This, along with the fact that *L. major* has already been demonstrated to be efficacious for protection against reinfection (leishmanization) (Khamesipour et al. 2005), makes this strain an excellent candidate to develop a genetically modified safe vaccine for human use.

13.5 CRISPR-Cas9-Generated *L. major Cen*$^{-/-}$ as an Alternative to *LdCen*$^{-/-}$

Using the CRISPR-Cas9 technique, our group has generated an antibiotic-resistant marker-free *L. major centrin* knockout (*LmCen*$^{-/-}$) mutant (Zhang et al. 2020b). As demonstrated for *LdCen*$^{-/-}$ parasites, *LmCen*$^{-/-}$ mutants did not cause disease in susceptible mouse and hamster models (Zhang et al. 2020b; Ismail et al. 2017), Furthermore, *LmCen*$^{-/-}$ parasites conferred protection against intradermal needle challenge with their homologous WT strain (Zhang et al. 2020b). While needle injection is a widely used technique to infect preclinical models with *Leishmania*, sandfly challenge is a more physiologically relevant route of infection. Previous studies have shown that a vaccine effective against a needle challenge may not necessarily be effective against sandfly challenge (Peters et al. 2009, 2012). These observations highlight the importance of testing vaccine-induced immunity in a sandfly infection model. In order to advance a *Leishmania* vaccine to clinical trials, it is imperative to demonstrate protection against sandfly challenge—a more stringent screening criteria that mimics natural human infection in endemic areas.

In our preclinical studies, immunization with *LmCen*$^{-/-}$ parasites resulted in protection against sandfly challenge with virulent *L. major*, in a manner comparable to leishmanization (Zhang et al. 2020b). This was facilitated by the recruitment of a population of multifunctional effector cells at the infection site, initiating a rapid immune response after sandfly challenge (Zhang et al. 2020b). Upon restimulation with *L. major* antigen, immunized and healed (Leishmanization model) challenged mice generated significantly higher percentages of IFN-γ-producing $CD4^{+}CD44^{Hi}Ly6C^{+}T\text{-}bet^{+}$ T effector cells at the lesion site, compared to nonimmunized challenged mice (Zhang et al. 2020b). However, the contribution of other immune cell types, such as tissue-resident and central memory T-cell populations to *LmCen*$^{-/-}$-mediated long-term protection, still remains to be assessed in mice and hamsters. Recently, we showed that immunization with *LmCen*$^{-/-}$ parasites generates skin resident memory T-cell immune response analogous to leishmanization and is protective against *L. major* challenge (Ismail et al. 2022). Overall, these results show that *LmCen*$^{-/-}$ parasites are safe and efficacious for protection against needle or sandfly challenge with their homologous virulent strain in preclinical models.

While efficacy against CL is important, an effective *Leishmania*-based vaccine should also elicit long-term protection against the more serious, and often fatal, VL form. Previous studies have shown that *L. major* can confer protection against visceral infection with *L. infantum* (Romano et al. 2015). Our group has demonstrated that immunization with *LmCen*$^{-/-}$ parasites results in long-term protection against sandfly challenge with *L. donovani* in a hamster model (Karmakar et al. 2021). This was mediated by a Th1-polarized immune response characterized by the elevated expression of *IFN-γ, TNF-α, T-bet, CXCR3,* and other pro-inflammatory genes, as well as diminished Th2 responses, characterized by IL-10 and IL-4 in the spleen (Karmakar et al. 2021). Taken together, these results show that *LmCen*$^{-/-}$ parasites elicit long-term protection against heterologous challenge in a VL model.

13.6 Toward Human Clinical Trials: A Pan-*Leishmania* Vaccine

Laboratory-grade *LmCen*$^{-/-}$ mutants have shown great promise in preclinical models; however, in order to advance to clinical trials, these parasites have to be produced following Good Manufacturing Practices (cGMP). As a first step toward this process, our group has recently grown under Good Laboratory Practices (GLP) *LmCen*$^{-/-}$ parasites in a bioreactor at a small industrial scale with cGMP quality control characteristics. We showed that immunization with GLP-grade *LmCen*$^{-/-}$ mutants also results in long-lasting protection against needle and sandfly challenge with *L. donovani* in a hamster model (Karmakar et al. 2021). Interestingly, GLP-grade *LmCen*$^{-/-}$ parasites also promoted the production of both IFN-γ and IL-10 in human PBMCs (Karmakar et al. 2021). The ratio between IFN-γ/IL-10 was significantly higher in PBMCs exposed to GLP-grade *LmCen*$^{-/-}$ parasites, compared to uninfected controls, highlighting a Th1-predominant response, consistent with the preclinical studies (Karmakar et al. 2021).

Due to attributes related to safety, immunogenicity, and efficacy against challenge with different *Leishmania* strains, *LmCen*$^{-/-}$ mutants are in the process of being manufactured under cGMP and tested for toxicology studies to be advanced to Phase I clinical trials. Nevertheless, toward the goal of developing a pan-*Leishmania* vaccine, more research needs to be conducted to assess the efficacy of these mutants against other forms of leishmaniasis, such as mucocutaneous leishmaniasis (MCL), as well as other Old World/New World causative strains of VL and CL. The fact that *LmCen*$^{-/-}$ parasites confer protection against *L. donovani* in hamsters (Karmakar et al. 2021), together with the observations of *LdCen*$^{-/-}$ providing cross-protection against *L. braziliensis* and *L. mexicana* challenge in mice (Selvapandiyan et al. 2009; Dey et al. 2014), suggests that *Cen*$^{-/-}$ *Leishmania* mutants present a broad range of antigens that are similar among different strains, including those endemic from different continents and causing VL, CL, and MCL. Furthermore, our group is currently developing other *centrin* knockout mutants, such as *L. mexicana Cen*$^{-/-}$

(*LmexCen*$^{-/-}$), which has been reported to be safe in immunocompromised mice and to be protective against challenge with its homologous wild-type strain in different mouse models (Volpedo et al. 2022), as well as heterologous VL causing *L. donovani* strain (Karmakar et al. 2022). Additional studies are needed to assess the cross-protection of *L. mexicana centrin*-deleted mutants against other *Leishmania* strains. Another group recently developed an *L. braziliensis centrin* deletion mutant (*LbCen*$^{-/-}$) using CRISPR/Cas9. *LbCen*$^{-/-}$ parasites also showed a growth defect in the amastigote stage and did not cause cutaneous lesions analogously to other *Cen*$^{-/-}$ mutants. However, immunogenicity and efficacy remain to be determined (Sharma et al. 2021). Lastly, despite their promise as prophylactic vaccines, the therapeutic potential of *Cen*$^{-/-}$ parasites has not yet been fully assessed even though our previous studies with *Ldcen−/−* parasites as immunogens showed that they induce immunity in asymptomatic animal models and protect against future infection (Bhattacharya et al. 2016). Therapeutic vaccines have been developed for chronic conditions to boost the immune system to fight infection in already sick individuals (Kutscher et al. 2012). Further research is needed to determine whether *LmCen*$^{-/-}$ mutants are also efficacious for therapeutic use.

Overall, *centrin* knockout *Leishmania* mutants have shown potential to become the first successful pan-*Leishmania* vaccine for human use. The development of a pan-*Leishmania* vaccine could alleviate the public health, as well as the socioeconomic burden of *Leishmania* endemic areas, especially for low- and middle-income countries (LMIC) (Volpedo et al. 2021a). Furthermore, climate change, travel, and migration have allowed the spread of endemic *Leishmania* strains to nonendemic regions (Oryan and Akbari 2016; Stamm and Human 2016). Due to a lack of effective strategies to control and manage leishmaniasis in endemic and nonendemic areas, it is imperative to develop an effective prophylactic vaccination strategy.

References

Adaui V et al (2020) Application of CRISPR/Cas9-based reverse genetics in leishmania braziliensis: conserved roles for HSP100 and HSP23. Genes (Basel) 11. https://doi.org/10.3390/genes11101159

Andrade-Narvaez FJ, Loría-Cervera EN, Sosa-Bibiano EI, Van Wynsberghe NR (2016) Asymptomatic infection with American cutaneous leishmaniasis: epidemiological and immunological studies. Mem Inst Oswaldo Cruz 111:599–604. https://doi.org/10.1590/0074-02760160138

Avishek K et al (2016) Gene deleted live attenuated Leishmania vaccine candidates against visceral leishmaniasis elicit pro-inflammatory cytokines response in human PBMCs. Sci Rep 6:33059. https://doi.org/10.1038/srep33059

Baker N et al (2021) Systematic functional analysis of Leishmania protein kinases identifies regulators of differentiation or survival. Nat Commun 12:1244. https://doi.org/10.1038/s41467-021-21360-8

Banerjee A et al (2018) Live attenuated leishmania donovani centrin gene-deleted parasites induce IL-23-dependent IL-17-protective immune response against visceral leishmaniasis in a murine model. J Immunol 200:163–176. https://doi.org/10.4049/jimmunol.1700674

Beneke T et al (2017) A CRISPR Cas9 high-throughput genome editing toolkit for kinetoplastids. R Soc Open Sci 4:170095. https://doi.org/10.1098/rsos.170095

Bern C et al (2007) The epidemiology of visceral leishmaniasis and asymptomatic leishmanial infection in a highly endemic Bangladeshi village. Am J Trop Med Hyg 76:909–914

Bhattacharya P et al (2015) Genetically modified live attenuated leishmania donovani parasites induce innate immunity through classical activation of macrophages that direct the Th1 response in mice. Infect Immun 83:3800–3815. https://doi.org/10.1128/IAI.00184-15

Bhattacharya P et al (2016) Live attenuated leishmania donovani centrin knock out parasites generate non-inferior protective immune response in aged mice against visceral leishmaniasis. PLoS Negl Trop Dis 10:e0004963. https://doi.org/10.1371/journal.pntd.0004963

Bhattacharya P et al (2020) Essential role of neutrophils in the protective immune response induced by a live attenuated. J Immunol 205:3333–3347. https://doi.org/10.4049/jimmunol.2000829

Bhaumik SK, Singh MK, Karmakar S, De T (2009) UDP-Gal: N-acetylglucosamine beta 1-4 galactosyltransferase expressing live attenuated parasites as vaccine for visceral leishmaniasis. Glycoconj J 26:663–673. https://doi.org/10.1007/s10719-008-9212-y

Bumb RA et al (2013) Long-term efficacy of single-dose radiofrequency-induced heat therapy vs. intralesional antimonials for cutaneous leishmaniasis in India. Br J Dermatol 168: 1114–1119. https://doi.org/10.1111/bjd.12205

Carrión J, Folgueira C, Soto M, Fresno M, Requena JM (2011) Leishmania infantum HSP70-II null mutant as candidate vaccine against leishmaniasis: a preliminary evaluation. Parasit Vectors 4: 150. https://doi.org/10.1186/1756-3305-4-150

CDC (2020) *Leishmaniasis: Epidemiology and Risk Factors*, https://www.cdc.gov/parasites/leishmaniasis/epi.html#:~:text=For%20cutaneous%20leishmaniasis%2C%20estimates%20of,to%20400%2C000%20or%20more%20cases

Chakravarty J et al (2019) Determinants for progression from asymptomatic infection to symptomatic visceral leishmaniasis: a cohort study. PLoS Negl Trop Dis 13:e0007216. https://doi.org/10.1371/journal.pntd.0007216

de Oliveira CI, Brodskyn CI (2012) The immunobiology of Leishmania braziliensis infection. Front Immunol 3:145. https://doi.org/10.3389/fimmu.2012.00145

Dey R et al (2010) Characterization of a Leishmania stage-specific mitochondrial membrane protein that enhances the activity of cytochrome c oxidase and its role in virulence. Mol Microbiol 77: 399–414. https://doi.org/10.1111/j.1365-2958.2010.07214.x

Dey R et al (2013) Live attenuated Leishmania donovani p27 gene knockout parasites are non-pathogenic and elicit long-term protective immunity in BALB/c mice. J Immunol 190:2138–2149. https://doi.org/10.4049/jimmunol.1202801

Dey R et al (2014) Characterization of cross-protection by genetically modified live-attenuated Leishmania donovani parasites against Leishmania mexicana. J Immunol 193:3513–3527. https://doi.org/10.4049/jimmunol.1303145

Domínguez-Bernal G et al (2015) HisAK70: progress towards a vaccine against different forms of leishmaniosis. Parasit Vectors 8:629. https://doi.org/10.1186/s13071-015-1246-y

Duthie MS, Reed SG (2014) The emergence of defined subunit vaccines for the prevention of leishmaniasis. Curr Trop Med Rep.:154–162. https://doi.org/10.1007/s40475-014-0024-9

Duthie MS et al (2017) A defined subunit vaccine that protects against vector-borne visceral leishmaniasis. NPJ Vaccines 2:23. https://doi.org/10.1038/s41541-017-0025-5

Espada CR et al (2021) Effective genome editing in Leishmania (Viannia) braziliensis stably expressing Cas9 and T7 RNA polymerase. Front Cell Infect Microbiol 11:772311. https://doi.org/10.3389/fcimb.2021.772311

Fiuza JA et al (2013) Induction of immunogenicity by live attenuated Leishmania donovani centrin deleted parasites in dogs. Vaccine 31:1785–1792. https://doi.org/10.1016/j.vaccine.2013.01.048

Fiuza JA et al (2015) Vaccination using live attenuated Leishmania donovani centrin deleted parasites induces protection in dogs against Leishmania infantum. Vaccine 33:280–288. https://doi.org/10.1016/j.vaccine.2014.11.039

Gannavaram S et al (2019) miR-21 expression determines the early vaccine immunity induced by. Front Immunol 10:2273. https://doi.org/10.3389/fimmu.2019.02273

Gedda MR et al (2020) Post kala-azar dermal leishmaniasis: a threat to elimination program. PLoS Negl Trop Dis 14:e0008221. https://doi.org/10.1371/journal.pntd.0008221

Gonçalves AAM et al (2019) An overview of immunotherapeutic approaches against canine visceral leishmaniasis: what has been tested on dogs and a new perspective on improving treatment efficacy. Front Cell Infect Microbiol 9:427. https://doi.org/10.3389/fcimb.2019.00427

Ishemgulova A et al (2018) CRISPR/Cas9 in Leishmania mexicana: a case study of LmxBTN1. PLoS One 13:e0192723. https://doi.org/10.1371/journal.pone.0192723

Ismail N, Kaul A, Bhattacharya P, Gannavaram S, Nakhasi HL (2017) Immunization with Live attenuated Leishmania donovani Centrin −/− parasites is efficacious in asymptomatic infection. Front Immunol 8:1788. https://doi.org/10.3389/fimmu.2017.01788

Ismail N et al (2022) *Leishmania Major* centrin gene-deleted parasites generate skin resident memory T-cell immune response analogous to leishmanization. Front Immunol 13. https://doi.org/10.3389/fimmu.2022.864031

Karmakar S et al (2021) Preclinical validation of a live attenuated dermotropic Leishmania vaccine against vector transmitted fatal visceral leishmaniasis. Commun Biol 4:929. https://doi.org/10.1038/s42003-021-02446-x

Karmakar S et al (2022) Centrin-deficient Leishmania mexicana confers protection against Old World visceral leishmaniasis. NPJ Vaccines 7:157. https://doi.org/10.1038/s41541-022-00574-x

Kaszak I, Planellas M, Dworecka-Kaszak B (2015) Canine leishmaniosis - an emerging disease. Ann Parasitol 61:69–76

Khamesipour A et al (2005) Leishmanization: use of an old method for evaluation of candidate vaccines against leishmaniasis. Vaccine 23:3642–3648. https://doi.org/10.1016/j.vaccine.2005.02.015

Kutscher S, Bauer T, Dembek C, Sprinzl M, Protzer U (2012) Design of therapeutic vaccines: hepatitis B as an example. Microb Biotechnol 5:270–282. https://doi.org/10.1111/j.1751-7915.2011.00303.x

Lari A, Lari N, Biabangard A (2022) Immunoinformatics approach to design a novel subunit vaccine against visceral leishmaniasis. Int J Pept Res Ther 28:34. https://doi.org/10.1007/s10989-021-10344-3

Le Rutte EA et al (2016) Feasibility of eliminating visceral leishmaniasis from the Indian subcontinent: explorations with a set of deterministic age-structured transmission models. Parasit Vectors 9:24. https://doi.org/10.1186/s13071-016-1292-0

Lestinova T, Rohousova I, Sima M, de Oliveira CI, Volf P (2017) Insights into the sand fly saliva: blood-feeding and immune interactions between sand flies, hosts, and Leishmania. PLoS Negl Trop Dis 11:e0005600. https://doi.org/10.1371/journal.pntd.0005600

Martel D, Beneke T, Gluenz E, Späth GF, Rachidi N (2017) Characterisation of casein kinase 1.1 in Leishmania donovani using the CRISPR Cas9 toolkit. Biomed Res Int 2017:4635605. https://doi.org/10.1155/2017/4635605

Medkour H et al (2019) Potential animal reservoirs (dogs and bats) of human visceral leishmaniasis due to Leishmania infantum in French Guiana. PLoS Negl Trop Dis 13:e0007456. https://doi.org/10.1371/journal.pntd.0007456

Moafi M, Rezvan H, Sherkat R, Taleban R (2019) Vaccines entered in clinical trials: a review of literature. Int J Prev Med 10:95. https://doi.org/10.4103/ijpvm.IJPVM_116_18

Okwor I, Uzonna J (2016) Social and economic burden of human leishmaniasis. Am J Trop Med Hyg 94:489–493. https://doi.org/10.4269/ajtmh.15-0408

Oryan A, Akbari M (2016) Worldwide risk factors in leishmaniasis. Asian Pac J Trop Med 9:925–932. https://doi.org/10.1016/j.apjtm.2016.06.021

Pacheco-Fernandez T et al (2021) Revival of Leishmanization and Leishmanin. Front Cell Infect Microbiol 11:639801. https://doi.org/10.3389/fcimb.2021.639801

Palatnik-de-Sousa CB, Nico D (2020) The delay in the licensing of protozoal vaccines: a comparative history. Front Immunol 11:204. https://doi.org/10.3389/fimmu.2020.00204

Papadopoulou B et al (2002) Reduced infectivity of a Leishmania donovani biopterin transporter genetic mutant and its use as an attenuated strain for vaccination. Infect Immun 70:62–68. https://doi.org/10.1128/IAI.70.1.62-68.2002

Peters NC et al (2008) In vivo imaging reveals an essential role for neutrophils in leishmaniasis transmitted by sand flies. Science 321:970–974. https://doi.org/10.1126/science.1159194

Peters NC et al (2009) Vector transmission of leishmania abrogates vaccine-induced protective immunity. PLoS Pathog 5:e1000484. https://doi.org/10.1371/journal.ppat.1000484

Peters NC et al (2012) Evaluation of recombinant Leishmania polyprotein plus glucopyranosyl lipid A stable emulsion vaccines against sand fly-transmitted Leishmania major in C57BL/6 mice. J Immunol 189:4832–4841. https://doi.org/10.4049/jimmunol.1201676

Petitdidier E et al (2019) Peptide-based vaccine successfully induces protective immunity against canine visceral leishmaniasis. NPJ Vaccines 4:49. https://doi.org/10.1038/s41541-019-0144-2

Pires M, Wright B, Kaye PM, da Conceição V, Churchill RC (2019) The impact of leishmaniasis on mental health and psychosocial well-being: a systematic review. PLoS One 14:e0223313. https://doi.org/10.1371/journal.pone.0223313

Rawat AK et al (2020) The CD200-CD200R cross-talk helps Leishmania donovani to down regulate macrophage and CD4. Int J Biol Macromol 151:394–401. https://doi.org/10.1016/j.ijbiomac.2020.02.189

Ready PD (2014) Epidemiology of visceral leishmaniasis. Clin Epidemiol 6:147–154. https://doi.org/10.2147/CLEP.S44267

Ribeiro-Gomes FL, Sacks D (2012) The influence of early neutrophil-Leishmania interactions on the host immune response to infection. Front Cell Infect Microbiol 2:59. https://doi.org/10.3389/fcimb.2012.00059

Rogers ME (2012) The role of leishmania proteophosphoglycans in sand fly transmission and infection of the Mammalian host. Front Microbiol 3:223. https://doi.org/10.3389/fmicb.2012.00223

Romano A, Doria NA, Mendez J, Sacks DL, Peters NC (2015) Cutaneous infection with Leishmania major mediates heterologous protection against visceral infection with Leishmania infantum. J Immunol 195:3816–3827. https://doi.org/10.4049/jimmunol.1500752

Salari S, Sharifi I, Keyhani AR, Ghasemi Nejad Almani P (2020) Evaluation of a new live recombinant vaccine against cutaneous leishmaniasis in BALB/c mice. Parasit Vectors 13:415. https://doi.org/10.1186/s13071-020-04289-7

Salehi Sangani G et al (2019a) Generation of a CRISPR/Cas9-based vector specific for gene manipulation in. Iran J Parasitol 14:78–88

Salehi Sangani G et al (2019b) Generation of a CRISPR/Cas9-based vector specific for gene manipulation in Leishmania major. Iran J Parasitol 14:78–88

Santi AMM et al (2018) Growth arrested live-attenuated Leishmania infantum KHARON1 null mutants display cytokinesis defect and protective immunity in mice. Sci Rep 8:11627. https://doi.org/10.1038/s41598-018-30076-7

Selvapandiyan A et al (2001) Expression of a mutant form of Leishmania donovani centrin reduces the growth of the parasite. J Biol Chem 276:43253–43261. https://doi.org/10.1074/jbc.M106806200

Selvapandiyan A et al (2004) Centrin gene disruption impairs stage-specific basal body duplication and cell cycle progression in Leishmania. J Biol Chem 279:25703–25710. https://doi.org/10.1074/jbc.M402794200

Selvapandiyan A et al (2009) Intracellular replication-deficient Leishmania donovani induces long lasting protective immunity against visceral leishmaniasis. J Immunol 183:1813–1820. https://doi.org/10.4049/jimmunol.0900276

Selvapandiyan A et al (2014) Generation of growth arrested Leishmania amastigotes: a tool to develop live attenuated vaccine candidates against visceral leishmaniasis. Vaccine 32:3895–3901. https://doi.org/10.1016/j.vaccine.2014.05.009

Seyed N, Peters NC, Rafati S (2018) Translating observations from leishmanization into non-living vaccines: the potential of dendritic cell-based vaccination strategies against. Front Immunol 9: 1227. https://doi.org/10.3389/fimmu.2018.01227

Sharma R et al (2021) Targeted deletion of centrin in leishmania braziliensis using CRISPR-Cas9-based editing. Front Cell Infect Microbiol 11:790418. https://doi.org/10.3389/fcimb.2021.790418

Shrivastava R, Tupperwar N, Drory-Retwitzer M, Shapira M (2019) Deletion of a single LeishIF4E-3 allele by the CRISPR-Cas9 system alters cell morphology and infectivity of. mSphere 4. https://doi.org/10.1128/mSphere.00450-19

Singh OP, Hasker E, Sacks D, Boelaert M, Sundar S (2014) Asymptomatic Leishmania infection: a new challenge for Leishmania control. Clin Infect Dis 58:1424–1429. https://doi.org/10.1093/cid/ciu102

Singh RK et al (2018) Centrin-deleted *Leishmania donovani* parasites Help CD4+ T cells to acquire Th1 phenotype and multi-functionality through downregulation of CD200-CD200R immune inhibitory axis. Front Immunol 9:1176. https://doi.org/10.3389/fimmu.2018.01176

Solana JC et al (2017) Vaccination with a Leishmania infantum HSP70-II null mutant confers long-term protective immunity against Leishmania major infection in two mice models. PLoS Negl Trop Dis 11:e0005644. https://doi.org/10.1371/journal.pntd.0005644

Solana JC et al (2020) Subcutaneous immunization of leishmania HSP70-II null mutant line reduces the severity of the experimental visceral leishmaniasis in BALB/c mice. Vaccines (Basel) 8. https://doi.org/10.3390/vaccines8010141

Soto M et al (2021) Inoculation of the leishmania infantum HSP70-II null mutant induces long-term protection against L. amazonensis infection in BALB/c mice. Microorganisms 9. https://doi.org/10.3390/microorganisms9020363

Stamm L, Human V (2016) Migration and leishmaniasis-on the move. JAMA Dermatol 152:373–374. https://doi.org/10.1001/jamadermatol.2015.4765

Sundar S, Singh OP, Chakravarty J (2018) Visceral leishmaniasis elimination targets in India, strategies for preventing resurgence. Expert Rev Anti-Infect Ther 16:805–812. https://doi.org/10.1080/14787210.2018.1532790

Tomiotto-Pellissier F et al (2018) Macrophage polarization in leishmaniasis: broadening horizons. Front Immunol 9:2529. https://doi.org/10.3389/fimmu.2018.02529

Topno RK et al (2010) Asymptomatic infection with visceral leishmaniasis in a disease-endemic area in bihar, India. Am J Trop Med Hyg 83:502–506. https://doi.org/10.4269/ajtmh.2010.09-0345

Viana KF et al (2016) Application of rapid in vitro co-culture system of macrophages and T-cell subsets to assess the immunogenicity of dogs vaccinated with live attenuated Leishmania donovani centrin deleted parasites (LdCen−/−). Parasit Vectors 9:250. https://doi.org/10.1186/s13071-016-1528-z

Volpedo G et al (2021a) From infection to vaccination: reviewing the global burden, history of vaccine development, and recurring challenges in global leishmaniasis protection. Expert Rev Vaccines 20:1431–1446. https://doi.org/10.1080/14760584.2021.1969231

Volpedo G et al (2021b) Mechanisms of immunopathogenesis in cutaneous leishmaniasis and post Kala-azar Dermal leishmaniasis (PKDL). Front Cell Infect Microbiol 11:685296. https://doi.org/10.3389/fcimb.2021.685296

Volpedo G et al (2022) Centrin-deficient Leishmania mexicana confers protection against New World cutaneous leishmaniasis. NPJ Vaccines 7:32. https://doi.org/10.1038/s41541-022-00449-1

WHO, W. H. O. in *WHO/HTM/NTD/IDM/2013.1* (eds World Health Organization WHO & J. Ruiz Postigo) 29 (2013)

Zhang WW, Lypaczewski P, Matlashewski G (2020a) Application of CRISPR/Cas9-mediated genome editing in leishmania. Methods Mol Biol 2116:199–224. https://doi.org/10.1007/978-1-0716-0294-2_14

Zhang WW et al (2020b) A second generation leishmanization vaccine with a markerless attenuated Leishmania major strain using CRISPR gene editing. Nat Commun 11:3461. https://doi.org/10.1038/s41467-020-17154-z

Understanding the Heterogeneity in Mast Cell Role in Host Defence During Leishmaniasis

14

Nilofer Naqvi, Rahul Srivastava, Angamuthu Selvapandiyan, and Niti Puri

Abstract

Mast cells (MCs) are involved in clearing invading pathogens at the skin surface and play an important role in innate and adaptive defences. Their role in the control of leishmaniasis due to the protozoa *Leishmania*, which is inoculated by infected sand fly vector on the skin surface, has been explored. Many investigations in *in vitro* studies, *in vitro* animal models and in clinical settings have been carried out in this field. Some reports show that MCs play a protective role against leishmaniasis, whereas some suggest that they exacerbate the infection, whereas a few reports also suggest that MCs have no role in leishmaniasis. The contradictory findings could be due to differences in the genetics of the pathogen or the host, but so far research in this area has led to the conclusion that MCs have an important participation in host defence against leishmaniasis in a differential manner. Understanding the heterogenetic role of MCs in leishmaniasis may lead to the development of novel therapeutics or vaccine interventions against this disease.

Nilofer Naqvi and Rahul Srivastava contributed equally to this work.

N. Naqvi
Cellular and Molecular Immunology Laboratory, School of Life Sciences, Jawaharlal Nehru University, New Delhi, India

Department of Microbiology, University of Chicago, Chicago, Illinois, IL, USA

R. Srivastava · N. Puri (✉)
Cellular and Molecular Immunology Laboratory, School of Life Sciences, Jawaharlal Nehru University, New Delhi, India
e-mail: purin@mail.jnu.ac.in

A. Selvapandiyan
Department of Molecular Medicine, Jamia Hamdard, New Delhi, India

© The Author(s), under exclusive license to Springer Nature Singapore Pte Ltd. 2023

A. Selvapandiyan et al. (eds.), *Challenges and Solutions Against Visceral Leishmaniasis*, https://doi.org/10.1007/978-981-99-6999-9_14

Keywords

MCs · Leishmaniasis · Innate immunity · Mast cell extracellular trap · Host defence

Abbreviations

BMMCs	Bone marrow–derived mast Cells
CBMCs	Cord-blood-derived mast cells
CL	Cutaneous Leishmaniasis
COX	Cyclooxygenase
Cpa3	Carboxypeptidase A3
CTMC	Connective tissue mast cell
DCs	Dendritic cells
ETs	Extracellular traps
IFN	Interferon
IL	Interleukin
LPG	Lipophosphoglycan
LPS	Lipopolysaccharide
LTs	Leukotrienes
MCETs	Mast cell extracellular traps
MCL	Mucocutaneous leishmaniasis
MCP	Mast cell protease
MCs	Mast cells
MIP	Macrophage inflammatory protein
MMC	Mucosal mast cell
MSP	Major surface protein
ND	Not determined
NETs	Neutrophil extracellular traps
PAMPs	Pathogen-associated molecular pattern
PAR	Protease activate receptor
PG	Prostaglandins
PGs	Proteoglycans
PMCs	Peritoneal mat cells
PMN	Polymorphonuclear leukocytes
PPGs	Proteophosphoglycans
PRR	Pathogen recognition receptor
PSP	Promastigote surface protease
RBL	rat basophilic leukaemia
SCF	Stem cell factor
TLR	Toll like receptor
TNF	Tumour necrosis factor
VL	Visceral Leishmaniasis

Vectors-injected microbes, when introduced beneath the skin surface, are known to face the host's innate immune response by interacting with the phagocytic cells, *viz.* neutrophils, macrophages and dendritic cells (DCs), mainly to colonize the macrophages. Mast cells (MCs) on the other hand are special kind of immune cells of hematopoietic origin that can secrete various mediators, which have a role in adaptive and innate immunity. They are also central effectors of asthma, anaphylaxis, and allergy. All these functions strongly rely on proinflammatory mediator's release, which is caused by the stimulation of cell surface receptors like FcγRI, FcεRI, CD48 and C5aR. Apart from exocytosis, and inflammation, MCs also have phagocytosis ability towards particulate materials. They are spread throughout the body, especially in those areas, which are first to interact with the invading pathogens. However, the details of the roles of the MCs during infection with such microbes as *Leishmania* are still obscure.

14.1 Introduction to Leishmaniasis

Leishmaniasis is a multifaceted protozoan parasitic infection caused by *Leishmania*. These parasites usually spread through the bite of infected hematophagous invertebrates and are capable of multiplying and living in host blood and tissues (de Freitas e Silva and von Stebut 2021). The different species causing leishmaniasis and their disease manifestations are briefly discussed in Fig. 14.1 In human, each *Leishmania* species causes a distinct clinical syndrome such as fatal visceral leishmaniasis (VL), cutaneous leishmaniasis (CL), post kala-azar dermal leishmaniasis (PKDL) and mucocutaneous leishmaniasis (MCL). VL is caused by *Leishmania infantum*, *Leishmania donovani*, *Leishmania chagasi* and *Leishmania tropica*, whereas CL is due to *Leishmania major*, *Leishmania amazonensis*, *Leishmania aethiopica*, *Leishmania braziliensis* complex and *Leishmania mexicana* complex. *Leishmania braziliensis* and *Leishmania panamensis/Leishmania guyanensis* may cause either CL or MCL (Thakur et al. 2018). It is reported that to date leishmaniasis has affected around 12 million people. In 2018, Brazil, Ethiopia, Iraq, China, India, Nepal, Kenya, Somalia and Sudan reported more than 95% of the global VL cases and annually 20,000 deaths have been reported (Gardinassi et al. 2017; Badaró et al. 1986). In 2018, over 85% of the global CL cases were reported in Afghanistan, Algeria, Bolivia, Brazil, Colombia, the Islamic Republic of Iran, Iraq, Pakistan, Peru, the Syrian Arab Republic and Tunisia (Leishmaniasis 2022). Nowadays, there are many reports on atypical leishmaniasis, which means that parasite species classically associated with the visceral disease can also cause cutaneous phenotype and *vice versa* (de Freitas e Silva and von Stebut 2021; Thakur et al. 2018). In total, 10–15% of the cured cases of VL have been reported to develop PKDL in the Indian subcontinent. Based on the modelling of the cases that have been reported, it has been suggested that it would be difficult to achieve the elimination of *Leishmania* in low VL endemic areas. Apart from PKDL being a reservoir, other challenges to the

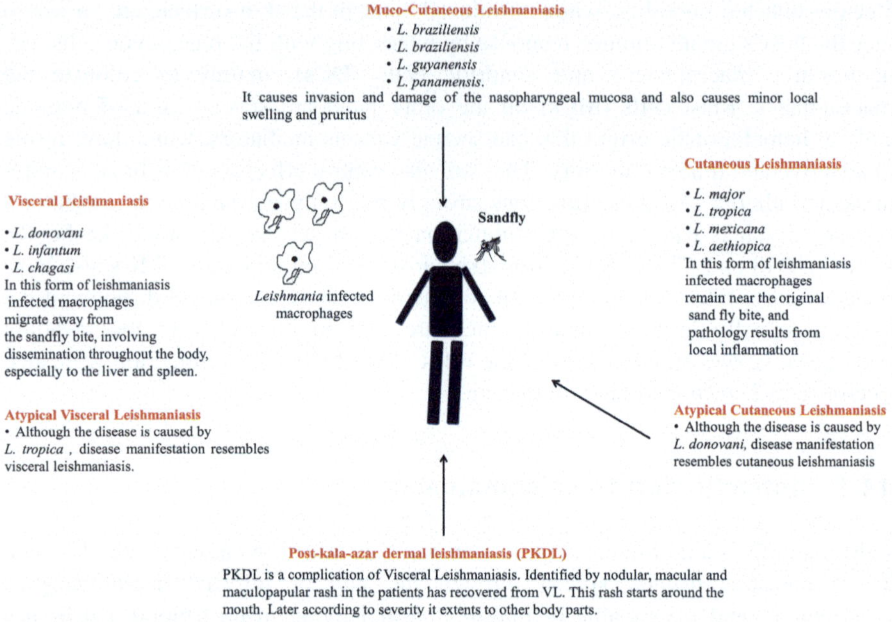

Fig. 14.1 Different forms of Leishmaniasis and the related causative species of *Leishmania*. Under each of the diseases, a brief description of the disease is also given

elimination of the disease include the emergence of human immunodeficiency virus (HIV)–VL coinfection and the asymptomatic to symptomatic conversion in 5–10% of the cases per annum (Selvapandiyan et al. 2019).

The flagellated promastigote form multiplies in Phlebotomine flies (insects), which belong to the genus *phlebotomus*. The vector of this disease belongs to the order Diptera, class Insecta, family Psychodidae (Teles et al. 2016), whereas the other rudimentary flagellated amastigote form of the parasite replicates in the mammalian macrophages (Bates 2007). The expression of surface glycolipid lipophosphoglycan (LPG) is the most important step in the developmental stage of disease progression. The expression of LPG makes virulent metacyclic *L. major* unable to attach to the gut wall, and as a result, they relocate to the insect proboscis (Rodríguez and Wilson 2014). When these female sand flies, which are infected by *Leishmania*, suck blood from a non-infected mammalian host, they inject these parasites into the dermis. The contents of the sand fly saliva along with extracellular components are transferred with the parasites and influence the local immune response and progression of the disease. The saliva of sand fly consists of a vasodilator peptide known as the maxadilan protein, which causes vasodilation as well as recruits mononuclear phagocytes to the region (Lerner et al. 2007; Morris et al. 2001). These phagocytes phagocytose the promastigotes; the process of conversion of promastigotes to amastigotes is initiated, and this cycle is completed in 2–5 days (Rodríguez and Wilson 2014). This conversion is indicated by various biochemical and morphological changes that include the disappearance of stage-specific proteophosphoglycans (PPGs), flagellum, and the major LPG, which is present at the surface of the parasites, and downregulation with relocalization of the surface protease. This transformation results in the replication of the parasites as well as dissemination to other macrophages ultimately causing the disease (Galvao-Quintao et al. 1990; McConville et al. 2007; Piani et al. 1999). This intracellular stage of the parasite is of greater interest to the research community in the development of vaccines and drugs.

14.2 Interactions of *Leishmania* with Immune Cells

Besides macrophages, there are other innate immune cells such as MCs, neutrophils, eosinophils and dendritic cells (DCs), which also serve as a host for the parasite. In one of the studies, it has been shown that during the initial infection (2 h), an increase in the neutrophil numbers occurs at the site of *L. major* inoculation, and they phagocytosed the *Leishmania* parasites (Peters et al. 2008). Such an environment persists till 18 h later. The population of macrophages is increased, between 2 and 7 days after infection (Peters et al. 2008), and they remain as the predominant infected host cells (Thalhofer et al. 2011). DCs are also activated and phagocytose *Leishmania* (Feijó et al. 2016). DCs displaying *Leishmania* antigens secrete interleukin-12 (IL-12). This IL-12 further helps in T-helper (Th1) type 1 cells differentiation, and these Th1 cells secrete interferon gamma (IFN-γ) (Lopes et al. 2015; Scheb-Wetzel et al. 2014; McLachlan et al. 2008). The production and secretion of IFN-γ by activated cell types, such as $CD8^+$ T cells, natural killer cells and $CD4^+$ Th1 cells is commonly known to correlate with protection in leishmaniasis (Carroll-Portillo et al. 2015; Gri et al. 2012). The protection against amastigote form of *Leishmania* occurs through the production of nitric oxide (NO) in macrophages (de Freitas e Silva and von Stebut 2021). Thus, it can be concluded that when a sand fly bites the local early inflammatory response takes place, which depicts the development of innate immunity, and the development of adaptive immunity is depicted by the formation of skin lesions (Peters et al. 2008; Thalhofer et al. 2011; Feijó et al. 2016; Belkaid et al. 2000). Neutrophils have been proposed to function as 'Trojan Horse' as neutrophils become apoptotic on phagocytosis of *Leishmania*, which they further transfer to macrophages (van Zandbergen et al. 2004), where these promastigote form of parasites develop into a flagellated amastigote form of the parasites (Tabbabi 2019). Another defence mechanism has been discovered in neutrophils, where neutrophils release their deoxyribonucleic acid (DNA) along with reactive oxygen species (ROS), neutrophil elastase, myeloperoxidase and histones. This process is known as the formation of neutrophil extracellular traps (NETs) (Brinkmann et al. 2004). This defence mechanism also gets activated in the case of *Leishmania* interaction with neutrophils. Different species of *Leishmania* respond differently to NET formation. *Leishmania amazonensis* is killed by NETs, whereas *L. donovani* evades the killing mechanism of NETs (Gabriel et al. 2010; Guimarães-Costa et al. 2009). This mechanism of NET formation by the release of a scaffold chromatin associated with intracellular proteins and different granular contents is known as NETosis (Brinkmann et al. 2004; Rochael et al. 2015; Fuchs et al. 2007). Upon interaction with *Leishmania*, NETosis is induced in neutrophils, which directly depends on ROS generation, which is a NADPH oxidase complex–dependent mechanism (Fuchs et al. 2007; Lim et al. 2011; Yost et al. 2009; Kirchner et al. 2012). Eosinophils are also recruited at the infection site and control the parasite load, which also further influences the other innate immune cells' functions by controlling the damage, by initiating an adaptive immune response along with local inflammatory responses (Rodríguez and Wilson 2014). At the site of parasite inoculation during the initial hours, eosinophils were observed close to degranulating (MCs), which also helps in the clearance of parasites (Grimaldi et al. 1984). In the present scenario, several researchers have focused on the better vaccine and drug development for leishmaniasis through the interaction of innate immune cells with *Leishmania* in both *in vitro* and *in vivo* studies. Hence, in this chapter, we are mainly focussing on the interaction of MCs with *Leishmania* and their effector responses.

14.3 Importance of Mast cells as Sentinels in Response to Pathogen

MCs have now been considered as the master regulator known for an array of immune responses. The role of MCs is well recognized in the initiation of allergic diseases and their activation during certain types of infections (Metz and Maurer 2007). MC is distinctive for 'armamentarium' of receptor systems and mediators for recognition and response to pathogen-associated signals. MCs are mostly distributed at the host–environment interface areas, such as the gastrointestinal tract, respiratory tract, genitourinary tract and mucosae, demise, which are mostly recognized as sites of infection throughout the host body. There is evidence that MCs are also found in the skin, mostly in the superficial dermis (Naqvi et al. 2017; Maurer et al. 2006). The first strong evidence that MCs protect against infectious diseases was depicted by studies on host–parasite interactions (Lee et al. 1986). Many studies have been conducted to understand MCs' responses for the benefit of the host on interaction with various pathogens like bacteria, fungus, virus, helminths and protozoa (Abraham and St John 2010). All MCs have common descent, granulated morphology and functions. The cytokine derived from stromal cells triggers MC/stem cell growth factor receptor (c-kit) signalling. This leads to the expression of dermal and connective tissue MCs (Dawicki and Marshall 2007). The most important peculiarity among MC populations is the heterogeneity in the content of granules as well as lipid mediator production (Puri and Roche 2008). It is also observed that distinct tissues possess specific subpopulations of MCs. For example, MCs found in the mucosa of the airway of rodents markedly differ from those found in the skin or peritoneal cavity. In rodents, the MC population has been classified as connective tissue type and mucosal type of MCs. Tryptase-only-positive MCs (MC_T) and tryptase- and chymase-positive MCs (MC_{TC}) are two subsets of human MCs classified based on protease content, that is, tryptase (T) and chymase (C) of their granule (Abraham and St John 2010). However, recently, it has been identified that there exists a broad spectrum of MC types with distinct characteristics in different tissues, which might change in the perspective of infection and inflammation. MCs have different receptors on their surface, which respond to different stimuli. During various pathological processes, MCs use different modes of receptor activation, leading to the release of various preformed mediators and the release of various cytokines as well as chemokines, which results in total degranulation (Gaudenzio et al. 2016).

14.3.1 Mast Cells—A Storehouse of Different Mediators

MCs are the storehouse of different groups of mediators that vary in their functionality. There are mainly three classes of MC mediators, specifically (i) granule-associated preformed mediators, (ii) *de novo* synthesized lipid mediators and (iii) a variety of cytokines as well as chemokines synthesized after receiving the trigger. The granule-associated mediators comprise biogenic amines, proteoglycans (PGs), non-MC-specific proteases and neutral proteases. Biogenic amines mainly include histamine, serotonin, dopamine and other polyamines. Histamine is present in all subtypes of MCs in all species and promotes bronchoconstriction, vasodilation and vascular permeability. Moreover, it plays a potential role in MC-mediated signalling to nerve endings. Unlike in human, rodent MCs express a high level of serotonin, which acts as a neurotransmitter and mediates MC signalling to nerve endings. Polyamines like spermidine and spermine regulate granule ultrastructure and storage of histamine, serotonin and enzymes like β-hexosaminidase (García-Faroldi et al. 2010). MC proteases consist of neutral MC-specific proteases and non-MC-specific proteases. MC-specific proteases are demarcated into different groups such as serine proteases, which comprise tryptases as well as chymases, and the zinc-dependent metalloproteases such as carboxypeptidase A (CPA) (Metz and Maurer 2007). All of these are essential for inflammatory response. It is reported that the distribution of MC protease amounts varies from species to species and among MC subtypes. Non-MC-specific proteases include lysosomal cathepsins, granzymes and matrix metalloproteinase-9 (MMP-9), which contribute to the antibacterial and pro-apoptotic effects of MCs on target cells, respectively. It is found that MCs mediate angiotensin II generation and regulation of blood vessels through aspartic acid protease called renin (Metz and Maurer 2007). The most well-known highly expressed proteoglycan in MC is serglycin, which plays a major role in promoting the storage of proteases and amines and metachromatic staining of MCs.

When MCs are activated, there is an initiation of *de novo* synthesized lipid mediators to stimulate the immune response. Important lipid mediators include prostaglandins (PG) and thromboxanes, belonging to the cyclooxygenase (COX) family and leukotrienes (LTs) and hydroepoxyeicosatrienoic acids (HPETEs), belonging to the lipoxygenase family (Galli and Tsai 2012). MCs predominately produce LTB4, which acts as a chemoattractant for MC progenitors but not mature MCs. LTB4 can enhance lysosomal enzyme release and superoxide anion production. PGD2 secreted during MC degranulation causes chemotaxis of eosinophils, promoting the development of allergy. Activated MCs also secrete platelet-activating factor (PAF) that acts as a chemoattractant (Urb and Sheppard 2012). In addition to these, MCs also synthesize and secrete cytokines as well as chemokines, which have an important role in the immune response, inflammation, cell repair, growth and infection (Moon et al. 2014). MCs can secrete both preformed and *de novo* synthesized proinflammatory mediators like tumour necrosis factor-alpha (TNF-α). They can release various cytokines such as granulocyte-macrophage colony-stimulating factor (GM-CSF), basic fibroblast growth factor-2 (bFGF-2), transforming growth factor (TGF)-β1, IL-6, IL-5, IL-4, IL-3, IL-33, IL-13 and IL-10 (Urb and Sheppard 2012), some of which are anti-inflammatory and some are proinflammatory. Thus, MCs are a storehouse of various mediators with diverse functions that are released upon activation as shown in Fig. 14.2.

14.3.2 Receptors on Mast Cells

MCs can be activated in various ways. The antigen cross-linking of IgE bound to high-affinity receptor (FcεRI) mediates the activation of MCs, which is the most extensively studied pathway. FcεRI is composed of three subunits FcεRIα, FcεRIβ and two FcεRIγ. Immunoglobulin G (IgG) receptors expressed on the surface can also be engaged for the activation of MCs. The γ-subunits of FcεRI and FcγRI are common, but the pathway for activation is not the same. The Fc receptors (FcεR and FcγR) can bind to pathogen-specific antibodies, which help MCs in recognition of pathogens and lead to stimulation of T-helper 1 type cells (Th1) or T-helper 2 type cells (Th2) immune response (Metz and Maurer 2007; Marshall and Jawdat 2004). The degranulation of MC can also take place due to complement components such as C3a, C5a and C4a, and they interact with cell surface G protein-coupled receptors C3aR, C4aR and C5aR (Schäfer et al. 2013) *in vivo*, which leads to anaphylactic reactions. It is reported that, in several instances, IgE-mediated MCs degranulation occurs for anaphylaxis. However, sometimes MC degranulation also occurs without the involvement of IgE. IgE-independent pathways can cause degranulation of MCs through the help of a large number of surface receptors such as mastocyte-related G-protein coupled receptor member X2 (MRGPRX2 and toll-like receptors (TLR). There are certain drugs that are capable of causing anaphylactic reactions and direct MC degranulation by activating the MRGPRX2 receptor (Porebski et al. 2018). The TLRs found on the surface of MCs get activated through pathogen-associated molecular patterns (PAMPs), which further sensitize the immune system (St John and Abraham 2013). MCs express TLR-2, TLR-3, TLR-4, TLR-6, TLR-7, TLR-8 and TLR-9 of the 11 members of the TLR family. TLRs are stimulated by different PAMPs. Lipopolysaccharide (LPS), a component of Gram-negative bacteria cell wall, stimulates rodent MCs *via* TLR4, whereas peptidoglycans (a constituent of gram-positive bacteria) induce cytokine production via TLR-2. Virus particles are identified *via* TLR-3 on MCs that recognize the viral dsRNA (Abraham and St John 2010).

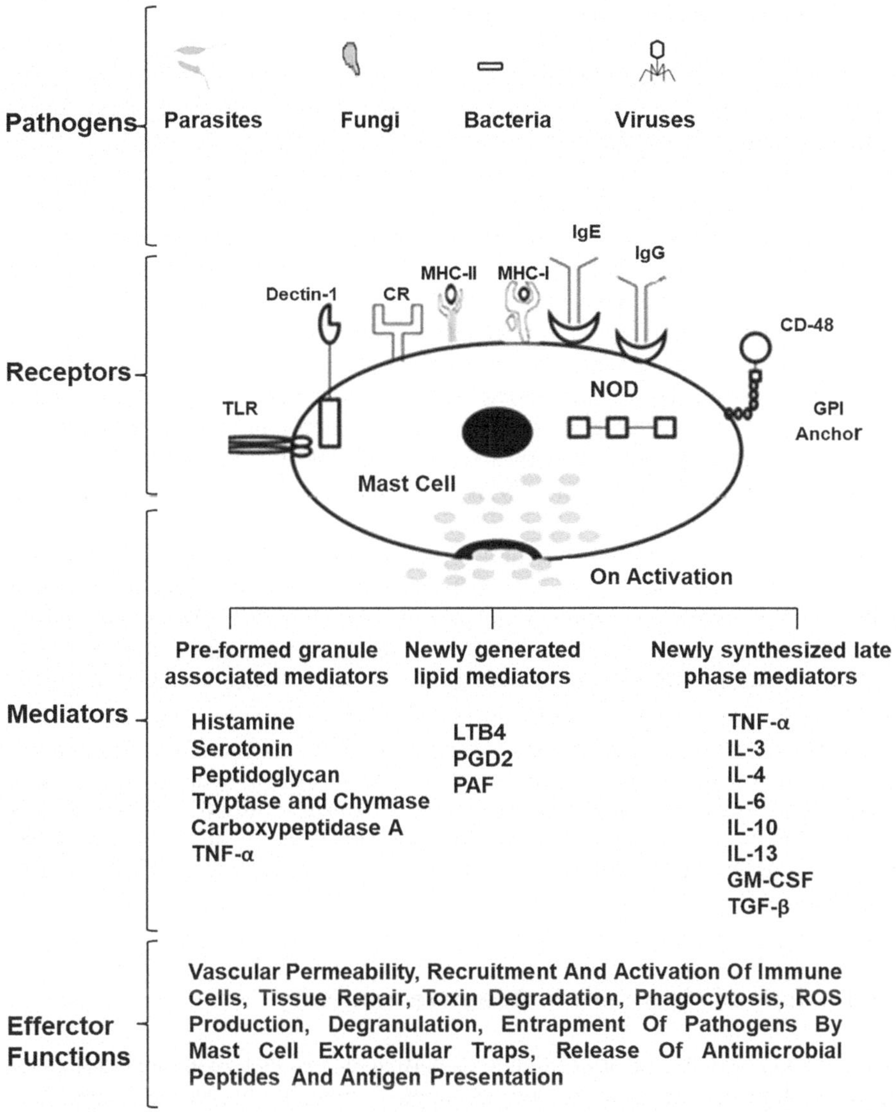

Fig. 14.2 Effector responses of MCs. MCs are equipped with various receptors, which recognize various pathogens, can do phagocytosis, generate ROS, undergo degranulation, can release DNA along with granules, antimicrobial peptides (AMPs) and histones by a process known as MCETs and can even present antigens to T cells. Receptor, when cross-linked with IgE+ antigen as well as other pathogens, when bound to the receptors, leads to activation of MCs, and further leads to degranulation, which lead to the release of various mediators, which having many functions that make them multifaceted

14.3.3 Antimicrobial Activity of Mast Cells

MCs can bind to opsonized bacteria. *Salmonella typhimurium* binds to complement receptor 3 (CR3) on the MC membrane, when coated with the iC3b fragment, which further leads to phagocytosis and the release of various mediators (Sher et al. 1979). As MCs express several IgG receptors such as FcγRI, FcγRII and FcγRIII, opsonized bacteria such as *Escherichia coli* expressing FimH, *Enterobacter cloacae* and *Klebsiella pneumoniae* bind to these IgG receptors and undergo phagocytosis. These pathogens are killed by the oxidative burst and acidification of phagocytic vacuoles in MCs (Malaviya et al. 1994). *Mycobacterium tuberculosis* uses a cholesterol-dependent pathway to infect MCs (Muñoz et al. 2003, 2009). *Mycobacterium bovis* Bacillus Calmette-Guerin (BCG) is also phagocytosed by MCs (Naqvi et al. 2021). Human cord-blood-derived MCs (CBMCs) and MCs from the human leukemic cell line HMC-1 bind and internalize several types of opsonized gram-negative and gram-positive bacteria (Arock et al. 1998). Thus, MCs also play a significant role in microbial phagocytosis and intracellular killing, as discussed in Fig. 14.2.

Degranulation and release of antimicrobial peptides such as proteases (chymase and tryptase), cathelicidins (LL-37 or CRAMP) and β-defensins play an important role in the extracellular antimicrobial activity of MCs. MCETs contain DNA, specific granule proteins of mast cell-like tryptase, histones and CRAMP/LL-37 and are released in a ROS-dependent pathway (von Köckritz-Blickwede et al. 2008). MCs on interaction with microorganisms such as *Streptococcus pyogenes*, *Staphylococcus pneumonia*, *Pseudomonas aeruginosa*, *Enterococcus faecalis*, *Listeria monocytogenes*, BCG and *Candida albicans* release extracellular traps (ETs) and these microbes are entrapped in these structures (Lopes et al. 2015; Scheb-Wetzel et al. 2014; Naqvi et al. 2021; von Köckritz-Blickwede et al. 2008; Campillo-Navarro et al. 2017). *S. pyogenes* are killed in these ETs (von Köckritz-Blickwede et al. 2008).

14.3.4 Role of Mast Cells in Adaptive Immunity

Some reports have demonstrated that MCs participate in adaptive immune responses through interaction with various cells of adaptive immunity by various means.

14.3.4.1 Dendritic Cells (DCs)

MCs secrete various mediators, which help in the maturation and function of DCs (McLachlan et al. 2008). Synapse formation takes place in the case of DCs and MCs to facilitate antigen transfer for T-cell activation (Carroll-Portillo et al. 2015). Generally, MCs produce TNF-α that helps in the maturation of DCs. It also enables the expression of costimulatory receptors on MCs and enhancement of MHC class I molecules, which causes activation of $CD8^+$ T cells in the tumour-draining lymph node (Pal et al. 2020). In the case of drug-resistant leishmaniasis, a combinatorial strategy of using intraperitoneal and heat-inactivated promastigotes is known to activate the dendritic cells that release IL-6 and aid in a T-helper 17 (Th17) type cells response, thus increasing the efficacy. This approach of stimulating the effector immunity by means of microbes could prove to be a viable approach for targeting multiple pathogens as a part of protective cellular immunity to intracellular pathogens (Selvapandiyan et al. 2023).

14.3.4.2 T Cells

T cells and MCs can form tight cell–cell interactions. As a result, they regulate their functions in a bidirectional manner (Gri et al. 2012; Mekori et al. 2016; Valitutti and Espinosa 2010). MCs can also participate in antigen presentation to T cells in both major histocompatibility complex (MHC) class II as well as class I pathways (Mekori et al. 2016; Kambayashi and Laufer 2014; Marshall 2004; Stelekati et al. 2009). *In vitro* experimentation has shown that antigen internalization into MCs occurs through FcεRI for activating antigen-specific T-cell responses, which is independent of the expression of MHC class II by MCs but requires that such MCs undergo apoptosis and later phagocytosis by antigen-presenting cells (Kambayashi et al. 2008). This is, therefore, an indirect presentation not a direct presentation by MCs using the MHC class II pathway. Both human and murine MCs after activation can release granule-associated mediators such as matrix MMP-9 and histamine, which on physical contact with activated T cells can stimulate the production of several cytokines like IL-8, IL-6, IL-4 and TNF-α. Moreover, it is also seen that when MCs are incubated with the isolated cell membrane of activated T cells, they can also induce expression and release of the above-said mediators. MCs and myristate 13-acetate (PMA) or anti-CD3-activated T cells can be attributed in T-cell-induced MC activation through surface molecules like lymphotoxin-β receptor and intercellular adhesion molecule 1 with their corresponding ligands. Hence, direct contact between surface molecules on MCs and activated T cells can induce stimulatory signals for MC degranulation and T-cell-independent cytokine release. This is well demonstrated when these two cell populations are separated by a semipermeable porous membrane, which prevents the above MC activation pathway (Mekori and Hershko 2012).

14.3.4.3 B Cells

Activated MCs can activate both naïve and activated B cells. After interaction with activated MCs, there was an increased blast formation, proliferation and expression of MHC class II, CD86 and CD19 expression in B cells. MCs stimulate the secretion of immunoglobulin M (IgM) and IgG in IgM^+ B cells. Thus, MCs promote class-switching in B cells (Palm et al. 2016). Plasma cells secrete IgE that binds to the high-affinity IgE receptor on MCs. When the antigen cross-links by surface-bound IgE, MCs secrete IL-4 and express CD40L, which upon binding with IL-4 receptors on the activated B cell results in IgE isotype switching by B cells. It is reported that when IL-4 binds to IL-4 surface receptors on B cells, signal transduction occurs, which leads to the activation of JAK-1 and JAK-3 pathways, which ultimately causes phosphorylation of transcriptional regulator STAT6. Second, there are several pieces of evidence, which reveal that, at the site of allergic reaction, the cross-linking of CD40 L with CD40 surface receptors on B cells can also cause IgE class-switching. Hence, MCs play an important role in driving IgE class-switching and IgE production by B cells (Charles et al. 2001).

14.3.5 Role of Mast Cells in Parasitic Infection

MCs have a significant role in host defence responses against parasitic infections. For instance, when *Toxoplasma gondii* parasite lysate was incubated with Bone Marrow-derived MCs (BMMCs), it is recognized by TLRs in a MyD88-dependent manner. Further, the *in vivo* studies have also shown strong evidence of Th1-cell responses, when W/Wv mice reconstituted with BMMCs are infected with *T. gondii* (Cruz et al. 2014). In nematode infection, the chymases mouse mast cell protease-1 (mMCP-1) and mouse mast cell protease-9 (mMCP-9) secreted by MCs contribute to expulsion (Sasaki et al. 2005). MCs also provide Th2 immunity against helminth infection. It is reported that MC-deficient mice ($Kit^{W}/Kit^{W\text{-}v}$ and $Kit^{W\text{-}Sh}$) show inhibition of Th2 cell priming upon helminth infection. Several studies demonstrated that at the early stage of helminth infection, such as during helminth larval invasion into small intestinal tissue, MCs degranulate and regulate the tissue-derived cytokines IL-33 and IL-25 in the intestine (Hepworth et al. 2012). In pathogenesis by malaria, MCs release histamine *via* an IgE-independent or IgE-mediated pathway (Mecheri 2012), which causes the release of nitric oxide as well as enhances vascular permeability (Van de Voorde and Leusen 1983), and hence, by triggering the MCs, malaria parasites can promote malaria pathogenesis (Lu and Huang 2017).

Further investigation is needed for understanding the role of MCs in leishmaniasis because it is found that at the site of *Leishmania* infection by sand flies on the skin, a number of MCs get localized (Maurer et al. 2006). Various previous reports have depicted that MCs play a role in *Leishmania* pathogenesis. It is also reported that, during granuloma formation, skin MCs are involved in the recruitment of macrophages (von Stebut et al. 2003). Thus, MCs also play a significant role in strengthening the immune responses against parasitic infection by phagocytosis or degranulation. In a few cases, they can also increase the pathophysiology of parasitic infection (Lu and Huang 2017). The role of MCs in parasitic infections has been briefly discussed in Table 14.1.

In this chapter, the role of MCs in leishmaniasis due to which 12 million people are affected worldwide is discussed in detail. As MCs are involved in clearing invading pathogens at the skin surface and play an important role in innate and adaptive defences, they may have a significant role in the leishmaniasis control, caused by *Leishmania* that is inoculated by infected sand fly vector on the skin surface. A large number of investigations in clinical studies, *in vitro* animal model systems and *in vitro* studies have been carried out in this field.

Table 14.1 Response of mast cells on interaction with various parasites

S. no.	Parasites (Species)	Tissue	Observation of function	Implicated factors	References
1	*Toxoplasma gondii*	Neurons	Suppresses phospholipase Cγ-mediated Ca^{2+} mobilization	Inhibition of mast cell–mediated degranulation	Smith et al. (2013)
2	Protozoa (*Leishmania major*)	Skin	Low cells recruitment, high parasite burden with large skin lesions in $MC^{-/-}$ mice.	Lesions associated with low level of IL-12	Maurer et al. (2006)
			Phagocytosis of *Leishmania* by BMMCs	Release of β-hexosaminidase and TNF-α	Bidri et al. (1997)
3	Larval tick (*Haemaphysalis longicornis*)	Skin	MCs-deficient skin grafts show lower resistance to ticks as compared to MCs-sufficient skin grafts	Not determined (ND)	Matsuda et al. (1990)
4	Protozoa (*Plasmodium berghei*)	Blood	High level of parasitaemia shows in $MC^{-/-}$ mice	TNF-α, MCs derived, is protective against Protozoa.	Furuta et al. (2006)
5	Nematode (*Strongyloides venezuelensis*)	Gut	MCs-promoted parasite expulsion	IL-3 contributes to protection	Abe et al. (1992)
6	Nematode (*Strongyloides ratti*)	Gut	Delay in parasite expulsion and increased larval numbers showed in MC-deficient mice.	ND	Abe et al. (1992)
7	Nematode (*Trichinella spiralis*)	Gut	Defects in MCs lead to an increase in deposition of larvae number in muscle cells.	Absence of mast cell protease (MCP)-6 causes delay in eosinophil recruitment, but not expulsion of parasite.	Shin et al. (2008)
			High level of MC precursors	ND	Dillon and MacDonald (1986)
			In rat, activated mucosal MC coincides with expulsion of parasite.	High level of MCP-2 coincides with parasite expulsion.	Dillon and MacDonald (1986); Woodbury et al. (1984)
			$MC^{-/-}$ mice show parasite expulsion when reconstituted with MCs.	MCP-1 lacking mice show delayed parasite expulsion.	Knight et al. (2000)

(continued)

Table 14.1 (continued)

S. no.	Parasites (Species)	Tissue	Observation of function	Implicated factors	References
8	Hookworm (*Nippostrongylus brasiliensis*)	Gut	Compared to MC-deficient mice, in MC-sufficient, there is a higher level of parasite expulsion seen under primary infection than under secondary infection.	ND	Ohnmacht and Voehringer (2010)
			Mobilization of MC precursor from blood to gut for proliferation.	IL-3 and IL-4	Ohnmacht and Voehringer (2010); Kasugai et al. (1995)
			In rat, activated mucosal MC coincides with expulsion of parasite.	Expulsion of worm coincides with peak systemic MCP-2 levels.	Woodbury et al. (1984)
9	Helminth (*Heligmosomoides polygyrus bakeri*)	Gut	The generation of Th2 immunity increased.	IL-25 and IL-33	Hepworth et al. (2012)

14.4 Clinical Studies of Leishmaniasis to Study a Role for Mast Cells

A clinical study was undertaken, in leishmaniasis to understand the relevance of MC function. This study revealed a variation in relative numbers of MCs, which depends on the species of *Leishmania* responsible for causing CL in these patients. A high level of MC number was observed in patients infected with *L. braziliensis*, whereas lesions due to *L. amazonensis* showed the lowest numbers of MCs (Naqvi et al. 2020). It was also found that the rate of healing depends on the number of MCs (Tuon et al. 2008). The reason behind the variations could be the genetic background of the patients as well as the pathogens. The age of the patients could also be taken into consideration as several reports indicate that, with aging, the number of MCs increases (Gunin et al. 2011). As the symptoms of CL appear as a result of local inflammatory response, this suggests the role of MCs in mediating the local inflammatory lesion that is induced by the *L. major* infection (Wershil et al. 1994). In CL, higher serum IgE level was seen in atopic patients than that in non-atopic patients (Al-Qadhi et al. 2015). In CL, FcεRII/CD23 (IgE, low-affinity receptor) expression and serum IgE level increase. Nitric oxide (NO) is produced when IgE binds to CD23 receptors, forming an immune complex that leads to leishmanicidal activity (Cabrera et al. 2003). Histidine decarboxylase (HDC) enzyme is known for the synthesis of histamine. The gene expression of HDC increases on the activation of MCs through their FcεRI receptors (Jutel et al. 2001; Takahashi et al. 2002). It has been hypothesized by De Oliveira and others (de Oliveira et al. 2005) that the MC and promastigote interaction releases histamine and IL-4, which could be responsible for promoting the survival of the parasite leading to a prolonged *L. major* infection. Thus, these observations have led to the conclusion that due to the higher number of MCs the pathogenesis of CL is reduced due to leishmanicidal activity.

14.5 Mast Cell–*Leishmania* Interactions *In Vivo*

Due to limitations in the availability of clinical samples and variations in age as well as genetic background in human patients, there is a need to study the role of MCs in standard animal models in greater detail. *In vivo* studies have been carried out in various animal models to understand leishmaniasis. Many aspects of the host immune response to leishmaniasis have been studied extensively in experimental animal models such as mice (BALB/c and C57BL/6), cotton rats, multimammate rats, Syrian golden hamsters, wild canines, domestic dogs and non-human primates such as Asian rhesus macaques, owl monkeys, squirrel monkeys and marmosets (Rodríguez and Wilson 2014; Fulton and Joyner 1948; Nolan and Farrell 1987). Many mouse models have been developed to understand specifically the functions of MCs. For the maturation of tissue MCs and their lineage progenitors, stem cell growth factor (SCF) binding with its Kit receptor is important. MC function was also explored in the mutated Kit gene in mouse strains commonly known as MC-deficient mice. $Kit^{W/Wv}$ mice having point mutation and truncating W and W^{v} alleles show a reduction in Kit gene expression, signal transmission and severe deficiency in MCs, but such mutation also leads to deficiency in pacemaker cells in the intestine, certain subpopulations of intraepithelial T cells, erythrocytes, neutrophils, melanocytes and germ cells. Mouse strain $Kit^{W\text{-}sh/W\text{-}sh}$ has an upstream inversion in the Kit gene showing a selective reduction in Kit expression, resulting in MC deficiency, but this also gave rise to other abnormalities such as megakaryocytic and splenic myeloid hyperplasia (Katz and Austen 2011). The mouse steel (SI) locus encodes the ligand (KL) for the c-Kit receptor. The KL gene in SI/SId mice is deleted due to which the transmembrane and intracellular domains are removed, and as a result, there is a deficiency in MCs (Flanagan et al. 1991). This also affects the development of other hematopoietic cells, germ cells and melanocytes (Flanagan et al. 1991). There is a recently developed MC-deficient mice model known as $Cpa3^{Cre}$ strain, which expresses Cre recombinase. The Cre recombinase expression depends on the knock-in allele of MC Cpa3 (carboxypeptidase A3) locus. The toxicity of Cre leads to a complete deficiency of MCs in mucosal as well as connective tissues (Feyerabend et al. 2011). Thus, various *in vivo* models have been developed to investigate MC's role in various diseases and immune responses.

To investigate the role of MCs during infection with *L. major*, a quantitative analysis of MCs was done at cutaneous sites in relatively resistant C57BL/6 mice and susceptible BALB/c mice after injecting *L. major*. There was extensive degranulation by dermal MCs at sites of *L. major* inoculation in both strains of mice. Kinetics of the course of *L. major* infection in genetically MC-deficient mice was studied. It was found that MCs elevated the intensity and size of the cutaneous lesions at sites of *L. major* infection. Through these types of studies, it is clear that MCs are responsible for the vigorous local inflammatory response on lesions induced by *L. major* infection (Wershil et al. 1994). *Saha et al.* reported that MCs have a pro-parasitic role in a susceptible host. In a relatively susceptible host (BALB/c), the parasite induces an early MC infiltration, to the site of infection, and they are unable to kill the parasite (*Saha et al.*2004). The type 2 response of MC occurs through IL-13, IL-4 and histamine release, which would antagonize the development of a type 1 immune response upon parasite infection (*Romão et al.* 2009). Romao et al. (2009) predicted that MCs could aggravate leishmaniasis in the murine model. To test this hypothesis, MCs were depleted by treating mice with G-protein activating agent Compound 48/80 to degranulate MCs (*Palomäki and Laitinen* 2006). Initial degranulation of MCs with Compound 48/80 led to a significant reduction in the lesion size of *L. major* developed in both resistant C57BL/6 mice and susceptible BALB/c mice (*Romão et al.* 2009).

MC degranulation resulted in reduced IL-4 production and enhanced mRNA expression such as inducible NO, IFN-γ, C-C motif chemokine ligand (CCL2) and CCL5. The degranulation of MCs also led to a decrease in parasite loads in IL-4 knockout (KO) animals, which is a clear indication that mediators other than IL-4 are involved in susceptibility *in vivo*. Thus, these results indicate that MC degranulation upon *L. major* infection not only causes a reduction in MC aggravation of inflammation but also reduces the capacity of promoting type-2 adaptive responses (*Romão et al.* 2009). An influx of MC was observed to peak after 1 week, which remained high until 8 weeks in genetically resistant C57BL/6 mice (*Villaseñor-Cardoso et al.* 2008). So, these are the reports that suggest that MCs exacerbate *Leishmania* infection.

There are also reports, which suggest that MCs have a protective role in leishmaniasis. The formation of granulomas around the parasite is important for effective immune response (Murray 1982). During skin inflammation, TNF-α plays an important role in recruiting white blood cells and localizing at cutaneous granulomas (Amiri *et al.* 1992; Kindler 1989). Macrophages play a pivotal role in cutaneous granuloma development. MCs regulate fibrosis and granuloma formulation in bone marrow and liver and hence restrict parasite dissemination (Jones *et al.* 1998). In polyacrylamide gel (PAG)-induced cutaneous granulomas, low level of neutrophils and macrophages recruitment was seen due to deficiency in either MC mediators like TNF-α or MCs in MC-deficient $Kit^{W}/Kit^{W/v}$ mice (von Stebut et al. 2003). Macrophage inflammatory protein (MIP)-1 α/β–producing neutrophils were required before MC-TNF-α–dependent macrophage influx as polymorphonuclear leukocytes (PMN) depletion before the induction of cutaneous granulomas completely inhibited macrophage recruitment, which was restored after reconstitution with PMN supernatant. Thus, it is concluded that the formation of granuloma is a sequential event of inflammatory response triggered by MC–derived TNF-α followed by neutrophils recruitment and release of MIP-1 α/β. Similarly, Maurer et al. also showed that MC-deficient mice have a reduction in the recruitment of DCs, neutrophils and macrophages into lesions on infection with *L. major* and also a reduction in the production of Th1-promoting cytokines such as IL-12. In these mice, the systemic immune response facilitated type 2 cytokines profile with an increment in IL-4 production and reduction in IFN-γ. When these mice were reconstituted with MCs, restoration of the phenotype of wild-type mice was seen (Maurer *et al.* 2006). Hence, MCs through the recruitment of DCs, neutrophils and macrophages trigger systemic immune response upon *L. major* infection. So, these are the reports that suggest that MCs play a protective role in leishmaniasis.

Upon infection with *L. amazonensis*, there was no significant difference observed in eosinophil number deposited in the lesions, in anti-*Leishmania* IgE antibody and in the size of the lesion, between the wild-type mice and their MC-deficient $Kit^{W}/Kit^{W/w}$ littermates (Katakura et al. 1993). Paul *et al.* used a strategy of developing MC-deficient $Cpa3^{Cre}$ mice on the protective C57BL/6 and the susceptible BALB/c genetic backgrounds. They subjected these mice to CL by injecting *L. major* using low and high dose of infection. In these infection models, the role of MCs in Th1 *versus* Th2-lineage specification was not seen (Paul *et al.* 2016). Thus, the deficiency of MCs did not influence disease progression or parasite loads, which negated the reports of either the protective or the aggravating role of MCs during murine cutaneous *Leishmania* infections. Hence, the *in vivo* studies on MCs interaction with leishmaniasis have revealed contrasting outcomes from no contribution to the protective and pro-pathogenic role of MCs in leishmaniasis, mainly because of a multitude of experimental mouse models used, different *Leishmania* species and a variety of experimental strategies used to study the role of MCs during *Leishmania* infection *in vivo*. These reports are still inconclusive and beg further investigation

Table 14.2 Role of MCs in leishmaniasis

Mast cell's role	Mouse strain	*Leishmania species*	Observations/outcome	References
Exacerbate	BALB/c mice (Susceptible host)	*L. major* *L. tropica*	• An early onset of MC infiltration fails to kill *Leishmania* at the site of infection. • Before *Leishmania major*, degranulation of MC decreases the ability of inflammation, exacerbates through histamine release and promotes type 2 immune responses; also due to this reason, in BALB/c mice, the development of type 1 immune response is suppressed. • Both subcutaneous (in the footpad) and intradermal (in ear pinna) *in vivo* infections with 10^5 *L. major* promastigotes cause dermal MC degranulation.	Saha et al. (2004),de Oliveira et al. (2005),Romão et al. (2009)
Protective	C57BL/6 mice (Resistant host)	*L. major* *L. tropica*	• MCs can cause killing of *Leishmania.* • In PAG-induced cutaneous granulomas, low level of neutrophil and macrophage was seen in mice with deficiency of either TNF-α (an MC product) or MCs. • MC deficiency leads to lowering of DCs, neutrophils and macrophages recruitment into the infected lesions, which further results in decreased production of Th1 promoting cytokines such as IL-12. Hence, MCs are important for guiding the systemic immune response against *Leishmania* spp. *via* subsequent recruitment	Jones et al. (1998), von Stebut et al. (2003), Maurer et al. (2006)

(continued)

Table 14.2 (continued)

Mast cell's role	Mouse strain	*Leishmania species*	Observations/outcome	References
			of DCs, neutrophils and macrophages.	
No role	Mast cell–deficient mice ($Cpa3^{cre}$) C57BL/6 or BALB/c	*L. amazonensis.*	• Between the MC-deficient and normal littermates, upon *L. amazonensis* infection, no significance differences were observed with respect to accumulation of eosinophil in the lesion, size of lesion and anti-*Leishmania* IgE antibody. • The genetic background against CL under high- and low-dose infection protocol with $Cpa3^{cre}$ mice, which is MC–deficient mice on the protective C57BL/6 and on the susceptible BALB/c, shows no role of MCs in Th1 versus Th2 lineage specification.	Katakura et al. (1993), Paul et al. (2016)

with better characterized models and improved experimental designs. Various reports of the interaction of MCs with *Leishmania* through *in vivo* studies have been briefly summarized in Table 14.2.

14.6 MC–*Leishmania* Interaction Studies *In Vitro*

As we found that *in vivo* studies provided many contradictory results regarding the role of MCs in leishmaniasis due to the presence of other immune cells and genetic variations in animal models, actual mechanisms of MCs' interaction with *Leishmania* could not be elucidated. *In vivo* studies were done to understand the mechanism of direct interaction of MCs with *Leishmania*. Our group has found that *L. tropica* is internalized by rat basophilic leukaemia (RBL-2H3) MCs, whereas *L. donovani* is not. Moreover, the treatment of cytochalasin D to peritoneal mast cells (PMCs) isolated from BALB/c mice with *L. tropica* coculture brought a reduction in the killing of *L. tropica*. There was no change in the survival of *L. donovani* on coculture with cytochalasin D-treated MCs, which further supports the finding that *L. tropica* is phagocytosed, whereas *L. donovani* is not. RBL-2H3

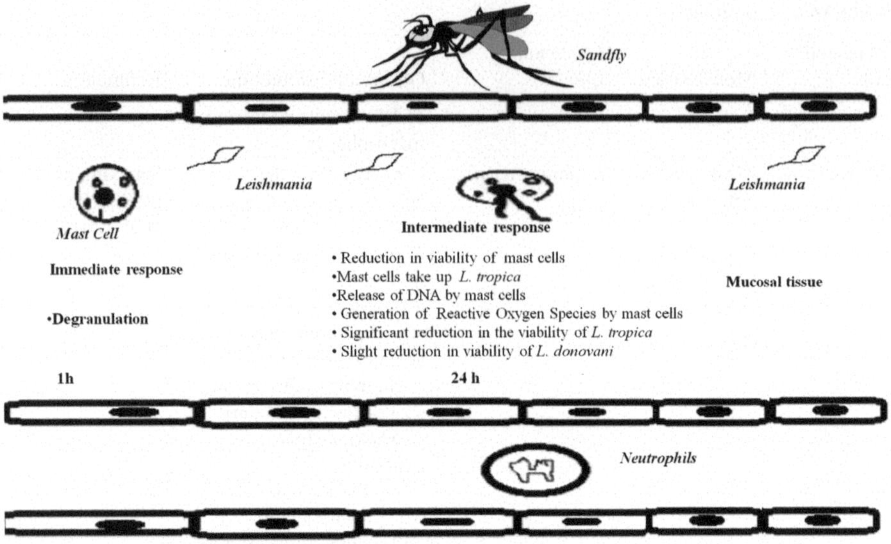

Fig. 14.3 Mast cell's interaction with *Leishmania in vivo.* When sand flies that are infected by *Leishmania* suck blood from a non-infected mammalian host, *Leishmania* promastigotes are injected into the dermis of the mammalian host. There are MCs below the dermis, which become activated and degranulate within an hour. Different species of *Leishmania* behave differently with MCs. There is a reduction in the recovery of MCs after 24 h, and they release out DNA and trap *L. tropica* and *L. donovani. L. tropica* are phagocytosed by MCs and their recoveries are very low, whereas *L. donovani* do not behave in this manner

MCs also show similar results (Naqvi et al. 2017). On interaction with these parasites, β-hexosaminidase is released. Coculture of these parasites led to ROS generation in MCs. Moreover, MCs also formed MCETs in which DNA along with tryptase and histones were released (Naqvi et al. 2017). PMCs isolated from susceptible BALB/c mice also show similar results. The interaction of MCs with *Leishmania* is briefly shown in Fig. 14.3 Bidri et al. investigated whether the interaction of mouse bone marrow–derived MCs (BMMC) and stationary metacyclic *L. infantum* or *L. major* promastigotes could lead to MC mediator release and parasites' phagocytosis. It was found that promastigotes were able to bind the BMMCs within 1 h. It also led to the release of prestored mediators like β-hexosaminidase and TNF-α (Bidri et al. 1997). BMMCs also phagocytosed the amastigotes form of *L. infantum* or *L. major*. Their uptake increased with an increase in multiplicity of infection and increasing period, but the amastigotes were not killed by BMMCs. Amastigotes were able to replicate in the BMMCs and further led to the lysis of BMMCs (Bidri et al. 1997). Thus, these results clearly show that MCs can play a role in controlling the pathogenesis of leishmaniasis (briefly summarized in Naqvi et al. (2020)).

14.7 Factors Affecting Mast Cells' Role in Leishmaniasis

14.7.1 Host Genetics—A Factor That Affects Disease Progression in Leishmaniasis

There are various reports, which suggest that disparity in host background has an influence in the functional role and recruitment of MCs with *Leishmania* infection. The spleen infiltrate shows variation in the percentage of MCs after *L. donovani* infection in relatively resistant C57BL/6 mice and susceptible BALB/c mice. The infiltration of MCs is low in the case of C57BL/6 mice and high in the case of BALB/c mice. The proliferation and increase in the number of MCs in BALB/c are IL-3 and stem cell factor (SCF) dependent (Saha et al. 2004). These results further show the confirmation of the hypothesis through studies, which show that, upon *L. major* infection in susceptible BALB/c mice, IL-3 level elevates (Rodríguez and Wilson 2014; Saha et al. 1999). To further study the interaction between cytokines and MC functions, macrophages were generated from infected BALB/c mice infected with *L. donovani* and subjected to incubation with supernatants collected from either C57BL/6 or BALB/c MCs. A high level of TNF-α secretion was found in supernatants of C57BL/6 MCs, which when treated with macrophages of BALB/c cause a reduction in parasite load. Similarly, when BALB/c MCs supernatants were treated in the same way, it caused an increase in parasite load of macrophages through IL-3- and IL-4-dependent manner (Rodríguez and Wilson 2014). Such differences were also perceived in recruitment as well as the role of MCs after infection with *L. major*. In susceptible BALB/c mice, the upper dermis showed high infiltration of MCs for a prolonged period, whereas low infiltration of MCs was seen in relatively resistant CBA/T6T6 mice. Although the numbers of degranulating MCs were high, infiltration of MCs was observed to be lower in CBA/T6T6 mice (Rodríguez and Wilson 2014). Similar results were also observed on infection with *L. mexicana* (Villaseñor-Cardoso et al. 2008). Upon stimulation or immunization with Lipophosphoglycan (LPG) of *L. mexicana* and as well as in basal condition, the expression of TNF-α, IL-10 and TLR-2 by MCs is lower in the case of BALB/c than that in C57BL/6 mice. Moreover, there was a difference in the degranulation of mast cells as well as in the numbers of MCs infiltrating at the site of lesion among the susceptible and resistant mice. In susceptible BALB/c mice around the third day of infection, the infiltration of MCs culminates, whereas, after 1 week in the case of genetically resistant C57BL/6 mice, MC infiltration culminates and stayed high for around 8 weeks. Thus, high MC influx helps in disease progression during initial stages of infection (Villaseñor-Cardoso et al. 2008). Hence, it is concluded that upon *Leishmania* infection there is a difference in MC gene expression and infiltration kinetics seen in the relatively susceptible and resistant mice strains. It will be interesting to investigate the mechanisms responsible for the difference in the influx of MCs as well as differences in gene expression between the resistant and susceptible mice strains.

14.7.2 Importance of Pathogen Genetics as a Factor Affecting Leishmaniasis

The function of MCs during *Leishmania* infection is quite complex. Host background is not the only factor that plays a role in this complexity. In a clinical study, it was seen that *L. amazonensis* and *L. braziliensis* showed a differential modulation in MC numbers (Tuon et al. 2008). Both *in vivo* animal model studies and human studies depict that, during leishmaniasis, MCs show inflammatory response, but the infiltration level and MC function depend upon both host origin and *Leishmania* species. *L. tropica* is phagocytosed by RBL MCs and causes the release of MCETs, which kill *L. tropica*, and there is a decrease in cell death upon DNase treatment, which was also confirmed by PMCs isolated from BALB/c mice. RBL MCs are not able to phagocytose *L. donovani* promastigote and cause MCETs release, but *L. donovani* are not killed (Naqvi et al. 2017). Hence, the interaction of MCs with *L. donovani* can cause killing evasion. These evasion strategies are also commonly seen between other *Leishmania* spp. and other immune cells interaction. There is a report that suggests that neutrophil extracellular traps (NETs) do not affect the viability of *L. donovani* because of a high amount of surface virulence glycolipid (Späth et al. 2003). There are also other reports that suggest that *L. donovani* uses many evasion strategies involving change of various signalling pathways in macrophages (Murray et al. 1982; Nandan and Reiner 1995; Privé and Descoteaux 2000; Shadab and Ali 2011). Thus, MCs can respond differently on interaction with different *Leishmania* spp. through various effector responses. Moreover, various *Leishmania* spp. also show different evasion strategies to protect themselves from different MC effector responses.

14.7.3 Importance of Environmental Factors Affecting Leishmaniasis

There is an expansion of several *Leishmania* species in different continents of the world. These species are transmitted and affect the populations in different environmental conditions. There are several pieces of evidence, which suggest that the incidence of leishmaniasis is highly influenced by environmental factors (humidity, weather conditions, temperature and rainfall/precipitation) (Mohammadbeigi et al. 2021). As sand flies are the intermediate host (vector) of *Leishmania*, the aforesaid environmental conditions (temperature: 7–37 °C, humidity >70%) affect the survival and developmental speed of different stages in the sand fly life cycle. It is estimated through population study that people staying near waterbodies (marshy lands, swamps, ponds and ephemeral canals) are highly affected by Leishmaniasis (Abdullah et al. 2017). Factors such as coinfection of *Leishmania* with *Leptomonas* in the Indian subcontinent can also alter patients' immunity, because the antigenicity and structure are mostly similar for both parasites (Selvapandiyan et al. 2015). Thus, these environmental and coinfection factors can be a cause for differences in the outcome when combined with differences in genotype and immunity status.

14.8 Conclusions and Future Prospects

In this chapter, we have highlighted the MCs' roles in the outcome and immune response to *Leishmania* spp. MCs that were traditionally known for their role in allergy are now known to be multifaceted. MCs are the primary cells that interact with the invading pathogens. The ability of MCs to release preformed and *de novo* synthesized inflammatory mediators upon contact with *Leishmania* has suggested that they can recruit and activate other innate immune cells such as macrophages, DCs, neutrophils and even those cells that are involved in the adaptive immunity. Several research studies were undertaken to determine the role of MCs in leishmaniasis. These studies have shown heterogeneity in responses of MCs on interaction with *Leishmania* as some studies depict that MCs play a protective role against Leishmaniasis, some studies depict that MCs exacerbate *Leishmania* infection, whereas others depict that MCs have no role in *Leishmania* infection. It can be deduced that the genetics of the host as well as *Leishmania* play a pivotal role in regulating the responses of MCs against Leishmaniasis. Many areas are required to be explored further such as what happens to the phagocytosed as well as MCET-trapped *Leishmania*. How does the initial interaction of MCs with *Leishmania* play out before and after the arrival of neutrophils? What are the evasion strategies of *L. donovani,* when faced with MCs? Is there involvement of TLRs or other receptors on MCs involved in *Leishmania* uptake or response? What are the signalling or changes in MCs on encounters with *Leishmania*? Is there any role of MCs in PKDL, MCL and atypical leishmaniasis? Understanding the mechanisms by which MCs regulate pathogenesis during *Leishmania* infection may potentially lead to the development of new and unique therapeutic targets. MCs being part of the initial innate response and their direct interaction with *Leishmania* may set the stage for the development of effective vaccines as well as activate the immune responses. Therefore, further studies can evaluate MC's importance in *Leishmania* infection; keeping in mind strain-specific parasites, all the genetics and coinfection/differences in immune status will be important for the control of leishmaniasis. Factors such as coinfection of *Leishmania* with *Leptomonas* in the Indian subcontinent (Kindler et al. 1989) can also alter patients' immunity, because the antigenicity and structure of both the parasites are mostly similar Thus all these be a cause for difference in the outcome due to changes in the different genotypes and immunity statuses.

Acknowledgments NP was supported by research grant from Department of Science and Technology-Science and Engineering Board (DST-SERB), Government of India—Project (No. CRG/2019/003651). NN and RS were supported by a grant from University Grants Commission (UGC), India. AS was supported by a research grant from Indian Council of Medical Research (ICMR) (No. GIA/2/VBD/2021/ECD-II) and Biotechnology Industry Research Assistance Council (BIRAC) (No. BT/CRS0214/CRS -10/16).

References

Abdullah AYM, Dewan A, Shogib MRI, Rahman MM, Hossain MF (2017) Environmental factors associated with the distribution of visceral leishmaniasis in endemic areas of Bangladesh: modeling the ecological niche. Trop Med Health 45(1):13

Abe T, Sugaya H, Yoshimura K, Nawa Y (1992) Induction of the expulsion of Strongyloides ratti and retention of Nippostrongylus brasiliensis in athymic nude mice by repetitive administration of recombinant interleukin-3. Immunology 76(1):10–14

Abraham SN, St John AL (2010) Mast cell-orchestrated immunity to pathogens. Nat Rev Immunol 10(6):440–452

Al-Qadhi BN, Musa IS, Al-Mulla Hummadi YMK (2015) Comparative immune study on cutaneous leishmaniasis patients with single and multiple sores. J Parasit Dis 39(3):361–370

Amiri P, Locksley RM, Parslow TG, Sadick M, Rector E, Ritter D et al (1992) Tumour necrosis factor alpha restores granulomas and induces parasite egg-laying in schistosome-infected SCID mice. Nature 356(6370):604–607

Arock M, Ross E, Lai-Kuen R, Averlant G, Gao Z, Abraham SN (1998) Phagocytic and tumor necrosis factor alpha response of human mast cells following exposure to gram-negative and gram-positive bacteria. Infect Immun 66(12):6030–6034

Badaró R, Jones TC, Lorenço R, Cerf BJ, Sampaio D, Carvalho EM et al (1986) A prospective study of visceral leishmaniasis in an endemic area of Brazil. J Infect Dis 154(4):639–649

Bates PA (2007) Transmission of Leishmania metacyclic promastigotes by phlebotomine sand flies. Int J Parasitol 37(10):1097–1106

Belkaid Y, Mendez S, Lira R, Kadambi N, Milon G, Sacks D (2000) A natural model of Leishmania major infection reveals a prolonged "silent" phase of parasite amplification in the skin before the onset of lesion formation and immunity. J Immunol Baltim Md 1950. 165(2):969–977

Bidri M, Vouldoukis I, Mossalayi MD, Debré P, Guillosson JJ, Mazier D et al (1997) Evidence for direct interaction between mast cells and Leishmania parasites. Parasite Immunol 19(10): 475–483

Brinkmann V, Reichard U, Goosmann C, Fauler B, Uhlemann Y, Weiss DS et al (2004) Neutrophil extracellular traps kill bacteria. Science 303(5663):1532–1535

Cabrera M, Rodriguez O, Monsalve I, Tovar R, Hagel I (2003) Variations in the serum levels of soluble CD23, nitric oxide and IgE across the spectrum of American cutaneous leishmaniasis. Acta Trop 88(2):145–151

Campillo-Navarro M, Leyva-Paredes K, Donis-Maturano L, González-Jiménez M, Paredes-Vivas-Y, Cerbulo-Vázquez A et al (2017) Listeria monocytogenes induces mast cell extracellular traps. Immunobiology 222(2):432–439

Carroll-Portillo A, Cannon JL, te Riet J, Holmes A, Kawakami Y, Kawakami T et al (2015) Mast cells and dendritic cells form synapses that facilitate antigen transfer for T cell activation. J Cell Biol 210(5):851–864

Charles A, Janeway J, Travers P, Walport M, Shlomchik MJ. The production of IgE. Immunobiol Immune Syst Health Dis 5th Ed [Internet]. 2001 [cited 2022 Feb 20]; Available from: https://www.ncbi.nlm.nih.gov/books/NBK27117/

Cruz A, Mendes E, Andrade M, Nascimento V, Cartelle C, Arantes R et al (2014) Mast cells are crucial in the resistance against Toxoplasma gondii oral infection. Eur J Immunol 1:44

Dawicki W, Marshall JS (2007) New and emerging roles for mast cells in host defence. Curr Opin Immunol 19(1):31–38

de Freitas e Silva R, von Stebut E. Unraveling the Role of Immune Checkpoints in Leishmaniasis. Front Immunol [Internet]. 2021 [cited 2022 Feb 20];12. Available from: https://www.frontiersin.org/article/10.3389/fimmu.2021.620144

de Oliveira MP, Lima MCR, Calheiros AS, Martins MA, Antas PRZ, De Luca PM et al (2005) Leishmania (Viannia) braziliensis: human mast cell line activation induced by logarithmic and stationary promastigote derived-lysates. Exp Parasitol 109(2):72–79

Dillon SB, MacDonald TT (1986) Limit dilution analysis of mast cell precursor frequency in the gut epithelium of normal and Trichinella spiralis infected mice. Parasite Immunol 8(5):503–511

Feijó D, Tibúrcio R, Ampuero M, Brodskyn C, Tavares N (2016) Dendritic cells and leishmania infection: adding layers of complexity to a complex disease. J Immunol Res 2016:3967436

Feyerabend TB, Weiser A, Tietz A, Stassen M, Harris N, Kopf M et al (2011) Cre-mediated cell ablation contests mast cell contribution in models of antibody- and T cell-mediated autoimmunity. Immunity 35(5):832–844

Flanagan JG, Chan DC, Leder P (1991) Transmembrane form of the kit ligand growth factor is determined by alternative splicing and is missing in the Sld mutant. Cell 64(5):1025–1035

Fuchs TA, Abed U, Goosmann C, Hurwitz R, Schulze I, Wahn V et al (2007) Novel cell death program leads to neutrophil extracellular traps. J Cell Biol 176(2):231–241

Fulton JD, Joyner LPY (1948) Infections by Leishmania donovani in the Cotton Rat. Microbiology 2(1):103–109

Furuta T, Kikuchi T, Iwakura Y, Watanabe N (2006) Protective roles of mast cells and mast cell-derived TNF in murine malaria. J Immunol Baltim Md 1950 177(5):3294–3302

Gabriel C, WR MM, Girard D, Descoteaux A (2010) Leishmania donovani promastigotes evade the antimicrobial activity of neutrophil extracellular traps. J Immunol Baltim Md 1950. 185(7): 4319–4327

Galli SJ, Tsai M (2012) IgE and mast cells in allergic disease. Nat Med 18(5):693–704

Galvao-Quintao L, Alfieri SC, Ryter A, Rabinovitch M (1990) Intracellular differentiation of Leishmania amazonensis promastigotes to amastigotes: presence of megasomes, cysteine proteinase activity and susceptibility to leucine-methyl ester. Parasitology 101(Pt 1):7–13

García-Faroldi G, Rodríguez CE, Urdiales JL, Pérez-Pomares JM, Dávila JC, Pejler G et al (2010) Polyamines are present in mast cell secretory granules and are important for granule homeostasis. PLoS One 5(11):e15071

Gardinassi LG, DeSouza-Vieira TS, da Silva NO, Garcia GR, Borges VM, Campos RNS et al (2017) Molecular signatures of neutrophil extracellular traps in human visceral leishmaniasis. Parasit Vectors 10(1):285

Gaudenzio N, Sibilano R, Marichal T, Starkl P, Reber LL, Cenac N et al (2016) Different activation signals induce distinct mast cell degranulation strategies. J Clin Invest 126(10):3981–3998

Gri G, Frossi B, D'Incà F, Danelli L, Betto E, Mion F, et al (2012) Mast Cell: An Emerging Partner in Immune Interaction. Front Immunol [Internet]. [cited 2022 Feb 20];3. Available from: https://www.frontiersin.org/article/10.3389/fimmu.2012.00120

Grimaldi G, Soares MJ, Moriearty PL (1984) Tissue eosinophilia and Leishmania mexicana mexicana eosinophil interactions in murine cutaneous leishmaniasis. Parasite Immunol 6(5): 397–408

Guimarães-Costa AB, Nascimento MTC, Froment GS, Soares RPP, Morgado FN, Conceição-Silva F et al (2009) Leishmania amazonensis promastigotes induce and are killed by neutrophil extracellular traps. Proc Natl Acad Sci U S A 106(16):6748–6753

Gunin AG, Kornilova NK, Vasilieva OV, Petrov VV (2011) Age-related changes in proliferation, the numbers of mast cells, eosinophils, and cd45-positive cells in human dermis. J Gerontol A Biol Sci Med Sci 66(4):385–392

Hepworth MR, Maurer M, Hartmann S (2012) Regulation of type 2 immunity to helminths by mast cells. Gut Microbes 3(5):476–481

Jones DE, Elloso MM, Scott P (1998) Host susceptibility factors to cutaneous leishmaniasis. Front Biosci 3:D1171–D1180

Jutel M, Watanabe T, Klunker S, Akdis M, Thomet OA, Malolepszy J et al (2001) Histamine regulates T-cell and antibody responses by differential expression of H1 and H2 receptors. Nature 413(6854):420–425

Kambayashi T, Laufer TM (2014) Atypical MHC class II-expressing antigen-presenting cells: can anything replace a dendritic cell? Nat Rev Immunol 14(11):719–730

Kambayashi T, Baranski JD, Baker RG, Zou T, Allenspach EJ, Shoag JE et al (2008) Indirect involvement of allergen-captured mast cells in antigen presentation. Blood 111(3):1489–1496

Kasugai T, Tei H, Okada M, Hirota S, Morimoto M, Yamada M et al (1995) Infection with Nippostrongylus brasiliensis induces invasion of mast cell precursors from peripheral blood to small intestine. Blood 85(5):1334–1340

Katakura K, Saito S, Hamada A, Matsuda H, Watanabe N (1993) Cutaneous leishmaniasis in mast cell-deficient W/Wv mice. Infect Immun 61(5):2242–2244

Katz HR, Austen KF (2011) Mast cell deficiency, a game of kit and mouse. Immunity 35(5): 668–670

Kindler V, Sappino AP, Grau GE, Piguet PF, Vassalli P (1989) The inducing role of tumor necrosis factor in the development of bactericidal granulomas during BCG infection. Cell 56(5):731–740

Kirchner T, Möller S, Klinger M, Solbach W, Laskay T, Behnen M (2012) The impact of various reactive oxygen species on the formation of neutrophil extracellular traps. Mediat Inflamm 2012:849136

Knight PA, Wright SH, Lawrence CE, Paterson YY, Miller HR (2000) Delayed expulsion of the nematode Trichinella spiralis in mice lacking the mucosal mast cell-specific granule chymase, mouse mast cell protease-1. J Exp Med 192(12):1849–1856

Lee TD, Swieter M, Befus AD (1986) Mast cell responses to helminth infection. Parasitol Today Pers Ed 2(7):186–191

Leishmaniasis [Internet]. [cited 2022 Feb 20]. Available from: https://www.who.int/westernpacific/health-topics/leishmaniasis

Lerner EA, Iuga AO, Reddy VB (2007) Maxadilan, a PAC1 receptor agonist from sand flies. Peptides 28(9):1651–1654

Lim MBH, Kuiper JWP, Katchky A, Goldberg H, Glogauer M (2011) Rac2 is required for the formation of neutrophil extracellular traps. J Leukoc Biol 90(4):771–776

Lopes JP, Stylianou M, Nilsson G, Urban CF (2015) Opportunistic pathogen Candida albicans elicits a temporal response in primary human mast cells. Sci Rep 5:12287

Lu F, Huang S (2017) The roles of mast cells in parasitic protozoan infections. Front Immunol 8: 363

Malaviya R, Ross EA, MacGregor JI, Ikeda T, Little JR, Jakschik BA et al (1994) Mast cell phagocytosis of FimH-expressing enterobacteria. J Immunol Baltim Md 1950 152(4): 1907–1914

Marshall JS (2004) Mast-cell responses to pathogens. Nat Rev Immunol 4(10):787–799

Marshall JS, Jawdat DM (2004) Mast cells in innate immunity. J Allergy Clin Immunol 114(1): 21–27

Matsuda H, Watanabe N, Kiso Y, Hirota S, Ushio H, Kannan Y et al (1990) Necessity of IgE antibodies and mast cells for manifestation of resistance against larval Haemaphysalis longicornis ticks in mice. J Immunol Baltim Md 1950 144(1):259–262

Maurer M, Lopez Kostka S, Siebenhaar F, Moelle K, Metz M, Knop J et al (2006) Skin mast cells control T cell-dependent host defense in Leishmania major infections. FASEB J 20(14): 2460–2467

McConville MJ, de Souza D, Saunders E, Likic VA, Naderer T (2007) Living in a phagolysosome; metabolism of Leishmania amastigotes. Trends Parasitol 23(8):368–375

McLachlan JB, Shelburne CP, Hart JP, Pizzo SV, Goyal R, Brooking-Dixon R et al (2008) Mast cell activators: a new class of highly effective vaccine adjuvants. Nat Med 14(5):536–541

Mecheri S (2012) Contribution of allergic inflammatory response to the pathogenesis of malaria disease. Biochim Biophys Acta 1822(1):49–56

Mekori Y, Hershko A. T cell-mediated modulation of mast cell function: heterotypic adhesion-induced stimulatory or inhibitory effects. Front Immunol [Internet]. 2012 [cited 2022 Feb 21];3. Available from: https://www.frontiersin.org/article/10.3389/fimmu.2012.00006

Mekori YA, Hershko AY, Frossi B, Mion F, Pucillo CE (2016) Integrating innate and adaptive immune cells: mast cells as crossroads between regulatory and effector B and T cells. Eur J Pharmacol 778:84–89

Metz M, Maurer M (2007) Mast cells--key effector cells in immune responses. Trends Immunol 28(5):234–241

Mohammadbeigi A, Khazaei S, Heidari H, Asgarian A, Arsangjang S, Saghafipour A et al (2021) An investigation of the effects of environmental and ecologic factors on cutaneous leishmaniasis in the old world: a systematic review study. Rev Environ Health 36(1):117–128

Moon TC, Befus AD, Kulka M (2014) Mast Cell Mediators: Their Differential Release and the Secretory Pathways Involved. Front Immunol [Internet]. [cited 2022 Feb 20];5. Available from: https://www.frontiersin.org/article/10.3389/fimmu.2014.00569

Morris RV, Shoemaker CB, David JR, Lanzaro GC, Titus RG (2001) Sandfly maxadilan exacerbates infection with Leishmania major and vaccinating against it protects against L. major infection. J Immunol Baltim Md 1950. 167(9):5226–5230

Muñoz S, Hernández-Pando R, Abraham SN, Enciso JA (2003) Mast cell activation by Mycobacterium tuberculosis: mediator release and role of CD48. J Immunol Baltim Md 1950 170(11): 5590–5596

Muñoz S, Rivas-Santiago B, Enciso JA (2009) Mycobacterium tuberculosis entry into mast cells through cholesterol-rich membrane microdomains. Scand J Immunol 70(3):256–263

Murray HW (1982) Cell-mediated immune response in experimental visceral leishmaniasis. II. Oxygen-dependent killing of intracellular Leishmania donovani amastigotes. J Immunol Baltim Md 1950 129(1):351–357

Murray HW, Masur H, Keithly JS (1982) Cell-mediated immune response in experimental visceral leishmaniasis. I. Correlation between resistance to Leishmania donovani and lymphokine-generating capacity. J Immunol Baltim Md 1950 129(1):344–350

Nandan D, Reiner NE (1995) Attenuation of gamma interferon-induced tyrosine phosphorylation in mononuclear phagocytes infected with Leishmania donovani: selective inhibition of signaling through Janus kinases and Stat1. Infect Immun 63(11):4495–4500

Naqvi N, Ahuja K, Selvapandiyan A, Dey R, Nakhasi H, Puri N (2017) Role of Mast Cells in clearance of Leishmania through extracellular trap formation. Sci Rep 7(1):13240

Naqvi N, Srivastava R, Selvapandiyan A, Puri N (2020) Host mast cells in leishmaniasis: friend or foe? Trends Parasitol 36(12):952–956

Naqvi N, Srivastava R, Naskar P, Puri N (2021) Mast cells modulate early responses to Mycobacterium bovis Bacillus Calmette-Guerin by phagocytosis and formation of extracellular traps. Cell Immunol 365:104380

Nolan TJ, Farrell JP (1987) Experimental infections of the multimammate rat (Mastomys natalensis) with Leishmania donovani and Leishmania major. Am J Trop Med Hyg 36(2): 264–269

Ohnmacht C, Voehringer D (2010) Basophils protect against reinfection with hookworms independently of mast cells and memory Th2 cells. J Immunol Baltim Md 1950 184(1):344–350

Pal S, Nath S, Meininger CJ, Gashev AA. Emerging roles of mast cells in the regulation of lymphatic immuno-physiology. Front Immunol [Internet]. 2020 [cited 2022 Feb 20];11. Available from: https://www.frontiersin.org/article/10.3389/fimmu.2020.01234

Palm AKE, Garcia-Faroldi G, Lundberg M, Pejler G, Kleinau S (2016) Activated mast cells promote differentiation of B cells into effector cells. Sci Rep 6(1):20531

Palomäki VAB, Laitinen JT (2006) The basic secretagogue compound 48/80 activates G proteins indirectly via stimulation of phospholipase D-lysophosphatidic acid receptor axis and 5-HT1A receptors in rat brain sections. Br J Pharmacol 147(6):596–606

Paul C, Wolff S, Zapf T, Raifer H, Feyerabend TB, Bollig N et al (2016) Mast cells have no impact on cutaneous leishmaniasis severity and related Th2 differentiation in resistant and susceptible mice. Eur J Immunol 46(1):114–121

Peters NC, Egen JG, Secundino N, Debrabant A, Kimblin N, Kamhawi S et al (2008) In vivo imaging reveals an essential role for neutrophils in leishmaniasis transmitted by sand flies. Science 321(5891):970–974

Piani A, Ilg T, Elefanty AG, Curtis J, Handman E (1999) Leishmania major proteophosphoglycan is expressed by amastigotes and has an immunomodulatory effect on macrophage function. Microbes Infect 1(8):589–599

Porebski G, Kwiecien K, Pawica M, Kwitniewski M. Mas-Related G Protein-Coupled Receptor-X2 (MRGPRX2) in Drug Hypersensitivity Reactions. Front Immunol [Internet]. 2018 [cited 2022 Feb 20];9. Available from: https://www.frontiersin.org/article/10.3389/fimmu.2018.03027

Privé C, Descoteaux A (2000) Leishmania donovani promastigotes evade the activation of mitogen-activated protein kinases p38, c-Jun N-terminal kinase, and extracellular signal-regulated kinase-1/2 during infection of naive macrophages. Eur J Immunol 30(8):2235–2244

Puri N, Roche PA (2008) Mast cells possess distinct secretory granule subsets whose exocytosis is regulated by different SNARE isoforms. Proc Natl Acad Sci 105(7):2580–2585

Rochael NC, Guimarães-Costa AB, Nascimento MTC, DeSouza-Vieira TS, Oliveira MP, Garcia e Souza LF et al (2015) Classical ROS-dependent and early/rapid ROS-independent release of Neutrophil Extracellular Traps triggered by Leishmania parasites. Sci Rep. 5(1):18302

Rodríguez NE, Wilson ME (2014) Eosinophils and mast cells in leishmaniasis. Immunol Res 59(1–3):129–141

Romão PRT, Da Costa SH, Ramos CDL, De Oliveira CFE, Monteiro MC, De Queiroz CF et al (2009) Mast cell degranulation contributes to susceptibility to Leishmania major. Parasite Immunol 31(3):140–146

Saha B, Saini A, Germond R, Perrin PJ, Harlan DM, Davis TA (1999) Susceptibility or resistance to Leishmania infection is dictated by the macrophages evolved under the influence of IL-3 or GM-CSF. Eur J Immunol 29(7):2319–2329

Saha B, Tonkal AMDJ, Croft S, Roy S (2004) Mast cells at the host-pathogen interface: host-protection versus immune evasion in leishmaniasis. Clin Exp Immunol 137(1):19–23

Sasaki Y, Yoshimoto T, Maruyama H, Tegoshi T, Ohta N, Arizono N, Nakanishi K, et al. IL-18 with IL-2 protects against Strongyloides venezuelensis infection by activating mucosal mast cell-dependent type 2 innate immunity. J Exp Med [Internet]. 2005 Sep 5 [cited 2022 Feb 21];202(5). Available from: https://pubmed.ncbi.nlm.nih.gov/16129701/

Schäfer B, Piliponsky AM, Oka T, Song CH, Gerard NP, Gerard C, et al (2013) Mast cell anaphylatoxin receptor expression can enhance IgE-dependent skin inflammation in mice. J Allergy Clin Immunol 131(2):541–8.e1–9

Scheb-Wetzel M, Rohde M, Bravo A, Goldmann O (2014) New insights into the antimicrobial effect of mast cells against Enterococcus faecalis. Infect Immun 82(11):4496–4507

Selvapandiyan A, Ahuja K, Puri N, Krishnan A (2015) Implications of co-infection of Leptomonas in visceral leishmaniasis in India. Parasitology 142(14):1657–1662

Selvapandiyan A, Croft SL, Rijal S, Nakhasi HL, Ganguly NK (2019) Innovations for the elimination and control of visceral leishmaniasis. PLoS Negl Trop Dis 13(9):e0007616

Selvapandiyan A, Puri N, Kumar P, Alam A, Ehtesham NZ, Griffin G et al (2023) Zooming in on common immune evasion mechanisms of pathogens in phagolysosomes: potential broad-spectrum therapeutic targets against infectious diseases. FEMS Microbiol Rev. 47(1):fuac041

Shadab M, Ali N (2011) Evasion of host defence by Leishmania donovani: subversion of signaling pathways. Mol Biol Int 2011:e343961

Sher A, Hein A, Moser G, Caulfield JP (1979) Complement receptors promote the phagocytosis of bacteria by rat peritoneal mast cells. Lab Investig J Tech Methods Pathol 41(6):490–499

Shin K, Watts GFM, Oettgen HC, Friend DS, Pemberton AD, Gurish MF et al (2008) Mouse mast cell tryptase mMCP-6 is a critical link between adaptive and innate immunity in the chronic phase of Trichinella spiralis infection. J Immunol Baltim Md 1950 180(7):4885–4891

Smith NL, Abi Abdallah DS, Butcher BA, Denkers EY, Baird B, Holowka D (2013) Toxoplasma gondii inhibits mast cell degranulation by suppressing phospholipase Cγ-mediated Ca(2+) mobilization. Front Microbiol 4:179

Späth GF, Garraway LA, Turco SJ, Beverley SM (2003) The role(s) of lipophosphoglycan (LPG) in the establishment of Leishmania major infections in mammalian hosts. Proc Natl Acad Sci U S A 100(16):9536–9541

St John AL, Abraham SN (2013) Innate immunity and its regulation by mast cells. J Immunol Baltim Md 1950. 190(9):4458–4463

Stelekati E, Bahri R, D'Orlando O, Orinska Z, Mittrücker HW, Langenhaun R et al (2009) Mast cell-mediated antigen presentation regulates CD8+ T cell effector functions. Immunity 31(4): 665–676

Tabbabi A (2019) Review of leishmaniasis in the Middle East and North Africa. Afr Health Sci 19(1):1329

Takahashi HK, Yoshida A, Iwagaki H, Yoshino T, Itoh H, Morichika T et al (2002) Histamine regulation of interleukin-18-initiating cytokine cascade is associated with down-regulation of intercellular adhesion molecule-1 expression in human peripheral blood mononuclear cells. J Pharmacol Exp Ther 300(1):227–235

Teles CBG, Santos AP d A d, Freitas RA, de Oliveira AFJ, Ogawa GM, Rodrigues MS et al (2016) Phlebotomine sandfly (Diptera: Psychodidae) diversity and their *Leishmania* DNA in a hot spot of American Cutaneous Leishmaniasis human cases along the Brazilian border with Peru and Bolivia. Mem Inst Oswaldo Cruz 111:423–432

Thakur L, Singh KK, Shanker V, Negi A, Jain A, Matlashewski G et al (2018 Sep 27) Atypical leishmaniasis: a global perspective with emphasis on the Indian subcontinent. PLoS Negl Trop Dis 12(9):e0006659

Thalhofer CJ, Chen Y, Sudan B, Love-Homan L, Wilson ME (2011) Leukocytes infiltrate the skin and draining lymph nodes in response to the protozoan Leishmania infantum chagasi. Infect Immun 79(1):108–117

Tuon FF, Duarte MIS, Amato VS (2008) A quantitative and morphometric study of mast cells in cutaneous leishmaniasis. Parasite Immunol 30(11–12):641–645

Urb M, Sheppard DC. The Role of Mast Cells in the Defence against Pathogens. PLoS Pathog [Internet]. 2012 Apr [cited 2022 Feb 20];8(4). Available from: https://www.ncbi.nlm.nih.gov/pmc/articles/PMC3343118/

Valitutti S, Espinosa E (2010) Cognate interactions between mast cells and helper T lymphocytes. Self/Nonself 1(2):114–122

Van de Voorde J, Leusen I (1983) Role of the endothelium in the vasodilator response of rat thoracic aorta to histamine. Eur J Pharmacol 87(1):113–120

van Zandbergen G, Klinger M, Mueller A, Dannenberg S, Gebert A, Solbach W et al (2004) Cutting edge: neutrophil granulocyte serves as a vector for Leishmania entry into macrophages. J Immunol Baltim Md 1950 173(11):6521–6525

Villaseñor-Cardoso MI, Salaiza N, Delgado J, Gutiérrez-Kobeh L, Pérez-Torres A, Becker I (2008) Mast cells are activated by Leishmania mexicana LPG and regulate the disease outcome depending on the genetic background of the host. Parasite Immunol 30(8):425–434

von Köckritz-Blickwede M, Goldmann O, Thulin P, Heinemann K, Norrby-Teglund A, Rohde M et al (2008) Phagocytosis-independent antimicrobial activity of mast cells by means of extracellular trap formation. Blood 111(6):3070–3080

von Stebut E, Metz M, Milon G, Knop J, Maurer M (2003) Early macrophage influx to sites of cutaneous granuloma formation is dependent on MIP-1alpha /beta released from neutrophils recruited by mast cell-derived TNFalpha. Blood 101(1):210–215

Wershil BK, Theodos CM, Galli SJ, Titus RG (1994) Mast cells augment lesion size and persistence during experimental Leishmania major infection in the mouse. J Immunol Baltim Md 1950 152(9):4563–4571

Woodbury RG, Miller HR, Huntley JF, Newlands GF, Palliser AC, Wakelin D (1984) Mucosal mast cells are functionally active during spontaneous expulsion of intestinal nematode infections in rat. Nature 312(5993):450–452

Yost CC, Cody MJ, Harris ES, Thornton NL, McInturff AM, Martinez ML et al (2009) Impaired neutrophil extracellular trap (NET) formation: a novel innate immune deficiency of human neonates. Blood 113(25):6419–6427

15 Feasibility of Therapeutic Vaccine for the Management and Control of VL

Alok K. Yadav, Niharika Gupta, Amogh A. Sahasrabuddhe, and Anuradha Dube

Abstract

Visceral leishmaniasis (VL), a neglected tropical disease, is caused by the parasite *Leishmania donovani* complex. Whereas the transmission of *L. donovani* infection is anthroponotic in the Indian subcontinent and East Africa, the spread of *Leishmania infantum*, as well as *Leishmania chagasi*, is zoonotic, with dogs serving as the reservoir host throughout Europe, North Africa, and regions of Latin America. Although zoonotic VL is becoming less common, anthroponotic VL continues to cause epidemics periodically. These parasites infect the host's macrophages and damage the immune system, leading to pronounced immunosuppression in the affected person. In general, a small percentage of *L. donovani*-infected population develop full-blown clinical disease and are treated, but the majority of them remain asymptomatic. In addition, the treated VL patients develop post-kala-azar dermal leishmaniasis (PKDL) conditions, and these populations contribute to spreading the VL disease via the sandfly vector. This condition is one of the main obstacles to the World Health Organization's efforts to eradicate this fatal disease. Designing therapeutic vaccines that can only target infected macrophages has been attempted in the absence of any reliable and efficient vector control measures. Three vaccines—Leishmune®, Leish-Tec®, and Leish-111f® with monophosphoryl lipid A (MPLA) plus squalene emulsion in combination with glucantime—have thus far effectively undergone immunotherapeutic evaluation against canine VL (CVL). A third-generation vaccine—

A. K. Yadav · N. Gupta · A. A. Sahasrabuddhe
Biochemistry and Structural Biology, Division, CSIR-Central Drug Research Institute (CSIR-CDRI), Lucknow, India

A. Dube (✉)
Division of Molecular Microbiology and Immunology, CSIR-Central Drug Research Institute (CSIR-CDRI), Lucknow, India

© The Author(s), under exclusive license to Springer Nature Singapore Pte Ltd. 2023

A. Selvapandiyan et al. (eds.), *Challenges and Solutions Against Visceral Leishmaniasis*, https://doi.org/10.1007/978-981-99-6999-9_15

ChAd63 KH has also shown its immunotherapeutic potential against PKDL. Nevertheless, it is still necessary to translate positive findings from a number of experimental studies against human VL into therapeutic applications. This chapter summarizes the numerous approaches that have been attempted and evaluated for generating therapeutic vaccines to stave off this dreaded disease.

Keywords

VL · PKDL · Canine VL · Immunosuppression · Therapeutic vaccines · Leishmune® · Leish-Tec® · ChAd 63KH

15.1 Introduction

Visceral leishmaniasis (VL), often known as "kala-azar" or "black fever," is a progressively systemic and fatal disease caused by *Leishmania donovani* complex (which includes *L. donovani* and *L. infantum* or *L. chagasi*), an obligate intracellular protozoan parasite (WHO 2010). In the absence of any appropriate treatment, it may lead to death in 90% of VL cases. Children under the age of 5 years are the primary victims of this disease, which is associated with malnutrition and other immunosuppressive disorders such as human immunodeficiency virus (HIV)–acquired immunodeficiency syndrome (AIDS).

According to its modes of transmission, this disease can either be zoonotic or be anthroponotic (WHO, 2010). Approximately 90% of the 100,000 recorded cases of VL each year are found in Asia (India and Bangladesh), South America (Brazil), and Africa (Ethiopia, Sudan, and South Sudan) (WHO 2010). Although the real number of VL cases is much greater due to underreporting, it predominantly affects the underprivileged section of people living in secluded areas with inadequate basic health-care facilities (Singh et al. 2006, 2010; Alvar et al. 2012). *L. donovani*, the agent responsible for the disease in East Africa and the Indian subcontinent, is primarily transmitted through *Phlebotomus argentipes* and *Phlebotomus orientalis*, respectively (Ready 2014); however, *L. infantum*, which is primarily causing VL throughout the Mediterranean area and the American subcontinent, is spread by *Lutzomyia* spp. The parasites exist either as intracellular, nonflagellated amastigotes within the mammalian host or as extracellular flagellated promastigotes within the insect vector. This digenetic way of life ensures the parasite's better survival inside the vector and host (Rodrigues et al. 2016).

Leishmania parasites primarily target the mononuclear phagocytic cells of the liver, the spleen, and the bone marrow. They, however, frequently spread to the skin and other visceral organs (such as the gut, lung, etc.), especially seen in the areas affected by *L. infantum*. (Rodrigues et al. 2016). The defining characteristics of human VL are prolonged fever, hepatosplenomegaly, pancytopenia, and hypergammaglobulinemia (WHO 2010). The disease has a 100% fatality rate, as the patient usually dies in the absence of any medication. The use of available chemotherapeutics is limited due to their several demerits, and besides, they are administered to patients having clinical symptoms only.

Human VL occurs more in a heterogeneous form with diverse chronicity. Only a small portion of infected individuals develop clinical symptoms; hence, only these individuals are subjected to antileishmanial treatment. It is unknown whether the vast majority of the endemic population—who accounts for 6–10 times as many people as VL patients—carries the disease asymptomatically or with subclinical infection and has any role in disease transmission (Singh et al. 2014). Furthermore, VL patients are more prone to secondary infections due to their immunocompromised state. Another significant issue is the emergence of parasite resistance following drug therapy and a rise in reports of disease relapse. Furthermore, it is challenging to determine the total eradication of the disease as the endpoint of the treatment is not defined, and patients frequently fail to follow up 6 months after therapy.

Emerging Challenge of *Leishmania*/HIV Coinfection: HIV coinfection is increasingly posing a significant issue in the era of VL elimination initiatives. As a result of the geographic overlap between VL- and HIV-affected areas, more people with *Leishmania*/HIV coinfection are exhibiting atypical VL symptoms (Lindoso et al. 2014) and play an essential role in parasite transmission. Both infections are opportunistic, which causes them to have a comparable negative impact on the immune system and raise the likelihood of relapse and recurrence cases. However, antiretroviral therapy greatly slows down disease development and relapses, improving the prognosis for individuals who are also coinfected (Lindoso et al. 2014). In 35 nations around the world, including India, *Leishmania* infection has spread due to an increase in HIV cases (Cruz et al. 2006; Diro et al. 2014). Splenomegaly, one of the defining characteristics of VL, may not be present in people with leishmaniasis and HIV (Mondain-Miton et al. 1995). On the other hand, there is an association with a typical organ, such as the lungs or the gastrointestinal system (Monge-Maillo et al. 2014).

Post-Kala-Azar Dermal Leishmaniasis (PKDL): PKDL is a well-recognized complication (a stigmatizing skin condition or cutaneous syndrome) of VL. It often develops in patients who are successfully treated for VL. Occasionally, between 10% and 50% of VL patients will experience PKDL, which is characterized by maculopapular or nodular skin rashes, during or after treatment (Zijlstra et al. 2003.) *L. donovani* strains isolated from patients with VL and parasites obtained from patients with PKDL are biochemically equivalent. As a result, it is primarily limited to regions where *L. donovani* is the causative parasite. PKDL due to *L. infantum* parasites may also be seen in immunocompromised persons. In Sudan, PKDL follows VL by 0–6 months and by 2–3 years, respectively. Treatment for PKDL is typically time-consuming and expensive. Such individuals may likely play a major role in interepidemic peaks of VL transmission since the lesions of PKDL may persist for up to 20 years, serving as a possible reservoir for VL parasites (Desjeux et al. 2013). In HIV coinfection, PKDL occurs more frequently and is more severe (Abongomera et al. 2019). In HIV coinfection, skin lesions may come before, after, or in addition to VL, some of which may be referred to as PKDL (Zijlstra 2014).

Canine VL: Canine VL is an important disease caused by *L. infantum*. Dogs play a crucial part in the transmission cycle of urban VL, which also affects humans, as they are the main reservoir hosts of *L. infantum*. The course of infection may be different from one individual dog to another, ranging from spontaneous cure to acute evolution if not attended properly. Although the treatment with antileishmanial successfully reduces the infection, the treatment must be continued for the rest of life. Drugs explicitly intended for human purposes are not allowed to be used to treat Canine VL due to concerns about the spread of parasites that are resistant to treatment.

Some dogs with parasite infections are able to control the parasite and do not develop the disease in the short term, sometimes for years or for the rest of their lives, whereas other infected dogs exhibit progressive diseases. Dogs with the progressive infection experience weight loss, cachexia, difficulty with mobility, conjunctivitis, epistaxis, lymphadenopathies, dermatitis, onychogryphosis, strong antibody response, and reduced cell-mediated immune response after an incubation period of 2–4 months. (Semião-Santos et al. 1995). However, a long-term study following several cohorts of diseased dogs in the Priorat area of Tarragona, Spain, divulged that 15% of all infected dogs could avoid the onset of the illness and the disease has resolved it on its own (Fisa et al. 1999; Moreno and Alvar 2002).

15.2 Immunopathology of VL, PKDL, and Canine VL

15.2.1 Human VL

Patients with active VL exhibit suppressed cell-mediated immunity as they exhibit negative leishmanin skin test (Gidwani et al. 2009). In addition, their peripheral blood mononuclear cells (PBMCs) do proliferate or produce interferon (IFN)-γ not in the presence of *Leishmania* antigen (Sacks et al. 1987; Caldas et al. 2005). On the contrary, stimulating PBMCs of treated VL patients with parasite antigen increases the production of IFN-γ, indicating that the antigen-specific T-cell response may not mediate the negative clinical effects of immunosuppression but may instead be caused by a number of other factors (Singh et al. 2012). Several studies on human VL showed that there was also a rise in the production of Th1 type of cytokines and chemokines, indicating that the immune response did not clearly favor T-helper cell type 2 (Th2 type) (Faleiro et al. 2014). Clinical investigations clearly show that individuals with active VL have raised serum interleukin (IL)-10 levels, which are supported by elevated messenger ribonucleic acid (mRNA) transcript levels in the lymph nodes, the spleen, and the bone marrow (BM; Singh et al. 2012). In addition, regulatory T (T-reg) cells that have been discovered to secrete IL-10 in the bone marrow of VL patients may be able to inhibit antiparasitic immunity (Rai et al. 2012). Further, the mRNA expression of transforming growth factor-beta (TGF-β) was also found to be increased in the splenic tissue of patients with active VL (Nylén et al. 2007). Moreover, a characteristic VL patient had a higher level of antibodies in the plasma that resulted in the development of immunological complexes (Buxbaum and Scott 2005).

Studies on human VL–HIV coinfection have revealed that extremely complicated pathways involving the deregulation of host immune responses contribute to the pathogenesis of HIV and *Leishmania*-mediated immune activation (Okwor and Uzonna 2013). *Leishmania* and HIV-1 both infect and grow in cells of myeloid or lymphoid origin, providing the ideal conditions for reciprocal control of the pathogenesis of both the diseases. In addition, in immunocompromised patients with PKDL, which often presents as skin lesions, there is a very real chance that HIV-mediated immunosuppression (induced by CD4(+) T-cell depletion) would trigger the reactivation of latent infections (reactivation leishmaniasis). This is crucial because the long-term survival of parasites at the sites of initial infection and their draining lymph nodes are related to the recovery from leishmaniases. This is either because VL cannot be effectively treated or because persistent parasites have suppressed immunity, and this syndrome is immunologically different from VL in a number of ways (Zijlstra et al. 2003). In PKDL lesions, higher concentrations of IFN-γ, IL-4, and IL-10 were found, along with a greater infiltration of CD3 + T cells (Ismail et al. 1999). According to other studies, the development of PKDL was accompanied by a more pronounced parasite-specific T-cell response (Gasim et al. 2000). In addition, PKDL lesions revealed that T-reg cell aggregation was favorably correlated with parasite burden. Although IL-10-mediated immunosuppression was evident in the majority of PKDL patients, additional unregulated inflammatory responses must also be present (Katara et al. 2011).

15.2.2 PKDL

PKDL develops as an immune reconstitution inflammatory syndrome (IRIS), which happens from a loss of immune suppression following drug therapy for VL (Chatterjee et al. 2020). The evidence so far reviewed in this study suggests that, similar to VL, the pathophysiology of PKDL involves an enhanced Th1/Th2 response with a Th2 bias, as evidenced by the increased levels of IL-4, IL-5, IL-13, IL-10, and TGF-β. However, data are fragmentary, and no systematic picture of the immune status at presentation or after initiating drug therapy in PKDL has yet emerged. An increased population of antigen-specific, IL-10-producing anergic T cells in peripheral blood, a decline in the presence of dendritic cells at lesion sites, a significant infiltration of CD68+ alternatively activated macrophages, and a dermal pathology dominated by IL-10 and FoxP3, which individually or more likely collectively contribute to the establishment of a systemic and dermal immunosuppressive milieu, are all present. In addition, cellular infiltration is characterized by the near-complete absence of CD4+ T cells and increased CD8+ T cells, which display signs of exhaustion. Because *L. donovani* parasites have ingeniously developed immune escape mechanisms, it is crucial to develop immunotherapeutic strategies to restore effector responses.

15.2.3 Canine VL (CVL)

It is well known that type 1 immune response, which is characterized by the production of IFN-γ, TNF-α, and IL-2 in relation to the resistance profile, triggers the majority of the immune response against the parasite (Gonçalves et al. 2019). This type of immunological response is related to an upregulation in macrophage's antileishmanial activity (Koutinas and Koutinas 2014). This is the primary effector mechanism of the intracellular death of *Leishmania* amastigotes (Baneth et al. 2008). In this way, type 1 immune response causes the production of cytokines like IFN-γ and TNF-α, which are common in dogs, who are asymptomatic and exhibit disease-preventive potential (Costa-Pereira et al. 2015). Solano-Gallego et al. (2016) showed that infected dogs that displayed high levels of IFN-γ had decreased parasite burdens as compared to those infected dogs, who did not produce this cytokine, and in them, the clinical symptoms were more severe, and the parasite burden is higher (Martínez-Orellana et al. 2017). Likewise, Th17 cells induce *L. infantum* to control growth (Nascimento et al. 2015; Rodriguez-Cortes et al. 2017).

Contrarily, CVL susceptibility is associated with type 2 immune response, which is characterized by IL-4, IL-5, IL-10, and TGF-β cytokines (Sanches et al. 2014; Rodrigues et al. 2016; Rodriguez-Cortes et al. 2017; Rossi et al. 2016; Solano-Gallego et al. 2016; Solcà et al. 2016; Tonin et al. 2016; de Martini et al. 2018). The clinical indications of these vulnerable dogs progress in a predictable pattern, with severity and diversity rising as the disease advances and the majority of clinicopathological alterations being apparent after a year of infection (Foglia Manzillo et al. 2013). The cellular immune response to *L. infantum* infection is deactivated by type 2 immune response, which offers an anti-inflammatory cytokine microenvironment (Rodriguez-Cortes et al. 2017). In addition, a strong anti-*Leishmania* humoral response results in significant amounts of nonimmunoprotective antibodies (Barbiéri 2006; Gradoni 2015), highlighting the polyclonal B-cell response, which is indicative of CVL vulnerability (Koutinas and Koutinas 2014). It is still unclear which IgG subclass is linked to CVL resistance or susceptibility (Lima et al. 2017; Chaabouni et al. 2018). Moreover, increased activation of humoral immunity may result in the formation of autoantibodies (Koutinas and Koutinas 2014), such as anti-actin and anti-tubulin (Pateraki et al. 1983), antinuclear (Smith et al. 2004; Ginel et al. 2008), and anti-transferrin (Chaabouni et al. 2018).

Although the measures of cellular and humoral immunity help in understanding the development of CVL and the mechanisms underlying resistance or vulnerability, integrated investigations of several biomarkers are required for a deeper comprehension of the disease (Solcà et al. 2016). Hematological and biochemical markers in asymptomatic dogs often remain unaltered, whereas changes may take place in symptomatic dogs (Maia and Campino 2018). Dogs with symptoms had significantly lower levels of platelets, lymphocytes, eosinophils, and red blood cells (Lopes et al. 2018). The general health state of CVL can be evaluated using biochemical parameters. Hyperproteinemia, hypoalbuminemia, and variations in the amounts of aspartate aminotransferase, alanine aminotransferase, alkaline phosphatase, urea, and creatinine are all characteristics of ongoing CVL (Heidarpour et al. 2012;

Ribeiro et al. 2018). Since kidney impairment associated with the disease is practically inevitable, these parameters are particularly intriguing indications for treatment monitoring (Ribeiro et al. 2018).

15.3 Preclinical Models Employed for the Evaluation of Therapeutic Vaccines

15.3.1 Rodents

Syrian hamsters and mice models, particularly BALB/c mice, are frequently employed for experimental studies for VL, but each animal model has its limitations (Garg and Dube 2006). Susceptible mouse strains initially permit parasite multiplication in the liver, and the disease resolves in the later phases of infection (Kumar and Nylén 2012); as such, they offer a more appropriate model of acute infection than the developing chronic VL. Although they do not represent the ideal clinicopathological symptoms, the parasites cause splenomegaly, with the degeneration of the lymphoid follicles, which proceeds for a longer duration in the spleen and is absent in the liver (Engwerda et al. 2004). Similar to the human immunological response, exemplified by a mixed Th1/Th2 type, the spleen of mice demonstrated enhanced levels of TNF-α as well as IL-10. Production of IL-10 initiated by TNF-α enhanced the progression of the disease by making the macrophages unresponsive to activation signals and also prevented the priming of T-cell responses by acting on dendritic cells (Nylén and Gautam 2010). Therefore, BALB/c mice are used as appropriate models of self-healing or subclinical VL infection (Garg and Dube 2006).

The Syrian golden hamster is yet another rodent model that closely reflects the clinical and pathological features of active canine and human VL, which eventually progresses to a fatal disease, leading to the host's death (Garg and Dube 2006). In hamsters infected with *Leishmania*, there is decreased expression of inducible nitric oxide synthase and increased levels of Th1 cytokines and significant amounts of IL-10 and TGF-β mRNA transcripts. This promotes the parasite's survival and growth, which is similar to the condition in humans (Melby et al. 2001). But distinct to humans, hamsters exhibit the onset of acute ascites and glomerulonephritis (Sartori et al. 1992). Even though the hamster is a very relevant model, it is not widely used as no commercial reagents are available for immunological and molecular studies (Garg and Dube 2006; Kumar and Nylén 2012).

15.3.2 Dogs

Dogs serve as an important natural reservoir for the visceralizing form of *Leishmania*, making them the ideal model for research on zoonotic VL as it is a natural host of *L. infantum*. (Alvar et al. 2004; Garg and Dube 2006). Canine VL is a multisystemic disease associated with poor body condition scores, lymph adenomegaly, and characteristics of granulomatous inflammatory reactions (Alvar et al. 2004). Infected dogs exhibit a balance between Th1 and Th2 cytokines, with enhanced synthesis of IL-10 and INF-γ (Lage et al. 2007). Nevertheless, there are detectable levels of IL-4 mRNA transcripts associated with both clinical symptoms and the parasite burden (Lage et al. 2007; Quinnell et al. 2001). Moreover, the CD4+ and CD8+ populations decrease during infection but are restored following drug therapy (Bourdoiseau et al. 1997). On the other hand, compared with symptomatic and uninfected dogs, there is substantial production of IL-2 and TNF-α by the asymptomatic dogs' PBMCs (Baneth et al. 2008). Despite being the relevant model for canine VL, dogs are only employed in a small amount of research since they are more expensive and raise ethical concerns (Kumar and Nylén 2012).

15.3.3 Nonhuman Primates or Monkeys

Primates are less fashionable in vaccine development because they are expensive laboratory animals that are difficult to obtain and handle and immunological black boxes. For VL, efforts were made to establish infection in New and Old World monkeys, wherein *Aotus trivirgatus* (owl monkeys) and *Saimiri sciureus* (squirrel monkeys) developed a fulminating but short-lived infection. The Old World monkeys such as *Macaca spp.*, viz. *Macaca mulatta*, *Macaca fascicularis*, and *Macaca nemestrina*, and African vervet monkeys developed low and/or inconsistent infections (Dube et al. 2004), whereas *Presbytis entellus* (Indian langurs) was highly susceptible to intravenous inoculation of *L. donovani* amastigotes and invariably produced the consistent and progressive acute fatal infection, leading to death in less than 150 days post-infection. The infected animals presented all the clinicoimmunopathological features as observed in human kala-azar (Anuradha, et al. 1990; Dube et al. 1999). Although the Indian langurs were employed for preclinical evaluation of potential vaccine (Dube et al. 1998; Misra et al. 2001; Srivastava et al. 2003), their further use in research has been restrained due to ethical issues.

In summary, the hamster is the most appropriate model to study progressive VL for both when compared to the mouse as the latter exhibits different clinicoimmunopathological characteristics, compared with humans. Nevertheless, due to the lack of immunological markers, the usage of hamsters is a constraint for the evaluation of vaccine candidates; however, some recent research has suggested that other techniques, such as real-time polymerase chain reaction, may be employed to examine the immune status after vaccination (Melby et al. 2001; Basu et al. 2005; Samant et al. 2009). Among the higher models, the dog serves as a pertinent model for research on zoonotic VL (Garg and Dube 2006). The dynamics of the disease will be better understood as a result of ongoing research to create methods that could imitate natural transmission by using either natural reservoir hosts or lower infectious doses of parasites or bioactive saliva. This will eventually help produce more informative data on the immune response that mimics human disease, greatly improving the experimental models used to evaluate potential vaccine candidates.

15.4 Probability of a Vaccine against VL

Chemotherapy has been the lone treatment option for VL (WHO 2010). Since 2005, the WHO has started eradication efforts in areas where VL is endemic through the detection of early cases and full treatment to lower the number of infected people. Through the use of antileishmanial drugs, this campaign has successfully saved many lives. Despite the fact that these chemotherapeutics work, they have several drawbacks, including limited availability, high treatment costs (drugs and hospital stays), low efficacy, side effects including toxicity, and the need for extended regimens with an intrusive mode of administration (parenteral) (Savoia 2015; Hendrickx et al. 2015). Furthermore, it is challenging to control this disease with chemotherapy alone due to the emergence of drug resistance in parasites (Sundar et al. 2014).

The aim of the current VL control program is also defeated by the presence of asymptomatic individuals, especially in hyperendemic regions of the world. To maintain the effectiveness of the existing kala-azar elimination effort, there is a pressing need for alternative therapeutic methods that are accessible (Engwerda and Matlashewski 2015). In order to completely eradicate this dreadful disease, researchers have been striving for the past few decades to develop or produce a vaccine against VL as an economically viable therapeutic strategy (Engwerda and Matlashewski 2015). It was clear that individuals with recovered VL acquire robust immune systems, making them resistant to reinfections, suggesting that VL can be prevented through either preventive or therapeutic vaccination (Chappuis et al. 2007). Identifying potential vaccine candidates through an accurate understanding of Leishmania disease's immunobiology and pathophysiology is crucial for realizing the vaccine development program (Kumar and Engwerda 2014). Choosing an appropriate adjuvant, immunomodulator, or delivery method can enhance immune responses and produce long-lasting immunity (Mohan et al. 2013). Therefore, there is a pressing need to improve currently available treatment and encourage funding to develop novel treatment approaches like immunotherapy or vaccines, which can immediately stimulate powerful and protective immunity against the parasite (Musa et al. 2010).

Prophylactic vaccines were considered an evident priority, and for which a number of molecules were identified and assessed as vaccine targets, as reviewed by Joshi et al. (2014), Gillespie et al. (2016), Moafi et al. (2019), and Velez and Gállego (2020). Only a small number of them advanced to the point of clinical trials and were commercialized (Table 15.1). Whereas one vaccine, CaniLeish®, was developed in Europe to combat canine VL, some others, viz. LeishTec® and Leishmune®, were developed and registered in Brazil. While Leishmune® and CaniLeish® afforded notable protection in naturally infected dogs, LeishTec® provided just 40% protection (Gradoni 2015). In addition to these trials, a separate vaccine candidate known as Leish-F1, a polyprotein produced in bacteria, has progressed to Phase I and Phase II clinical studies (Skeiky et al. 2002). Furthermore, it seemed to have reduced the time to cure when used along with chemotherapy (Nascimento et al. 2010; Llanos-Cuentas et al. 2010). In light of the promising

Table 15.1 Components of Leishmania vaccines under clinical/field trial

S. No.	Vaccines	Component(s)
1	ALM + BCG	Alum-precipitated autoclaved *Leishmania major* (alum/ALM) vaccine + *Bacillus Calmette-Guérin* (BCG)
2	Leish-F1 or Leish-111f or MML	*L. major* thiol-specific antigen (TSA) + *L. major* stress-inducible protein 1 (LmSTI1) + *L. braziliensis* homolog of eukaryotic ribosomal elongation and translation initiation factor 4A (LeIF)
3	Leish-F2 + MPL-SE or Leish 110F	TSA, LmSTI1, LeIF + monophosphoryl lipid in squalene emulsion (MPL-SE); modification in Leish-111F six-His sequence near the amino terminus was removed to eliminate a potential regulatory concern, and an apparent proteolytic hot spot was eliminated by replacing Lys274 with Gln to potentially improve the manufacture of the fusion protein. The new 110-kDa construct was named Leish-110f.
4	Leish F3 + MPL-SE/GLA-SE	NH36 (*L. donovani* nucleoside hydrolase 36 kDa) + SMT (*L. infantum* sterol 24-c-methyltransferase) + MPL-SE/glucopyranosyl lipid A-stable oil-in-water nano emulsion, a Toll-like receptor 4 Th1 (T helper 1) promoting nano emulsion adjuvant
5	Leish F3+	NH36 + SMT + *L. mexicana* truncated cysteine protease B (delta CPB)
6	**Leish-Tec®**	Recombinant *L. donovani* A2 protein
7	**Leishmune®**	FML (*L. donovani* fucose-mannose ligand + Quill A) *Leishmune vaccine is the first licensed vaccine against CVL.*
8	**CaniLeish®**	Purified excreted–secreted proteins of *L. infantum* associated with adjuvant–purified fraction of saponin (QA-21)
9	LaSap	Total antigens of *L. amazonensis* plus saponin
10	LaMPL	Antigens of *L. amazonensis* associated with MPL adjuvant
11	LBMPL	Antigens of *L. braziliensis* associated with MPL adjuvant
12	ChAd-63-KH (Leish-2a)	Simian adenovirus backbone (ChAd63) + *L. donovani* Kinetoplastid membrane protein-11 (KMP-11), *L. donovani* hydrophilic acylated surface protein B (HASPB)
13	Chimera E-T	A chimera of *L. donovani* enolase and *L. donovani* triose phosphate isomerase (TPI)
14	LEISHDNAVAX	*L. donovani* KMP11 + *L. major* TSA + *L. tarentolae* (CPA + truncated CPB) + *L. donovani* elongation factor-α p74. *The five-antigen MIDGE-TH1 DNA vaccine is referred to as LEISHDNAVAX.*

outcomes from Leish-F1, a new construct, Leish-F2, was designed, lacking a histidine tag due to regulatory considerations, and it also entered clinical trials (Phase I and Phase II) (Bertholet et al. 2009). Moreover, Leish-F3, which contains the nucleoside hydrolase from *L. donovani* and the sterol 24-c-methyltransferase from *L. infantum*, formulated with a toll-like receptor 4 (TLR4)-based adjuvant, glucopyranosyl lipid A, was found to be both immunogenic and safe (Coler et al. 2015). In addition to these trials, the latest study called Multivalent Vaccination for

Human VL (MuLeVaClin) investigated newer vaccine approaches against VL in preclinical models (Fernández et al. 2021). In this study, the efficacy of a novel vaccine consisting of the *Leishmania* membrane protein Kinetoplastid membrane protein-11 (KMP11), Leish-F3+ (a recombinant fusion protein, composed of epitopes of the parasite proteins nucleoside hydrolase, sterol-24-c-methyltransferase, and cysteine protease B), and the sandfly salivary protein LJL143, with the addition of the TLR4 agonist glucopyranosyl lipid adjuvant (GLA)- squalene emulsion (SE) as an adjuvant and the use of virosomes (VS) as a delivery system, was successfully investigated in a hamster model against *L. infantum.* There was a significantly lower splenic parasite burden in hamsters immunized with the combination of all antigens in VS + GLA-SE as compared to control animals. This state-of-the-art vaccine formulation conferring protection against *L. infantum* infection supports the advancement of the vaccine formulation into process development and manufacturing along with the conduction of toxicity studies toward future Phase I human clinical trials.

15.5 Therapeutic Vaccines: Current Status

The creation of therapeutic vaccines has received some attention, although it would have been more helpful for those with active infections to control their immune responses in the direction of a cure. Therapeutic vaccinations are believed to be vital, especially in the case of incessant chronic VL infections, as the majority of the endemic population exhibits either subclinical or asymptomatic infection, which often progresses into an active disease state. The intervention with therapeutic vaccines seems to be an important avenue for combating infections by stimulating the natural defense system of infected individuals (Sela and Hilleman 2004; Engwerda and Matlashewski 2015). It has been demonstrated that immunotherapy, a therapy focused on activating the host immune system, can replace the meager effectiveness of conventional chemotherapy in managing VL (Ratnapriya et al. 2021). Such circumstances emphasize the exploration of new therapeutic strategies for improving the immune status of the infected person with better resolution of infection because the primary site of the *Leishmania* parasite is the host, macrophages, resulting in compromised immunity and in marked T-cell immunosuppression (Gupta et al. 2013). In this situation, immunotherapy and combined therapies (immunochemotherapy) are gaining popularity as a logical strategy for the secure upkeep of currently available medications despite parasite resistance, the restoration of immunity in nonresponsive animals, and faster and more reliable therapeutic success with shorter treatment times and fewer side effects (Joshi et al. 2014; Roatt et al. 2017; Viana et al. 2018). Therefore, in these endemic populations, a need for a therapeutic vaccination is a prerequisite that can be used to successfully stimulate the patient's own immune defense system. This would be a major asset in controlling the spread of the disease (Gupta et al. 2013). These therapeutic vaccinations are reported to be quite effective against a number of chronic disorders (Bachmann and Dyer 2004) such as HIV infection (Mylvaganam et al. 2015), tuberculosis (Gröschel et al. 2014), Chagas disease (Beaumier et al. 2016), human papillomavirus infection (Kim et al. 2014), as well as cancer (Melero et al. 2014; Melief et al. 2015), demonstrating the advent of the successful therapeutic vaccination era.

Immune regulation by the use of immunomodulatory drugs aids in the control of chronic diseases that exhibit either an inadequate or an overactive immune response (Kumar and Nylén 2012). The therapy approach entails modifying the immune responses of *Leishmania*-infected people toward the protective type by using biological substances (whole or their components) along with either adjuvants or medicines (Roatt et al. 2014). These components are promising adjuvants for regulating VL because they are necessary for eliciting a very potent and sustained immune response (humoral and cellular). To date, the immunoprophylaxis and immunotherapy of VL have made extensive use of plant-based immunomodulators (such as saponin, garlic extract, and others) and nonvirulent bacteria-based adjuvants (such as *Bacillus Calmette-Guérin* [BCG], *Propionibacterium acnes*, and others). Therefore, this approach, which does not cause adverse effects and either restores or creates an efficient immune response, could be a promising substitute for traditional chemotherapeutics (Roatt et al. 2014; Okwor and Uzonna 2009). In addition, there is a critical need for immunostimulatory formulations with additives that can strongly stimulate the CMI response to already-administered, insufficiently immunogenic antigens in order to provide the best possible therapeutic effectiveness against this disease. Liposomes, nondegradable nanoparticles, viral vectors, and virus-like particles have all been demonstrated to produce cellular immunity among the different vaccine delivery vehicles currently being explored, making them crucial carrier systems.

This chapter also focuses on those therapeutic vaccines that are either under preclinical trials or under clinical trials against various forms of VL-CVL and PKDL, as summarized in Tables 15.2 and 15.3.

15.5.1 Therapeutic Vaccines Under Preclinical Studies

Therapeutic vaccinations containing the entire parasite, when used alone or in combination with immunomodulators or antileishmanial drugs, may be more successful in managing a range of VL forms. Here we described the different vaccine candidates under the categories tested for the immunotherapeutic use against VL, such as live attenuated vaccines, killed or inactivated whole pathogens or their fractions (first-generation vaccine); full single recombinant proteins, peptides, and polyproteins or fusion proteins (second-generation vaccine); and deoxyribonucleic acid (DNA) vaccines in the form of a plasmid or with viral vectors as the third-generation vaccines (Table 15.2).

Table 15.2 Summary of therapeutic vaccines under preclinical trial

Vaccine	Dose/route of administration	Host and parasite (causative organism/stage)	Drugs (dose, schedule/route of administration)	Intervention on day (D) post-infection	Day of Assessment post-vaccination	Assessment criteria	Remarks/references
1. First-generation vaccine whole/attenuated/killed parasite or parasite fractions, nondefined composition							
UR6 (live/sonicated), (1999)	10^8 parasites × 3 (monthly)/s.c.	Hamster *L. donovani*/ amastigote (AG83)	–	D30	D135	Splenic and hepatic parasite burden by LDU	Reduced parasite burden in the spleen (95/93%) and the liver (94/93%) (Mukhopadhyay et al. 1999)
	UR6 (live/sonicated), 10^9 parasites × 3 (monthly)/s.c.	BALB/c mice/ *L. donovani*/ amastigote (AG83)	–	D30	D165	Splenic and hepatic parasite burden by LDU	Reduced parasite burden in the spleen (97/95%) and the liver (82/84%) (Mukhopadhyay et al. 2000)
Attenuated *L. donovani* promastigotes irradiated at 100 Gy and 150 Gy	5×10^6 parasites × 3 (15-day intervals)/i.m.	BALB/c mice/ *L. donovani*/ promastigote (AG83)	–	D75	D120	Splenic and hepatic burden by LDU NO, ROS, as well as cytokine and antibody response proliferation and cytokine response	Reduced parasite burden in the spleen (77–80%) and the liver (84–93%) NO and ROS ↑, IFN-γ, IL-10 ↑, and IgG2/IgG1 ↑ (Datta et al. 2012), T-cell proliferation ↑, with inducible nitric oxide synthase (iNOS) ↑ and TGF-β level ↑ (Datta et al. 2015)

(continued)

Table 15.2 (continued)

Vaccine	Dose/route of administration	Host and parasite (causative organism/stage)	Drugs (dose, schedule/route of administration)	Intervention on day (D) post-infection	Day of Assessment post-vaccination	Assessment criteria	Remarks/references
Autoclaved *Leishmania major* (ALM) + *M . vaccae*	100 µg × 3 (monthly)/ i.d.	Dogs (experimental; Iran)/*L. infantum/* amastigote	Meglumine antimoniate, 100 mg/kg/d × 30/i.m.	D60	D270	Parasite burden in dab smear of the bone marrow, the spleen, and the liver, survival	Parasite number reduced and animal becomes serologically negative relapsed (Jamshidi et al. 2011)
Live attenuated *L. donovani* centrin knockout (KO) (LdCen−/−)	3 × 10^6 LdCen−/−/ (i.v.)	C57Bl/6 mice representing asymptomatic infection with low dose (10^3) of *L. donovani* wild-type (LdWT) 10 weeks' post-immunization with LdCen−/− mice challenged i.v. with 10^5 virulent *L. donovani* metacyclic promastigote (Ld1S)	–	D 21	8 weeks post-immunization and 8 weeks post-challenge	Analysis of CD4+ T cells in the spleens, inguinal, axillary, cervical, submandibular, and parotid lymph nodes; parasite load, and analysis of immune response (production of pro-inflammatory cytokines by effector T cells as well as proliferation)	Elimination of parasites from the prior experimental asymptomatic infection with low-dose *LdWT* upon immunization with the *LdCen−/−* parasites (Ismail et al. 2017)
Chondroitin sulfate A (CSA) (from attenuated parasite)	100 µg × 2 (15-day interval)/s.c.	BALB/c mice *L. donovani/* promastigote (AG83)	Sodium antimony gluconate, 25 mg/kg/d × 5/i.m.	D 15 (acute) and D120 (chronic)	D60 (acute) and D165 (chronic)	Hepatic and splenic parasite burdens by LDU/limiting dilution, cytokine response	Complete and 99% protection in acute and chronic stages, IFNγ ↑ and IL-10 ↓ (Bhaumik et al. 2009)

SLA	(25 μg/mL)-pulsed DC, 10^5 cells × 3 (weekly)/i.v.	BALB/c mice/ *L. donovani*/ amastigote	Sodium stibogluconate 50 mg/kg × 3	D30	D70	Splenic and liver parasite burden by LDU	Complete clearance (Ghosh et al. 2003)
L. braziliensis antigen + saponin LBSap vaccine	(60 μg) + (100 μg) in 2/3 series of 5 days each, with an interval of 5 days between the series	Hamster/ *L. infantum* promastigote (strain MCAN/ BR/2008/OP46)	Miltefosine 2 mg/kg × 14 consecutive uninterrupted days (half course of treatment)	D60	D100	Splenic parasite burden by quantitative polymerase chain reaction (qPCR)	Hematobiochemical condition restored serum levels of IgG-anti-*Leishmania* ↓; ↑ in the number of CD4+ lymphocytes producers of IFN-γ, and TNF-α ↑. IL-10 ↓, a splenic parasitic burden low (Carvalho et al. 2021)
Leishmune® (FML + saponin)	150 μg + 100 μg × 3 (weekly)/s.c.	BALB/c mice/ *L. donovani*/ amastigote (LD-1S)	–	D15	D45	Liver parasite load by LDU delayed-type hypersensitivity (DTH), cell proliferation, and antibody response	Reduced burden (94.7%) DTH ↑, IgG1 ↑, IgG2a ↑, IgG2b levels ↑, and cell proliferation ↑ (Santos et al. 2003)
	1.5 mg + 1.0 mg × 3 (20–30-day interval)/s.c.	Mongrel dogs (experimental; Brazil)/ *L. donovani*/ Amastigote (LD-1S)	–	D127	D365	Clinical symptoms, parasitic burden in dab smears of the skin, spleen, liver, lymph nodes, and bone marrow; DTH, antibody response, and PBMC analysis	Asymptomatic amastigotes are absent; DTH↑, IgG2 levels↑, normal levels of CD4, CD8, and CD21 (Borja-Cabrera et al. 2004)

(continued)

Table 15.2 (continued)

Vaccine	Dose/route of administration	Host and parasite (causative organism/stage)	Drugs (dose, schedule/route of administration)	Intervention on day (D) post-infection	Day of Assessment post-vaccination	Assessment criteria	Remarks/references
	1.5 mg + 1.0 mg × 3 (20–30-day interval)/s.c.	Mongrel dogs (experimental; Brazil)/*L. chagasi*/amastigote (BH46)	–	D180	D450	Clinical symptoms, parasite load in the liver by LDU, survival, DTH, antibody level, and cytokine response	↓ Clinical score, ↓ parasite load, IgG2 ↑, IgG ↓, and normal CD4+ count (Santos et al. 2007)
2. Second-generation vaccine membrane/soluble proteins (native/recombinant) and fusion proteins							
GRP78 + MPLA	10 μg + 40 μg × 2 (15-day interval)/s.c.	BALB/c mice/*L. donovani*/promastigote (Dd8)	Cisplatin (0.5 mg/kg × 5 days) × 2 (14-day interval)/i.p.	D30	D55, D70, and D85	Parasitic burden in the liver and the spleen by LDU, DTH, antibody level, and cytokine response	Reduction in the liver (84–93%) and the spleen (87–93%), burden, DTH ↑, IgG2a ↑, IgG1 ↑, IFN-γ ↑, IL-2 ↑, IL-10 ↓, and IL-4 ↓ levels (Joshi and Kaur 2014)
rLAld/rEno + BCG	25 and 12.5 μg × 3 (15-day interval) i.d.	Hamsters/*L. donovani* amastigote (Dd8)	–	D15	D60 and D90	Parasite burden, DTH, lymphocyte transformation test (LTT), Th1 and Th2 cytokines, antileishmanial antibody response	~60% ↓parasitic load, DTH and LTT responses ↑, IFN-γ and TNF-α high, IL-10 and TGF-β ↓ (Keerti et al. 2018)

Leishmania-derived rAld (aldolase)	12.5 μg × 3 (15-day interval) i.d.	Hamsters/ *L. donovani* amastigote (Dd8)	AmBisome 2.5 mg/kg/ animal after 28 days post-infection	D15	D60 and D90	Parasite burden, DTH, LTT, Th1 and Th2 cytokines, antileishmanial antibody response	~80% reduction in parasitic load, DTH and LTT responses ↑,IFN-γ and TNF-α high, IL-10 and TGF-β ↓ (Keerti et al. 2022)
rLdT-E chimeric protein + BCG	100 μg × 3 (15-day interval) + 2 × 10^5 CFU (100 μL) i.d.	Hamster/ *L. donovani* amastigote (Dd8)	–	D15	D60 and D90	Parasite burden, DTH, LTT, Th1 and Th2 cytokines, antileishmanial antibody response	70% inhibition of splenic parasitic multiplication, ↑ IFN-γ and TNF-α and ↑ IL-10, IgG2-DTH and LTT responses ↑ (Ratnapriya et al. 2021)
3. Third-generation vaccine—DNA vaccines							
LEISHDNAVAX	100 μg/200 μg/i.d.	C57BL/6Jmice/ *L. donovani*/ amastigote (HU3)	AmBisome 0.8 mg/kg × 2 (14-day interval)/i.v.	D7	D31	Parasite burden in the spleen and the liver by LDU	↓ Spleen (75%) and liver (68%) parasite burden (Seifert et al. 2015)
4. Delivery of therapeutic vaccines using carrier system/vector-derived vaccine							
Adenovirus-based vaccine (Ad5-KH) comprising a synthetic *haspb* gene linked to a *kmp11* gene via a viral 2A sequence		BALB/c mice/ *L. donovani* amastigotes (LV9)	–	D21	D31		Antibody responses were boosted, and IFNc1CD81 T-cell responses, particularly to HASPB, became apparent (Maroof et al. 2012).

(continued)

Table 15.2 (continued)

Vaccine	Dose/route of administration	Host and parasite (causative organism/stage)	Drugs (dose, schedule/route of administration)	Intervention on day (D) post-infection	Day of Assessment post-vaccination	Assessment criteria	Remarks/references
Positively charged liposome containing SLA	15 μg × 3 (14-day interval)/i.p.	BALB/c mice/ *L. donovani* amastigote (AG83)	–	D60	D100	Parasite burden cytokines	90% elimination of parasites from the liver and the spleen, IFN-γ and IL-12 ↑, IL-4 and IL-10 ↓ (Bhowmick et al. 2007)

s.c. subcutaneous, *i.m.* intramuscular, *i.d.* intradermal, *i.p.* intraperitoneal, *i.v.* intravenous, *LDU* Leishman-Donovan unit, *NO* nitric oxide, *ROS* reactive oxygen species, *iNOS* inducible nitric oxide synthase, *DC* dendritic cell, *SLA* soluble Leishmania antigen, *DTH* delayed-type hypersensitivity, *LTT* lymphocyte transformation test, *IgG* immunoglobulin G, *mg* milligram, *Kg* kilogram, *μg* microgram, *μL* microliter, *CFU* colony-forming unit, *Gy* gamma irradiated

Table 15.3 Summary of therapeutic vaccines under clinical/field trials against VL (PKDL and canine VL)

Disease	Tested against strain of *L. donovani*/ location of study	Therapeutic vaccines	Dose/route of administration	Assessment criteria (number of participants)	Outcome/reference	Remarks (phase start/ completion date)
Human PKDL	*L. donovani* Sudanese strain, Sudan	Alum/autoclaved *Leishmania major* (ALM) + *Bacillus Calmette-Guérin* (BCG) + sodium stibogluconate (SSG)	100 μg × 4 weekly)/i. m. + 20 mg/kg/d × 40/i.v.	Parasite burden, clinical symptoms, cytokine response ($N = 15$)	87% of patients cured by D60 post-treatment, ↑ IFN-γ/IL-10 (Musa et al. 2008)	Phase II, completed in 2008
		Leish-F2 + MPL-SE + sodium stibogluconate (SSG)	3 × 10 μg + 25 μg (14-day interval) + 20 mg/kg/day for 40 days)	Clinical course of PKDL antibody and T-cell response. (after 1 year) ($N = 42$)	Completed	Phase I (ClinicalTrial. gov) (November 2008–April 2011)
		Leish2a (ChAd63-KH), simian adenovirus)-KH	Single dose of 1×10^{10} viral particles (v.p.; adults only) or 7.5×10^{10} v.p. (adults and adolescents) i.m. in arm followed up for 120 days post-vaccination	Clinical changes in PKDL disease (2 year), presence of IFN-γ-producing T cells and serum antibodies against Leishmania peptides ($N = 100$)	Safety and immunogenicity studies completed (Younis et al. 2021)	Phase II (ClinicalTrial. gov), ongoing (April, 2020–December 31, 2022)
		Leish2b (ChAd63-KH), simian adenovirus)-KH	Single dose of 1×10^{10} viral particles (v.p.; adults only) or 7.5×10^{10} v.p. (adults and adolescents) i.m. in the arm	Clinical changes in PKDL disease (2 years), presence of IFN-γ-producing T cells and serum antibodies against	Safety and immunogenicity studies completed (Lacey et al. 2022); efficacy studies are ongoing.	Phase II (ClinicalTrial. gov) ongoing

(continued)

Table 15.3 (continued)

Disease	Tested against strain of *L. donovani*/ location of study	Therapeutic vaccines	Dose/route of administration	Assessment criteria (number of participants)	Outcome/reference	Remarks (phase start/ completion date)
			followed up for 120 days post-vaccination	Leishmania peptides ($N = 100$)		
		Leish3 (ChAd63-KH)	Single dose of vaccine in 1 mL injection into the deltoid region	For prevention of PKDL in Sudanese VL patients previously treated with SSG/paromomycin sulfate (PM), parasite load prevaccination and at PKDL onset, T cells response and transcriptome, B cells and antibody level ($N = 262$)	Yet to be started	Phase II (ClinicalTrial.gov) (October 1, 2022–July 2025), yet to be started
Canine VL (naturally infected dogs)	*L. chagasi* Brazil	Leishmune™ (fucose-mannose ligand [FML]–QuilA) alone	1.5 mg + 0.5 mg × 3 doses at 21-day interval/s.c., a booster at 12-month interval	Parasitic burden in BM by qPCR, clinical, biochemical, hematological and immunological response ($N = 34$)	Vaccinated dogs develop the mild disease with low clinical score. IL-10 and IL-4 levels were high (Almeida et al. 2021).	Vaccine was not effective.

		Leishmune® + allopurinol or amphotericin B	1.5 mg + 1.0 mg × 3 (20–30-day interval)/s.c. ($n = 31$) + 10 mg/kg/12 h for 15 months)/i.v. ($n = 24$) or 0.5 mg/kg × 16 (3-day interval)/ i.v. ($n = 11$)	Assessment on D90 and D240, clinical symptoms, parasitic burden in dab smears of aspirates of popliteal lymph nodes, PCR for Leishmania DNA, DTH response, IgG level ($N = 91$)	Reduced symptomatic cases with decreased CVL deaths only after immunochemotherapy, DTH ↑ (Borja-Cabrera et al. 2010)	An additional advantage of immunochemotherapy along with therapeutic vaccine in the control and cure of CVL
		Leish-F2 + SLA-SE + allopurinol	3 × 20 μg × 6 (3-week interval) + 20 mg/kg/ day for 90 days) ($N = 8$)	Assessment on D90, D180, D360 clinical course, parasite burden in bone marrow ($N = 28$)	Clinical condition improved, Long-term clearance of parasite clearance (Nascimento et al. 2019)	Same as above
		rLeish-110f + MPL-SE + glucantime (N-methyl meglumine antimoniate)	20 μg + 25 μg × 3 (21-day interval)/s. c. ($n = 6$) + 100 mg/ kg × 10 days × 2 cycles (10-day interval)/i.m.	Assessment on D180, clinical symptoms, parasitic burden in BM aspirate, skin biopsy, survival ($N = 30$)	Improvement in clinical response, reduced number of deaths, higher survival, increased antibody titers (Miret et al. 2008)	Same as above
		Leish-111f + MPL-SE + glucantime	20 μg + 25 μg × 4 (weekly intervals)/s. c. + 20 mg/kg × 30 days/i.m.	Assessment after 3 years, clinical symptoms, antibody response ($N = 59$)	92% of vaccinated dogs were cured with the treatment of a vaccine, antibody titer ↑, (Trigo et al. 2010) d	Vaccine alone was effective
		Recombinant cysteine proteinase of *L. infantum* (rLdccys1) + *Propionibacterium acnes*,	500 μg × 3 (monthly)/ s.c. ($N = 10$) + 500 μg × 3 (monthly)/s.c. ($N = 10$)	Assessment on D90, clinical symptoms, splenic parasite burden by limiting dilution, DTH, IgG,	Lower clinical scores, reduced splenic burden, IgG2 high, IFN-γ high, IL-10 low, survival up to 12 months (Ferreira et al. 2014)	Vaccine alone was effective.

(continued)

Table 15.3 (continued)

Disease	Tested against strain of *L. donovani*/ location of study	Therapeutic vaccines	Dose/route of administration	Assessment criteria (number of participants)	Outcome/reference	Remarks (phase start/ completion date)
				lymphokine, survival ($N = 30$)		
		L. braziliensis associated with MPL (LBMPL)	Three treatment series, each series composed of 10 doses with increasing concentrations of the vaccine antigen (60–300 μg antigen protein +5–25 μg of MPL (days 1–5) and 300 μg of antigen protein +25 μg of MPL (days 6–10) in 1 mL of sterile 0.9% saline in first series; 300 μg antigen protein +25 μg of MPL in 1 mL of sterile 0.9% saline in the second and third series by subcutaneous route and a 10-day interval between each series)	Assessment on D90, biochemical/ hematological, immunological, clinical, and parasitological ($N = 16$), hemoglobin, hematocrit, and platelets parameters restored	Disease signs ↓, body weight ↑, splenomegaly ↓, urea, aspartate transaminase (AST), alkaline phosphatase (ALP), and bilirubin and erythrocytes ↑, CD3+ T lymphocytes and their subpopulations (TCD4+ and TCD8+), ↓ CD21+ B lymphocytes ↑, NK cells (CD5− CD16+) and CD14+ monocytes ag-specific LTT ↑ TNF-α ↑, IL-10 ↓. (Roatt et al. 2017)	A good alternative therapeutic vaccine

		LaSap (*L. amazonensis* antigen + saponin).	5 s.c. doses at 7-day intervals/8 animals ($N = 6$)	Assessment on D90 and D180, hematology, liver and renal biochemical analyses, serology, lymphoproliferation, and parasite load by qPCR ($N = 14$)	Improvement in clinical status, IgG ↓, lymphoproliferative capacity ↑ (Viana et al. 2018)	A better alternative vaccine formulation
	L. infantum France	LiF2 + glucantime	(50 μg × 3 (weekly)/i.m. + 300 mg/kg × 20 (2-day interval) i.m. ($N = 8$)	D90 and D180, cytokine response, clinical symptoms, parasite burden in dab smear of bone marrow ($N = 24$)	Symptoms disappeared, 100% parasitological cure on combined therapy (Neogy et al. 1994)	Radical cure for CVL in rural endemic areas where post-treatment surveillance is difficult
	L. infantum Spain	Soluble *Leishmania* antigen (SLA) of L. infantum + glucantime	300 μL of SLA × 3 s.c. at 14-day intervals (7 days before Sb treatment) + 100 mg/kg × 21 s.c. ($N = 5$)	Assessment on D60 and D120, clinical symptoms, parasite burden in popliteal lymph node aspirate, cell proliferation assay, lymphoid subsets in PBMCs by flow cytometry, ($N = 10$)	Clinical improvement in some dogs with negative parasite burden, proportion of T-lymphocytes in PBMCs ↑, cell proliferation ↓ (Guarga et al. 2002)	Poor response to therapeutic vaccine
	L. infantum United States	Leish-Tec (*Leishmania* A2 protein + saponin)	100 μg +500 μg × 3 s.c. doses at 14-day intervals ($N = 250$ animals)	Assessment of safety and clinical progression in infected, asymptomatic dogs ($N = 557$)	Clinical progression ↓, mortality ↓ (Toepp et al. 2018)	A better alternative vaccine formulation

s.c. subcutaneous, *i.m.* intramuscular, *i.d.* intradermal, *i.p.* intraperitoneal, *i.v.* intravenous, *LDU* Leishman-Donovan unit, *NO* nitric oxide, *DTH* delayed-type hypersensitivity, *LTT* lymphocyte transformation test, *IgG* immunoglobulin G, *mg* milligram, *Kg* kilogram, *μg* microgram, *μL* microliter, *CFU* colony-forming unit, *Gy* gamma irradiated

15.5.2 Immunotherapy with First-Generation Vaccines

Since significant protection was developed in patients who had recovered from cutaneous leishmaniasis, leishmanization (LZ), a form of immunization employing live parasites, was successfully utilized in Western and Southwestern Asia (CL) (Mutiso et al. 2013). These parasites were then used in many prophylactic vaccination trials with a greater success rate (Joshi et al. 2014). Similarly, the immunotherapeutic potential of these live parasites was also evaluated by Mukhopadhyay et al. (Mukhopadhyay et al. 1999; Mukhopadhyay et al. 2000). Datta et al. (2012) evaluated the immunotherapeutic effects of live attenuated UR6 leishmanial parasite for the first time, wherein it prevents infected hamsters and BALB/c mice from developing the disease. It was observed that this particular parasite alone provides significant protection by reducing the parasite load in the liver and spleen of these animal models (Table 15.1). Later on, Datta et al. (2012) evaluated the therapeutic efficacy of attenuated (gamma-irradiated) *L. donovani* parasite in infected mice and found that, in the 100 Gy and 150 Gy treated infected mice, the restoration of T-cell response and higher production of nitric oxide and reactive oxygen species facilitated the clearance of the parasites. The immunotherapeutic potential of autoclaved *L. major* (ALM) antigen, one of the recognized vaccines of the first generation, was evaluated by Jamshidi et al. (2011), against experimental canine VL. They observed decreased parasite counts in dogs administered with ALM + *Mycobacterium vaccae* as an adjuvant in combination with or without meglumine antimoniate, an antileishmanial drug.

Currently, Leishmania vaccine research is focused on live attenuated or mutant vaccines, which include genetically altering a parasite's genes responsible for its virulence and/or survival. A growth-regulated gene called centrin, when removed from the amastigote stage of *L. donovani*, the resulting parasite, known as LDCen−/−, (Selvapandiyan et al. 2009), when evaluated for its prophylactic potential in mice and hamsters, was found to be both safe and protective against virulent challenge (Selvapandiyan et al. 2009). Testing this live attenuated LdCen−/− vaccination in asymptomatic individuals would be of epidemiological relevance given that asymptomatic people are thought to be reservoirs for the parasite and facilitate the continuous transmission of the disease. Mice infected with wild-type (WT) *L. donovani* were used as a model in the absence of a perfect experimental model representing an asymptomatic *L. donovani* infection. These infected mice exhibited immunological conditions similar to those of asymptomatic people, including high percentages of IFN-γ-producing cells and lower numbers of IL-10-producing cells, as well as low parasite burdens, which are signs of host defense. The central memory response after vaccination in asymptomatic and naive mice was found to be remarkably similar, demonstrating that LdCen−/− is immunogenic in individuals with asymptomatic conditions as well. Moreover, these mutants induce protection against WT *L. donovani* challenge in asymptomatic mice, similar to naïve mice vaccinated with LdCen−/− (Ismail et al. (2017).

The knowledge of VL pathogens has offered potential explanations regarding the failure of vaccinations with the whole parasite (killed or live attenuated) as either they do not confer lifetime protection or their attenuation may change their immunogenicity (Joshi et al. 2014). However, when Bhaumik et al. (2009) assessed the complete soluble antigen of attenuated parasites along with a suboptimal dose of sodium antimony gluconate in *Leishmania*-infected BALB/c mice (both antimony-sensitive and antimony-resistant), this led to parasitic clearance probably due to improved T-cell response. Taking a different approach, Ghosh et al. (2003) evaluated the effectiveness of soluble Leishmania antigen (SLA) trapped in dendritic cells (DC) along with a drug (sodium stibogluconate [SSG]) to treat mouse VL. The results revealed total clearance of parasite burden from both the liver and the spleen, suggesting that the antileishmanial efficacy of SLA was increased by DC-based immunotherapy or vice versa, as they are inefficient at treating established infection when used alone. In another strategy, phospholipid vesicle, called a liposome-based delivery system, was used for SLA-based vaccines. The liposomes have been studied as adjuvants and carriers of antigens. The main characteristics of vaccine delivery methods based on liposomes are their flexibility and plasticity (Schwendener 2014. Cationic liposomes, in particular, have a potent Th1- and Th17-mediated immune stimulatory effect at the injection site and provide a long-acting depot effect. There are few investigations on therapeutic vaccinations, even though liposomized administration of preventive antigens against VL has been the subject of numerous studies (Das and Ali 2012). In one such study, using cationic liposome as a vehicle, SLA was used to treat *Leishmania*-infected BALB/c mice, which showed the generation of Th1 cytokines (IFN-γ and IL-12) as well as 90% parasite reduction from the liver and the spleen (Bhowmick et al. 2007). These investigations demonstrated that liposomes SLA-based immunotherapy greatly reduced the parasite burden due to shifting immune responses to the Th1 subtype.

Furthermore, choosing suitable vaccine candidates is challenging due to the parasite's complexity (Joshi et al. 2014). Therefore, throughout the past few years, numerous proteomic methods have been employed for the evaluation of crude, soluble, and other parasite fractions, leading to the identification of various proteins that have been investigated as potential VL vaccine candidates. Therefore, in order to reach any definitive conclusions, the prophylactic potential of these different parasite fractions was evaluated, and as a result, the therapeutic potential of a few of these fractions was also examined (Joshi et al. 2014). One such fraction-derived vaccine is a saponin formulation of fucose-mannose ligand (FML), Leishmune. It is a glycoprotein complex, expressed throughout the life cycle of *L. donovani* parasite, used in experimental rodents (mouse and hamster) and dog models. Leishmune was found to be safe, immunogenic, and protective in experimental rodent (mouse and hamster) and dog models with significant prophylactic efficacy (Palatnik-de-Sousa et al. 1994; Santos et al. 2003, 2007). The vaccine was evaluated for its immunotherapeutic potential in mice, and it led to a strong lymphoproliferative response and a decrease in liver parasite burden (Santos et al. 2003). Later, it was successfully evaluated therapeutically against experimentally (*L. donovani*) infected dogs (Borja-Cabrera et al. 2004). Santos et al. (2007) also studied the therapeutic efficacy of Leishmune with higher saponin concentrations in infected, seropositive, and symptomatic dogs with experimental canine VL where they found a decline in clinical outcomes and a shift in the responsiveness toward Th1 type. In a different study, nucleoside hydrolase NH36, a 36-kDa glycoprotein identified in the FML complex, demonstrated to be protective given either as the native recombinant protein or as a bivalent DNA vaccine, when combined with garlic extract exhibited immunotherapeutic efficacy in BALB/c mice infected with *L. chagasi*, with a significant decrease in liver parasitic load and 100% survival (Gamboa-León et al. 2006).

A recent study by Carvalho et al. (2021) demonstrated the immunochemotherapy of a heterologous *L. braziliensis* complete crude extract-based antigens (LB) combined with saponin (LBSap) along with the half-course of miltefosine (Milt+LBSap), in hamsters for VL treatment. It was observed that the status of the hematobiochemical parameters has improved, and the serum levels of immunoglobulin G (IgG)-anti-*Leishmania* have decreased. Beyond that, a decrease in the splenic parasite burden was noted, as well as an increase in the number of CD4+ lymphocytes that produce IFN-γ and a decrease in the production of IL-10. The findings point to the immunochemotherapy protocol as a potentially effective option for treating VL since it can elicit an expressive cellular response that is sufficient to reduce spleen parasitism.

15.5.3 Immunotherapy with Second-Generation Vaccine

With the evolution of molecular biology, the capability of scaling up the procedure to industrial production of second-generation vaccines that consist of specific antigens and adjuvants, the synthesis of a single antigen or a polyprotein is simply attainable. It has the advantage of easy standardization of the vaccine production as well as rational designing of the vaccine. In a study by Nagill and Kaur (2010), immunization with a glucose-regulated protein GRP78 antigen (a 78-kDa Ca2+-binding chaperone molecule or binding immunoglobulin protein [BiP]), a defined native antigen of *Leishmania* in combination with various adjuvants, has offered optimum protection with monophosphoryl lipid A (MPLA). Encouraged by these findings, the researchers used the recombinant GRP78 with MPLA for assessing the therapeutic potential in mice infected with *L. donovani* along with cisplatin, a platinum-based anticancer medication with antileishmanial action (Kaur et al. 2010). The therapy significantly inhibited the parasite, increased the delayed-type hypersensitivity response, and boosted the production of IFN-γ (Joshi and Kaur 2014).

The possibility of creating a vaccine against VL is made possible by the fact that people with *Leishmania* infection have lifelong immunity. In addition, during the past few years, the soluble and other parasite fractions have been assessed using a variety of proteomic methods, leading to the identification of a number of proteins that have been examined as potential VL vaccine candidates (Coler and Reed 2005). Previous studies (Kumari et al. 2008; Gupta et al. 2007, 2012, 2014; Kushawaha et al. 2011; Kushawaha et al. 2012; Joshi et al. 2019) indicated that six recombinant proteins isolated from the soluble fraction of *L. donovani* promastigote, that is, triose phosphate isomerase (TPI), elongation factor-2 (elF-2), enolase, protein disulfide isomerase (PDI), aldolase, and p45 were responsible for T-helper 1 (Th1) type of cellular response, in VL-treated patients as well as in Syrian hamsters. All of these six proteins have exhibited considerable prophylactic potential against experimental VL.

Due to their immunostimulatory properties, the Th1 stimulatory recombinant proteins were further examined for their immunotherapeutic potential in *Leishmania*-infected hamsters. The therapeutic potency of four recombinant proteins—aldolase, enolase, p45, and triose phosphate isomerase—was examined. The results demonstrated that these proteins are moderately effective against experimental VL. Immunization with lesser doses of aldolase and enolase, that is, 25 and 12.5 mg, showed a significant decline (60%) in parasite load along with an increased cellular immune response. To enhance the therapeutic efficacy of these two proteins—rLdAld and rLdEno—were further evaluated in conjunction with either BCG or AmBisome in *L. donovani*-infected hamsters. The study showed that rLdAld + BCG-treated hamsters offered a ~ 75% inhibition of parasitic load (Keerti et al. 2018) and were strongly indicative of rLdAld being the potential therapeutic vaccine candidate. Further, when rLdAld was used with a suboptimal dose of AmBisome, it showed a substantial reduction in parasitic load in *L. donovani*-infected hamsters, and animals developed immunity against the parasite (Keerti et al. 2022).

Based on the above-stated promising results, the researchers further worked on the generation of recombinant chimeric antigens involving the two most immunodominant *L. donovani* proteins (triosephosphate isomerase and enolase). This chimera was evaluated for its immunotherapeutic efficacy associated with adjuvant BCG. The results showed a remarkable 70% inhibition of splenic parasitic burden and boosted Th1-dominant immune response against *L. donovani* in hamsters. These findings demonstrated the chimera due to its potential as a candidate for successful vaccination against visceral leishmaniasis (Ratnapriya et al. 2021).

15.5.4 Immunotherapy with Third-Generation Vaccines

Although it is easy to synthesize peptide-based vaccines, it is challenging to formulate them in a way that addresses the complicated antigenicity of infectious agents and the diverse immunogenetics of the target populations. The aforementioned criteria appear to be met in DNA and RNA vaccines as they have shown the capacity to elicit humoral and cellular immune responses (Zepp 2010). Clinical experiments using DNA vaccines have shown favorable safety profiles, good tolerability, and no signs of an autoimmune reaction. The levels of vaccine-induced immunity, however, were largely insufficient, most likely as a result of inefficient transfection and transgene expression (Kutzler and Weiner 2008). More complex candidates other than plasmid in the form of heterologous expression systems like viral vectors offer in situ antigen expression as well as intrinsic adjuvant activity. In a study, Maroof et al. (2012) developed a vaccine (Ad5-KH) based on an adenovirus and evaluated the vaccine's therapeutic effectiveness in BALB/c mice infected with *L. donovani*.

Adenovirus provides a common viral-vector platform for several vaccines against HIV infection, tetanus, influenza, and Alzheimer's disease (Mohan et al. 2013). Results indicated that a single immunization boosted humoral and cellular (CD8 + T-cell) responses while slightly inhibiting splenic parasite burden. These findings demonstrated the usefulness of the adenoviral vector, which can more effectively guard against leishmaniasis by delivering leishmanial antigens into *Leishmania*-infected animals (Maroof et al. 2012).

In addition to these approaches, the immunotherapeutic effects of a new T-cell epitope-based DNA vaccine (LEISHDNAVAX composed of five antigens, that is, KMP11, TSA, CPA, truncated CPB, and P74) and AmBisome in experimental VL were examined by Das et al. (2014). This DNA vaccine in clinical trials has been well tolerated and has demonstrated good safety. They discovered that there was a considerable decrease in the load of splenic parasites, but there was no change in the number of CD3+ cells (Seifert et al. 2015).

15.6 Therapeutic Vaccines Under Clinical/Field trials

15.6.1 PKDL

The advancement of VL and progression to PKDL are driven by lack of efficient or correctly focused cell-mediated immunity, especially CD8+ T-cell responses, which can also restrict the therapeutic efficiency of anti-leishmanial drugs. In light of this, in addition to the need for prophylactic vaccines against Leishmaniasis, the development of therapeutic vaccines for use either alone or in conjunction with immunochemotherapy has been identified as an unmet clinical need. Treatment options for PKDL patients using only antileishmanial drugs are still challenging. For instance, 20 days of intravenous therapy is needed for East African PKDL, and recurrent VL is frequently observed in patients with HIV and immunodeficiency. Therefore, for PKDL patients, a successful therapeutic immunization administered alone would have a very large clinical benefit by minimizing the need for prolonged hospitalization and chemotherapy. Immuno-chemotherapy, which combines vaccinations and drugs, may significantly increase the life of new drugs while limiting the development of drug resistance with lower doses and shorter regimens.

A formulation of autoclaved *L. major* precipitated with alum (alum–ALM), which has shown to be immunogenic and induced Th1 response in an animal model, was tested along with the drug against PKDL. Alum–ALM coupled with BCG was safe to use and produced a potent immune response in healthy volunteers and patients with PKDL (Khalil et al. 2006) (Table 15.3). The therapeutic effectiveness of the ALM vaccination in combination with the antileishmanial drug SSG was next examined in PKDL patients. The patients who received alum-ALM + BCG with SSG once a week experienced considerable healing, most likely as a result of increased IFN-γ production, which shifted the patients' immune systems to the Th1 type (Musa et al. 2008). Combined therapy was significantly more effective than chemotherapy alone (Musa et al. 2008; Modabber 2010). Therefore, the enhanced effectiveness of combined therapy, which has been demonstrated in both experimental and human studies, may be attributable to the drug-induced decrease in parasite load, which was aided by the stimulation of sensitized T-lymphocytes by particular mitogenic antigens. It eventually led to the production of lymphokines, which in turn further prompted the activation of macrophages for the purpose of removing any remaining parasites in the host (Neogy et al. 1994).

Very recently, a new third-generation therapeutic vaccine, ChAd63-KH for VL/CL/PKDL, has been developed, which has been demonstrated to be safe and immunogenic in Phase 1 clinical trials in the UK. A Phase 2a (Leish 2a) (open-label three-phase) clinical trial was conducted for the safety and immunogenicity of the ChAd63-KH vaccine in Sudanese patients involving adult and adolescent patients with persistent PKDL (ClinicalTrials.gov: NCT02894008; Younis et al. 2021). Patients received a single intramuscular injection of 1×10^{10} viral particles (v.p.; adults only) or 7.5×10^{10} v.p. (adults and adolescents), with 42–120 days of follow-up used to evaluate the primary (safety) and secondary (clinical response and immunogenicity) endpoints. Few unfavorable side effects and potent innate and cell-mediated immune responses were observed in PKDL patients following vaccination with ChAd63-KH, as shown by whole-blood transcriptomics and ELISpot. Five of 23 patients (21.7%) demonstrated only partial clinical improvement, while seven of 23 patients (30.4%) who were followed up to the end of the study showed 90% clinical improvement. To evaluate the therapeutic effectiveness and safety of ChAd63-KH's treatment in Sudanese patients with persistent PKDL, a randomized, double-blind placebo-controlled Phase 2b trial is underway (ClinicalTrials.gov: NCT03969134; Lacey et al. 2022).

15.6.2 Canine VL

The development of vaccines with a highly protective power to disrupt the parasite transmission cycle has been a primary objective in the management of canine visceral leishmaniasis (CVL). However, in addition to promising vaccination research, scientists have worked to develop new drugs that can get rid of parasites in both humans and dogs. There is little proof that the current control methods for canine leishmaniasis (CanL) are effective. This is critical because dogs are the main source of *L. infantum* infection and a major source of parasite transmission to humans. Because of the concern about the spread of parasites that are resistant to treatment, drugs designed primarily for human use are not allowed to be used to treat CanL. Although the antileishmanial drugs effectively decrease the parasite burden in CanL, the treatment needs to be maintained for life.

Two canine leishmaniosis vaccines that have been licensed for use in Brazil since 2004 (Leishmune, whose manufacture and marketing license was revoked in 2014, and Leish-Tec) have been shown to possess therapeutic efficacy (Table 15.3). FML, the first commercial Leishmune veterinary vaccine, was found to be safe, protective, and immunogenic in experimental rodent (mouse and hamster) models and against canine VL with significant prophylactic efficacy, as reviewed by Palatnik-de-Sousa

(2008). This formulation was licensed after a series of canine VL field studies (Palatnik-de-Sousa 2012). It was successfully evaluated therapeutically and also against naturally (*L. chagasi*) infected dogs way back in 2004 (Borja-Cabrera et al. 2004). When taken in combination with allopurinol or amphotericin B against naturally infected dogs, Leishmune supplemented with saponin was able to control the disease and make most of the dogs asymptomatic. This suggested that administering chemotherapy along with vaccines helps eradicate the disease (Borja-Cabrera et al. 2010).

Leish-Tec (Hertape Calier Saude Animal, later Ceva, Brazil) is formulated with a recombinant protein A2 from *L. donovani* amastigotes and saponin as a vaccine adjuvant. It was licensed in Brazil in 2007, and it is currently the only CanL vaccine being used in that country. In the latest randomized, double-blinded field trial, the safety of Leish-Tec was assessed in infected and asymptomatic hunting dogs from the United States that were *L infantum* seropositive and were monitored for 9 months. It was observed that vaccination with Leish-Tec has no adverse reactions; therefore, the vaccine is likely to be safe (Toepp et al. 2018).

Apart from Leishmune and Leish-Tec vaccines, various single-component/poly-protein vaccines were developed as prophylactic vaccines. The potential of *L. infantum*-derived fraction 2 (LiF2) antigen was tested in naturally infected dogs, either alone or in combination with chemotherapy. In dogs receiving combination therapy, complete parasitological recovery was seen, although the absence of clinical consequences was unrelated to it (Neogy et al. 1994). Further, it was assumed that a polyprotein vaccine would elicit an enhanced protective response as it overcomes the genetic variability of the mammalian immune system (Goto et al. 2011). Moreover, it would have lower manufacturing costs and more straightforward quality control testing (Skeiky et al. 2002). These polyproteins were further evaluated for their therapeutic potential against natural infection of *L. infantum* in dogs. Leish-F1, also known as Leish-111f or MML, is a fusion of three proteins, viz. TSA, *L. major* stress-inducible protein 1 (LmSTI1), and LeIF proteins, was the first subunit vaccine to enter the clinical trial. Leish-111f, in formulation with monophosphoryl lipid A plus squalene emulsion (MPL-SE) (Leish-111f + MPL-SE), has been reported to fail to prevent naturally infected canine VL cases (Coler et al. 2007). However, its immunotherapeutic potential in combination with N-methyl meglumine antimoniate (glucantime) studied in natural *L. chagasi*-infected dogs revealed improvement in the clinical picture in dogs, both in the chemotherapy alone and immunochemotherapy cohorts, with reduced death (Miret et al. 2008). In another study, using two separate trials (open and blinded), the therapeutic efficacy of Leish-111f + MPL-SE with or without glucantime was evaluated in naturally infected canine VL cases, wherein there was 75% efficacy in canine VL cases receiving a combination of Leish-111f + MPL-SE + glucantime, when monitored up to 36 months post-therapy (open trial), and this efficacy was subsequently confirmed in a blinded trial. The results of these trials suggested that well-characterized polyprotein vaccines, as adjuncts, can proficiently boost the current chemotherapy efficiency against canine VL (Trigo et al. 2010). Due to potential regulatory concerns from the Leish-111f protein, His-tag was removed, and an apparent proteolytic hot spot was replaced with Lys274 with Gln to potentially improve the manufacture of the fusion protein and was renamed as Leish-110f or Leish-F2. Nascimento et al. (2020) tested the therapeutic efficacy of Leish-F2 in conjunction with drug therapy to improve the clinical and parasitological outcome of CanL. *L. infantum*-infected dogs were treated with allopurinol and Leish-F2 antigen vaccination combined with the administration of a stable emulsion of a second-generation lipid adjuvant (SLA) that is a TLR4 agonist (SE; SLA-SE). Concomitant immunization with Leish-F2 + SLA-SE provided long-term clearance of *L. infantum* from systemic organs and lymphoid tissues and improved the clinical status of dogs.

Likewise, Ferreira et al. (2014) explored the therapeutic potential of a recombinant cysteine proteinase from *L. infantum chagasi* (rLdccys1), in conjunction with *P. acnes* as an adjuvant, which was earlier found to be protective in both canine and murine models. Dogs that underwent the treatment showed reduced parasitic load in the spleen without an increase in clinical symptoms, accompanied by enhanced levels of IFN-γ and immunoglobulin G2 (IgG2) antibodies (Ferreira et al. 2014).

In another study, Roatt et al. (2017) evaluated the effect of using a heterologous therapeutic vaccine—LBMPL—composed of *L. braziliensis* antigens associated with MPLA adjuvant in symptomatic dogs naturally infected by *L. infantum*. The results showed that the immunotherapy with LBMPL vaccine promoted the normalization of hematological and biochemical parameters, the restoration, activation, and polarization of immune response with a protective profile, clinical improvement, and reduction in parasitic burden with the potential to block transmission to sandflies.

Similarly, Viana et al. (2018) investigated the therapeutic efficacy of killed *L. amazonensis* plus saponin (LaSap) in naturally infected dogs with *L. infantum*. The results were encouraging because it improved the clinical status, preserved hematological and biochemical parameters, lowered serum IgG levels, promoted lymphoproliferative activity against soluble *L infantum* antigens, and significantly reduced skin parasite load.

15.7 Future Perspectives and Conclusion

Given the complexity and heterogeneity of VL and the way in which its pathophysiology is closely related to multiple innate and adaptive immunological systems, it is necessary to develop a cohesive approach that targets a wide range of mediators and pathways. This calls for a thorough understanding of the immunological processes engaged in disease control, which can ultimately help generate potent therapeutic vaccines and contribute to the eradication of VL (Khamesipour 2014).

An effective therapeutic vaccine against leishmaniasis could swiftly result in broad-ranging direct and indirect health benefits. None of the therapeutic vaccination regimens that have been tested with promising outcomes in experimental human VL have been translated clinically. As shown in Fig. 15.1, the following issues must be addressed for therapeutic vaccines against human VL to be effective:

- In order to find innovative immunotherapeutic vaccine targets against particularly affected regions of the host in an established infection, it is necessary to have a

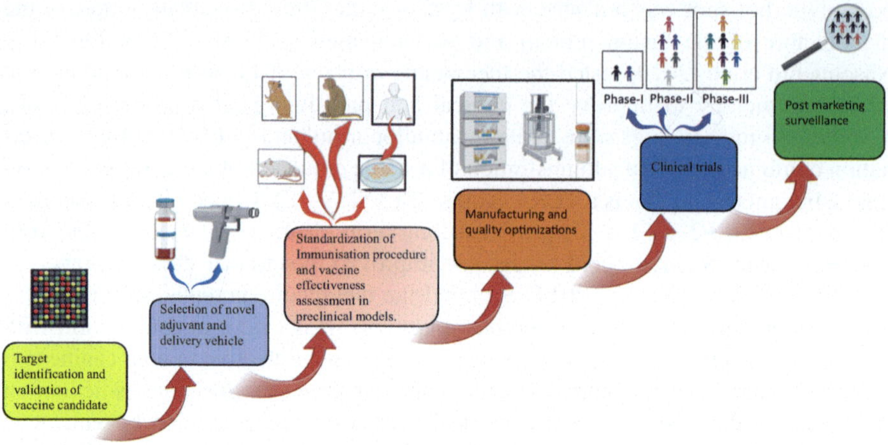

Fig. 15.1 Schematic representation of therapeutic vaccine development

basic grasp of the biology of VL disease as well as the underlying fundamental processes of immune suppression. Identifying and developing possible proteins or epitopes as vaccine targets would require the use of genomics, proteomics, and computer modeling. The ideal immunogens would be those that may enhance a Th1-type cellular immune response (Joshi et al. 2014; Seyed et al. 2016). Moreover, a therapeutic vaccine may need a significant amount of effector CD8 T cells that are trained to kill *Leishmania* parasites already present in the infected host. Optimization of the immunization schedule would further increase the efficacy of the therapeutic vaccine.

- Adjuvants and particular formulations or vector system(s) with enhanced immunogenicity must be created in order to produce potent immune responses for low immunogenic molecular vaccines.
- A precise experimental model is necessary to minimize the variability in experimental results, such as conventional cytokine quantification and infection load, which can be extrapolated to human scenarios (patients).
- It is necessary to create a standard operating method that takes into account pharmacokinetic factors, such as dose and therapy duration, in order to assess and compare the immunotherapeutic advantages of all cutting-edge approaches. Ultimately, this would reduce the vast range of readouts from different immunological assays.
- It is necessary to identify immunological biomarkers to evaluate the effectiveness of immunizations. There are several difficulties that need to be resolved about how therapeutic vaccinations work, as well as concerns regarding their safety and efficacy.
- Last but not the least, more research is needed on the use of therapeutic vaccines to treat asymptomatic PKDL and cases of HIV–VL coinfection, because they serve as silent reservoirs in endemic areas and would, thus, jeopardize the efficacy

of current VL eradication programs. For the complete eradication of VL, lessons must be gained from the experiences of smallpox and polio eradication.

Today, there are more opportunities than ever for therapeutic vaccinations against infectious diseases. Ten therapeutic vaccines for humans are currently in development, mostly against cancer (Guo et al. 2013) (brain, cervix, skin, breast, lung, head and neck, and pancreatic cancers), celiac disease (Bakshi et al. 2012), and recurrent vulvovaginal candidiasis (De Bernardis et al. 2015). In addition, some persistent viral infections, particularly those caused by the hepatitis B virus, human papillomavirus, HIV, and others, theoretically may be a good candidate for active, targeted immunotherapy. Future chronic infections can be effectively managed with the help of these immunotherapeutic measures (Moingeon et al. 2003). Therefore, the development of an effective therapeutic vaccine(s) against VL will ultimately result from ongoing, coordinated efforts in academic research.

References

Abongomera C, Battaglioli T, Adera C, Ritmeijer K (2019) Severe post-kala-azar dermal leishmaniasis successfully treated with miltefosine in an Ethiopian HIV patient. Int J Infect Dis 81:221–224

Almeida GG, Coura FM, de Melo Barbieri J, Moura A, de Oliveira Paes-Leme F, da Costa-Val AP (2021) FML/QuilA-vaccinated dogs naturally infected with *Leishmania infantum*: serum cytokines, clinicopathological profile, and parasitological parameters. Biomed Res Int 2021: 3192960. https://doi.org/10.1155/2021/3192960

Alvar J, Cañavate C, Molina R, Moreno J, Nieto J (2004) Canine leishmaniasis. Adv Parasitol 57:1–88. https://doi.org/10.1016/S0065-308X(04)57001-X

Alvar J, Vélez ID, Bern C, Herrero M, Desjeux P, Cano J, Jannin J, den Boer M, Leishmaniasis Control Team WHO (2012) Leishmaniasis worldwide and global estimates of its incidence. PLoS One 7(5):e35671

Anuradha, Pal R, Guru PY, Katiyar JC, Sethi N (1990) Successful transmission of *Leishmania donovani* in Indian langurs (Presbytis entellus). Curr Sci 59(12):612–615

Bachmann MF, Dyer MR (2004) Therapeutic vaccination for chronic diseases: a new class of drugs in sight. *Nature reviews*. Drug Discov 3(1):81–88

Bakshi A, Stephen S, Borum ML, Doman DB (2012) Emerging therapeutic options for celiac disease: potential alternatives to a gluten-free diet. Gastroenterol Hepatol 8(9):582–588

Baneth G, Koutinas AF, Solano-Gallego L, Bourdeau P, Ferrer L (2008) Canine leishmaniosis - new concepts and insights on an expanding zoonosis: part one. Trends Parasitol 24(7):324–330

Barbiéri CL (2006) Immunology of canine leishmaniasis. Parasite Immunol 28(7):329–337

Basu R, Bhaumik S, Basu JM, Naskar K, De T, Roy S (2005) Kinetoplastid membrane protein-11 DNA vaccination induces complete protection against both pentavalent antimonial-sensitive and -resistant strains of *Leishmania donovani* that correlates with inducible nitric oxide synthase activity and IL-4 generation: evidence for mixed Th1- and Th2-like responses in visceral leishmaniasis. J Immunol (Baltimore, Md.: 1950) 174(11):7160–7171

Beaumier CM, Gillespie PM, Strych U, Hayward T, Hotez PJ, Bottazzi ME (2016) Status of vaccine research and development of vaccines for Chagas disease. Vaccine 34(26):2996–3000

Bertholet S, Goto Y, Carter L, Bhatia A, Howard RF, Carter D, Coler RN, Vedvick TS, Reed SG (2009) Optimized subunit vaccine protects against experimental leishmaniasis. Vaccine 27(50): 7036–7045

Bhaumik SK, Naskar K, De T (2009) Complete protection against experimental visceral leishmaniasis with complete soluble antigen from attenuated *Leishmania donovani* promastigotes involves Th1-immunity and down-regulation of IL-10. Eur J Immunol 39(8):2146–2160

Bhowmick S, Ravindran R, Ali N (2007) Leishmanial antigens in liposomes promote protective immunity and provide immunotherapy against visceral leishmaniasis via polarized Th1 response. Vaccine 25(35):6544–6556

Borja-Cabrera GP, Cruz Mendes A, Paraguai de Souza E, Hashimoto Okada LY, de A Trivellato FA, Kawasaki JK, Costa AC, Reis AB, Genaro O, Batista LM, Palatnik M, Palatnik-de-Sousa CB (2004) Effective immunotherapy against canine visceral leishmaniasis with the FML-vaccine. Vaccine 22(17–18):2234–2243

Borja-Cabrera GP, Santos FN, Santos FB, Trivellato FA, Kawasaki JK, Costa AC, Castro T, Nogueira FS, Moreira MA, Luvizotto MC, Palatnik M, Palatnik-de-Sousa CB (2010) Immunotherapy with the saponin enriched-Leishmune vaccine versus immunochemotherapy in dogs with natural canine visceral leishmaniasis. Vaccine 28(3):597–603

Bourdoiseau G, Bonnefont C, Hoareau E, Boehringer C, Stolle T, Chabanne L (1997) Specific IgG1 and IgG2 antibody and lymphocyte subset levels in naturally *Leishmania infantum*-infected treated and untreated dogs. Vet Immunol Immunopathol 59(1–2):21–30

Buxbaum LU, Scott P (2005) Interleukin 10- and Fc gamma receptor-deficient mice resolve Leishmania mexicana lesions. Infect Immun 73(4):2101–2108

Caldas A, Favali C, Aquino D, Vinhas V, van Weyenbergh J, Brodskyn C, Costa J, Barral-Netto M, Barral A (2005) Balance of IL-10 and interferon-gamma plasma levels in human visceral leishmaniasis: implications in the pathogenesis. BMC Infect Dis 5:113

Carvalho LM, Ferreira FC, Gusmão MR, Costa A, de Brito R, Aguiar-Soares R, Reis AB, Cardoso J, Carneiro CM, Roatt BM (2021) Heterologous vaccine therapy associated with half course of Miltefosine promote activation of the proinflammatory response with control of splenic parasitism in a hamster model of visceral leishmaniasis. Curr Res Immunol 2:194–201

Chaabouni A, Boubaker Elandoulsi R, Mhadhbi M, Gharbi M, Sassi A (2018) Comparative analysis of the *Leishmania infantum*-specific antibody repertoires and the autoantibody repertoires between asymptomatic and symptomatic dogs. Vet Parasitol 261:9–17

Chappuis F, Sundar S, Hailu A, Ghalib H, Rijal S, Peeling RW, Alvar J, Boelaert M (2007) Visceral leishmaniasis: what are the needs for diagnosis, treatment and control? Nat Rev Microbiol 5(11): 873–882

Chatterjee M, Sengupta R, Mukhopadhyay D, Mukherjee S, Dighal A, Moulik S, Sengupta S (2020) Immune responses in post Kala-azar Dermal Leishmaniasis. Indian J Dermatol 65(6): 452–460

Coler RN, Reed SG (2005) Second-generation vaccines against leishmaniasis. Trends Parasitol 21(5):244–249

Coler RN, Goto Y, Bogatzki L, Raman V, Reed SG (2007) Leish-111f, a recombinant polyprotein vaccine that protects against visceral Leishmaniasis by elicitation of CD4+ T cells. Infect Immun 75(9):4648–4654

Coler RN, Duthie MS, Hofmeyer KA, Guderian J, Jayashankar L, Vergara J, Rolf T, Misquith A, Laurance JD, Raman VS, Bailor HR, Cauwelaert ND, Reed SJ, Vallur A, Favila M, Orr MT, Ashman J, Ghosh P, Mondal D, Reed SG (2015) From mouse to man: safety, immunogenicity and efficacy of a candidate leishmaniasis vaccine LEISH-F3+GLA-SE. Clin Transl Immunol 4(4):e35

Costa-Pereira C, Moreira ML, Soares RP, Marteleto BH, Ribeiro VM, França-Dias MH, Cardoso LM, Viana KF, Giunchetti RC, Martins-Filho OA, Araújo MS (2015) One-year timeline kinetics of cytokine-mediated cellular immunity in dogs vaccinated against visceral leishmaniasis. BMC Vet Res 11:92

Cruz I, Nieto J, Moreno J, Cañavate C, Desjeux P, Alvar J (2006) Leishmania/HIV co-infections in the second decade. Indian J Med Res 123(3):357–388

Das A, Ali N (2012) Vaccine development against *Leishmania donovani*. Front Immunol 3:99

Das S, Freier A, Boussoffara T, Das S, Oswald D, Losch FO, Selka M, Sacerdoti-Sierra N, Schönian G, Wiesmüller KH, Seifert K, Schroff M, Juhls C, Jaffe CL, Roy S, Das P, Louzir H, Croft SL, Modabber F, Walden P (2014) Modular multiantigen T cell epitope-enriched DNA vaccine against human leishmaniasis. Sci Transl Med. 6(234):234ra56

Datta S, Manna M, Khanra S, Ghosh M, Bhar R, Chakraborty A, Roy S (2012) Therapeutic immunization with radio-attenuated Leishmania parasites through i.m. route revealed protection against the experimental murine visceral leishmaniasis. Parasitol Res 111(1):361–369

Datta S, Roy S, Manna M (2015) Therapy with radio-attenuated vaccine in experimental murine visceral leishmaniasis showed enhanced T cell and inducible nitric oxide synthase levels, suppressed tumor growth factor-beta production with higher expression of some signaling molecules. Braz J Infect Dis 19(1):36–42

De Bernardis F, Arancia S, Sandini S, Graziani S, Norelli S (2015) Studies of immune responses in Candida vaginitis. Pathogens (Basel, Switzerland) 4(4):697–707

de Martini CC, de Andrade JT, de Almeida S, Oliveira Silva KL, de Rezende Eugenio F, Dos Santos P, de Lima V (2018) Cellular apoptosis and nitric oxide production in PBMC and spleen from dogs with visceral leishmaniasis. Comp Immunol Microbiol Infect Dis 57:1–7

Desjeux P, Ghosh RS, Dhalaria P, Strub-Wourgaft N, Zijlstra EE (2013) Report of the Post Kala-azar Dermal Leishmaniasis (PKDL) Consortium Meeting, New Delhi, India, 27-29 June 2012. Parasit Vectors 6:196

Diro E, Lynen L, Ritmeijer K, Boelaert M, Hailu A, van Griensven J (2014) Visceral Leishmaniasis and HIV coinfection in East Africa. PLoS Negl Trop Dis 8(6):e2869

Dube A, Sharma P, Srivastava JK, Misra A, Naik S, Katiyar JC (1998) Vaccination of langur monkeys (*Presbytis entellus*) against *Leishmania donovani* with autoclaved L. major plus BCG. Parasitology 116(3):219–221

Dube A, Srivastava JK, Sharma P, Chaturvedi A, Katiyar JC, Naik S (1999) *Leishmania donovani*: cellular and humoral immune responses in Indian langur monkeys, *Presbytis entellus*. Acta Trop 73(1):37–48

Dube A, Murthy PK, Puri SK, Misra-Bhattacharya S (2004) *Presbytis entellus*: a primate model for parasitic disease research. Trends Parasitol 20(8):358–360

Engwerda CR, Matlashewski G (2015) Development of Leishmania vaccines in the era of visceral leishmaniasis elimination. Trans R Soc Trop Med Hyg 109(7):423–424

Engwerda CR, Ato M, Kaye PM (2004) Macrophages, pathology and parasite persistence in experimental visceral leishmaniasis. Trends Parasitol 20(11):524–530

Faleiro RJ, Kumar R, Hafner LM, Engwerda CR (2014) Immune regulation during chronic visceral leishmaniasis. PLoS Negl Trop Dis 8(7):e2914

Fernández L, Solana JC, Sánchez C, Jiménez MÁ, Requena JM, Coler R, Reed SG, Valenzuela JG, Kamhawi S, Oliveira F, Fichera E, Glueck R, Bottazzi ME, Gupta G, Cecilio P, Pérez-Cabezas B, Cordeiro-da-Silva A, Gradoni L, Carrillo E, Moreno J (2021) Protective efficacy in a hamster model of a multivalent vaccine for human visceral leishmaniasis (MuLeVaClin) consisting of the KMP11, LEISH-F3+, and LJL143 antigens in virosomes, plus GLA-SE adjuvant. Microorganisms 9(11):2253

Ferreira JH, Silva L, Longo-Maugéri IM, Katz S, Barbiéri CL (2014) Use of a recombinant cysteine proteinase from *Leishmania (Leishmania) infantum chagasi* for the Immunotherapy of canine visceral Leishmaniasis. PLoS Negl Trop Dis 8(3):e2729

Fisa R, Gállego M, Castillejo S, Aisa MJ, Serra T, Riera C, Carrió J, Gállego J, Portús M (1999) Epidemiology of canine leishmaniosis in Catalonia (Spain) the example of the Priorat focus. Vet Parasitol 83(2):87–97

Foglia Manzillo V, Di Muccio T, Cappiello S, Scalone A, Paparcone R, Fiorentino E, Gizzarelli M, Gramiccia M, Gradoni L, Oliva G (2013) Prospective study on the incidence and progression of clinical signs in naïve dogs naturally infected by Leishmania infantum. PLoS Negl Trop Dis. 7(5):e2225

Gamboa-León R, Paraguai de Souza E, Borja-Cabrera GP, Santos FN, Myashiro LM, Pinheiro RO, Dumonteil E, Palatnik-de-Sousa CB (2006) Immunotherapy against visceral leishmaniasis with the nucleoside hydrolase-DNA vaccine of *Leishmania donovani*. Vaccine 24(22):4863–4873

Garg R, Dube A (2006) Animal models for vaccine studies for visceral leishmaniasis. Indian J Med Res 123(3):439–454

Gasim S, Elhassan AM, Kharazmi A, Khalil EA, Ismail A, Theander TG (2000) The development of post-kala-azar dermal leishmaniasis (PKDL) is associated with acquisition of Leishmania reactivity by peripheral blood mononuclear cells (PBMC). Clin Exp Immunol 119(3):523–529

Ghosh M, Pal C, Ray M, Maitra S, Mandal L, Bandyopadhyay S (2003) Dendritic cell-based immunotherapy combined with antimony-based chemotherapy cures established murine visceral leishmaniasis. J Immunol (Baltimore, Md.: 1950) 170(11):5625–5629

Gidwani K, Rai M, Chakravarty J, Boelaert M, Sundar S (2009) Evaluation of leishmanin skin test in Indian visceral leishmaniasis. Am J Trop Med Hyg 80(4):566–567

Gillespie PM, Beaumier CM, Strych U, Hayward T, Hotez PJ, Bottazzi ME (2016) Status of vaccine research and development of vaccines for leishmaniasis. Vaccine 34(26):2992–2995

Ginel PJ, Camacho S, Lucena R (2008) Anti-histone antibodies in dogs with leishmaniasis and glomerulonephritis. Res Vet Sci 85(3):510–514

Gonçalves A, Leite JC, Resende LA, Mariano R, Silveira P, Melo-Júnior O, Ribeiro HS, de Oliveira DS, Soares DF, Santos T, Marques AF, Galdino AS, Martins-Filho OA, Dutra WO, da Silveira-Lemos D, Giunchetti RC (2019) An overview of immunotherapeutic approaches against canine visceral leishmaniasis: what has been tested on dogs and a new perspective on improving treatment efficacy. Front Cell Infect Microbiol 9:427

Goto Y, Bhatia A, Raman VS, Liang H, Mohamath R, Picone AF, Vidal SE, Vedvick TS, Howard RF, Reed SG (2011) KSAC, the first defined polyprotein vaccine candidate for visceral leishmaniasis. Clin Vaccine Immunol 18(7):1118–1124

Gradoni L (2015) Canine Leishmania vaccines: still a long way to go. Vet Parasitol 208(1–2): 94–100

Gröschel MI, Prabowo SA, Cardona PJ, Stanford JL, van der Werf TS (2014) Therapeutic vaccines for tuberculosis--a systematic review. Vaccine 32(26):3162–3168

Guarga JL, Moreno J, Lucientes J, Gracia MJ, Peribáñez MA, Castillo JA (2002) Evaluation of a specific immunochemotherapy for the treatment of canine visceral leishmaniasis. Vet Immunol Immunopathol 88(1–2):13–20. https://doi.org/10.1016/s0165-2427(02)00128-9

Guo C, Manjili MH, Subjeck JR, Sarkar D, Fisher PB, Wang XY (2013) Therapeutic cancer vaccines: past, present, and future. Adv Cancer Res 119:421–475

Gupta SK, Sisodia BS, Sinha S, Hajela K, Naik S, Shasany AK, Dube A (2007) Proteomic approach for identification and characterization of novel immunostimulatory proteins from soluble antigens of *Leishmania donovani* promastigotes. Proteomics 7(5):816–823

Gupta R, Kushawaha PK, Tripathi CD, Sundar S, Dube A (2012) A novel recombinant *Leishmania donovani* p45, a partial coding region of methionine aminopeptidase, generates protective immunity by inducing a Th1 stimulatory response against experimental visceral leishmaniasis. Int J Parasitol 42(5):429–435

Gupta G, Oghumu S, Satoskar AR (2013) Mechanisms of immune evasion in leishmaniasis. Adv Appl Microbiol 82:155–184

Gupta R, Kumar V, Kushawaha PK, Tripathi CP, Joshi S, Sahasrabuddhe AA, Mitra K, Sundar S, Siddiqi MI, Dube A (2014) Characterization of glycolytic enzymes--rAldolase and rEnolase of *Leishmania donovani*, identified as Th1 stimulatory proteins, for their immunogenicity and immunoprophylactic efficacies against experimental visceral leishmaniasis. PLoS One 9(1): e86073

Heidarpour M, Soltani S, Mohri M, Khoshnegah J (2012) Canine visceral leishmaniasis: relationships between oxidative stress, liver and kidney variables, trace elements, and clinical status. Parasitol Res 111(4):1491–1496

Hendrickx S, Mondelaers A, Eberhardt E, Delputte P, Cos P, Maes L (2015) In vivo selection of paromomycin and miltefosine resistance in *Leishmania donovani* and *L. infantum* in a Syrian hamster Model. Antimicrob Agents Chemother 59(8):4714–4718

Ismail A, El Hassan AM, Kemp K, Gasim S, Kadaru AE, Moller T, Kharazmi A, Theander TG (1999) Immunopathology of post kala-azar dermal leishmaniasis (PKDL): T-cell phenotypes and cytokine profile. J Pathol 189(4):615–22. https://doi.org/10.1002/(SICI)1096-9896 (199912)189:4<615::AID-PATH466>3.0.CO;2-Z.

Ismail N, Kaul A, Bhattacharya P, Gannavaram S, Nakhasi HL (2017) Immunization with live attenuated *Leishmania donovani* Centrin−/− parasites is efficacious in asymptomatic infection. Front Immunol 8:1788

Jamshidi S h, Avizeh R, Mohebali M, Bokaie S (2011) Immunotherapy using autoclaved *L. major* antigens and M. vaccae with meglumine antimoniate, for the treatment of experimental canine visceral Leishmaniasis. Iran J Parasitol 6(3):26–34

Joshi J, Kaur S (2014) To investigate the therapeutic potential of immunochemotherapy with cisplatin + 78 kDa + MPL-A against *Leishmania donovani* in BALB/c mice. Parasite Immunol 36(1):3–12

Joshi S, Rawat K, Yadav NK, Kumar V, Siddiqi MI, Dube A (2014) Visceral Leishmaniasis: advancements in vaccine development via classical and molecular approaches. Front Immunol 5.380

Joshi S, Yadav NK, Rawat K, Kumar V, Ali R, Sahasrabuddhe AA, Siddiqi MI, Haq W, Sundar S, Dube A (2019) Immunogenicity and protective efficacy of T-cell epitopes derived from potential Th1 stimulatory proteins of *Leishmania (Leishmania) donovani.* Front Immunol 10: 288

Katara GK, Ansari NA, Verma S, Ramesh V, Salotra P (2011) Foxp3 and IL-10 expression correlates with parasite burden in lesional tissues of post kala azar dermal leishmaniasis (PKDL) patients. PLoS Negl Trop Dis 5(5):e1171

Kaur S, Sachdeva H, Dhuria S, Sharma M, Kaur T (2010) Antileishmanial effect of cisplatin against murine visceral leishmaniasis. Parasitol Int 59(1):62–69

Keerti, Yadav NK, Joshi S, Ratnapriya S, Sahasrabuddhe AA, Dube A (2018) Immunotherapeutic potential of Leishmania (Leishmania) donovani Th1 stimulatory proteins against experimental visceral leishmaniasis. Vaccine 36(17):2293–2299

Keerti, Yadav NK, Joshi S, Ratnapriya S, Sahasrabuddhe AA, Dube A (2022) Combined immunotherapeutic effect of Leishmania-derived recombinant aldolase and Ambisome against experimental visceral leishmaniasis. J Microbiol Immunol Infect 56(1):163–171. https://doi.org/10.1016/j.jmii.2022.06.003.

Khalil EAG, Musa AM, Modabber F, El-Hassan AM (2006) Safety and immunogenicity of a candidate vaccine for visceral leishmaniasis (Alum-precipitated autoclaved Leishmania major+ BCG) in children: an extended phase II study. Ann Trop Paediatr 26(4):357–361

Khamesipour A (2014) Therapeutic vaccines for leishmaniasis. Expert Opin Biol Ther 14(11): 1641–1649

Kim TJ, Jin HT, Hur SY, Yang HG, Seo YB, Hong SR, Lee CW, Kim S, Woo JW, Park KS, Hwang YY, Park J, Lee IH, Lim KT, Lee KH, Jeong MS, Surh CD, Suh YS, Park JS, Sung YC (2014) Clearance of persistent HPV infection and cervical lesion by therapeutic DNA vaccine in CIN3 patients. Nat Commun 5(1):1–14

Koutinas AF, Koutinas CK (2014) Pathologic mechanisms underlying the clinical findings in canine leishmaniosis due to *Leishmania infantum*/chagasi. Vet Pathol 51(2):527–538

Kumar R, Engwerda C (2014) Vaccines to prevent leishmaniasis. Clin Transl Immunol 3(3):e13

Kumar R, Nylén S (2012) Immunobiology of visceral leishmaniasis. Front Immunol 3:251

Kumari S, Samant M, Misra P, Khare P, Sisodia B, Shasany AK, Dube A (2008) Th1-stimulatory polyproteins of soluble *Leishmania donovani* promastigotes ranging from 89.9 to 97.1 kDa offers long-lasting protection against experimental visceral leishmaniasis. Vaccine 26(45): 5700–5711

Kushawaha PK, Gupta R, Sundar S, Sahasrabuddhe AA, Dube A (2011) Elongation factor-2, a Th1 stimulatory protein of *Leishmania donovani*, generates strong IFN-γ and IL-12 response in cured Leishmania-infected patients/hamsters and protects hamsters against Leishmania challenge. J Immunol 187(12):6417–6427

Kushawaha PK, Gupta R, Tripathi CDP, Khare P, Jaiswal AK, Sundar S, Dube A (2012) *Leishmania donovani* triose phosphate isomerase: a potential vaccine target against visceral leishmaniasis. PLoS One 7(9):e45766. https://doi.org/10.1371/journal.pone.0045766.

Kutzler MA, Weiner DB (2008) DNA vaccines: ready for prime time? Nat Rev Genet 9(10): 776–788

Lacey C, Musa A, Khalil ET, Younis B, Osman M, Wiggins R, Kaye P (2022) LEISH2b-A phase 2b study to assess the safety, efficacy, and immunogenicity of the Leishmania vaccine ChAd63-KH in post-kala azar dermal leishmaniasis. Wellcome Open Res 7:200

Lage RS, Oliveira GCD, Busek SU, Guerra LL, Giunchetti RC, Corrêa-Oliveira R, Reis AB (2007) Analysis of the cytokine profile in spleen cells from dogs naturally infected by Leishmania chagasi. Vet Immunol Immunopathol 115(1–2):135–145

Lima ÁL, de Lima ID, Coutinho JF, de Sousa ÚP, Rodrigues MA, Wilson ME et al (2017) Changing epidemiology of visceral leishmaniasis in northeastern Brazil: a 25-year follow-up of an urban outbreak. Trans R Soc Trop Med Hyg 111(10):440–447

Lindoso JA, Cota GF, da Cruz AM, Goto H, Maia-Elkhoury ANS, Romero GAS et al (2014) Visceral leishmaniasis and HIV coinfection in Latin America. PLoS Negl Trop Dis 8(9):e3136

Llanos-Cuentas A, Calderón W, Cruz M, Ashman JA, Alves FP, Coler RN et al (2010) A clinical trial to evaluate the safety and immunogenicity of the LEISH-F1+ MPL-SE vaccine when used in combination with sodium stibogluconate for the treatment of mucosal leishmaniasis. Vaccine 28(46):7427–7435

Lopes VV, Belo VS, Pereira DA, Coelho MB, Pena HP, Alves NR et al (2018) IgG avidity index and complete blood count as biomarkers of clinical disease in naturally infected dogs with *Leishmania infantum*. Vet Parasitol 261:96–103

Maia C, Campino L (2018) Biomarkers associated with *Leishmania infantum* exposure, infection, and disease in dogs. Front Cell Infect Microbiol 8:302

Maroof A, Brown N, Smith B, Hodgkinson MR, Maxwell A, Losch FO et al (2012) Therapeutic vaccination with recombinant adenovirus reduces splenic parasite burden in experimental visceral leishmaniasis. J Infect Dis 205(5):853–863

Martínez-Orellana P, Quirola-Amores P, Montserrat-Sangrà S, Ordeix L, Llull J, Álvarez-Fernández A, Solano-Gallego L (2017) The inflammatory cytokine effect of Pam3CSK4 TLR2 agonist alone or in combination with *Leishmania infantum* antigen on ex-vivo whole blood from sick and resistant dogs. Parasit Vectors 10(1):1–7

Melby PC, Chandrasekar B, Zhao W, Coe JE (2001) The hamster as a model of human visceral leishmaniasis: progressive disease and impaired generation of nitric oxide in the face of a prominent Th1-like cytokine response. J Immunol 166(3):1912–1920

Melero I, Gaudernack G, Gerritsen W, Huber C, Parmiani G, Scholl S, Thatcher N, Wagstaff J, Zielinski C, Faulkner I, Mellstedt H (2014) Therapeutic vaccines for cancer: an overview of clinical trials. Nat Rev Clin Oncol 11(9):509–524

Melief CJ, van Hall T, Arens R, Ossendorp F, van der Burg SH (2015) Therapeutic cancer vaccines. J Clin Invest 125(9):3401–3412

Miret J, Nascimento E, Sampaio W, França JC, Fujiwara RT, Vale A et al (2008) Evaluation of an immunochemotherapeutic protocol constituted of N-methyl meglumine antimoniate (Glucantime®) and the recombinant Leish-110f®+ MPL-SE® vaccine to treat canine visceral leishmaniasis. Vaccine 26(12):1585–1594

Misra A, Dube A, Srivastava B, Sharma P, Srivastava JK, Katiyar JC, Naik S (2001) Successful vaccination against *Leishmania donovani* infection in Indian langur using alum-precipitated autoclaved Leishmania major with BCG. Vaccine 19(25–26):3485–3492

Moafi M, Rezvan H, Sherkat R, Taleban R (2019) Leishmania vaccines entered in clinical trials: a review of literature. Int J Prev Med. 10:95

Modabber F (2010) Leishmaniasis vaccines: past, present and future. Int J Antimicrob Agents 36: S58–S61

Mohan T, Verma P, Rao DN (2013) Novel adjuvants & delivery vehicles for vaccines development: a road ahead. Indian J Med Res 138(5):779–795

Moingeon P, Almond J, de Wilde M (2003) Therapeutic vaccines against infectious diseases. Curr Opin Microbiol 6(5):462–471

Mondain-Miton V, Toussaint-Gari M, Hofman P, Marty P, Carles M, De Salvador F, Miton F, Le Fichoux Y, Dellamonica P (1995) Atypical leishmaniasis in a patient infected with human immunodeficiency virus. Clin Infect Dis 21(3):663–665

Monge-Maillo B, Norman FF, Cruz I, Alvar J, López-Vélez R (2014) Visceral leishmaniasis and HIV coinfection in the Mediterranean region. PLoS Negl Trop Dis 8(8):e3021

Moreno J, Alvar J (2002) Canine leishmaniasis: epidemiological risk and the experimental model. Trends Parasitol 18(9):399–405

Mukhopadhyay S, Sen P, Bhattacharyya S, Majumdar S, Roy S (1999) Immunoprophylaxis and immunotherapy against experimental visceral leishmaniasis. Vaccine 17(3):291–300

Mukhopadhyay S, Bhattacharyya S, Majhi R, De T, Naskar K, Majumdar S, Roy S (2000) Use of an attenuated leishmanial parasite as an immunoprophylactic and immunotherapeutic agent against murine visceral leishmaniasis. Clin Diagnostic Lab Immunol 7(2):233–240

Musa AM, Khalil EA, Mahgoub FA, Elgawi SH, Modabber F, Elkadaru AE, Aboud MH, Noazin S, Ghalib HW, El-Hassan AM, Leishmaniasis Research Group/Sudan (2008) Immunochemotherapy of persistent post-kala-azar dermal leishmaniasis: a novel approach to treatment. Trans R Soc Trop Med Hyg. 102(1):58–63

Musa AM, Noazin S, Khalil EA, Modabber F (2010) Immunological stimulation for the treatment of leishmaniasis: a modality worthy of serious consideration. Trans R Soc Trop Med Hyg. 104(1):1–2

Mutiso JM, Macharia JC, Kiio MN, Ichagichu JM, Rikoi H, Gicheru MM (2013) Development of Leishmania vaccines: predicting the future from past and present experience. J Biomed Res 27(2):85–102

Mylvaganam GH, Silvestri G, Amara RR (2015) HIV therapeutic vaccines: moving towards a functional cure. Curr Opin Immunol 35:1–8

Nagill R, Kaur S (2010) Enhanced efficacy and immunogenicity of 78kDa antigen formulated in various adjuvants against murine visceral leishmaniasis. Vaccine 28(23):4002–4012

Nascimento E, Fernandes DF, Vieira EP, Campos-Neto A, Ashman JA, Alves FP, Coler RN, Bogatzki LY, Kahn SJ, Beckmann AM, Pine SO, Cowgill KD, Reed SG, Piazza FM (2010) A clinical trial to evaluate the safety and immunogenicity of the LEISH-F1+MPL-SE vaccine when used in combination with meglumine antimoniate for the treatment of cutaneous leishmaniasis. Vaccine 28(40):6581–6587

Nascimento MS, Albuquerque TD, Nascimento AF, Caldas IS, Do-Valle-Matta MA, Souto JT, Talvani A, Bahia MT, Galvão LM, Câmara AC, Guedes PM (2015) Impairment of interleukin-17A expression in canine visceral leishmaniosis is correlated with reduced interferon-γ and inducible nitric oxide synthase expression. J Comp Pathol 153(4):197–205

Nascimento L, Miranda D, Moura LD, Pinho FA, Werneck GL, Khouri R, Reed SG, Duthie MS, Barral A, Barral-Netto M, Cruz M (2019) Allopurinol therapy provides long term clinical improvement, but additional immunotherapy is required for sustained parasite clearance, in *L. infantum*-infected dogs. Vaccine X, 4:100048

Nascimento LFM, Miranda DFH, Moura LD, Pinho FA, Werneck GL, Khouri R, Reed SG, Duthie MS, Barral A, Barral-Netto M, Cruz MSP. Allopurinol therapy provides long term clinical improvement, but additional immunotherapy is required for sustained parasite clearance, in L. infantum-infected dogs. Vaccine X. 2020 Nov 20;4:100048. https://doi.org/10.1016/j.jvacx.2019.100048.

Neogy AB, Vouldoukis I, da Costa JM, Monjour L (1994) Exploitation of parasite-derived antigen in therapeutic success against canine visceral leishmaniosis. Veterinary Group of Lupino. Vet Parasitol 54(4):367–373

Nylén S, Gautam S (2010) Immunological perspectives of leishmaniasis. J Global Infect Dis 2(2): 135–146

Nylén S, Maurya R, Eidsmo L, Manandhar KD, Sundar S, Sacks D (2007) Splenic accumulation of IL-10 mRNA in T cells distinct from CD4+CD25+ (Foxp3) regulatory T cells in human visceral leishmaniasis. J Exp Med 204(4):805–817

Okwor I, Uzonna JE (2009) Immunotherapy as a strategy for treatment of leishmaniasis: a review of the literature. Immunotherapy 1(5):765–776

Okwor I, Uzonna JE (2013) The immunology of Leishmania/HIV co-infection. Immunol Res 56(1): 163–171

Palatnik-de-Sousa CB (2008) Vaccines for leishmaniasis in the fore coming 25 years. Vaccine 26(14):1709–1724

Palatnik-de-Sousa CB (2012) Vaccines for canine leishmaniasis. Front Immunol 3:69

Palatnik-de-Sousa CB, Moreno MMB, Paraguai-de-Souza E, Borojevic R (1994) The FML vaccine (fucose-mannose ligand) protects hamsters from experimental kala-azar Ciênc. cult. (Säo Paulo) 46(4):290–296

Pateraki E, Portocala R, Labrousse H, Guesdon JL (1983) Antiactin and antitubulin antibodies in canine visceral leishmaniasis. Infect Immunity 42(2):496–500

Quinnell RJ, Courtenay O, Shaw MA, Day MJ, Garcez LM, Dye C, Kaye PM (2001) Tissue cytokine responses in canine visceral leishmaniasis. J Infect Dis 183(9):1421–1424

Rai AK, Thakur CP, Singh A, Seth T, Srivastava SK, Singh P, Mitra DK (2012) Regulatory T cells suppress T cell activation at the pathologic site of human visceral leishmaniasis. PLoS One 7(2): e31551

Ratnapriya S, Keerti, Yadav NK, Dube A, Sahasrabuddhe AA (2021) A chimera of Th1 stimulatory proteins of Leishmania donovani offers moderate immunotherapeutic efficacy with a Th1-inclined immune response against visceral leishmaniasis. BioMed Res Int 2021:8845826

Ready PD (2014) Epidemiology of visceral leishmaniasis. Clin Epidemiol 6:147–154

Ribeiro RR, Michalick M, da Silva ME, Dos Santos C, Frézard F, da Silva SM (2018) Canine Leishmaniasis: an overview of the current status and strategies for control. BioMed Res Int. 2018:3296893

Roatt BM, Aguiar-Soares RD, Coura-Vital W, Ker HG, Moreira ND, Vitoriano-Souza J, Giunchetti RC, Carneiro CM, Reis AB (2014) Immunotherapy and immunochemotherapy in visceral leishmaniasis: promising treatments for this neglected disease. Front Immunol 5:272

Roatt BM, Aguiar-Soares RD, Reis LE, Cardoso JM, Mathias FA, de Brito RC, da Silva SM, Gontijo NF, Ferreira SA, Valenzuela JG, Corrêa-Oliveira R, Giunchetti RC, Reis AB (2017) A vaccine therapy for canine visceral leishmaniasis promoted significant improvement of clinical and immune status with reduction in parasite burden. Front Immunol 8:217

Rodrigues V, Cordeiro-da-Silva A, Laforge M, Silvestre R, Estaquier J (2016) Regulation of immunity during visceral Leishmania infection. Parasites Vectors. 9:118

Rodriguez-Cortes A, Martori C, Martinez-Florez A, Clop A, Amills M, Kubejko J, Llull J, Nadal JM, Alberola J (2017) Canine leishmaniasis progression is associated with vitamin D deficiency. Sci Rep 7(1):3346

Rossi CN, Tomokane TY, Batista LF, Marcondes M, Larsson CE, Laurenti MD (2016) In situ CUTANEOUS CELLULAR IMMUNE RESPONSE IN DOGS NATURALLY AFFECTED BY VISCERAL LEISHMANIASIS. Rev Inst Med Trop de Sao Paulo 58:48

Sacks DL, Lal SL, Shrivastava SN, Blackwell J, Neva FA (1987) An analysis of T cell responsiveness in Indian kala-azar. J Immunol (Baltimore, Md.: 1950). 138(3):908–913

Samant M, Gupta R, Kumari S, Misra P, Khare P, Kushawaha PK, Sahasrabuddhe AA, Dube A (2009) Immunization with the DNA-encoding N-terminal domain of proteophosphoglycan of Leishmania donovani generates Th1-type immunoprotective response against experimental visceral leishmaniasis. J Immunol (Baltimore, Md.: 1950). 183(1):470–479

Sanches FP, Tomokane TY, Da Matta VL, Marcondes M, Corbett CE, Laurenti MD (2014) Expression of inducible nitric oxide synthase in macrophages inversely correlates with parasitism of lymphoid tissues in dogs with visceral leishmaniasis. Acta Vet Scand 56(1):57

Santos WR, Aguiar IA, Paraguai de Souza E, de Lima VM, Palatnik M, Palatnik-de-Sousa CB (2003) Immunotherapy against murine experimental visceral leishmaniasis with the FML-vaccine. Vaccine 21(32):4668–4676

Santos FN, Borja-Cabrera GP, Miyashiro LM, Grechi J, Reis AB, Moreira MA, Martins Filho OA, Luvizotto MC, Menz I, Pessôa LM, Gonçalves PR, Palatnik M, Palatnik-de-Sousa CB (2007) Immunotherapy against experimental canine visceral leishmaniasis with the saponin enriched-Leishmune vaccine. Vaccine 25(33):6176–6190

Sartori A, Roque-Barreira MC, Coe J, Campos-Neto A (1992) Immune complex glomerulonephritis in experimental kala-azar. II: Detection and characterization of parasite antigens and antibodies eluted from kidneys of Leishmania donovani-infected hamsters. Clin Exp Immunol. 87(3):386–392

Savoia D (2015) Recent updates and perspectives on leishmaniasis. J Infect Dev Countries 9(6): 588–596

Schwendener RA (2014) Liposomes as vaccine delivery systems: a review of the recent advances. Ther Adv Vaccines 2(6):159–182

Seifert K, Juhls C, Salguero FJ, Croft SL (2015) Sequential chemoimmunotherapy of experimental visceral leishmaniasis using a single low dose of liposomal amphotericin B and a novel DNA vaccine candidate. Antimicrob Agents Chemother. 59(9):5819–5823

Sela M, Hilleman MR (2004) Therapeutic vaccines: realities of today and hopes for tomorrow. Proc Natl Acad Sci U S A. 101 Suppl 2(Suppl 2):14559

Selvapandiyan A, Debrabant A, Duncan R, Muller J, Salotra P, Sreenivas G, Salisbury JL, Nakhasi HL (2009) Centrin gene disruption impairs stage-specific basal body duplication and cell cycle progression in Leishmania. J Biol Chem 279(24):25703–25710

Selvapandiyan A, Dey R, Nylen S, Duncan R, Sacks D, Nakhasi HL (2009) Intracellular replication-deficient *Leishmania donovani* induces long lasting protective immunity against visceral leishmaniasis. J Immunol. 183(3):1813–1820

Semião-Santos SJ, el Harith A, Ferreira E, Pires CA, Sousa C, Gusmão R (1995) Evora district as a new focus for canine leishmaniasis in Portugal. Parasitol Res 81(3):235–239

Seyed N, Taheri T, Rafati S (2016) Post-genomics and vaccine improvement for leishmania. Front Microbiol. 7:467

Singh SP, Reddy DC, Rai M, Sundar S (2006) Serious underreporting of visceral leishmaniasis through passive case reporting in Bihar, India. Trop Med Int Health. 11(6):899–905

Singh VP, Ranjan A, Topno RK, Verma RB, Siddique NA, Ravidas VN, Kumar N, Pandey K, Das P (2010) Estimation of under-reporting of visceral leishmaniasis cases in Bihar. India. Am J Trop Med Hyg. 82(1):9–11

Singh OP, Gidwani K, Kumar R, Nylén S, Jones SL, Boelaert M, Sacks D, Sundar S (2012) Reassessment of immune correlates in human visceral leishmaniasis as defined by cytokine release in whole blood. Clin Vaccine Immunolo. 19(6):961–966

Singh OP, Hasker E, Sacks D, Boelaert M, Sundar S (2014) Asymptomatic Leishmania infection: a new challenge for Leishmania control. Clin Infect Dis. 58(10):1424–1429

Skeiky YA, Coler RN, Brannon M, Stromberg E, Greeson K, Crane RT, Webb JR, Campos-Neto A, Reed SG (2002) Protective efficacy of a tandemly linked, multi-subunit recombinant leishmanial vaccine (Leish-111f) formulated in MPL adjuvant. Vaccine 20(27–28):3292–3303

Smith BE, Tompkins MB, Breitschwerdt EB (2004) Antinuclear antibodies can be detected in dog sera reactive to Bartonella vinsonii subsp. berkhoffii, Ehrlichia canis, or Leishmania infantum antigens. J Vet Intern Med 18(1):47–51

Solano-Gallego L, Montserrat-Sangrà S, Ordeix L, Martínez-Orellana P (2016) Leishmania infantum-specific production of IFN-γ and IL-10 in stimulated blood from dogs with clinical leishmaniosis. Parasites Vectors 9(1):317

Solcà MS, Andrade BB, Abbehusen MM, Teixeira CR, Khouri R, Valenzuela JG, Kamhawi S, Bozza PT, Fraga DB, Borges VM, Veras PS, Brodskyn CI (2016) Circulating biomarkers of immune activation, oxidative stress and inflammation characterize severe canine visceral leishmaniasis. Sci Rep 6:32619

Srivastava JK, Misra A, Sharma P, Srivastava B, Naik S, Dube A (2003) Prophylactic potential of autoclaved *Leishmania donovani* with BCG against experimental visceral leishmaniasis. Parasitology 127(2):107–114

Sundar S, Singh A, Singh OP (2014) Strategies to overcome antileishmanial drugs unresponsiveness. J Trop Med. 2014:646932

Toepp A, Larson M, Wilson G, Grinnage-Pulley T, Bennett C, Leal-Lima A, Anderson B, Parrish M, Anderson M, Fowler H, Hinman J, Kontowicz E, Jefferies J, Beeman M, Buch J, Saucier J, Tyrrell P, Gharpure R, Cotter C, Petersen C (2018) Randomized, controlled, double-blinded field trial to assess Leishmania vaccine effectiveness as immunotherapy for canine leishmaniosis. Vaccine 36(43):6433–6441

Tonin AA, Calado AM, Bottari NB, Dalenogare D, Thomé GR, Duarte T, Duarte MM, Morsch VM, Schetinger MR, Alves LC, Tinucci-Costa M, Da Silva AS (2016) Novel markers of inflammatory response and hepatic dysfunction in canine leishmaniasis. Comp Immunol Microbiol Infect Dis 44:61–64

Trigo J, Abbehusen M, Netto EM, Nakatani M, Pedral-Sampaio G, de Jesus RS, Goto Y, Guderian J, Howard RF, Reed SG (2010) Treatment of canine visceral leishmaniasis by the vaccine Leish-111f+MPL-SE. Vaccine 28(19):3333–3340

Velez R, Gállego M (2020) Commercially approved vaccines for canine leishmaniosis: a review of available data on their safety and efficacy. Trop Med Int Health. 25(5):540–557. https://doi.org/10.1111/tmi.13382

Viana KF, Lacerda G, Teixeira NS, Rodrigues Cangussu AS, Sousa Aguiar RW, Giunchetti RC (2018) Therapeutic vaccine of killed *Leishmania amazonensis* plus saponin reduced parasite burden in dogs naturally infected with *Leishmania infantum*. Vet Parasitol. 254:98–104

WHO Expert Committee on the Control of the Leishmaniases & World Health Organization (2010) Control of the leishmaniases: report of a meeting of the WHO Expert Committee on the Control of Leishmaniases, Geneva, 22–26 March 2010. World Health Organization

Younis BM, Osman M, Khalil E, Santoro F, Furini S, Wiggins R, Keding A, Carraro M, Musa A, Abdarahaman M, Mandefield L, Bland M, Aebischer T, Gabe R, Layton AM, Lacey C, Kaye PM, Musa AM (2021) Safety and immunogenicity of ChAd63-KH vaccine in post-kala-azar dermal leishmaniasis patients in Sudan. Mol Ther. 29(7):2366–2377

Zepp F (2010) Principles of vaccine design - Lessons from nature. Vaccine 28(Suppl 3):C14–C24

Zijlstra EE (2014) PKDL and other dermal lesions in HIV co-infected patients with Leishmaniasis: review of clinical presentation in relation to immune responses. PLoS Negl Trop Dis. 8(11): e3258. https://doi.org/10.1371/journal.pntd.0003258

Zijlstra EE, Musa AM, Khalil EA, el-Hassan IM, el-Hassan AM (2003) Post-kala-azar dermal leishmaniasis. Lancet Infect Dis. 3(2):87–98

Worldwide Efforts for the Prevention of Visceral Leishmaniasis Using Vaccinations

16

Angamuthu Selvapandiyan, Niti Puri, Enam Reyaz, Mirza A. Beg, Poonam Salotra, Hira L. Nakhasi, and Nirmal K. Ganguly

Abstract

The potentially fatal visceral leishmaniasis (VL) disease in human is endemic to tropical world and caused by protozoan parasites *Leishmania donovani* (in the Indian subcontinent and African countries) and/or *Leishmania infantum* (in the Mediterranean region and Latin America). Post kala-azar dermal leishmaniasis (PKDL), a sequel of VL, occurs among certain healed VL cases in India as well as in Africa, and the VL-infested dogs serve as disease reservoirs. No licensed vaccine for leishmaniasis is available for human use, although several candidates are globally in development pipeline. They are either antigens LEISH-F3 recombinant protein formulated with glucopyranosyl lipid adjuvant in a squalene-in-water emulsion (GLA-SE) or multiantigen deoxyribonucleic acid (DNA) type or simply the sand fly saliva proteins as antigens or even gene (centrin)-deleted whole live attenuated *Leishmania donovani/Leishmania major* parasite as probable vaccine elements. A few among these have reached the clinical trial stage.

A. Selvapandiyan (✉) · E. Reyaz · M. A. Beg
Department of Molecular Medicine, Jamia Hamdard, New Delhi, India
e-mail: selvapandiyan@jamiahamdard.ac.in

N. Puri
School of Life Sciences, Jawaharlal Nehru University, New Delhi, India

P. Salotra
ICMR-National Institute of Pathology, New Delhi, India

H. L. Nakhasi
Division of Emerging and Transfusion Transmitted Diseases, Office of Blood Research and Review, Center for Biologics Evaluation and Research, Food and Drug Administration, Silver Spring, MD, USA

N. K. Ganguly
Institute of Liver and Biliary Sciences, New Delhi, India

© The Author(s), under exclusive license to Springer Nature Singapore Pte Ltd. 2023

A. Selvapandiyan et al. (eds.), *Challenges and Solutions Against Visceral Leishmaniasis*, https://doi.org/10.1007/978-981-99-6999-9_16

Also, some protein vaccines are already available mainly as canine vaccines (Leish-Tec, CaniLeish and LetiFend) in Europe and Brazil to target transmission blocking of VL in domestic dogs. These vaccines could easily become commercially feasible, as the canine population is largely at high risk for VL in such areas. For such canine vaccines to reach human, it must cross the rigorous clinical screenings/regulations. This chapter reviews current developments in *Leishmania* vaccine area, an essential aspect in the disease elimination process, to achieve the target of less than one case per 10,000 people at block level in the endemic regions.

Keywords

Visceral leishmaniasis · Live attenuated vaccine · *Leishmania donovani* · Canine *Leishmania* vaccine · *Leishmania* protein vaccine · *Leishmania* elimination

16.1 Introduction

Parasitic diseases are common in the tropical world with significant morbidity and mortality. Among such diseases, leishmaniasis due to the unicellular parasite *Leishmania* sp. in the poverty-stricken countries exerts severe socioeconomic burden. Among different clinical forms of leishmaniasis, visceral leishmaniasis (VL)/kala-azar is widespread majorly in the Indian subcontinent and African countries, where it is caused by *Leishmania donovani*, and in Latin America and the Mediterranean region, caused by another close member, *L. infantum.* Being a potential fatal infection, VL affects the visceral organs of mammals. This disease occurs annually among one million people as of May 2021 report by World Health Organization (WHO; https://www.who.int/news-room/fact-sheets/detail/leishmaniasis). Unfortunately, VL affects many tropical and subtropical regions regardless of numerous local, national and international endeavours in the direction of its eradication over the past several years (Selvapandiyan et al. 2019). A sequel of VL, Post kala-azar dermal leishmaniasis (PKDL), occurs among 5–20% of the apparently cured population in India and Africa. Hence, PKDL patients act as a natural reservoir of *Leishmania* parasites. In addition, VL-infested dogs/canines in Europe and South American countries serve as a reservoir of VL. *Leishmania*–human immunodeficiency virus (HIV) coinfection also increases relapse and mortality rates. The VL elimination target is to achieve less than 1 case per 10,000 annually at the district level (Sundar et al. 2018). Missing several target deadlines recently, WHO has reset 2025 as an elimination target.

The other less severe major clinical forms are cutaneous leishmaniasis (CL) caused by *L. major*, *Leishmania mexicana*, *Leishmania tropica* or *Leishmania amazonensis* mainly in the Middle East, Mediterranean region and certain South American and Asian countries and mucocutaneous leishmaniasis (MCL) due to *Leishmania braziliensis* in the South American countries. All forms of leishmaniasis are disseminated by the vector, sand fly. Leishmaniasis endemicity is also based on the existence of the vector population. Since the medication is expensive and the duration of treatment is long, poor and neglected populations are affected more. *Leishmania* is a digenetic parasite with two life forms: one in the vector gut, as flagellated promastigotes, and the second in the mammalian host as rudimentary-flagellated amastigotes. Researchers focus mainly on the later stage of the parasite for the development of drugs or vaccines (Ahuja et al. 2018; Avishek et al. 2018). The commonly used drugs in the world against leishmaniasis are certain antimonials, Amphotericin B, Miltefosine, Paromomycin, Pentamidine, Azoles, etc. with variable cure rates (Haldar et al. 2011; Chakravarty and Sundar 2019). However, resistance to these drugs and adverse side effects necessitate looking for alternative therapy (Ponte-Sucre et al. 2017).

A vaccination strategy could serve as the most efficient tool towards achieving the elimination target set by the WHO. Owing to the long-term ineffectiveness of various drug therapies and reported relapse of disease, there is an urgent need for a scale-up of research in vaccine development.

16.2 The Critical Aspect of Immunity and Vaccines Against Leishmaniasis

16.2.1 Required Immunity Against Leishmaniasis

In general, recruitment of inflammatory monocytes, neutrophils and macrophages as first-line host defence, also called as 'innate response', is required for the *Leishmania* sp. to establish infection in mammals and also to set stage for host to clear infection. Macrophages are the primary host cells for *Leishmania* (Volpedo et al. 2021), although the initial immune cells that encounter the parasites are mast cells and neutrophils, acting as a Trojan Horse that passes on parasites to macrophages (Peters et al. 2008; Naqvi et al. 2017). Hence, the cure in most cases is the activation of cellular immune response in such infected macrophages activated by the T cells (the second lines of host defence or the 'adaptive response') to eliminate the parasites. Such cell-mediated immunity can be achieved by stimulating Th1 response in those cells, that is, by eliciting proinflammatory cytokines like interferon gamma (IFN-γ), interleukin (IL)-2, tumour necrosis factor-alpha (TNF-α), IL-12 and IL-6 and reducing anti-inflammatory cytokines, viz. IL-4, IL-5, IL-10 and transforming growth factor-beta (TGF-β) (Selvapandiyan et al. 2009; Dey et al. 2013; Fiuza et al. 2013; Bhattacharya et al. 2015; Volpedo et al. 2021, 2022; Ismail et al. 2022; Karmakar et al. 2022; Tandon et al. 2023). On the other hand, distinct B-cell populations involved in a protective role by antibody generation against the parasite, via induction of Th2 response, as the other form of adaptive response, are also observed (Sacks 1988; Silva-Barrios et al. 2016). Vaccine-instigated innate immunity is crucial for the development of protective immune response due to its intense role in establishing adaptive immunity and subsequent generation of memory T cells (Volpedo et al. 2021; Bhattacharya et al. 2022).

16.2.2 Recombinant Antigens as Vaccines or Leishmanization

With all the concerns and hardships associated with current *Leishmania* therapy and its management, elimination remains crucial to sustainable eradication of the disease. In the absence of a licensed vaccine against any clinical form of leishmaniases, several vaccine strategies are being pursued by different laboratories worldwide (Table 16.1). A recent report stressed on vaccines for the prevention of four main diseases, viz. VL, PKDL, MCL and CL (Malvolti et al. 2021). Among these, the first three are involved in or will lead to the elimination of kala-azar. Among different types of vaccines, live attenuated vaccines are more promising, as they provide a plethora of parasite proteins in an active form as antigens to stimulate innate and adaptive immunity.

Towards such a strategy, *L. donovani* centrin (parasite's basal body associated and cell division–specific) gene-deleted live attenuated parasite as vaccine candidate was developed using funds from Indo-US Vaccine Action Program (VAP) provided to the US Food and Drug Administration (US FDA) and the Indian Council of Medical Research (ICMR; India) (Selvapandiyan et al. 2004). Similarly, another *L. donovani* attenuated parasite, deleted for p27 (a component of an active cytochrome c oxidase complex) gene, was developed (Dey et al. 2013). These parasites have been reported to produce an enhanced protective immunity in the experimental animals (mice, hamsters and/or dogs) and/or in human cells ex vivo against VL (Selvapandiyan et al. 2009; Fiuza et al. 2013; Dey et al. 2014; Fiuza et al. 2015; Avishek et al. 2016; Bhattacharya et al. 2016). Out of these, centrin-deleted *L. donovani* vaccine candidate has also recently completed its preclinical safety and toxicity tests in the animals in India, showing that such live attenuated vaccine was well tolerated by hamsters and dogs without adverse effects with even threefold higher vaccine dose (manuscript under preparation). To enhance the safety aspects of vaccine use in humans and based on the age-old success stories of leishmanization (that used less fatal cutaneous leishmaniasis causing whole live *Leishmania* sp.), different encouraging centrin gene knockout *L. major* and *L. mexicana* vaccine lines have been created by Dr. H.L. Nakhasi group (US FDA) and collaborators (from McGill University, Montreal, Canada, and The Ohio State University, Columbus, OH, USA) using marker-free CRISPR-Cas9 methodology (Zhang et al. 2020). These parasites are safe and are reported to generate potent innate and adaptive protective immune responses against both CL and VL (cross-protection) by in vitro and in vivo studies in the animals, including the sand fly–based infection/challenge model (Zhang et al. 2020; Karmakar et al. 2021; Ismail et al. 2022; Volpedo et al. 2022). Currently, the *L. major* vaccine line is being evaluated in canine population in the endemic region and manufactured under Good Laboratory Practice (GLP) and Current Good Manufacturing Product (cGMP) approaches for the human trials.

The third-generation DNA and also therapeutic vaccines are under development by other researchers globally. Especially from the group of Dr. P. Kaye, ChAd63-KH, an adenoviral vaccine encoding kinetoplastid membrane protein 11 (KMP-11) and hydrophilic acylated surface protein B (HASPB) proteins as antigens, has been

Table 16.1 List of current and pipeline vaccines against visceral leishmaniasis

S. no.	Vaccine name	Status	Chemistry	Company/institute	Funding	References
1	$LdCen^{-/-}$	Experimentally protects against VL and CL. Preclinical toxicity completed against VL	Centrin gene-deleted *L. donovani* live attenuated vaccine candidate	Food and Drug Administration, Silver Spring, MD, USA; National Institute of Pathology, New Delhi, India; Jamia Hamdard, New Delhi, India	INDO US Vaccine Action Program; Department of Biotechnology (DBT), India; Science and Engineering Research Board, India	Selvapandiyan et al. (2009), Selvapandiyan et al. (2014), Fiuza et al. (2015), Avishek et al. (2016), Bhattacharya et al. (2016)
2	$LmCen^{-/-}$	Preclinical studies completed against VL and CL	Centrin gene-deleted *L. major* marker-free live attenuated vaccine candidate	Food and Drug Administration, Silver Spring, MD, USA; McGill University, Montreal, Canada; Ohio State University, Columbus, OH, USA	Global Health Innovative Technology Fund; The Canadian Institutes of Health Research; Intramural funding from CBER, FDA	Zhang et al. (2020), Karmakar et al. (2021), Ismail et al. (2022)
3	$LmxCen^{-/-}$	Experimental studies conducted to confirm safety and protection	Centrin gene-deleted *L. mexicana* marker-free live attenuated vaccine candidate	Food and Drug Administration, Silver Spring, MD, USA; McGill University, Montreal, Canada; Ohio State University, Columbus, OH, USA	Global Health Innovative Technology (GHIT) Fund and NIH grant	Karmakar et al. (2022), Volpedo et al. (2022)
4	ChAd63-KH	In clinical trials as a therapeutic vaccine for persistent PKDL in Sudan; new phase IIa/b trial to assess whether ChAd63-KH as vaccine can avert the progression of VL to PKDL in previously	Adenoviral vaccine encoding KMP-11 and HASPB proteins as antigens	University of York, Heslington, York, United Kingdom	Welcome Trust, TRANSVAC funding	Osman et al. (2017), Younis et al. (2021)

(continued)

Table 16.1 (continued)

S. no.	Vaccine name	Status	Chemistry	Company/institute	Funding	References
		treated VL patients is in progress in Sudan				
5	F3+/GLA-SE	Preclinical evaluations	LEISH-F2-expressing RNA vaccine followed later by subunit vaccine	Infectious Research Institute, Seattle, WA, USA	National Institute of Allergy and Infectious Diseases of the National Institutes of Health; Bill and Melinda Gates Foundation	Duthie et al. (2017)
6	Leishmune	Licence withdrawn due to lack of effectiveness	Fucose-mannose ligand of Ld + Saponin	Fort Dodge Wyeth, later Zoetis, Brazil	–	Borja-Cabrera et al. (2002), Aguiar-Soares et al. (2020), Velez and Gallego (2020)
7	Leish-Tec	In use as vaccine against VL in dogs	A2 protein of *L. donovani* and Saponin adjuvant	Hertape Calier Saúde Animal, later Ceva, Brazil	–	Aguiar-Soares et al. (2020), Santos et al. (2021)
8	CaniLeish	In use as a vaccine against VL in dogs	*L. infantum*'s purified excreted–secreted proteins with saponin adjuvant	Virbac, France	–	Moreno et al. (2012)
9	LetiFend	In use as a vaccine against VL in dogs	Five antigenic peptides (called protein Q) from four proteins of *L. infantum*, viz. ribosomal proteins LiP2a, LiP2b and LiP0 and the histone H2A without an adjuvant	Laboratorios LETI, Spain	–	Parody et al. (2004)

in clinical trials as a therapeutic vaccine for continual PKDL in Sudan (Osman et al. 2017). Also, the latest Phase II a/b trials to assess whether ChAd63-KH as a cure can avoid the progression of VL to PKDL in previously cured VL population are under progress in Sudan, with funds from Welcome Trust. This vaccine displayed reduced adverse symptoms in PKDL cases and elicited both potent innate and cell-mediated immunity (Younis et al. 2021; Mohan et al. 2022). Further, randomized controlled trials to assess whether the clinical responses are vaccine-related and such vaccine has clinical effectiveness are underway. In addition, this group secured TRANSVAC funding to identify biomarkers linked with PKDL cure and vaccine response, which would allow patient stratification for vaccine or drug intervention (personal communication).

In addition, the course of technology handover for making F3+/GLA-SE (LEISH-F3 recombinant subunit vaccine formulated with glucopyranosyl lipid adjuvant [GLA] in a squalene-in-water emulsion [SE]) vaccine against VL by Infectious Research Institute, Seattle, WA, USA, after massive antigen discovery and adjuvant development is also in focus (Duthie et al. 2017). Further preclinical generation of a multiantigen DNA vaccine (KMP11, thiol-specific antioxidant [TSA], cysteine proteases A [CPA], cysteine proteases B [CPB], P74 [elongation factor], hydrophilic acidic surface protein B [HASPB] and A2) (Das et al. 2014) continues. Recently, a delivery system called cationic solid lipid nanoparticle (cSLN) has been used to deliver a DNA vaccine containing *L. donovani* A2 gene besides *L. infantum* cysteine proteinase. The result exhibited protection to Balb/c mice against *L. infantum* and an elevated level of IFN-γ (Saljoughian et al. 2013). In addition, the ability of salivary proteins of sand fly as DNA antigen vaccine by various groups has also been reported against both VL and CL (Abbehusen et al. 2018; Gholami et al. 2019; Lajevardi et al. 2022a, b; Souissi et al. 2023). This is based on the initial findings of Dr. S. Kamhawi and Dr. J.G. Valenzuela (Gomes et al. 2008; Oliveira et al. 2015) that sand fly salivary proteins contribute largely towards host immune modulation for leishmaniasis disease progression.

In addition, to understand the disease pathogenesis, to determine correlates of protection and to deliver a conduit for hastening vaccine generation, Dr. P. Kaye's group has recently unveiled a controlled human infection model (CHIM) (Ashwin et al. 2021). Here infectivity, drug sensitivity and sand fly-based transmission competence of two new *L. major* strains from CL patients of Israel were used in the CHIM optimization. Such an approach discourses a major barrier in the making of vaccines for leishmaniasis, giving a crucial resource for the CHIM of the vector-transmitted CL. On the other hand, the usage of Toll-like receptors agonists, for example, Toll-like receptor (TLR)-4 ligand (monophosphoryl lipid A [MPL])-based formulations as adjuvants in the vaccine development is common, mostly to enhance the antibody responses to subunit antigens. However, of late the discovery and inclusion of ribonucleic acid (RNA)-based molecules in the immunization schedule, to stimulate innate and adaptive immunity, specifically to enhance CD8 T cell responses, have opened avenues for the development of new generation of vaccines (Duthie et al. 2018; Reed et al. 2020).

A recent article by Paul Kaye's group through a unique simulation study suggests that the total number of doses of vaccine needed for the prevention of VL would be 300–830 million and for CL would be 557–1400 million over a period of 10 years (Malvolti et al. 2021). In such a strategy for VL vaccination, the doses needed to curtail PKDL would be 167,000. Nevertheless, in the absence of effective vaccine (s) to VL, the doses needed for PKDL over 10 years would rise to 330,000. Overall, these studies contribute to the field of vaccine development by providing estimates on the needed number of vaccine units for the elimination of different forms of leishmaniasis and the importance of developing appropriate vaccines for disease control in an economical and a most feasible manner.

16.2.3 Canine Vaccines

Leishmaniasis in domestic dogs is another important vector-borne zoonotic disease (main reservoir host) predominant in Europe and South American countries, caused mainly by *L. infantum* (Almeida et al. 2022). So far, four prophylactic vaccines against dog visceral leishmaniasis have been commercialized, two from Brazil (Leishmune® and Leish-Tec®) and two from Europe (CaniLeish® and LetiFend®) (Velez and Gallego 2020). Leishmune is a vaccine, made up of the fucose-mannose ligand (FML) of *L. donovani* and the soapbark tree (*Quillaja saponaria*)-derived saponin adjuvant (Borja-Cabrera et al. 2002). With this vaccination, three vaccine doses were injected subcutaneously every 3 weeks to the dogs from the age of 4 months, with annual boosters. However, due to its absence of effective testimony in phase III trials, both the vaccine production and marketing (by Fort Dodge Wyeth, later Zoetis, Brazil) were revoked by the Brazilian Ministry of Agriculture (Aguiar-Soares et al. 2020) in 2014. The other Brazilian vaccine, Leish-Tec by Hertape Calier Saúde Animal, later Ceva, Brazil, a combination of *L. donovani* A2 protein and saponin as adjuvant, is the only available vaccine for dogs against VL. The vaccination procedure is similar to the one followed for Leishmune (https://www.ceva.com.br/Produtos/Lista-de-Produtos/LEISH-TEC). On the other hand, CaniLeish, marketed by Virbac, France, is composed of *L. infantum*'s purified excreted–secreted proteins with saponin adjuvant (Moreno et al. 2012). The immunization practice comprises of subcutaneously administered three doses to dogs aged 6 months and older, with a gap of 21 days between doses, followed by a single booster dose annually. The fourth canine vaccine currently in use in Europe against VL is Laboratorios LETI, Spain-marketed, LetiFend, since 2016. It is made up of five antigenic peptides (called protein Q) from four proteins of *L. infantum*, viz. ribosomal proteins LiP2a, LiP2b and LiP0 and the histone H2A not including an adjuvant (Parody et al. 2004). To the dogs aged 6 months and beyond, the vaccine is administered with a single dose followed by a single dose (booster) annually. These vaccines were not continued towards industrial clinical development as human vaccines for two major reasons. In canine vaccines, the profit margin is very high, since millions of doses are sold due to the high dog population in these regions and having a high risk of contracting VL. But for the development as human vaccines, better industrial clinical development procedures are required, particularly after a rigorous clinical regulatory screening.

16.3 Conclusion

The World Health Organization has reset the visceral leishmaniasis disease elimination target as 2025, after being unsuccessful with several of the earlier assigned targets. The development of a vaccine is a critical need in disease elimination, especially where drug therapy faces challenges with cost-effectiveness, long-term treatment regimen, side effects and more importantly development of resistance against the drugs. The current vaccines in the pipeline are either subunit or DNA vaccines or whole live parasites that are gene-deleted and attenuated, targeting directly against VL-causing parasites or vaccines that were originally developed against CL but anticipated to also target VL as cross-protection (summarized in the cartoon: Fig. 16.1). The frontrunners here are the already-in-commercial use canine visceral leishmaniasis vaccines in Europe and Brazil, generating confidence that someday a licensed vaccine against human VL is possible. The accessibility of vaccines against one or other forms of leishmaniasis may well offer an easy way to reduce morbidity and mortality. Preventive vaccines (prophylactic) may be used to prevent the disease or used as substitutes to or in combination with the current drugs (i.e. as therapeutic vaccines) for the elimination of primary (VL) or the sequelae (PKDL) leishmaniasis (Malvolti et al. 2021). Several factors would determine the relative value administration of preventive or therapeutic vaccines (Selvapandiyan et al. 2023), for example, type of vaccines (e.g. live attenuated or recombinant protein antigens), prevailing comorbidity at the endemic region, among the vaccinated population at risk, and alternative treatment options. Through the opinions and reports of vaccine stakeholders, their research will progress in their attempts towards the sustained control of VL globally by continued networking, publications, development of cost-effective, safe and highly efficacious vaccine to ensure elimination of VL, based upon justifiable disruption of transmission.

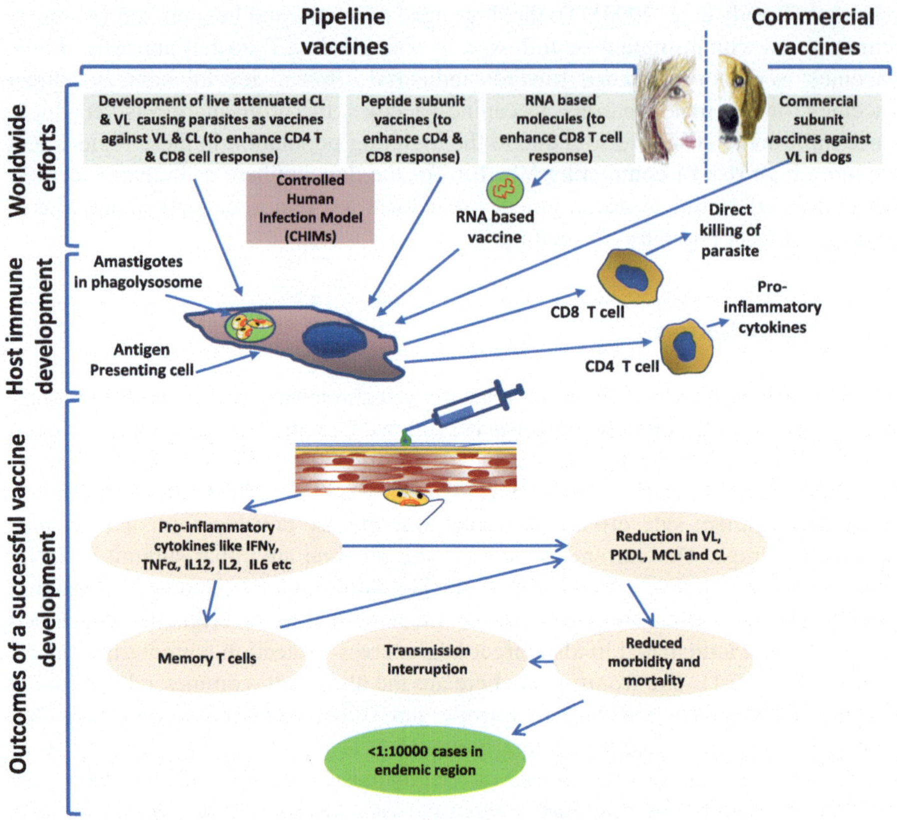

Fig. 16.1 Worldwide efforts and probable outcomes of successful vaccine development and administration strategies towards the elimination of leishmaniases. The scheme describes both the vaccines that are currently in use against canine leishmaniasis and the vaccines that are being developed/tested experimentally for human use

Acknowledgment This review research was funded by Indian Council of Medical Research (63/8/2013-BMS and GIA/2/VBD/2021/ECD-II), Biotechnology Industry Research Assistance Council (BIRAC/BT/CRS0214/CRS-10/16) and Science and Engineering Research Board (SERB) (EMR/2015/000874).

Conflict of Interest None of the authors has any conflict of interest to disclose.

References

Abbehusen MMC, Cunha J, Suarez MS, Teixeira C, Almeida VDA, Pereira LDS, Bordoni M, Gil-Santana L, Solca MDS, Fraga DBM, Fischer L, Bozza PT, Veras PST, Valenzuela JG, Kamhawi S, Andrade BB, Brodskyn CI (2018) Immunization of experimental dogs with salivary proteins from lutzomyia longipalpis, using DNA and recombinant canarypox virus induces immune responses consistent with protection against Leishmania infantum. Front Immunol 9:2558

Aguiar-Soares RDO, Roatt BM, Mathias FAS, Reis LES, Cardoso JMO, Brito RCF, Ker HG, Correa-Oliveira R, Giunchetti RC, Reis AB (2020) Phase I and II clinical trial comparing the LBSap, Leishmune((R)), and Leish-Tec((R)) vaccines against canine visceral leishmaniasis. Vaccine 8:690

Ahuja K, Beg MA, Sharma R, Saxena A, Naqvi N, Puri N, Rai PK, Chaudhury A, Duncan R, Salotra P, Nakhasi H, Selvapandiyan A (2018) A novel signal sequence negative multimeric glycosomal protein required for cell cycle progression of Leishmania donovani parasites. Biochim Biophys Acta, Mol Cell Res 1865:1148–1159

Almeida M, Maia C, Cristovao JM, Morgado C, Barbosa I, Ibars RF, Campino L, Goncalves L, Cortes S (2022) Seroprevalence and risk factors associated with leishmania infection in dogs from Portugal. Microorganisms 10:2262

Ashwin H, Sadlova J, Vojtkova B, Becvar T, Lypaczewski P, Schwartz E, Greensted E, Van Bocxlaer K, Pasin M, Lipinski KS, Parkash V, Matlashewski G, Layton AM, Lacey CJ, Jaffe CL, Volf P, Kaye PM (2021) Characterization of a new Leishmania major strain for use in a controlled human infection model. Nat Commun 12:215

Avishek K, Kaushal H, Gannavaram S, Dey R, Selvapandiyan A, Ramesh V, Negi NS, Dubey US, Nakhasi HL, Salotra P (2016) Gene deleted live attenuated Leishmania vaccine candidates against visceral leishmaniasis elicit pro-inflammatory cytokines response in human PBMCs. Sci Rep 6:33059

Avishek K, Ahuja K, Pradhan D, Gannavaram S, Selvapandiyan A, Nakhasi HL, Salotra P (2018) A Leishmania-specific gene upregulated at the amastigote stage is crucial for parasite survival. Parasitol Res 117:3215–3228

Bhattacharya P, Dey R, Dagur PK, Kruhlak M, Ismail N, Debrabant A, Joshi AB, Akue A, Kukuruga M, Takeda K, Selvapandiyan A, McCoy JP Jr, Nakhasi HL (2015) Genetically modified live attenuated leishmania donovani parasites induce innate immunity through classical activation of macrophages that direct the Th1 response in mice. Infect Immun 83:3800–3815

Bhattacharya P, Dey R, Dagur PK, Joshi AB, Ismail N, Gannavaram S, Debrabant A, Akue AD, KuKuruga MA, Selvapandiyan A, McCoy JP Jr, Nakhasi HL (2016) Live attenuated leishmania donovani centrin knock out parasites generate non-inferior protective immune response in aged mice against visceral leishmaniasis. PLoS Negl Trop Dis 10:e0004963

Bhattacharya P, Ismail N, Saxena A, Gannavaram S, Dey R, Oljuskin T, Akue A, Takeda K, Yu J, Karmakar S, Dagur PK, McCoy JP Jr, Nakhasi HL (2022) Neutrophil-dendritic cell interaction plays an important role in live attenuated Leishmania vaccine induced immunity. PLoS Negl Trop Dis 16:e0010224

Borja-Cabrera GP, Correia Pontes NN, da Silva VO, Paraguai de Souza E, Santos WR, Gomes EM, Luz KG, Palatnik M, Palatnik de Sousa CB (2002) Long lasting protection against canine kala-azar using the FML-QuilA saponin vaccine in an endemic area of Brazil (Sao Goncalo do Amarante, RN). Vaccine 20:3277–3284

Chakravarty J, Sundar S (2019) Current and emerging medications for the treatment of leishmaniasis. Expert Opin Pharmacother 20:1251–1265

Das S, Freier A, Boussoffara T, Das S, Oswald D, Losch FO, Selka M, Sacerdoti-Sierra N, Schonian G, Wiesmuller KH, Seifert K, Schroff M, Juhls C, Jaffe CL, Roy S, Das P, Louzir H, Croft SL, Modabber F, Walden P (2014) Modular multiantigen T cell epitope-enriched DNA vaccine against human leishmaniasis. Sci Transl Med 6:234ra256

Dey R, Dagur PK, Selvapandiyan A, McCoy JP, Salotra P, Duncan R, Nakhasi HL (2013) Live attenuated Leishmania donovani p27 gene knockout parasites are nonpathogenic and elicit long-term protective immunity in BALB/c mice. J Immunol 190:2138–2149

Dey R, Natarajan G, Bhattacharya P, Cummings H, Dagur PK, Terrazas C, Selvapandiyan A, McCoy JP Jr, Duncan R, Satoskar AR, Nakhasi HL (2014) Characterization of cross-protection by genetically modified live-attenuated Leishmania donovani parasites against Leishmania mexicana. J Immunol 193:3513–3527

Duthie MS, Pereira L, Favila M, Hofmeyer KA, Reed SJ, Metangmo S, Townsend S, Laurance JD, Picone A, Misquith A, Hossain F, Ghosh P, Khan MAA, Guderian J, Bailor HR, Liang H, Vergara J, Oliveira F, Howard RF, Kamhawi S, Mondal D, Coler RN, Valenzuela JG, Reed SG (2017) A defined subunit vaccine that protects against vector-borne visceral leishmaniasis. NPJ Vaccines 2:23

Duthie MS, Van Hoeven N, MacMillen Z, Picone A, Mohamath R, Erasmus J, Hsu FC, Stinchcomb DT, Reed SG (2018) Heterologous immunization with defined RNA and subunit vaccines enhances T cell responses that protect against Leishmania donovani. Front Immunol 9:2420

Fiuza JA, Santiago Hda C, Selvapandiyan A, Gannavaram S, Ricci ND, Bueno LL, Bartholomeu DC, Correa-Oliveira R, Nakhasi HL, Fujiwara RT (2013) Induction of immunogenicity by live attenuated Leishmania donovani centrin deleted parasites in dogs. Vaccine 31:1785–1792

Fiuza JA, Gannavaram S, Santiago Hda C, Selvapandiyan A, Souza DM, Passos LS, de Mendonca LZ, Lemos-Giunchetti Dda S, Ricci ND, Bartholomeu DC, Giunchetti RC, Bueno LL, Correa-Oliveira R, Nakhasi HL, Fujiwara RT (2015) Vaccination using live attenuated Leishmania donovani centrin deleted parasites induces protection in dogs against Leishmania infantum. Vaccine 33:280–288

Gholami E, Oliveira F, Taheri T, Seyed N, Gharibzadeh S, Gholami N, Mizbani A, Zali F, Habibzadeh S, Bakhadj DO, Meneses C, Kamyab-Hesari K, Sadeghipour A, Taslimi Y, Khadir F, Kamhawi S, Mazlomi MA, Valenzuela JG, Rafati S (2019) DNA plasmid coding for Phlebotomus sergenti salivary protein PsSP9, a member of the SP15 family of proteins, protects against Leishmania tropica. PLoS Negl Trop Dis 13:e0007067

Gomes R, Teixeira C, Teixeira MJ, Oliveira F, Menezes MJ, Silva C, de Oliveira CI, Miranda JC, Elnaiem DE, Kamhawi S, Valenzuela JG, Brodskyn CI (2008) Immunity to a salivary protein of a sand fly vector protects against the fatal outcome of visceral leishmaniasis in a hamster model. Proc Natl Acad Sci U S A 105:7845–7850

Haldar AK, Sen P, Roy S (2011) Use of antimony in the treatment of leishmaniasis: current status and future directions. Mol Biol Int 2011:571242

Ismail N, Karmakar S, Bhattacharya P, Sepahpour T, Takeda K, Hamano S, Matlashewski G, Satoskar AR, Gannavaram S, Dey R, Nakhasi HL (2022) Leishmania major centrin gene-deleted parasites generate skin resident memory T-cell immune response analogous to leishmanization. Front Immunol 13:864031

Karmakar S, Ismail N, Oliveira F, Oristian J, Zhang WW, Kaviraj S, Singh KP, Mondal A, Das S, Pandey K, Bhattacharya P, Volpedo G, Gannavaram S, Satoskar M, Satoskar S, Sastry RM, Oljuskin T, Sepahpour T, Meneses C, Hamano S, Das P, Matlashewski G, Singh S, Kamhawi S, Dey R, Valenzuela JG, Satoskar A, Nakhasi HL (2021) Preclinical validation of a live attenuated dermotropic Leishmania vaccine against vector transmitted fatal visceral leishmaniasis. Commun Biol 4:929

Karmakar S, Volpedo G, Zhang WW, Lypaczewski P, Ismail N, Oliveira F, Oristian J, Meneses C, Gannavaram S, Kamhawi S, Hamano S, Valenzuela JG, Matlashewski G, Satoskar AR, Dey R, Nakhasi HL (2022) Centrin-deficient Leishmania mexicana confers protection against Old World visceral leishmaniasis. NPJ Vaccines 7:157

Lajevardi MS, Gholami E, Taheri T, Sarvnaz H, Habibzadeh S, Seyed N, Mortazavi Y, Rafati S (2022a) Leishmania tarentolae as potential live vaccine co-expressing distinct salivary gland proteins against experimental cutaneous leishmaniasis in BALB/c mice model. Front Immunol 13:895234

Lajevardi MS, Taheri T, Gholami E, Seyed N, Rafati S (2022b) Structural analysis of PpSP15 and PsSP9 sand fly salivary proteins designed with a self-cleavable linker as a live vaccine candidate against cutaneous leishmaniasis. Parasit Vectors 15:377

Malvolti S, Malhame M, Mantel CF, Le Rutte EA, Kaye PM (2021) Human leishmaniasis vaccines: use cases, target population and potential global demand. PLoS Negl Trop Dis 15:e0009742

Mohan S, Revill P, Malvolti S, Malhame M, Sculpher M, Kaye PM (2022) Estimating the global demand curve for a leishmaniasis vaccine: a generalisable approach based on global burden of disease estimates. PLoS Negl Trop Dis 16:e0010471

Moreno J, Vouldoukis I, Martin V, McGahie D, Cuisinier AM, Gueguen S (2012) Use of a LiESP/QA-21 vaccine (CaniLeish) stimulates an appropriate Th1-dominated cell-mediated immune response in dogs. PLoS Negl Trop Dis 6:e1683

Naqvi N, Ahuja K, Selvapandiyan A, Dey R, Nakhasi H, Puri N (2017) Role of Mast Cells in clearance of Leishmania through extracellular trap formation. Sci Rep 7:13240

Oliveira F, Rowton E, Aslan H, Gomes R, Castrovinci PA, Alvarenga PH, Abdeladhim M, Teixeira C, Meneses C, Kleeman LT, Guimaraes-Costa AB, Rowland TE, Gilmore D, Doumbia S, Reed SG, Lawyer PG, Andersen JF, Kamhawi S, Valenzuela JG (2015) A sand fly salivary protein vaccine shows efficacy against vector-transmitted cutaneous leishmaniasis in nonhuman primates. Sci Transl Med. 7:290ra290

Osman M, Mistry A, Keding A, Gabe R, Cook E, Forrester S, Wiggins R, Di Marco S, Colloca S, Siani L, Cortese R, Smith DF, Aebischer T, Kaye PM, Lacey CJ (2017) A third generation vaccine for human visceral leishmaniasis and post kala azar dermal leishmaniasis: First-in-human trial of ChAd63-KH. PLoS Negl Trop Dis 11:e0005527

Parody N, Soto M, Requena JM, Alonso C (2004) Adjuvant guided polarization of the immune humoral response against a protective multicomponent antigenic protein (Q) from Leishmania infantum. A CpG + Q mix protects Balb/c mice from infection. Parasite Immunol 26:283–293

Peters NC, Egen JG, Secundino N, Debrabant A, Kimblin N, Kamhawi S, Lawyer P, Fay MP, Germain RN, Sacks D (2008) In vivo imaging reveals an essential role for neutrophils in leishmaniasis transmitted by sand flies. Science 321:970–974

Ponte-Sucre A, Gamarro F, Dujardin JC, Barrett MP, Lopez-Velez R, Garcia-Hernandez R, Pountain AW, Mwenechanya R, Papadopoulou B (2017) Drug resistance and treatment failure in leishmaniasis: a 21st century challenge. PLoS Negl Trop Dis 11:e0006052

Reed SG, Tomai M, Gale MJ Jr (2020) New horizons in adjuvants for vaccine development. Curr Opin Immunol 65:97–101

Sacks DL (1988) B cell dependent T lymphocyte responses in leishmaniasis. Mem Inst Oswaldo Cruz 83(Suppl 1):506–513

Saljoughian N, Zahedifard F, Doroud D et al (2013) Cationic solid–lipid nanoparticles are as efficient as electroporation in DNA vaccination against visceral leishmaniasis in mice. Parasite Immunol 35(12):397–408

Santos TTO, Ramos FF, Goncalves IAP, Tavares GSV, Ludolf F, Bandeira RS, Silva AM, Oliveira-da-Silva JA, Reis TAR, Machado AS, Lage DP, Freitas CS, Vale DL, Martins VT, Alves LA, Guimaraes NS, Chaves AT, Chavez-Fumagalli MA, Cota GF, Silveira JAG, Tupinambas U, Goncalves DU, Christodoulides M, Coelho EAF (2021) Potential of recombinant LiHyQ, a novel Leishmania infantum protein, for the diagnosis of canine visceral leishmaniasis and as a diagnostic and prognostic marker for human leishmaniasis and human immunodeficiency virus co-infection: a preliminary study. Acta Trop 224:106126

Selvapandiyan A, Debrabant A, Duncan R, Muller J, Salotra P, Sreenivas G, Salisbury JL, Nakhasi HL (2004) Centrin gene disruption impairs stage-specific basal body duplication and cell cycle progression in Leishmania. J Biol Chem 279:25703–25710

Selvapandiyan A, Dey R, Nylen S, Duncan R, Sacks D, Nakhasi HL (2009) Intracellular replication-deficient Leishmania donovani induces long lasting protective immunity against visceral leishmaniasis. J Immunol 183:1813–1820

Selvapandiyan A, Dey R, Gannavaram S, Solanki S, Salotra P, Nakhasi HL (2014) Generation of growth arrested Leishmania amastigotes: a tool to develop live attenuated vaccine candidates against visceral leishmaniasis. Vaccine 32:3895–3901

Selvapandiyan A, Croft SL, Rijal S, Nakhasi HL, Ganguly NK (2019) Innovations for the elimination and control of visceral leishmaniasis. PLoS Negl Trop Dis 13:e0007616

Selvapandiyan A, Puri N, Kumar P, Alam A, Ehtesham NZ, Griffin G, Hasnain SE (2023) Zooming in on common immune evasion mechanisms of pathogens in phagolysosomes: potential broad-spectrum therapeutic targets against infectious diseases. FEMS Microbiol Rev 47:fuac041

Silva-Barrios S, Smans M, Duerr CU, Qureshi ST, Fritz JH, Descoteaux A, Stager S (2016) Innate immune B cell activation by Leishmania donovani exacerbates disease and mediates hypergammaglobulinemia. Cell Rep 15:2427–2437

Souissi C, Marzouki S, Elbini-Dhouib I, Jebali J, Oliveira F, Valenzuela JG, Srairi-Abid N, Kamhawi S, Ben Ahmed M (2023) PpSP32, the Phlebotomus papatasi immunodominant salivary protein, exerts immunomodulatory effects on human monocytes, macrophages, and lymphocytes. Parasit Vectors 16:1

Sundar S, Singh OP, Chakravarty J (2018) Visceral leishmaniasis elimination targets in India, strategies for preventing resurgence. Expert Rev Anti-Infect Ther 16:805–812

Tandon R, Reyaz E, Roshanara, Jadhav M, Gandhi M, Dey R, Salotra P, Nakhasi HL, Selvapandiyan A (2023) Identification of protein biomarkers of attenuation and immunogenicity of centrin or p27 gene deleted live vaccine candidates of Leishmania against visceral leishmaniasis. Parasitol Int 92:102661

Velez R, Gallego M (2020) Commercially approved vaccines for canine leishmaniosis: a review of available data on their safety and efficacy. Trop Med Int Health 25:540–557

Volpedo G, Huston RH, Holcomb EA, Pacheco-Fernandez T, Gannavaram S, Bhattacharya P, Nakhasi HL, Satoskar AR (2021) From infection to vaccination: reviewing the global burden, history of vaccine development, and recurring challenges in global leishmaniasis protection. Expert Rev Vaccines 20:1431–1446

Volpedo G, Pacheco-Fernandez T, Holcomb EA, Zhang WW, Lypaczewski P, Cox B, Fultz R, Mishan C, Verma C, Huston RH, Wharton AR, Dey R, Karmakar S, Oghumu S, Hamano S, Gannavaram S, Nakhasi HL, Matlashewski G, Satoskar AR (2022) Centrin-deficient Leishmania mexicana confers protection against New World cutaneous leishmaniasis. NPJ Vaccines 7:32

Younis BM, Osman M, Khalil EAG, Santoro F, Furini S, Wiggins R, Keding A, Carraro M, Musa AEA, Abdarahaman MAA, Mandefield L, Bland M, Aebischer T, Gabe R, Layton AM, Lacey CJN, Kaye PM, Musa AM (2021) Safety and immunogenicity of ChAd63-KH vaccine in post-kala-azar dermal leishmaniasis patients in Sudan. Mol Ther 29:2366–2377

Zhang WW, Karmakar S, Gannavaram S, Dey R, Lypaczewski P, Ismail N, Siddiqui A, Simonyan V, Oliveira F, Coutinho-Abreu IV, DeSouza-Vieira T, Meneses C, Oristian J, Serafim TD, Musa A, Nakamura R, Saljoughian N, Volpedo G, Satoskar M, Satoskar S, Dagur PK, McCoy JP, Kamhawi S, Valenzuela JG, Hamano S, Satoskar AR, Matlashewski G, Nakhasi HL (2020) A second generation leishmanization vaccine with a markerless attenuated Leishmania major strain using CRISPR gene editing. Nat Commun 11:3461

Emerging Concepts in *Leishmania* Vaccine Adjuvants

17

Amrita Das and Nahid Ali

Abstract

Leishmaniasis refers to a plethora of vector-borne protozoan diseases clinically manifested as self-healing cutaneous, mucocutaneous and most dangerous visceral forms, caused by *Leishmania* sp. It accounts for over 70,000 mortalities worldwide among 1.5–2 million cases reported annually. Considering the challenges of acute toxicities, chemo-resistance and relapse with the current chemotherapeutic regimes against leishmaniasis, there is an urgent need to develop a safe and effective pan-*Leishmania* vaccine. Despite decades of extensive research, no anti-leishmanial vaccine is still available for human use, partly due to lack of an appropriate adjuvant. Leishmanization, the gold standard of acquired immunity by live vaccination with virulent *Leishmania* parasites is obsolete due to safety and reactivity issues. Subunit vaccines, although safe, are poorly immunogenic when given alone. Thus, adjuvants are key components of both inactivated and modern subunit vaccines to augment specificity, magnitude and duration of immune response against pathogens. Although antibody generation is a well-established phenomenon in several human licensed adjuvants, very few adjuvants can elicit dominant cell-mediated immune (CMI) response against intracellular pathogens. Unlike several viral and bacterial pathogens, immunity against intracellular parasites like *Leishmania* requires potent $CD4^+$ and $CD8^+$ T-cell response. Hence, the adjuvants like MF59 and virosomes, which predominantly generate antibodies, are non-protective against intracellular pathogens failing to generate adequate CMI response. The history of anti-leishmanial vaccines evaluated so far demonstrates that only a few adjuvants including

A. Das · N. Ali (✉)
Infectious Diseases and Immunology Division, CSIR-Indian Institute of Chemical Biology, Kolkata, West Bengal, India
e-mail: nali@iicb.res.in

© The Author(s), under exclusive license to Springer Nature Singapore Pte Ltd. 2023

A. Selvapandiyan et al. (eds.), *Challenges and Solutions Against Visceral Leishmaniasis*, https://doi.org/10.1007/978-981-99-6999-9_17

saponins, immune-stimulating complexes (ISCOMs), interleukin (IL)-12, monophosphoryl lipid A (MPLA), glucopyranosyl lipid adjuvant (GLA) and liposomes could induce cellular immunity against *Leishmania* required for successful vaccination. Further, the physicochemical properties of emerging nanocarriers have great potential to overcome the limitations of conventional adjuvant systems posed to anti-leishmanial vaccines. This chapter intends to advance our understanding in the role of toll-like receptors (TLRs), deoxyribonucleic acid (DNA) sensors, C-type lectins and metabolic, cell death and epigenetic adjuvants in *Leishmania* vaccines. Emerging high-throughput multi-omic technologies like epigenomics, metabolomics, transcriptomics and mass cytometry can rapidly delineate correlates of vaccine-mediated immunity for adjuvant screening today. Such approach combining human models with system vaccinology during the primary stages of adjuvant screening has the translational potential for revitalizing anti-leishmanial prophylaxis.

Keywords

Adjuvant · Immunotherapy · Vaccine · TLR agonist · Nanoparticles

17.1 Introduction

As a neglected tropical disease, leishmaniasis accounts for 700,000 to 1 million cases annually (World Health Organization) with a broad clinical spectrum ranging from self-healing cutaneous lesions (caused by *Leishmania major*, *Leishmania tropica* and *Leishmania mexicana*), mucocutaneous (caused by *Leishmania panamensis*, *Leishmania guyanensis* and *Leishmania braziliensis*) to fatal visceral form (caused by *Leishmania infantum* and *Leishmania donovani*) of the disease. Leishmaniases, both anthroponotic and zoonotic, are transmitted from one host to another chiefly through the bite of infected phlebotomine sandfly vectors during blood meal. Although endemic to the Indian subcontinent, Africa, Mediterranean regions and Latin America, increased global travels, malnutrition, rising human immunodeficiency virus (HIV) co-infection and various malignancies pose a serious worldwide threat in the absence of any commercially available vaccine against *Leishmania*. Leishmaniasis management completely restricts itself to few drugs reportedly having systemic toxicities (Tiwari et al. 2018). However, lifelong immunity conferred in cured individuals against leishmaniases holds promise for vaccine-mediated protection against *Leishmania*. Vaccines capable of eliciting strong cell-mediated immune (CMI) responses are required to eradicate intracellular pathogens like *Leishmania*. Use of adjuvants is an integral part of modern vaccinology due to poor immunogenicity of subunit proteins/DNA, particularly against intracellular parasites like *Leishmania*, which require strong cell-mediated immunity.

In both animals and humans, anti-leishmanial immunity is mostly determined by the activation of antigen-specific T lymphocytes. The $CD4^+$ Th1 cells producing pro-inflammatory cytokines like interferon gamma (IFN-γ) and tumour necrosis factor-alpha (TNF-α) along with IL-2 and IL-12 activate the microbicidal activity of antigen-presenting cells (APCs), resulting in amastigote killing within macrophages (Mann et al. 2021). The $CD4^+$ Th2 subset, however, is associated with disease exacerbation through increased production of IL-4, IL-5, IL-10 and IL-13 and antibody. The role of IL-10, IL-27 and IL-21 have been implicated in active human visceral leishmaniasis (VL) helping in disease progression (Samant et al. 2021). Exhaustion of $CD8^+$ T response appears to result in parasite persistence and disease progression in these patients (Gautam et al. 2014). In addition, IL-10-producing $CD8^+$ T cells found in patients with cutaneous leishmaniasis (CL), mucocutaneous leishmaniasis (MCL) and post-kala-azar dermal leishmaniasis (PKDL) suggest their role in disease progression. Although the role of $CD8^+$ T cells is controversial in eliciting innate immune response in leishmaniasis, their role is well established in vaccine-induced protection and generation of long-lasting memory (Ikeogu et al. 2020). The fact that 85% of cured individuals are resistant to re-infection by *L. donovani* (Zijlstra et al. 2003) and restoration of depleted $CD8^+$ T cells after cure indicates the importance of $CD8^+$ T cells in disease resolution and lifelong immunity against VL (Nikolich-Zugich 2008; Kaushal et al. 2014). Many adjuvants like saponins, interleukin (IL)-12, IL-2, granulocyte-macrophage colony-stimulating factor (GM-CSF), Bacillus Calmette-Guerin (BCG), Montanide and CpG oligodeoxynucleotides (CpG-ODNs) have been evaluated for anti-leishmanial vaccine development with limited success (Das and Ali 2012). Recently emerging particulate adjuvants cum delivery systems based on lipids, minerals and polymer-based nanoparticles including co-polymer poly(lactide-co-glycolide) (PLGA), solid lipid nanoparticles (SLNs), liposomes, virosomes, micelles and dendrimers have shown promising results in immunoprophylaxis against *Leishmania*, as well as other intracellular pathogens (Tada et al. 2018a; Bhalla et al. 2021a). The recent concept of trained immunity, however, predicts the importance of innate immunity controlled through metabolic and epigenetic imprinting of the host. Here lie the future prospects of toll-like receptor (TLR), C-type lectin receptor (CLR) and NOD-like receptor (NLR) agonists in association with proper delivery systems. Such adjuvants can lead to innate reprogramming of immunological pathways and durable immunity not only against leishmaniasis but also against several other pathogens.

17.2 Adjuvants: The Non-Specific Immune Enhancers

Adjuvants ('*adjuvare*' means help in Latin) have been defined as non-specific immune activators that enhance the cellular and humoral immunity to the associated antigens for vaccine-mediated protection. Based on their mechanism of action, adjuvants can be classified as delivery systems, immunopotentiators and combination adjuvants. Delivery vehicles are specialized antigen carriers intended for protection, intracellular delivery and pro-inflammatory response or depot effect at the site of injection recruiting immune cells, cytokines and chemokines as mediators of innate immunity. Nanoparticles, emulsions, conventional liposomes and virosomes are some of the well-studied delivery systems. Immunopotentiators like mineral salts, TLR agonists (e.g. monophosphoryl lipid A, CpG, imidazoquinolines and flagellin) and cholera toxins are usually pathogen-associated molecular patterns (PAMPs) that stimulate the innate immune cells via pattern recognition receptors (PRRs) to augment cytokine and chemokine production in the host. Freund's complete adjuvant, derived from bacterial cell wall, is one of the earliest effective immunostimulants. However, this adjuvant is unsuitable for human use due to its associated toxicities (Oscherwitz et al. 2006). TLRs present on the plasma membranes (TLRs 1, 2, 4, 5, 6 and 11) and endolysosomes (3, 7, 8 and 9) of professional APCs recognize specific PRRs on pathogens (Kawasaki and Kawai 2014). Except TLR3, all other TLRs stimulate MyD88-dependent nuclear factor (NF)-kβ, interferon regulatory factors (IRF) or mitogen-activated protein kinase (MAPK) signalling, thereby upregulating pro-inflammatory cytokines crucial for adaptive immune response (Kawasaki and Kawai 2014). Unlike other TLRs, TLR3 recognizes viral double-stranded RNA (dsRNA) and its synthetic analogues and acts through TIR-domain-containing adaptor-inducing interferon-β (TRIF)-dependent and MyD88-independent signalling pathways. Combination adjuvants usually combine two or more immunopotentiators with or without delivery systems to activate multiple innate signalling pathways or PRRs along with safe delivery. An attractive strategy to induce optimum CMI response is combining particulate adjuvant/delivery systems with immunostimulant. Such adjuvant formulations can synergistically increase both innate and adaptive immune responses along with efficient antigen delivery to the target cells. Liposomes with TLR4, combination of TLR4 and TLR7/8 agonists, are some of the synergistically effective combinations being studied against various infectious diseases with various success rates (Kawasaki and Kawai 2014; Fischetti et al. 2017). Nevertheless, only few adjuvants like aluminium salts, virosomes, MF59, CpG1018, Matrix M™, AS01B and AS04 have been clinically approved for human administration.

17.3 Delivery Vehicles and Particulate Adjuvants

17.3.1 Nanoparticles

Conventional treatment options for leishmaniasis restrict themselves to injectable toxic drugs like sodium stibogluconate, pentavalent antimonial and amphotericin B, with emerging parasite resistance worsening the situation in endemic regions. Although nanomedicines have originally emerged as a highly promising approach in reducing drug-related toxicities due to better cellular absorption of existing drugs at lower doses, it is equally applicable in vaccine delivery (Das and Ali 2021). Polymeric nanoparticles (PNPs) are biodegradable, sophisticated, self-assembled synthetic or natural polymers (1–1000 nm in size), which can be nanosphere or nanocapsular type, with active ingredients entrapped within the core (Zielińska et al. 2020). In addition to their simple fabrication, functionalization and biocompatibility, polymeric nanoparticles revealed their potential for pH stability, tissue-targeted delivery and sustained release. Among various polymeric delivery vehicles like chitosan, alginate, poly(D,L-lactide) (PLA) and poly(D,L-glycolide) (PLG), the co-polymer poly(lactide-co-glycolide) (PLGA) is the US Food and Drug Administration (FDA)-approved nanoparticle successfully applied for drug delivery against various pathogens (Anselmo and Mitragotri 2019). They can carry both encapsulated and surface-attached antigens for efficient antigen delivery to APCs via simple phagocytosis or through receptor-mediated endocytosis. Chitosan and PLGA nanoparticles have been used in several experimental vaccines against *Mycobacterium tuberculosis*, tetanus, hepatitis B and influenza in mice resulting in enhanced cellular immunity (Pati et al. 2018). However, polymeric nanoparticles (PNPs) are often associated with poor vaccine/drug payloads and formation of toxic residues after degradation (Zielińska et al. 2020). Functionalized gold nanoparticles curtailed drug-related cytotoxicity (Kumar et al. 2019) and target MAPK of *Leishmania* to reduce the parasite burden in mammalian macrophages (Raj et al. 2022). Solid lipid nanoparticles (SLNs) on the other hand are less toxic, stable and cost-effective adjuvants to be used in anti-leishmanial vaccine. Cationic solid lipid nanoparticles encapsulating cysteine protease DNA vaccines against *L. major* (Das and Ali 2012) and *L. infantum* (Saljoughian et al. 2013) showed marked protection in BALB/c mice with increased IFN-γ and low IL-10 production in immunized animals.

17.3.2 Immune-Stimulating Complexes (ISCOMs)

ISCOMs are highly dense, buckminsterfullerene-shaped complexes with small porous particles of 40–60 nm diameter, developed in 1984 (Morein et al. 1984). They are composed of partially purified saponins from Quil-A, others lipids and cholesterol to form a hollow, cage-like structure with strong negative charge. Originally, antigen incorporation has been a major concern while working with ISCOM adjuvants. However, simple co-administration of ISCOMs with antigen is adequate to form an immune complex instead of tedious processes of conjugation, entrapment or chemical interactions. ISCOMATRIX and ISCOPREP readily mixed with the antigen, developed by CSL (Australia) (Baz Morelli et al. 2012; Silva et al. 2015), were less reactogenic than the parent saponin, yet retained strong adjuvant effects. Many trials used ISCOM matrices purely as adjuvants and formulated with vaccine antigens rather than incorporating them, which elicited immune responses as beneficial as those elicited by classical ISCOMs (Stertman et al. 2023). Owing to their particulate nature, ISCOMs and ISCOMATRIX are readily taken up by dendritic cells (DCs) and macrophages by endocytosis for efficient antigen presentation to T cells. ISCOMs induce both Th1 and Th2 immunity, production of pro-inflammatory cytokines, along with a strong cytotoxic T lymphocyte (CTL) response against a wide array of antigenic peptides in several animal models (Sun et al. 2009).

Since unmodified ISCOMs usually lacks the benefits of antigen encapsulation, which is important for the oral vaccines, ISCOM matrix adjuvants, when used for enteric vaccine antigens, are administered through intranasal or parenteral routes after oral priming with antigens alone or with other adjuvants (Chen et al. 2023; Silveira et al. 2023). Although preferably high $CD8^+$ major histocompatibility complex (MHC) class I-restricted adjuvant activity of ISCOMs has been reported for several pathogens, immune response probably depends on the choice of antigen formulated with ISCOMs in *Leishmania*. Intraperitoneal immunization with *L. major* gp63 formulated with ISCOM-induced antigen-specific IFN-γ, IL-2 and IL-10 (but no IL-4) for Th1 type protective immunity against cutaneous leishmaniasis (CL) (Sun et al. 2009). However, soluble leishmanial antigen (SLA) formulated with ISCOMATRIX injected subcutaneously generated a mixed Th1/Th2 response in mice which failed to protect against *L. major* (Mehravaran et al. 2016). Improved ISCOM adjuvant termed Matrix-M has shown successful induction of multifunctional $CD4^+$ and $CD8^+$ T cell responses against deadly human diseases like coronavirus disease (COVID-19), Ebola, malaria and influenza with good safety profiles (Stertman et al. 2023).

17.3.3 Virosomes

Virosomes are reconstituted, non-replicating virus-like particles (VLPs) without viral genes, recently approved by the FDA as safe vaccine carriers (Asadi and Gholami 2021). They are spherical, unilamellar phospholipid vesicles approximately 150–200 nm in diameter. Virosomes chiefly target lymph node DCs to enhance antigen uptake and induce MHC class I restricted $CD8^+$ T-cell response (Bungener et al. 2002). In addition, they are involved in DC maturation and activate $CD4^+$ T cells, crucial for CTL response and long-term memory. Similar to live virus, virosomes enter cells through receptor-mediated endocytosis using sialic acid receptors on host APCs (Huckriede et al. 2003). Endosomal fusion with virosomal membrane is usually triggered by acidic pH of the endosomes, facilitating the release of associated antigen into the cytoplasm. Virosome-encapsulated antigens are subsequently presented to T cells via MHC class I pathway leading to antigen-specific CTL responses (Bungener et al. 2002). Although many VLP-based vaccines for human papillomavirus (HPV) and hepatitis B virus (HBV) also use MPLA or aluminium salts, influenza virosomes are used alone to generate both cellular and humoral responses against the embedded antigen. Protective immunity has been reported in knockout C57/BL6 mice receiving α-Gal associated with Qβ VLPs against *L. infantum* and *Leishmania amazonensis* (Moura et al. 2017). Hence, virosomes represent an ideal delivery vehicle for entrapped or conjugated antigens to induce strong CMI response. However, rapid disintegration in serum and unwanted humoral responses by viral glycoproteins on its surface are two potential drawbacks of virosomes as adjuvants. Nevertheless, they are considered as biocompatible and efficient prophylactic and therapeutic delivery systems due to their targeted antigen delivery and immune activation. Choice of the right virosome type, quality control and optimization of cryoprotectants can be helpful in overcoming its disadvantages.

17.3.4 Liposomes

Liposomes, hitherto known as magic bullets, are one of the most unique delivery vehicles for a wide range of biomolecules, most importantly drugs and antigens used in infectious diseases and cancer. Liposomes were introduced in 1961 by Alec D. Bangham through a chance discovery while he was working with biomembranes. These multilamellar particles were later coined by Gerald Weissmann as "liposomes" (Sessa and Weissmann 1968; Bangham and Horne 1964). They are formed by the spontaneous integration of phospholipids making closed circular, unilamellar or multilamellar concentric layers of amphipathic lipids measuring 25–2500 nm in diameter (Akbarzadeh et al. 2013). This gives them the uniqueness to entrap both hydrophilic and hydrophobic materials, thus increasing their flexibility and versatility. In addition, liposomes also have dynamic properties of improving immunogenicity of the therapeutic agent (Fig. 17.2).

Liposomes can be categorized into conventional, charged and stealth liposomes. Conventional liposomes or the first generation liposomes are composed of simple phospholipids and cholesterol. These were designed and aimed for the delivery of the drugs to the target site. However, they suffered low retention time and limited stability. This was followed by charged and stealth liposomes or the second-generation liposomes, which were pure lipid, long circulating liposomes used for various infectious diseases including leishmaniasis. Ambisome™, Myocet™ and Daunoxome™ are examples of such conventional therapeutic liposomal drugs available in the market (Liu et al. 2022). Next came the surface-modified third-generation liposomes that aided in prolonged circulation, mostly by adding saturated or hydrogenated fatty acids and sphingomyelin for increased stability, for example, Marqibo™. The fourth-generation liposomes consist of polyethylene glycol (PEG) on their membrane, which helps to increase the circulation and retention time of the liposomes in the body, thereby increasing their serum stability and prolonged circulation. PEGylated stealth liposomes are the latest and currently used liposomes (Sharma and Agrawal 2021). Among numerous nanoparticle-based delivery systems, liposomes take the advantage of co-delivery of lipophilic antimicrobial drugs incorporated in their phospholipid bilayer and hydrophilic nucleic acid/peptide antigens in their closed aqueous core. The rationale of the nano-liposomal drug/vaccine against *Leishmania* parasites is that both parasites and liposomes are preferentially engulfed by the cells of reticuloendothelial system, hence accounting for superior efficacy. From its initial introduction since 1980s as army liposomal formulation (ALF), containing cholesterol, phospholipids and MPLA, liposomal adjuvant system has been extensively studied. Several liposome-based drug deliveries with the aim of prevention and treatment of diseases have been approved by the FDA for human use. To date, several liposomal vaccines are commercially available in the market against infectious diseases like malaria (Mosquirix®), influenza (Nasalflu®), COVID-19 (Moderna, Pfizer-BionTech), herpes (Shingrix®) and hepatitis-A (Inflexal® and Epaxal®), reviewed elsewhere (Krasnopolsky and Pylypenko 2022). While free antigens are susceptible to rapid clearance and low immunogenicity by getting diffused in the humoral system, liposome-encapsulated

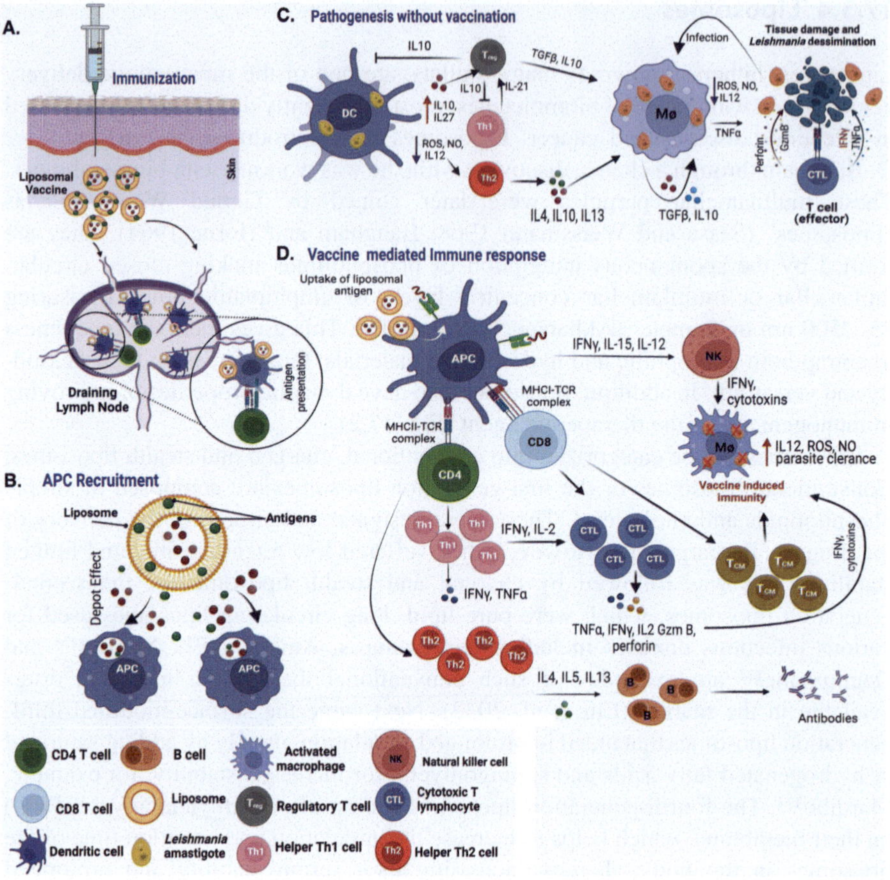

Fig. 17.1 Schematic diagram showing liposome-mediated antigen delivery and induction of anti-leishmanial immunity in mice. (**a**) Liposomal antigen delivery and tissue penetration ultimately accessing draining lymph nodes resulting in dendritic cell (DC) activation and DC–T cell cross-presentation. (**b**) Depot effect of liposome-associated antigens resulting in recruitment of antigen-presenting cells (APCs) at the site of injection, persistent and slow release of antigens ensuring better antigen presentation to T cells. (**c**) Disease progression without immunization is associated with impaired helper T (Th1) cell proliferation, recruitment of immunosuppressive helper T (Th2) and regulatory T cells (Treg) with enhanced IL-10 and TGF-β production leading to parasite multiplication in macrophages (MΦ). (**d**) The combined effect of multi-cytokine (IFN-γ, IL-2 and TNF-α) producing CD4$^+$and CD8$^+$ T cells along with natural killer (NK) cells results in successful parasite clearance from host macrophage

antigens give rise to local pro-inflammatory response and innate immune response at the site of inoculation due to depot effect. This in turn results in efficient antigen uptake by professional APCs and better innate stimulation via signalling through pattern recognition receptors (PRRs) or directly via cytokines. Immunogenicity of the soluble antigens is highly enhanced by entrapping them inside cationic liposomes facilitating their uptake by APCs (Fig. 17.1). This strategy of T-cell activation has been effectively used in designing highly immunogenic pneumococcal (Sessa and Weissmann 1968; Bangham and Horne 1964; Tada et al. 2018b; Bhalla et al. 2021b), anti-leishmanial (Das et al. 2018; Sabur et al. 2018a) and viral vaccines. Cationic liposomes greatly improve the vaccine immunogenicity, antigen uptake and generation of antigen-specific $CD4^+$ and $CD8^+$ T cells producing IFN-γ, IL-2 and TNF-α for robust protective immunity against experimental VL (Das and Ali 2021; Das et al. 2018). Despite several modifications, entrapment efficiency and rapid clearance in the blood were a persistent problem. To address this issue, PEG was incorporated into liposomes, which shielded liposomes, prevented opsonization and as a result increased the clearance time (Klibanov et al. 1990). Army liposomal formulation (ALF) containing cholesterol, phospholipids and most importantly MPLA, was the first immunomodulatory TLR agonist to be used in human vaccines (Alving et al. 2020). Later, Shingrix® and Mosquirix® vaccines were approved by the FDA against herpes zoster and malaria, respectively, containing AS01 adjuvant (liposomal formulation of MPLA and QS-21 saponin from *Quillaja saponaria*) (Didierlaurent et al. 2017). Amicable, non-toxic nature of liposomes and their nano range of size with their versatility to entrap any nature of therapeutic agent made them the choice for drug and antigen delivery. Interestingly, Latif and Bachhawat's (Latif and Bachhawat 1984) research on liposomes resulted in a paradigm shift from negatively charged to cationic liposomes as vaccine adjuvant (Korsholm et al. 2012), which was later reported by many others (Christensen et al. 2007; Nakanishi et al. 1997). Eventually cationic liposomes re-emerged as a potent delivery system reported to induce a strong humoral and cellular immunomodulatory response against various infectious diseases and cancer (De et al. 2020; Banerjee et al. 2008; Ishii et al. 1997; Attia et al. 2021; Joseph et al. 2006). Most recent success of cationic liposomes is those of the COVID-19 vaccines by Pfizer and Moderna, which utilize ionizable cationic lipids for the delivery of the surface spike protein of the virus and have been able to generate a substantial immune response in the mass, enough to combat the pandemic (Attia et al. 2021). Compared with large, unilamellar and unmodified liposomes, multilamellar positively charged or cationic liposomes have so far emerged as best adjuvant cum delivery system against leishmaniasis (Joseph et al. 2006). Cationic liposomes containing phosphatidylcholine and stearylamine specifically interact with the negatively charged phosphatidylserine (PS) on the parasite surface leading to *Leishmania* killing (Banerjee et al. 2008; Das et al. 2021). Other than nanoliposomes targeting PS, several studies show increasing importance of parasite kinases; metacaspases and folate transport pathways are gaining liposomal targets against leishmaniases. Among these, cationic mannosylated liposomes targeting mannose receptors of infected macrophages have shown promising results in leishmaniasis (Rathore et al. 2011). Upregulation of co-stimulatory molecules and increased production of pro-inflammatory cytokines have been reported in DCs, exposed to cationic nanoparticles (Vangasseri et al. 2006; Pizzuto et al. 2018; Lonez et al. 2014). Furthermore, several recent studies reported that positively charged lipids in cationic nanoparticles can augment antigen delivery to the cytosol and improve cross-presentation of exogenous antigens in MHC class I molecules to $CD8^+$ T cells

(Ishii et al. 1997; Gao et al. 2017; Vermeulen et al. 2018; Sabnis et al. 2018). Efficient lymphatic transport of nanoparticles of diameters ~100–150 nm, surface modifications and co-delivery of immunomodulators like TLR agonists with antigen are some of the advantages of nanoparticle-based delivery systems. Polymeric nanoparticles and liposomes have been shown to stimulate MHC class I antigen presentation (Das et al. 2018; Maji et al. 2016), making them extremely attractive for vaccine delivery. In experimental VL, distearoyl phosphatidylcholine (DSPC)-based cationic liposomal formulation of gp63 (Mazumder et al. 2011a) and cysteine protease (Das et al. 2018) with MPLA was shown to boost high IFN-γ-mediated parasite killing by $CD4^+$ and CD8 T^+ cells in VL. In another study, protective immunity was generated against CL by using CpG-ODN along with liposomal soluble leishmanial antigen (SLA) in BALB/c mice model (Heravi Shargh et al. 2012). Several immunoliposomes targeting professional APCs have also been reported to achieve strong CMI response against *Leishmania*. Mannose 5-DPPE-coated immunoliposomes entrapping SLA that target mannose macrophages were shown to protect against CL (Shimizu et al. 2007). Immune stimulatory action of such liposomes arises from robust carbohydrate–protein interaction via mannose receptors on APCs, leading to better antigen presentation. Table 17.1 presents some recently developed liposomal prophylactic formulations against *Leishmania*. More studies are warranted to overcome the challenges of dose optimization, toxicity and parasite resistance to these immunochemotherapy combinations.

17.4 Immunomodulators

17.4.1 Aluminium Salts

Aluminium phosphate and aluminium oxyhydroxide, commonly called alum, are the oldest adjuvants licensed for human use against several infectious diseases and allergic conditions. Cationic aluminium salts attract negatively charged protein antigens that are surface-adsorbed on alum by hydrophobic and van der Waals forces (Hem and Hogenesch 2007). Upon injection, alum reportedly induces high Th2 response towards vaccine antigen recruiting eosinophils, neutrophils, dendritic cells, monocytes and natural killer cells at the inoculation site. However, alum is a weak mediator of CMI response (Hogenesch 2013). Activation of NLR signalling pathway is an important aspect of adjuvancy of alum. Alum adjuvant is known to activate inflammatory dendritic cells through NLRP3 inflammasome pathway, resulting in the release of IL-18 and IL-1β (Kool et al. 2008). Recently, alum adjuvant like Alhydrogel has been modified to microparticles and nanoparticles (called nanoalum) to trigger CMI response in combination with TLR4 agonist (Orr et al. 2019). Clinically approved vaccines like Cervarix and Gardasil against HPV uses MPLA–alum and amorphous aluminium hydroxyphosphate sulphate as adjuvants, respectively. Although alum was unable to induce Th1 response upon systemic administration against *Leishmania*, it has been widely used intradermally against experimental CL (Askarizadeh et al. 2020). Complete protection against CL

Table 17.1 A summary of notable liposomal vaccine formulations evaluated against *Leishmania* in recent years

Formulation	Model system/ administration route	Organism	Protection	Reference
Prophylaxis				
SLA + dioleoyl trimethylammonium-propane (DOTAP) liposomes	BALB/c mice, subcutaneous	*L. major*	Th1 type immune response, protection	Firouzmand et al. (2013)
SLA + DOTAP/imiquimod liposomes	BALB/c mice, subcutaneous	*L. major*	Th1 type immune response, protection	Mehravaran et al. (2019)
SLA + DOTAP + CpG-ODN liposomes	BALB/c mice, subcutaneous	*L. major*	Th1 type immune response, protection	Shargh et al. (2012)
Chimera T synthetic Ag + dipalmitoyl phosphatidylcholine (DPPC) liposome	BALB/c mice, subcutaneous	*L. infantum*	Th1 type immune response	Lage et al. (2020)
CPC DNA + DSPC/ stearylamine (SA)/MPLA liposomes	BALB/c mice, intramuscular	*L. donovani*	Mixed Th1/Th2 immune response	Das et al. (2018)
Elongation factor (EF)1-α + DSPC/SA liposomes	BALB/c mice, intraperitoneal	*L. donovani*	Mixed Th1/Th2 immune response	Sabur et al. (2018b)
Cysteine protease (CP) A/ CPB/CPC cocktail antigen + DSPC/SA liposomes + MPL–trehalose dicorynomycolate (TDM)	Syrian golden hamster, subcutaneous	*L. donovani*	Mixed Th1/Th2 immune response	Das and Ali (2014)
LiHyp1 + DPPC/DOTAP liposome + CpG-ODN	BALB/c mice, subcutaneous DMN patch	*L. donovani*	Mixed Th1/Th2 immune response	Lanza et al. (2020)
Immunotherapy				
Glucantime + DSPC/DCP/ distearoyl phosphoethanolamine (DSPE)–PEG liposomes + IL-10 blocking	Mongrel dogs, intravenous	*L. infantum*	No protection	Cardoso et al. (2022)
DOTAP liposomes + protamine + CpG-ODN	BALB/c mice subcutaneous	*L. major*	Partial protection	Alavizadeh et al. (2012)
AmBisome + LEISHDNAVAX	C57BL/6 mice, intravenous	*L. donovani*	Protection	Seifert et al. (2015)

has been achieved by injecting live *L. major* mixed with CpG-ODN plus whole cell lysate of *L. major* bound to alum, in mice (Okwor and Uzonna 2009). A single dose of BCG mixed with autoclaved *L. major* with alum resulted in almost 70% protection against canine VL in Iran (Mohebali et al. 2004). Notably, aluminium phosphate enhances transfection of DNA vaccine and has shown to induce protection against *L. mexicana* when used with a DNA vaccine encoding leishmanial NH36, GP46 and GP63 antigens, in BALB/c mice (Rosado-Vallado et al. 2005). Although beneficial for inducing humoral immunity, significant production of IL-10 from macrophages and DCs by alum adjuvant is commonly associated with the suppression of protective Th1 response (Oleszycka et al. 2018), making it less suitable as an adjuvant against intracellular pathogens requiring strong CMI response. Although safe, alum-adjuvanted vaccines are water-insoluble and cannot be sterile-filtered, frozen and lyophilised. As alum appears to trigger several molecular mechanisms to enhance the immunogenicity of an antigen, more work is needed to evaluate its adjuvant potential to activate Th1 pathway when combined with other immunostimulants. Despite good safety profile of alum adjuvants, recent reports of autoimmunity and inflammatory disorders induced by alum, termed as 'Shoenfeld's syndrome' (Bragazzi et al. 2020) in genetically susceptible individuals, mediated by human leukocyte antigens (HLAs) has raised an important concern for its future use.

17.4.2 IL-12

Originally identified by Trinchieri and his colleagues (Kobayashi et al. 1989) for its ability to boost IFN-γ production from both natural killer (NK) cells and DCs. IL-12 is recognized as a strong inducer of Th1 response when used as an adjuvant (Sypek et al. 1993; Jalah et al. 2012). IL-12 alone has shown to be a promising adjuvant in enhancing the antigen-mediated CMI response against *Leishmania* and can be used successfully along with other adjuvants. Protective immunity was observed in rhesus monkeys immunized with heat-killed *L. amazonensis* combined with recombinant IL-12 and alum, with enhanced IFN-γ response without any increase in IL-4 or IL-10 (Kenney et al. 1999). Further, protection in immunized animals receiving DNA vaccines with gene-encoding IL-12 correlated with Th-1-biased immune response and generation of antigen-specific $CD4^+$ and $CD8^+$ T cells (Tewary et al. 2006; Hugentobler et al. 2012; Maspi et al. 2016). In leishmaniasis, continuous expression of IL-12 DNA is a prerequisite for its adjuvant action, which is demonstrated by a DNA vaccine encoding leishmanial LACK gene and IL-12 (Gurunathan et al. 1997; Méndez et al. 2001) compared with short-term protection induced by recombinant LACK antigen with IL-12 in protein form (Stobie et al. 2000; Raman et al. 2012). Despite its significant role as an adjuvant in the activation of both innate (NK cells) and adaptive (CMI) responses against leishmaniasis and cancer, severe side effects like hepatotoxicity, leukopenia and bone marrow dysfunction hindered the systemic administration of IL-12 in humans (Raman et al. 2012) (Fig. 17.2).

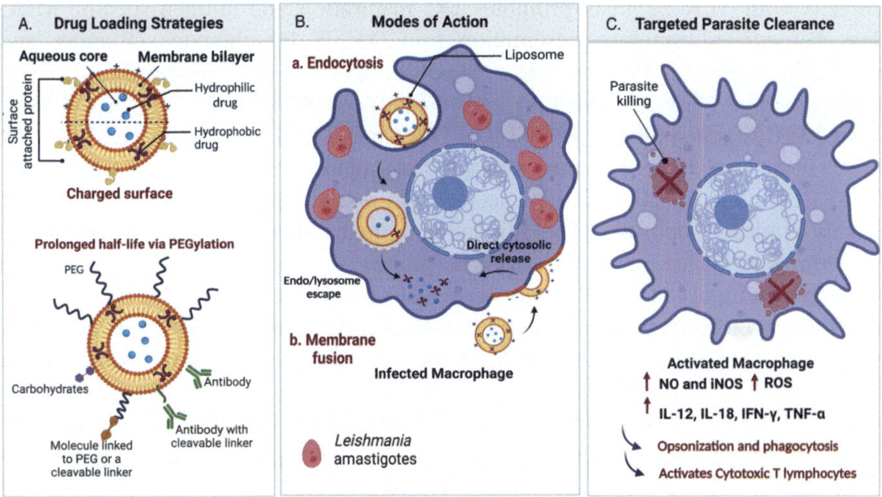

Fig. 17.2 Therapeutic delivery of liposomal drugs and their mode of action against leishmaniasis. (**a**) drug-loaded liposomal vesicles with different surface modifications. (**b**) Schematic representation of release of liposomal drug inside *Leishmania*-infected macrophage. Liposome uptake is considerably higher in infected macrophages than in normal ones, through receptor-mediated endocytosis or membrane fusion. (**c**) Nitric oxide (NO)- and reactive oxygen species (ROS)-mediated parasite killing inside macrophage along with the release of pro-inflammatory cytokines like IL-12, IL-18, IFN-γ and TNF-α from activated macrophage for disease resolution

17.4.3 TLR Agonists

Pattern recognition receptors (PRRs) are transmembrane receptors abundantly expressed on innate immune cells as well as fibroblasts and epithelial and endothelial cells for detecting pathogens. Among six different types of PRRs, TLR, NLR and CLRs are of particular interest for stimulating the innate activation of pro-inflammatory cytokines in response to infection. Almost all TLRs act through MyD88 pathway, except TLR3 that acts via TRIF pathway, leading to the activation of nuclear factor κB (NF-κB) and mitogen-activated protein kinases (MAPKs) (Luchner et al. 2021). IL-12 and IFN-γ produced from activated macrophages and DCs upon binding to TLR ligands are critically important for Th1-biased immune response against intracellular pathogens like *Leishmania*. Similarly, CLRs like DC-SIGN and L-SIGN help in interaction of *Leishmania* promastigotes with host macrophages and DCs (Caparrós et al. 2005), which may be of great importance in designing future vaccines. Carbohydrate-based CLR adjuvants like β-glucan (NCT04936529), mannan, chitosan (NCT00806962) and δ-inulin (Advax™) yielding satisfactory preclinical results are under clinical trials for different infectious diseases and cancer Garcia-Vello (2020). The goal of modern vaccine research is to orchestrate the ability of TLRs, CLRs and NLRs to modulate both innate and adaptive responses for total parasite clearance. Co-delivery of immunogenic antigens with TLR agonists promotes DC maturation and antigen-specific MHC class II-mediated antigen presentation to $CD4^+$ T lymphocytes (Blander and Medzhitov 2006). Polyinosinic : polycytidylic acid (poly I:C), CpG, monophosphoryl lipid A (MPLA) and resiquimod (R848) are some of the well-studied PRR agonists for TLR3, TLR9, TLR4 and TLR7/8, respectively. Till date, most widely used TLRs for immunoprophylaxis against leishmaniasis are TLR9 and TLR4 (Raman et al. 2010, 2012). So far, MPLA and CpG-1018 are the two TLR ligands clinically approved for adjuvating human vaccines (Yang et al. 2022). Co-administration of CpG DNA with several *Leishmania* vaccines has improved protective immunity in both CL and VL models, mediated by enhanced IFN-γ, IL-12 and IL-6 produced from $CD4^+$ T cells, B cells and NK cells (Ratnapriya et al. 2019). Non-coding plasmid DNAs with bacterial CpG motif can function as Th1-stimulating adjuvant even without antigen and have anti-parasitic effects against both VL (Mazumder et al. 2007) and CL (Montakhab-Yeganeh et al. 2022). Notably, a heterologous prime boost vaccination with recombinant gp63 and CpG-ODN against *L. donovani* resulted in 10^{10}-fold reduction in parasite load for durable protection in mice (Mazumder et al. 2011b). Moreover, suboptimal dose of drug, miltefosine, when combined with CpG-ODN, led to almost 97% reduction in *L. donovani* multiplication in both mice and hamsters with increased expression of IFN-γ, IL-12 and TNF-α with downregulation of Th2-promoting cytokines (Shivahare et al. 2014). As naked CPG-ODNs are susceptible to poor intracellular uptake, binding to plasma proteins and low bioavailability, their formulation with nanoparticles, ISCOMATRIX and liposomes appears to be promising strategies to enhance their synergistic adjuvant action. Recent studies show that CpG-ODN35 increases efficacy and treatment outcome of a low-dose sodium stibogluconate therapy in a primate model of CL (Thacker et al. 2020) and may also be suitable for vaccine development (Lanza et al. 2020). CpG-ODN35 is currently in human trial conducted by DNDi against uncomplicated CL (De Rycker et al. 2023). Although different TLR agonists share a common mechanism of action, they are reported to induce quite distinct immunological signalling pathways in rhesus macaques and human (van Haren et al. 2016; Qiao et al. 2022). Intradermal injections of TLR7/8 (R848) and TLR9 (CpG-2006) ligands preferentially activate dendritic cells and produce IFN-γ, IL-10 and IL-12, while the TLR4 ligands like MPLA and GLA induce the production of pro-inflammatory Th1 cytokines like IL-12, TNF-α and CXCL10 (van Haren et al. 2016). Emerging adjuvants like gardiquimod (TLR7 agonist) (Goyal 2021) and Pam2CSK4 (TLR2/6 agonist) (Qiao et al. 2022) have shown promising results as vaccine adjuvants that need further preclinical evaluations against *Leishmania*. *Leishmania*-specific immune response can be highly improved by TLR synergy as evidenced in several studies (Raman et al. 2010; Karmakar et al. 2012). Co-delivery of MPLA and imiquimod formulated with liposomal SLA synergistically enhanced the level of protection and CMI response against CL, compared with single adjuvants (Emami et al. 2018). COVID-19 pandemic has seen the emergence of many novel TLR adjuvants that hold promise for future vaccinations. New synthetic immunostimulators like CV8102, simultaneously activating TLR7/8 and RIG-1 pathways, need to be evaluated for infectious diseases (Yang et al. 2022). Another strategy to improve the immunostimulation by TLR ligands is its co-formulation with nanoparticles, as in case of GlaxoSmithkline's AS01 (liposomes, saponin QS21 and MPLA) and AS02 (saponin QS21 and MPLA emulsion) adjuvants for malaria vaccine. Cationic DSPC liposomes formulated with MPLA have shown remarkable protection against *L. donovani*, with Th1-biased CMI response and multi-cytokine (IFN-γ, IL-2 and TNF-α) producing $CD4^+$ and $CD8^+$ T cells in mice (Das et al. 2018). Taken together, various TLR agonists need to be modified, synergistically combined and formulated with correct nanoparticles for designing next-generation vaccines against *Leishmania*.

17.4.4 Metabolic and Epigenetic Adjuvants

Nutrient sensors or metabolic adjuvants like vitamin D, iron and L-carnitine are common molecules of intracellular metabolisms that can serve as adjuvants for modulating immune cells for against various diseases. Metformin is such a metabolic adjuvant studied against cancer immunotherapy, known to boost T-cell function, simultaneously reducing the expression of immune-checkpoint receptors in cancer cells (Munoz et al. 2021). More studies are required to predict the future of such adjuvants to combat infectious diseases like leishmaniasis.

Epigenetics refer to changes in gene expression patterns beyond the genetic code, due to environmental factors, behaviour and infectious agents. Epigenetic adjuvants help to train our innate immune system against multiple infectious agents through epigenetic reprogramming, along with significant B-cell and T-cell stimulation. Recent reports suggest that AS03-adjuvanted H5N1 influenza vaccine can lead to epigenetic changes in innate immune cells and increased functioning of antiviral genes to raise strong innate immunity against other influenza strains, dengue and Zika viruses (Wimmers et al. 2021). This is possibly due to superimposition of both refractory state and antiviral state in the same immune cell at the same time balancing the inflammation and vigilance over other viruses. Generation of such epigenetic memory by AS03 emulsion adjuvant with enhanced expression of disease-suppressing cytokines can be equally beneficial for other infectious diseases as well. Several epigenetic alterations in host documented after *Leishmania* infection include modifications in non-coding micro-RNAs and histones and several gene silencing in host macrophages that can serve as important targets for vaccine development (Afrin et al. 2019). More studies on post-infection epigenetic alterations in host, epigenetic sensitization of immune cells and reprogramming of vector epigenetics can lead to improved vaccines against leishmaniasis.

17.5 Conclusion

The development of new vaccine against infectious diseases is a complex process and usually interdisciplinary as in the case of leishmaniasis. Antigen selection is currently guided by generation of specific antigen-deficient parasites by CRISPR/

cas9 genetic engineering, based on T-cell receptor sequencing to identify novel epitopes (Zhang et al. 2020; Moreira et al. 2023). Similarly, computational system biology can provide efficient in silico simulations to predict adjuvant effects of novel immunostimulators targeting specific pathogens prior to in vivo validation (Pezeshki et al. 2019; Karmakar et al. 2021). Emerging evidence has suggested that strong $CD4^{+}$ and $CD8^{+}$ T-cell responses are critical for generating protective immunity against *Leishmania*. To achieve this, appropriate adjuvants should be incorporated in the antigen formulation for optimum immunogenic profiles. Herein, liposomes, especially cationic liposomes, are the most explored moiety that provides a safe, targeted, specific and desired delivery of the drugs with additional benefits of improved circulation, retention and immunomodulation; reduced toxicities; and synergistic effect of the available drugs. Vaccine formulations based on cationic nanoparticles are very promising platforms for immunotherapy. Notably, liposomes can synergistically enhance the immunomodulatory properties of antigen when co-formulated with other immune stimulators like CpG-ODNs, MPLA, imiquimod and resiquimod, which are mostly pathogen-associated molecular patterns (PAMPs) easily recognized by PRRs on professional APCs. As drug delivery cargoes, liposomes combine the benefits of excellent tissue-targeted delivery of drugs along with biocompatibility. Due to the structural differences in different types of cationic lipids used in liposomes, their immune-stimulating capacity is mediated via several molecular pathways involving multiple receptors in their mechanism of action, mostly resulting in strong antigen-specific T-cell responses. There is still immense scope of improvement especially in designing of the liposomes to improve their pharmacokinetics and dynamics further and also established in vitro models to test the efficacies of these formulations. Co-delivery of TLR agonists can significantly improve the immunogenicity of the candidate vaccine antigen via PAMP signalling mimicking pathogens leading to maturation of helper T (Th) cells and activation of $CD8^{+}$ T-cell-mediated adaptive response. Notably, sustained type-I IFN-γ-independent recruitment of antigen-loaded APCs towards the lymph node has been observed after the administration of TLR7/8 agonist (3M-052) in cationic liposomes, thereby exerting its adjuvant activity in C57/BL6 mice (Auderset et al. 2020). In conclusion, the ongoing extensive work in the field keeps developing and leaves little doubt that cationic lipid-based delivery and therapy are the future of nanomedicine. However, it is extremely important to invigilate the safety issues associated with possible reactogenicity of immunoliposomes containing immunomodulators like PAMPs (pathogen-associated molecular patterns) including inflammation and toxicities. The lack of reproducibility of liposomal drug/vaccine delivery from animal to clinical trials can be overcome with more rational designing and minimum variation during scale-up process using simplified manufacturing. In the struggle of developing newer and more efficient liposomal adjuvant system, understanding of basic cellular mechanisms of antigen uptake and associated receptors is equally important. Equipped with our recent experiences from COVID-19 pandemic, emphasis should also be given on standardized technology transfer with an accessible adjuvant library facilitating collaboration between various research groups for streamlined vaccine generation against leishmaniasis.

Acknowledgments The authors would like to acknowledge Council of Scientific and Industrial Research (CSIR), UK Research and Innovation via the Global Challenges Research Fund (GCRF) under agreement 'A Global Network for Neglected Tropical Diseases' (grant number: MR/P027989/1) and Sir J C Bose National Fellowship, India, for their support.

Conflict of Interest The authors declare no conflict of interest.

References

Afrin F, Khan I, Hemeg HA (2019) *Leishmania*-host interactions-an epigenetic paradigm. Front Immunol 10:492

Akbarzadeh A, Rezaei-Sadabady R, Davaran S, Joo SW, Zarghami N, Hanifehpour Y, Samiei M, Kouhi M, Nejati-Koshki K (2013) Liposome: classification, preparation, and applications. Nanoscale Res Lett 8(1):102

Alavizadeh SH, Badiee A, Khamesipour A, Jalali SA, Firouzmand H, Abbasi A, Jaafari MR (2012) The role of liposome-protamine-DNA nanoparticles containing CpG oligodeoxynucleotides in the course of infection induced by *Leishmania major* in BALB/c mice. Exp Parasitol 132(3): 313–319

Alving CR, Peachman KK, Matyas GR, Rao M, Beck Z (2020) Army liposome formulation (ALF) family of vaccine adjuvants. Expert Rev Vaccines 19(3):279–292

Anselmo AC, Mitragotri S (2019) Nanoparticles in the clinic: an update. Bioeng Transl Med 4(3): e10143

Asadi K, Gholami A (2021) Virosome-based nanovaccines; a promising bioinspiration and biomimetic approach for preventing viral diseases: a review. Int J Biol Macromol 182:648–658

Askarizadeh A, Badiee A, Khamesipour A (2020) Development of nano-carriers for *Leishmania* vaccine delivery. Expert Opin Drug Deliv 17(2):167–187

Attia MA, Essa EA, Elebyary TT et al (2021) Brief on recent application of liposomal vaccines for lower respiratory tract viral infections: from influenza to COVID-19 vaccines. Pharmaceuticals (Basel). 14. https://doi.org/10.3390/PH14111173. Epub ahead of print November 1, 2021

Auderset F, Belnoue E, Mastelic-Gavillet B, Lambert PH, Siegrist CA (2020) A TLR7/8 agonist-including DOEPC-based cationic liposome formulation mediates its adjuvanticity through the sustained recruitment of highly activated monocytes in a Type I IFN-independent but NF-κB-dependent manner. Front Immunol 11:580974

Banerjee A, De M, Ali N (2008) Complete cure of experimental visceral leishmaniasis with amphotericin B in stearylamine-bearing cationic liposomes involves down-regulation of IL-10 and favorable T cell responses. J Immunol 181:1386–1398

Bangham AD, Horne RW (1964) Negative staining of phospholipids and their structural modification by surface-active agents as observed in the electron microscope. J Mol Biol 8:660–668

Baz Morelli A, Becher D, Koernig S, Silva A, Drane D, Maraskovsky E (2012) ISCOMATRIX: a novel adjuvant for use in prophylactic and therapeutic vaccines against infectious diseases. J Med Microbiol 61(Pt 7):935–943

Bhalla M, Nayerhoda R, Tchalla EYI, Abamonte A, Park D, Simmons SR, Pfeifer BA, Bou Ghanem EN (2021a) Liposomal encapsulation of polysaccharides (LEPS) as an effective vaccine strategy to protect aged hosts against *S. pneumoniae* infection. Front Aging. 2:798868

Bhalla M, Nayerhoda R, Tchalla EYI, Abamonte A, Park D et al (2021b) Liposomal encapsulation of polysaccharides (LEPS) as an effective vaccine strategy to protect aged hosts Against *S. pneumoniae* infection. Front Aging. 2:798868

Blander JM, Medzhitov R (2006) Toll-dependent selection of microbial antigens for presentation by dendritic cells. Nature 440(7085):808–812

Bragazzi NL, Hejly A, Watad A, Adawi M, Amital H, Shoenfeld Y (2020) ASIA syndrome and endocrine autoimmune disorders. Best Pract Res Clin Endocrinol Metab 34(1):101412

Bungener L, Serre K, Bijl L, Leserman L, Wilschut J, Daemen T, Machy P (2002) Virosome-mediated delivery of protein antigens to dendritic cells. Vaccine 20(17–18):2287–2295

Caparrós E, Serrano D, Puig-Kröger A, Riol L, Lasala F, Martinez I, Vidal-Vanaclocha F, Delgado R, Rodríguez-Fernández JL, Rivas L, Corbí AL, Colmenares M (2005) Role of the C-type lectins DC-SIGN and L-SIGN in *Leishmania* interaction with host phagocytes. Immunobiology 210(2–4):185–193

Cardoso JMO, Brito RCF, Mathias FAS, Reis LES, Vieira JFP et al (2022) Comparative evaluation of meglumine antimoniate encapsulated in a mixture of conventional and PEGylated liposomes and immunotherapy using an anti-canine IL-10 receptor-blocking monoclonal antibody on canine visceral leishmaniasis. Mol Immunol 141:70–78

Chen K, Wang N, Zhang X, Wang M, Liu Y, Shi Y (2023) Potentials of saponins-based adjuvants for nasal vaccines. Front Immunol 14:1153042

Christensen D, Korsholm KS, Rosenkrands I et al (2007) Cationic liposomes as vaccine adjuvants. Expert Rev Vaccines 6:785–796

Das A, Ali N (2012) Vaccine development against *Leishmania donovani*. Front Immunol 3:99

Das A, Ali N (2014) Combining cationic liposomal delivery with MPL-TDM for cysteine protease cocktail vaccination against Leishmania donovani: evidence for antigen synergy and protection. PLoS Negl Trop Dis 8(8):e3091

Das A, Ali N (2021) Nanovaccine: an emerging strategy. Expert Rev Vaccines 20(10):1273–1290

Das A, Asad M, Sabur A et al (2018) Monophosphoryl lipid A based cationic liposome facilitates vaccine induced expansion of polyfunctional T cell immune responses against visceral leishmaniasis. ACS Appl Bio Mater 1:999–1018

Das A, Kamran M, Ali N (2021) HO-3867 induces ROS-dependent stress response and apoptotic cell death in *Leishmania donovani*. Front Cell Infect Microbiol 11:774899

De Rycker M, Wyllie S, Horn D, Read KD, Gilbert IH (2023) Anti-trypanosomatid drug discovery: progress and challenges. Nat Rev Microbiol 21(1):35–50

De M, Ghosh S, Asad M et al (2020) Combining doxorubicin with stearylamine-bearing liposomes elicits Th1 cytokine responses and cures metastasis in a mouse model. Cancer Immunol Immunother. https://doi.org/10.1007/s00262-020-02578-9. Epub ahead of print

Didierlaurent AM, Laupèze B, Di Pasquale A, Hergli N, Collignon C, Garçon N (2017) Adjuvant system AS01: helping to overcome the challenges of modern vaccines. Expert Rev Vaccines 16(1):55–63

Emami T, Rezayat SM, Khamesipour A, Madani R, Habibi G et al (2018) The role of MPL and imiquimod adjuvants in enhancement of immune response and protection in BALB/c mice immunized with soluble *Leishmania* antigen (SLA) encapsulated in nanoliposome. Artif Cells Nanomed Biotechnol. 46(sup2):324–333

Firouzmand H, Badiee A, Khamesipour A, Heravi Shargh V, Alavizadeh SH, Abbasi A, Jaafari MR (2013) Induction of protection against leishmaniasis in susceptible BALB/c mice using simple DOTAP cationic nanoliposomes containing soluble *Leishmania* antigen (SLA). Acta Trop 128(3):528–535

Fischetti L, Zhong Z, Pinder CL, Tregoning JS, Shattock RJ (2017) The synergistic effects of combining TLR ligand based adjuvants on the cytokine response are dependent upon p38/JNK signalling. Cytokine 99:287–296

Gao J, Ochyl L, Yang E et al (2017) Cationic liposomes promote antigen cross-presentation in dendritic cells by alkalizing the lysosomal pH and limiting the degradation of antigens. Int J Nanomedicine 12:1251–1264

Garcia-Vello (2020) Carbohydrate-based adjuvants Drug Discovery Today: Technologies: 35-36 57–68 https://doi.org/10.1016/j.ddtec.2020.09.005

Gautam S, Kumar R, Singh N, Singh AK, Rai M, Sacks D, Sundar S, Nylén S (2014) CD8 T cell exhaustion in human visceral leishmaniasis. J Infect Dis 209(2):290–299

Goyal (2021) Adjuvant effects of TLR agonist gardiquimod admixed with Leishmania vaccine in mice model of visceral leishmaniasis. Infect Genet Evol 93:104947. https://doi.org/10.1016/j.meegid.2021.104947

Gurunathan S, Sacks DL, Brown DR, Reiner SL, Charest H, Glaichenhaus N, Seder RA (1997) Vaccination with DNA encoding the immunodominant LACK parasite antigen confers protective immunity to mice infected with *Leishmania major*. J Exp Med 186(7):1137–1197

Hem SL, Hogenesch H (2007) Relationship between physical and chemical properties of aluminum-containing adjuvants and immunopotentiation. Expert Rev Vaccines 6(5):685–698

Heravi Shargh V, Jaafari MR, Khamesipour A, Jalali SA, Firouzmand H, Abbasi A, Badiee A (2012) Cationic liposomes containing soluble *Leishmania* antigens (SLA) plus CpG ODNs induce protection against murine model of leishmaniasis. Parasitol Res 111(1):105–114

Hogenesch H (2013) Mechanism of immunopotentiation and safety of aluminum adjuvants. Front Immunol 3:406

Huckriede A, Bungener L, ter Veer W, Holtrop M, Daemen T, Palache AM, Wilschut J (2003) Influenza virosomes: combining optimal presentation of hemagglutinin with immunopotentiating activity. Vaccine 21(9–10):925–931

Hugentobler F, Yam KK, Gillard J, Mahbuba R, Olivier M, Cousineau B (2012) Immunization against Leishmania major infection using LACK- and IL-12-expressing *Lactococcus lactis* induces delay in footpad swelling. PLoS One 7(2):e30945

Ikeogu NM, Akaluka GN, Edechi CA, Salako ES, Onyilagha C et al (2020) *Leishmania* immunity: advancing immunotherapy and vaccine development. Microorganisms. 8(8):1201

Ishii N, Fukushima J, Kaneko T et al (1997) Cationic liposomes are a strong adjuvant for a DNA vaccine of human immunodeficiency virus type 1. AIDS Res Hum Retrovir 13:1421–1428

Jalah R, Patel V, Kulkarni V, Rosati M, Alicea C et al (2012) IL-12 DNA as molecular vaccine adjuvant increases the cytotoxic T cell responses and breadth of humoral immune responses in SIV DNA vaccinated macaques. Hum Vaccin Immunother 8(11):1620–1629

Joseph A, Itskovitz-Cooper N, Samira S et al (2006) A new intranasal influenza vaccine based on a novel polycationic lipid - ceramide carbamoyl-spermine (CCS) I. Immunogenicity and efficacy studies in mice. Vaccine 24:3990–4006

Karmakar S, Bhaumik SK, Paul J, De T (2012) TLR4 and NKT cell synergy in immunotherapy against visceral leishmaniasis. PLoS Pathog 8(4):e1002646

Karmakar S, Ismail N, Oliveira F, Oristian J, Zhang WW et al (2021) Preclinical validation of a live attenuated dermotropic *Leishmania* vaccine against vector transmitted fatal visceral leishmaniasis. Commun Biol 4(1):929

Kaushal H, Bras-Gonçalves R, Negi NS, Lemesre JL, Papierok G, Salotra P (2014) Role of CD8(+) T cells in protection against *Leishmania donovani* infection in healed Visceral Leishmaniasis individuals. BMC Infect Dis 14:653

Kawasaki T, Kawai T (2014) Toll-like receptor signalling pathways. Front Immunol 5:461

Kenney RT, Sacks DL, Sypek JP, Vilela L, Gam AA, Evans-Davis K (1999) Protective immunity using recombinant human IL-12 and alum as adjuvants in a primate model of cutaneous leishmaniasis. J Immunol 163(8):4481–4488

Klibanov AL, Maruyama K, Torchilin VP et al (1990) Amphipathic polyethyleneglycols effectively prolong the circulation time of liposomes. FEBS Lett 268:235–237

Kobayashi M, Fitz L, Ryan M, Hewick RM, Clark SC, Chan S, Loudon R, Sherman F, Perussia B, Trinchieri G (1989) Identification and purification of natural killer cell stimulatory factor (NKSF), a cytokine with multiple biologic effects on human lymphocytes. J Exp Med 170(3): 827–845

Kool M, Pétrilli V, De Smedt T, Rolaz A, Hammad H et al (2008) Cutting edge: alum adjuvant stimulates inflammatory dendritic cells through activation of the NALP3 inflammasome. J Immunol 181(6):3755–3759

Korsholm KS, Andersen PL, Christensen D (2012) Cationic liposomal vaccine adjuvants in animal challenge models: overview and current clinical status. Expert Rev Vaccines 11:561–577

Krasnopolsky Y, Pylypenko D (2022) Licensed liposomal vaccines and adjuvants in the antigen delivery system. BioTechnologia (Pozn) 103(4):409–423

Kumar P, Shivam P, Mandal S, Prasanna P, Kumar S et al (2019) Synthesis, characterization, and mechanistic studies of a gold nanoparticle-amphotericin B covalent conjugate with enhanced antileishmanial efficacy and reduced cytotoxicity. Int J Nanomedicine 14:6073–6101

Lage DP, Ribeiro PAF, Dias DS, Mendonça DVC, Ramos FF et al (2020) Liposomal formulation of ChimeraT, a multiple T-cell epitope-containing recombinant protein, is a candidate vaccine for human visceral leishmaniasis. Vaccines (Basel) 8(2):289

Lanza JS, Vucen S, Flynn O, Donadei A, Cojean S et al (2020) A TLR9-adjuvanted vaccine formulated into dissolvable microneedle patches or cationic liposomes protects against leishmaniasis after skin or subcutaneous immunization. Int J Pharm 586:119390

Latif N, Bachhawat BK (1984) The effect of surface charges of liposomes in immunopotentiation. Biosci Rep 4:99–107

Liu P, Chen G, Zhang J (2022) A review of liposomes as a drug delivery system: current status of approved products, regulatory environments, and future perspectives. Molecules 27(4):1372

Lonez C, Bessodes M, Scherman D et al (2014) Cationic lipid nanocarriers activate Toll-like receptor 2 and NLRP3 inflammasome pathways. Nanomedicine 10(4):775–782

Luchner M, Reinke S, Milicic A (2021) TLR agonists as vaccine adjuvants targeting cancer and infectious diseases. Pharmaceutics. 13(2):142

Maji M, Mazumder S, Bhattacharya S, Choudhury ST, Sabur A et al (2016) A lipid based antigen delivery system efficiently facilitates MHC class-I antigen presentation in dendritic cells to stimulate CD8(+) T cells. Sci Rep 6:27206

Mann S, Frasca K, Scherrer S, Henao-Martínez AF, Newman S, Ramanan P, Suarez JA (2021) A review of leishmaniasis: current knowledge and future directions. Curr Trop Med Rep 8(2): 121–132

Maspi N, Ghaffarifar F, Sharifi Z, Dalimi A (2016) Codelivery of DNA vaccination encoding LeIF gene and IL-12 increases protection against *Leishmania major* infection in BALB/c mice. Parasite Immunol 38(4):228–235

Mazumder S, Ravindran R, Banerjee A, Ali N (2007) Non-coding pDNA bearing immunostimulatory sequences co-entrapped with leishmanial antigens in cationic liposomes elicits almost complete protection against experimental visceral leishmaniasis in BALB/c mice. Vaccine 25(52):8771–8781

Mazumder S, Maji M, Ali N (2011a) Potentiating effects of MPL on DSPC bearing cationic liposomes promote recombinant GP63 vaccine efficacy: high immunogenicity and protection. PLoS Negl Trop Dis 5(12):e1429

Mazumder S, Maji M, Das A, Ali N (2011b) Potency, efficacy and durability of DNA/DNA, DNA/protein and protein/protein based vaccination using gp63 against *Leishmania donovani* in BALB/c mice. PLoS One 6(2):e14644

Mehravaran A, Jaafari MR, Jalali SA, Khamesipour A, Ranjbar R, Hojatizade M, Badiee A (2016) The role of ISCOMATRIX bilayer composition to induce a cell mediated immunity and protection against leishmaniasis in BALB/c mice. Iran J Basic Med Sci 19(2):178–186

Mehravaran A, Rezaei Nasab M, Mirahmadi H, Sharifi I, Alijani E, Nikpoor AR, Akhtari J, Hojatizade M (2019) Immunogenicity and protection effects of cationic liposome containing imiquimod adjuvant on leishmaniasis in BALB/c mice. Iran J Basic Med Sci 22(8):922–931

Méndez S, Gurunathan S, Kamhawi S, Belkaid Y, Moga MA, Skeiky YA, Campos-Neto A, Reed S, Seder RA, Sacks D (2001) The potency and durability of DNA- and protein-based vaccines against *Leishmania major* evaluated using low-dose, intradermal challenge. J Immunol. 166(8): 5122–5128

Mohebali M, Khamesipour A, Mobedi I, Zarei Z, Hashemi-Fesharki R (2004) Double-blind randomized efficacy field trial of alum precipitated autoclaved *Leishmania major* vaccine mixed with BCG against canine visceral leishmaniasis in Meshkin-Shahr district, I.R. Iran. Vaccine 22(29–30):4097–4100

Montakhab-Yeganeh H, Shafiei R, Najm M, Masoori L, Aspatwar A, Badirzadeh A (2022) Immunogenic properties of empty pcDNA3 plasmid against zoonotic cutaneous leishmaniasis in mice. PLoS One 17(2):e0263993

Morein B, Sundquist B, Höglund S, Dalsgaard K, Osterhaus A (1984) Iscom, a novel structure for antigenic presentation of membrane proteins from enveloped viruses. Nature 308(5958): 457–460

Moreira POL, Nogueira PM, Monte-Neto RL (2023) Next-generation leishmanization: revisiting molecular targets for selecting genetically engineered live-attenuated *Leishmania*. Microorganisms 11(4):1043

Moura APV, Santos LCB, Brito CRN, Valencia E, Junqueira C et al (2017) Virus-like particle display of the α-Gal carbohydrate for vaccination against *Leishmania* infection. ACS Cent Sci 3(9):1026–1031

Munoz LE, Huang L, Bommireddy R, Sharma R, Monterroza L et al (2021) Metformin reduces PD-L1 on tumor cells and enhances the anti-tumor immune response generated by vaccine immunotherapy. J Immunother Cancer 9(11):e002614

Nakanishi T, Kunisawa J, Hayashi A et al (1997) Positively charged liposome functions as an efficient immunoadjuvant in inducing immune responses to soluble proteins. Biochem Biophys Res Commun 240(3):793–797

Nikolich-Zugich J (2008) Ageing and life-long maintenance of T-cell subsets in the face of latent persistent infections. Nat Rev Immunol 8(7):512–522

Okwor I, Uzonna J (2009) Vaccines and vaccination strategies against human cutaneous leishmaniasis. Hum Vaccin 5(5):291–301

Oleszycka E, McCluskey S, Sharp FA, Muñoz-Wolf N, Hams E, Gorman AL, Fallon PG, Lavelle EC (2018) The vaccine adjuvant alum promotes IL-10 production that suppresses Th1 responses. Eur J Immunol 48(4):705–715

Orr MT, Khandhar AP, Seydoux E, Liang H, Gage E et al (2019) Reprogramming the adjuvant properties of aluminum oxyhydroxide with nanoparticle technology. NPJ Vaccines 4:1

Oscherwitz J, Hankenson FC, Yu F, Cease KB (2006) Low-dose intraperitoneal Freund's adjuvant: toxicity and immunogenicity in mice using an immunogen targeting amyloid-beta peptide. Vaccine 24(15):3018–3025

Pati R, Shevtsov M, Sonawane A (2018) Nanoparticle vaccines against infectious diseases. Front Immunol 9:2224

Pezeshki A, Ovsyannikova IG, McKinney BA, Poland GA, Kennedy RB (2019) The role of systems biology approaches in determining molecular signatures for the development of more effective vaccines. Expert Rev Vaccines 18(3):253–267

Pizzuto M, Bigey P, Lachagès A-M et al (2018) Cationic lipids as one-component vaccine adjuvants: a promising alternative to alum. J Control Release 287:67–77

Qiao Y, Zhan Y, Zhang Y, Deng J, Chen A et al (2022) Pam2CSK4-adjuvanted SARS-CoV-2 RBD nanoparticle vaccine induces robust humoral and cellular immune responses. Front Immunol 13: 992062

Raj S, Sasidharan S, Tripathi T, Saudagar P (2022) Biofunctionalized Chrysin-conjugated gold nanoparticles neutralize *Leishmania* parasites with high efficacy. Int J Biol Macromol 205:211–219

Raman VS, Bhatia A, Picone A, Whittle J, Bailor HR et al (2010) Applying TLR synergy in immunotherapy: implications in cutaneous leishmaniasis. J Immunol 185(3):1701–1710

Raman VS, Duthie MS, Fox CB, Matlashewski G, Reed SG (2012) Adjuvants for *Leishmania* vaccines: from models to clinical application. Front Immunol 3:144

Rathore A, Jain A, Gulbake A, Shilpi S, Khare P, Jain A, Jain SK (2011) Mannosylated liposomes bearing Amphotericin B for effective management of visceral Leishmaniasis. J Liposome Res 21(4):333–340

Ratnapriya S, Keerti, Sahasrabuddhe AA, Dube A (2019) Visceral leishmaniasis: an overview of vaccine adjuvants and their applications. Vaccine 37(27):3505–3519

Rosado-Vallado M, Mut-Martin M, García-Miss Mdel R, Dumonteil E (2005) Aluminium phosphate potentiates the efficacy of DNA vaccines against *Leishmania mexicana*. Vaccine 23(46–47):5372–5379

Sabnis S, Kumarasinghe ES, Salerno T et al (2018) A novel amino lipid series for mRNA delivery: improved endosomal escape and sustained pharmacology and safety in non-human primates. Mol Ther 26:1509–1519

Sabur A, Bhowmick S, Chhajer R et al (2018a) Liposomal elongation factor-1α triggers effector CD4 and CD8 T cells for induction of long-lasting protective immunity against visceral leishmaniasis. Front Immunol 9:18

Sabur A, Bhowmick S, Chhajer R, Ejazi SA, Didwania N, Asad M, Bhattacharyya A, Sinha U, Ali N (2018b) Liposomal elongation factor-1α triggers effector CD4 and CD8 T cells for induction of long-lasting protective immunity against visceral leishmaniasis. Front Immunol 9:18

Saljoughian N, Zahedifard F, Doroud D, Doustdari F, Vasei M, Papadopoulou B, Rafati S (2013) Cationic solid-lipid nanoparticles are as efficient as electroporation in DNA vaccination against visceral leishmaniasis in mice. Parasite Immunol 35(12):397–408

Samant M, Sahu U, Pandey SC, Khare P (2021) Role of cytokines in experimental and human visceral leishmaniasis. Front Cell Infect Microbiol 11:624009

Seifert K, Juhls C, Salguero FJ, Croft SL (2015) Sequential chemoimmunotherapy of experimental visceral leishmaniasis using a single low dose of liposomal amphotericin B and a novel DNA vaccine candidate. Antimicrob Agents Chemother 59(9):5819–5823

Sessa G, Weissmann G (1968) Phospholipid spherules (liposomes) as a model for biological membranes. J Lipid Res 9(3):310–318

Shargh VH, Jaafari MR, Khamesipour A, Jaafari I, Jalali SA, Abbasi A, Badiee A (2012) Liposomal SLA co-incorporated with PO CpG ODNs or PS CpG ODNs induce the same protection against the murine model of leishmaniasis. Vaccine 30(26):3957–3964

Sharma VK, Agrawal MK (2021) A historical perspective of liposomes-a bio nanomaterial. Mater Today Proc 45:2963–2966

Shimizu Y, Takagi H, Nakayama T, Yamakami K, Tadakuma T, Yokoyama N, Kojima N (2007) Intraperitoneal immunization with oligomannose-coated liposome-entrapped soluble leishmanial antigen induces antigen-specific T-helper type immune response in BALB/c mice through uptake by peritoneal macrophages. Parasite Immunol 29(5):229–239

Shivahare R, Vishwakarma P, Parmar N, Yadav PK, Haq W, Srivastava M, Gupta S, Kar S (2014) Combination of liposomal CpG oligodeoxynucleotide 2006 and miltefosine induces strong cell-mediated immunity during experimental visceral leishmaniasis. PLoS One 9(4):e94596

Silva A, Mount A, Krstevska K, Pejoski D, Hardy MP, Owczarek C, Scotney P, Maraskovsky E, Baz MA (2015) The combination of ISCOMATRIX adjuvant and TLR agonists induces regression of established solid tumors in vivo. J Immunol 194(5):2199–2207

Silveira F, Rivera-Patron M, Deshpande N, Sienra S, Checa J et al (2023) *Quillaja brasiliensis* nanoparticle adjuvant formulation improves the efficacy of an inactivated trivalent influenza vaccine in mice. Front Immunol 14:1163858

Stertman L, Palm AE, Zarnegar B, Carow B, Lunderius Andersson C et al (2023) The Matrix-M™ adjuvant: a critical component of vaccines for the 21st century. Hum Vaccin Immunother 19(1): 2189885

Stobie L, Gurunathan S, Prussin C, Sacks DL, Glaichenhaus N, Wu CY, Seder RA (2000) The role of antigen and IL-12 in sustaining Th1 memory cells in vivo: IL-12 is required to maintain memory/effector Th1 cells sufficient to mediate protection to an infectious parasite challenge. Proc Natl Acad Sci U S A 97(15):8427–8432

Sun HX, Xie Y, Ye YP (2009) ISCOMs and ISCOMATRIX. Vaccine 27(33):4388–4401

Sypek JP, Chung CL, Mayor SE, Subramanyam JM, Goldman SJ, Sieburth DS, Wolf SF, Schaub RG (1993) Resolution of cutaneous leishmaniasis: interleukin 12 initiates a protective T helper type 1 immune response. J Exp Med 177(6):1797–1802

Tada R, Suzuki H, Takahashi S, Negishi Y, Kiyono H, Kunisawa J, Aramaki Y (2018a) Nasal vaccination with pneumococcal surface protein A in combination with cationic liposomes consisting of DOTAP and DC-chol confers antigen-mediated protective immunity against Streptococcus pneumoniae infections in mice. Int Immunopharmacol 61:385–393

Tada R, Suzuki H, Takahashi S et al (2018b) Nasal vaccination with pneumococcal surface protein A in combination with cationic liposomes consisting of DOTAP and DC-chol confers antigen-mediated protective immunity against Streptococcus pneumoniae infections in mice. Int Immunopharmacol 61:385–393

Tewary P, Saxena S, Madhubala R (2006) Co-administration of IL-12 DNA with rORFF antigen confers long-term protective immunity against experimental visceral leishmaniasis. Vaccine 24(13):2409–2416

Thacker SG, McWilliams IL, Bonnet B, Halie L, Beaucage S, Rachuri S, Dey R, Duncan R, Modabber F, Robinson S, Bilbe G, Arana B, Verthelyi D (2020) CpG ODN D35 improves the response to abbreviated low-dose pentavalent antimonial treatment in non-human primate model of cutaneous leishmaniasis. PLoS Negl Trop Dis 14(2):e0008050

Tiwari N, Gedda MR, Tiwari VK, Singh SP, Singh RK (2018) Limitations of current therapeutic options, possible drug targets and scope of natural products in control of leishmaniasis. Mini Rev Med Chem 18(1):26–41

van Haren SD, Ganapathi L, Bergelson I, Dowling DJ, Banks M, Samuels RC, Reed SG, Marshall JD, Levy O (2016) In vitro cytokine induction by TLR-activating vaccine adjuvants in human blood varies by age and adjuvant. Cytokine 83:99–109

Vangasseri DP, Cui Z, Chen W et al (2006) Immunostimulation of dendritic cells by cationic liposomes. Mol Membr Biol 23(5):385–395

Vermeulen LMP, de Smedt SC, Remaut K et al (2018) The proton sponge hypothesis: fable or fact? Eur J Pharm Biopharm 129:184–190

Wimmers F, Donato M, Kuo A, Ashuach T, Gupta S et al (2021) The single-cell epigenomic and transcriptional landscape of immunity to influenza vaccination. Cell 184(15):3915–3935.e21

Yang JX, Tseng JC, Yu GY, Luo Y, Huang CF, Hong YR, Chuang TH (2022) Recent advances in the development of toll-like receptor agonist-based vaccine adjuvants for infectious diseases. Pharmaceutics 14(2):423

Zhang WW, Lypaczewski P, Matlashewski G (2020) Application of CRISPR/Cas9-Mediated Genome Editing in *Leishmania*. Methods Mol Biol 2116:199–224

Zielińska A, Carreiró F, Oliveira AM, Neves A, Pires B, Venkatesh DN, Durazzo A, Lucarini M, Eder P, Silva AM, Santini A, Souto EB (2020) Polymeric nanoparticles: production, characterization, toxicology and ecotoxicology. Molecules 25(16):3731

Zijlstra EE, Musa AM, Khalil EA (2003) el-Hassan IM, el-Hassan AM. Post-kala-azar dermal leishmaniasis. Lancet Infect Dis 3(2):87–98

(2020) Carbohydrate-based adjuvants Drug Discovery Today: Technologies 35-3657-68 10.1016/j.ddtec.2020.09.005

(2021) Adjuvant effects of TLR agonist gardiquimod admixed with Leishmania vaccine in mice model of visceral leishmaniasis Infection Genetics and Evolution 93104947-10.1016/j.meegid.2021.104947

SPRINGER NATURE

GPSR Compliance

The European Union's (EU) General Product Safety Regulation (GPSR) is a set of rules that requires consumer products to be safe and our obligations to ensure this.

If you have any concerns about our products, you can contact us on ProductSafety@springernature.com

In case Publisher is established outside the EU, the EU authorized representative is:

Springer Nature Customer Service Center GmbH
Europaplatz 3
69115 Heidelberg, Germany

The manufacturer's authorised representative in the EU is Springer Nature Customer Service Centre GmbH, Europaplatz 3, 69115 Heidelberg, Germany. If you have any concerns regarding our products, please contact ProductSafety@springernature.com

Printed and bound by CPI Group (UK) Ltd, Croydon, CR0 4YY
01/10/2026
02225896-0016